Progress in Computational Physics (PiCP)

Volume 1: Wave Propagation in Periodic Media

Editor

Matthias Ehrhardt
Bergische Universität
Germany

CONTENTS

FOREWORD

Wave propagation in periodic media has attracted much attention from the scientists for a very long time. Forward or backward propagation regimes, refraction and diffraction properties, all these subjects are theoretically formalized and experimentally investigated since many centuries... Nevertheless, a renewed interest for this domain exists for a few decades with the emergence of the field of metamaterials and/or photonic crystals. The origin of this renewal can be found in the theoretical works of V. Veselago in the late 1960's, followed by the proposals of E. Yablonovitch for photonic crystals in the 1980's and of J. Pendry for metamaterials in the 1990's. The promise is the opportunity to design original devices built with artificial materials, metallic or dielectric, with extraordinary propagation properties (that is to say not found in nature). The most popular new effect is undoubtedly the "negative refraction" that can be obtained either by using specific propagation regimes in periodic structures or by creating double negative media in terms of permittivity and permeability. Such a concept has allowed generalizing the famous Snell-Descartes law for refraction at the interface of two propagating media to any arbitrary values of the refractive index, positive or negative, greater or lower than unity... Moreover, as it can be shown that most of the concepts do not depend on wavelength scale, they can be potentially applied from microwave to optical waves. The main limiting factor is the patterning scale of the supporting artificial materials which is proportional to the targeted wavelength of operation. Fortunately, owing to the progress in fabrication technology down to the nanometer scale, it is now possible to bring to reality most of the theoretical predictions, even in the visible.

At present, this research field is very attractive as shown by the huge literature devoted to metamaterials or photonic crystals. Perfect lenses, hyperlenses, cloaking or invisibility devices are widely investigated, exploiting the fact that intrinsic materials parameters as refraction index, impedance, permittivity and permeability can be locally modulated to any positive, near zero or negative values. Open theoretical problems also re-appear following some "advanced" proposals and "lively" debates occasionally occur within the scientific community. As the field gains in maturity, devices become more complex and new modeling approaches are now required to interpret and understand properly the underlying effects. That is why the field becomes more and more multidisciplinary with mathematicians, numericians and physicists for theory and concepts, applied physicists for fabrication and measurements in connection with chemists, telecommunication engineers or biologists for original applications.

The book edited by Prof. Matthias Ehrhardt provides some particularly interesting keys to enter in this vast and exciting research domain. By focusing on specific advanced subjects related to wave propagation in periodic media from the different viewpoints of analysis, numerical techniques and practical applications with chapters written by experts in their respective fields, the reader will find an overview of state-of-the-art research results over a wide range of approaches from theory and concepts to real devices. As such, the volume will be very useful to Ph.D. students and lecturers of computational physics and numerical mathematics, but also to applied physicists searching for accurate theoretical models to support their tremendous imagination.

Olivier Vanbésien, *Professor*
IEMN, University of Lille
Villeneuve d'Ascq
FRANCE

PREFACE

This is the first volume of a new e-book series that is devoted to very recent research trends in computational physics. Hereby, it focuses on the computational perspectives of current physical challenges, publishing new numerical techniques for the solution of mathematical equations including chapters describing certain real-world applications concisely. The goal of this series is to emphasize especially approaches that are of interdisciplinary nature.

Nowadays, with the help of powerful computers and sophisticated methods of numerical mathematics it is possible to simulate many ultramodern devices, e.g. photonic crystals structures, semiconductor nanostructures or fuel cell stacks devices, thus preventing expensive and longstanding design and optimization in the laboratories. This first volume treats both, mathematical analysis of periodic structure problems and state-of-the art numerical techniques, like frequency domain methods , beam propagation methods and eigenmode expansion methods. Several chapters are devoted to concrete applications of periodic media simulations, E.g. in optical applications these periodic media, have the special capability to select the ranges of frequencies of the waves that are allowed to pass or blocked in the waveguide and act as an efficient frequency filter.

This book consists of 8 invited chapters that are structured in the three parts analysis, numerical techniques and finally practical applications. In the first part analysis we deal with the analysis of periodic square lattices constructed of rhomboidal quantum wells interacting via narrow links. An accurate analysis of Bloch waves, based on DN–maps of the quantum wells is presented. In Chapters 2–3 so–called open periodic waveguides are investigated. First, the problem of resonant enhancement of fields in the waveguide and anomalous transmission of energy across it due to the interaction between guided electromagnetic or acoustic modes is considered. This mechanism for resonant scattering is studied analytically using the Floquet-Bloch decomposition of the periodic differential operator underlying the waveguide structure. Secondly, in Chapter 3, the author uses the spectrum of the Helmholtz operator on an infinite strip with quasiperiodic boundary conditions to describe the propagation of electromagnetic waves in dielectric slab waveguides with periodic corrugations. Hereby, the typical ingredients like guided modes, radiation modes and leaky modes are explained in detail. Furthermore, methods are presented to compute guided and leaky modes by matching the Dirichlet-to-Neumann operator on the corresponding interfaces.

This topic is a good bridge to the second part, the numerical techniques, consisting of Chapters 4–6. In Chapter 4 the authors review several numerical approaches for solving high-frequency scattering problems Most particularly, hereby focusing on the multiple scattering problem where rays are multiply bounced by a collection of separate objects. The next chapter describes a new efficient numerical method to simulate time harmonic wave propagation in infinite periodic media including a local perturbation. Here, the main challenge is the confinement of computations to a bounded region enclosing the perturbation using so–called Dirichlet-to-Neumann (DtN) operators. In Chapter 6 the authors explain how to solve problems with periodic coefficients of periodic geometry efficiently, if they are defined on an unbounded (or very large) domain. Hereby, the usual strategy is to introduce so-called artificial boundaries and impose suitable boundary conditions that mimic the perfect absorption of waves traveling out of the computational domain through the artificial boundaries.

In the last part we present some application, illustrating the impact of the mathematical ideas. In Chapter 7 several potential applications of negative refraction in artificial periodic media in the wavelength range from microwaves down to optics. Both, physical concepts to create such an abnormal propagation regime and practical examples of real devices, like a photonic crystal slab for optical waves, are presented. Finally, Chapter 8 considers electromagnetic waves propagating through periodically heterogeneous layer, involving dielectrics and conductors with the goal to obtain homogenized transmission conditions and to determine the optimal structure of the periodic cells with respect to desired material properties as in meta-materials. The influence of the material composition of the layer influences the reflection and transmission of the scattered fields is discussed concisely.

We would like to thank Prof. Olivier Vanbésien for writing the foreword and providing the figures for the title page and Bentham Science Publishers, particularly Manager Bushra Siddiqui, for their support and efforts.

CONTRIBUTORS

Xavier Antoine	Full Professor, Institut Elie Cartan Nancy (IECN), Université Henri Poincaré Nancy 1, Vandoeuvre-lès-Nancy Cedex, France
Nikolai Bagraev	Full Professor, A.F. Ioffe Physico-Tecnical Institute, St. Petersburg, Russia
Matthias Ehrhardt	Full Professor, Bergische Universität Wuppertal, Fachbereich C Mathematik und Naturwissenschaften, Lehrstuhl für Angewandte Mathematik und Numerische Analysis, Wuppertal, Germany
Sonia Fliss	Assistant Researcher, INRIA POEMS Project and Applied Mathematics Department, ENSTA, Paris, France
Christophe Geuzaine	Full Professor, University of Liège, Department of Electrical Engineering and Computer Science, Montefiore Institute, Liège, Belgium
Patrick Joly	Head of INRIA POEMS Project and Full Professor, Applied Mathematics Department, ENSTA, Paris, France
Günter Leugering	Full Professor, Lehrstuhl für Angewandte Mathematik II, Universität Erlangen–Nürnberg, Erlangen, Germany
Jing-Rebecca Li	Assistant Researcher, INRIA POEMS Project, Domaine de Voluceau – Rocquencourt, Le Chesnay Cedex, France
Gaven Martin	Full Professor, Massey University, Albany, Auckland, NZ Institute for Advanced study, New Zealand
Boris Pavlov	Full Professor, Massey University, Albany, Auckland, NZ Institute for Advanced study, New Zealand and V.A. Fock Institute for Physics of St.-Petersburg University, Petrodvorets, Russia
Karim Ramdani	Assistant Researcher, INRIA (CORIDA Team) & Institut Elie Cartan Nancy (IECN), Université Henri Poincaré Nancy 1, Vandoeuvre-lÃÍs-Nancy Cedex, France
Eduard Rohan	Head of MBS Department, University of West Bohemia, Research Centre New Technologies, Plzen, Czech Republic
Frantisek Seifrt	Associate Professor, Lehrstuhl für Angewandte Mathematik II, Universität Erlangen–Nürnberg, Erlangen, Germany
Stephen P. Shipman	Associate Professor, Department of Mathematics, Louisiana State University, Baton Rouge, USA
Johannes Tausch	Associate Professor, Department of Mathematics, Southern Methodist University, Dallas, Texas, USA
Olivier Vanbésien	Full Professor, Institut d'Electronique, de Microélectronique et de Nanotechnologie, Université des Sciences et Technologies de Lille, France
Chunxiong Zheng	Associate Professor, Department of Mathematical Sciences, Tsinghua University, Beijing, China

CHAPTER 1

Landau–Zener Phenomenon on a Double of Weakly Interacting Quasi-2d Lattices

N. Bagraev [1], G. Martin[2], B. Pavlov[2,3]

[1] *A.F. Ioffe Physico-Technical Institute, St. Petersburg, Russia.*
[2] *New Zealand Institute of Advanced Study, Massey University, New Zealand.*
[3] *V.A. Fock Institute for Physics of St.-Petersburg University, Petrodvorets, Russia.*

Abstract: The dynamics of a single electron is considered on a periodic square lattice constructed of rhomboidal quantum wells interacting via narrow links. The spectral structure of bands and gaps of the lattice is derived from an accurate analysis of Bloch waves, based on DN- maps of the quantum wells. For periodic lattice with rhomboidal periods, a solvable model, is constructed based on a rational approximation of DN- maps of the quantum wells by establishing a communication between them via partial boundary conditions emulating the covalent bonds. In the case of the corresponding double periodic lattice, the weak interaction of the two parallel periodic quasi-2d sub-lattices defines, due to the 2d Landau-Zener effect, a high mobility of the corresponding charge carriers in certain direction on the quasi-momenta plane.

1. INTRODUCTION

We start with comparison of the mysteriously efficient coupled cluster approach to few-body problems of solid-state physics with a modified analytic perturbation procedure based on selection of the first order approximation. Transport properties of periodic lattices are defined by the structure of the corresponding Bloch eigenfunctions. In the 1d case the Bloch eigenfunctions are found based on the transfer matrix constructed of the solutions of the relevant Cauchy problem. This approach fails in 2d, and, generally in the multi-dimensional case, because the Cauchy problem for the multi-dimensional Schrödinger equation is ill-posed. The approach based on "tight binding" ideas (Linear Combination of Atomic Orbitals - LCAO , see [18]) gives a reasonably good qualitative coincidence with experiment, but stays on a shaky mathematical basement.
The recent coupled cluster philosophy, see for instance [4] and references therein, gives a decisive hint for development an alternative approach to analysis of the Bloch functions in multidimensional periodic structures. Indeed, the periods of covalent crystals, sharing electrons, may play a role of *coupled clusters* - elementary blocks of the solid, connected by covalent bounds associated with certain partial summations of the perturbation series. It can be interpreted as substitution of the perturbed operator by the appropriate solvable model, which takes into account most essential spectral features of the perturbed operator, defined by the two-electron correlations.

In this paper we consider a simplest quasi-2d square lattice with rhomboidal periods weakly connected with each other by the relatively narrow links. The wave-functions components supported by the links correspond to covalent bonds. The whole 2d crystal is considered as a periodic quantum network, with the Fermi-level situated on the conductivity band. In the case when the products $K_- d_s$ of the exponents K_- of the evanescent waves and the lengths d_s of the connecting links are large, the matching of the evanescent waves generated by the neighboring clusters can be substituted, on the Fermi level, by the partial zero boundary conditions $P_- \psi\big|_\Gamma = 0$ imposed on the cross-sections Γ of the connecting links (the slots), with the projections P_- onto the entrance subspace E_- of the closed channels. Vice versa, the most essential part of the interaction defined by the covalent bonds is caused by the matching of the wave-functions in the open channels, which correspond to the oscillating component of the wave-function on the connecting links. We guess that a convenient and realistic solvable model of the interaction of the neighboring periods of the 2d lattice can be obtained based on the partial matching conditions $P_+ \psi\Big|_{\Gamma-0} = P_+ \psi\Big|_{\Gamma+0}$, with the complementary orthogonal projection P_+ onto $E_+ = L_2(\Gamma) \ominus E_-$, taking into account only the oscillating waves with the spectral parameter $\lambda = 2mE\hbar^{-2}$ situated inside the temperature interval $\Delta_T = [\Lambda - \frac{2m\kappa T}{\hbar^2}, \Lambda - \frac{2m\kappa T}{\hbar^2}]$ centered at the (scaled) Fermi level $\Lambda = 2mE_F\hbar^{-2}$

of the lattice. Thus the solvable model is defined by the choice of the entrance subspaces $E_{\pm}$ of the open and closed spectral channels. Selection of an appropriate entrance (cross-section) subspaces $E_{\pm}$ of the closed and open channels on the links is actually a freedom we can use to simplify the original problem of matching of all one-body orbitals of the neighboring clusters.

Though the above assumptions concerning the configuration of the links already allow to develop an appropriate analytic perturbation technique for the single-particle spectral analysis on the corresponding quantum network, based on filtering properties of the narrow channel, see for instance [6], we make one more step toward a simpler model, by neglecting the length of the links, but assuming that the one-body spectral problem on the periodic network is considered with the partial matching and zero boundary condition imposed on the common boundary $\Gamma_{\vec{l}} = \Gamma_{\vec{l}'} =: \Gamma$ of the neighboring periods $\Omega_{\vec{l}}$, $\Omega_{\vec{l}'}$:

$$-\Delta\psi_{\vec{l}} + V(x)\psi_{\vec{l}} = \lambda\psi_{\vec{l}},\ x \in \Omega_{\vec{l}},$$

$$\begin{aligned} P_{-}\psi_{\vec{l}}\Big|_{\Gamma} &= P_{-}\psi_{\vec{l}'}\Big|_{\Gamma} = 0, \\ P_{+}\psi_{\vec{l}}\Big|_{\Gamma} &= P_{+}\psi_{\vec{l}'}\Big|_{\Gamma}, \\ P_{+}\frac{\partial \psi_{\vec{l}}}{\partial n}\Big|_{\Gamma_{\vec{l}}} &+ P_{+}\frac{\partial \psi_{\vec{l}'}}{\partial n}\Big|_{\Gamma_{\vec{l}'}} = 0 \end{aligned} \tag{1}$$

and zero boundary condition on $\partial\Omega\backslash\Gamma$. The operator defined on the quantum network $\cup_{\vec{l}}\Omega_{\vec{l}}$ by (1) is self-adjoint and has a pure continuous spectrum. We guess that (1) takes into account an essential part of the interaction forming the Bloch-functions of the corresponding solvable perturbed Hamiltonian, defined by the selected from E_{+} cross-sections orbitals associated with the open channels on the lincs. Based on the model described, with use of the Dirichlet-to-Neumann map, see [16], we develop in this paper spectral analysis of the model 2d lattice from the first principles, and obtain the corresponding dispersion equations.

Based on results obtained for the 2d periodic lattice, we consider the corresponding double lattice assuming a weak interaction between the two sub-lattices. Considering the corresponding 2d Landau-Zener effect we conclude that the charge carriers dynamics on a two-layers periodic lattice may reveal some important features like high mobility in certain direction on the quasi-momenta plane.

Landau-Zener effect is the transformation of the intersection of terms $\lambda_1(p), \lambda_2(p)$, see Fig. (**1**) into

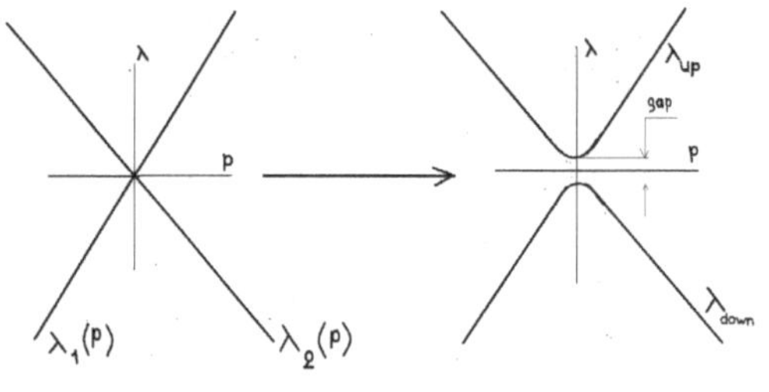

Fig. (**1**): One dimensional Landau-Zener effect.

quasi-intersection. It was observed first, see [17], in one-dimensional lattices, with use of the transfer-matrix as a main spectral tool for study of corresponding space or time- periodic structures, see [8]. It was noticed that the interaction of terms $\lambda_s(p)$ in solid-state quantum problems implies pseudo-relativistic properties of the corresponding particles / quasi-particles. Fresh interest for quasi-relativism in solid state physics arose in connection with discovery of high mobility of charge carriers in graphen, see for instance [13]. Recent discovery of quasi-relativistic behavior of terms in man-made bi-layer periodic quasi-2d lattices, see [2], allows to conjecture that the weak interaction of 2d periodic lattices may be used as a source of various artificial structures with useful transport properties. Study of the Landau-Zener transformation of 2d terms requires an adequate analytic machinery.

2. DISPERSION EQUATION OF THE 2D LATTICE VIA DN-MAP.

In [10] the Dirichlet-to-Neumann map was selected as an appropriate tool to substitute the transfer-matrix in analysis of perturbations of the two-dimensional terms. The standard DN-map is a linear transformation of the boundary "potential" $\psi\big|_{\Gamma}$, $\Gamma \subset \partial\Omega$ into the "boundary current" $\frac{\partial\psi}{\partial n}\big|_{\Gamma}$ of the solution ψ of the homogeneous Schrödinger equation on the domain Ω, with scaled spectral variable $\lambda = \frac{2mE}{\hbar^2}$.

$$-\Delta\psi + V\psi = \lambda\psi,\ \mathscr{D}\mathscr{N}(\lambda): \psi\big|_{\Gamma} \longrightarrow \frac{\partial\psi}{\partial n}\big|_{\Gamma}.$$

It this paper we consider a modified version of DN-map, restricted by an orthogonal projection P_{+} onto a contact subspace E_{+} of $L_2(\Gamma)$. We ignore the spin of electron and initially assume that the one-electron wave functions on the neighboring rhomboidal periods, see Fig. (**2**), communicate with each other via relatively narrow connecting channels, which filter the evanescent waves off, see an extended analysis of the filtering by narrow channels in [6].

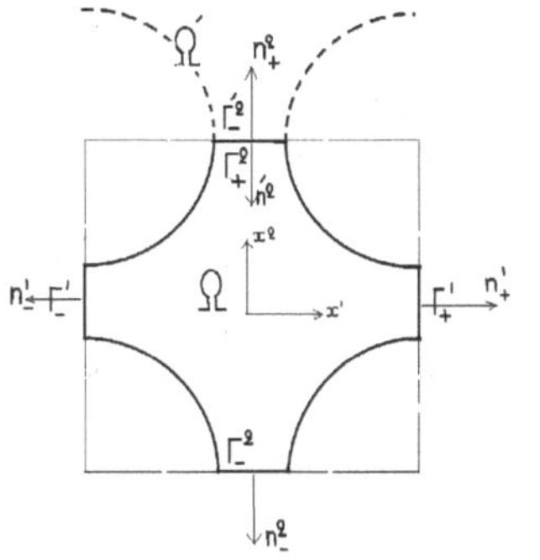

Fig. (**2**): A detail of a square lattice with rhomboidal periods. The connecting leads are not shown.

We simplified the spectral problem via replacement of the matching condition on Γ in closed channels by the partial zero boundary condition on the slots Γ, see (1):

$$P_+ \left[\psi_\Omega - \psi_{\Omega'}\right]\big|_{\Gamma_{\Omega,\Omega'}} = 0,$$

$$P_+ \left[\frac{\partial \psi_\Omega}{\partial n} + \frac{\partial \psi_{\Omega'}}{\partial n'}\right]\big|_{\Gamma_{\Omega,\Omega'}} = 0, \qquad (2)$$

$$P_- \psi\big|_\Gamma = 0$$

We consider the Schrödinger operator $-\Delta * +V(x)* =: L$ on the periodic 2d lattice with periods connected by the open channels only, with normals $n' = -n$. The Schrödinger operator with above boundary conditions is self-adjoint and can be analyzed based on a quasi-periodic problem on the period, with the partial matching boundary condition substituted by the quasi-periodicity on the pairs of opposite slots $\Gamma_\pm^s = \{x^s = \pm 1\}$ of the period Ω:

$$P_+ \left[\psi_\Omega\Big|_{\Gamma_-^s}\right] = e^{-2ip_s a} P_+ \left[\psi_\Omega\Big|_{\Gamma_+^s}\right],$$

$$P_+ \left[\frac{\partial \psi_\Omega}{\partial n}\Big|_{\Gamma_-^s}\right] = -e^{-2ip_s a} P_+ \left[\frac{\partial \psi_\Omega}{\partial n}\Big|_{\Gamma_+^s}\right], \qquad (3)$$

where the differentiation is done with respect to the outward normals on the boundary of the relevant periods, see Fig. (**1**). Hereafter we assume that the width $2a$ of the period is equal to 2, $\delta/2 << 1$, and the entrance subspace E_+^s of the open channel attached to each slot $\Gamma_{\pm a}^s$ is one-dimensional and spanned by $\frac{2}{\delta}\sin\frac{\pi y}{\delta} = e^s$, $P_+^s = e^s\rangle\langle e^s$ on each slot. The electron with the wave-function having a nontrivial boundary data on $\Gamma^s = \partial\Omega \cap \partial\Omega'$ from common contact subspaces $E^s = L_2(\Gamma^s)$ on the slots $\Gamma_{\pm a}^s$ belongs to both periods and forms a covalent bond between the blocks Ω, Ω', see [7].

We consider the *intermediate boundary problem* with the partial boundary data and introduce the corresponding DN-map, see [1], by formal setting the exponential in the closed channels as $K_- = \infty$ and correspondingly choosing the zero boundary conditions on the bottom sections of closed channels. Then the partial DN-map $\mathcal{DN}^\Lambda$ is defined as a restriction of the standard DN-map onto the slots Γ with subsequent framing by the projections $P_+ = \sum_{s=1,2,sgn} P_{sgn}^s$ onto $E_+ = \sum_{s=1,sgn} E_{sgn}^1$:

$$\mathcal{DN}^\Lambda = P_+ \mathcal{DN} P_+$$

$$= \sum_{s,t=1,2,sgn,sgn'} P_{sgn}^s\Big|_{\Gamma_{sgn}^s} \mathcal{DN} P_{sgn'}^t\Big|_{\Gamma_{sgn'}^t}.$$

Hence the partial DN-map $\mathcal{DN}^\Lambda$ is defined by the matrix elements of the standard DN-map of the period in the decomposition of the contact space $E = E_+ + E_- \equiv E = L_2(\Gamma)$ of the slot Γ into an orthogonal sum of the entrance subspaces of the open and closed channels. We characterize the period Ω on given spectral interval Δ_T by the rational approximation with an appropriate correcting term

$$\mathcal{DN}^\Lambda(\lambda) = \sum_{r=1}^{n} \frac{Q_r}{\lambda - \lambda_r} + P_+ K P_+, \ \lambda_r \in \Delta_T, \qquad (4)$$

where

$$\frac{Q_r}{\lambda - \lambda_r} =$$

$$= \sum_{s,sgn;t,sgn'} e_{sgn}^s\rangle \frac{\langle e_{sgn}^s, \frac{\partial \psi_r}{\partial n}\rangle \langle \frac{\partial \psi_r}{\partial n}, e_{sgn'}^t\rangle}{\lambda - \lambda_r} \langle e_{sgn'}^t,$$

$$P_+ K P_+ = \sum_{s,t,sgn,sgn'} e_{sgn}^s\rangle \langle e_{sgn}^s K e_{sgn'}^t\rangle \langle e_{sgn'}^t. \qquad (5)$$

Here λ_r are the eigenvalues of the Schrödinger operator with partial Dirichlet boundary conditions in closed channels on the essential spectral interval Δ_T and $P_+ K P_+$ - the restriction of the correcting (regular on Δ_T) part of the DN-map onto the open channels of the leads. The correcting term contains contributions to the DN - map from the complementary spectral subspace, corresponding to the eigenvalues on the complement of Δ_T.

The spectral structure of the Schrödinger operator on the 2d periodic lattice is established based on study of the quasi-periodic spectral problem on the period Ω, which is defined by the quasiperiodic boundary conditions connecting the projections of the boundary values and the boundary currents of the solutions of the Schrödinger equation $L\psi = \lambda\psi$ on the opposite slots $\Gamma^s_{\pm}$ of the period:

$$P_+\begin{pmatrix}\psi^1_-\\ \psi^1_+\\ \psi^2_-\\ \psi^2_+\end{pmatrix} = P_+\begin{pmatrix}e^{-2ip_1}\psi^1_+\\ \psi^1_+\\ e^{-2ip_2}\psi^2_+\\ \psi^2_+\end{pmatrix} = \psi^1_+ v^1 + \psi^2_+ v^2,$$

$$P_+\begin{pmatrix}\psi'^1_-\\ \psi'^1_+\\ \psi'^2_-\\ \psi'^2_+\end{pmatrix} = P_+\begin{pmatrix}-e^{-2ip_1}\psi'^1_+\\ \psi'^1_+\\ -e^{-2ip_2}\psi'^2_+\\ \psi'^2_+\end{pmatrix} = \psi'^1_+\mu^1 + \psi'^2_+\mu^2, \qquad (6)$$

where

$$v^1 = e^1\begin{pmatrix}e^{-2ip_1}\\ 1\\ 0\\ 0\end{pmatrix},\ v^2 = e^2\begin{pmatrix}0\\ 0\\ e^{-2ip_2}\\ 1\end{pmatrix},$$

$$\mu^1 = e^1\begin{pmatrix}-e^{-2ip_1}\\ 1\\ 0\\ 0\end{pmatrix},\ \mu^2 = e^2\begin{pmatrix}0\\ 0\\ -e^{-2ip_2}\\ 1\end{pmatrix},$$

and

$$\psi^s_+ = \langle \psi\Big|_{\Gamma^s_{+a}}, e^s\rangle,\ \psi'^s_+ = \langle \frac{\partial\psi}{\partial n}\Big|_{\Gamma^s_{+a}}, e^s\rangle.$$

Then the quasi-periodicity condition implies the equation:

$$\mathcal{DN}^{\Lambda}[\psi^1_+ v^1 + \psi^2_+ v^2] = \psi'^1_+\mu^1 + \psi'^2_+\mu^2, \qquad (7)$$

with scalar coefficients ψ^s_+, ψ'^2_+. Notice that $\langle v^s, \mu^t\rangle = 0$, which implies

$$\langle v^1\mathcal{DN}v^1\rangle\psi^1_+ + \langle v^1\mathcal{DN}v^2\rangle\psi^2_+ = 0,$$

$$\langle v^2\mathcal{DN}v^1\rangle\psi^1_+ + \langle v^2\mathcal{DN}v^2\rangle\psi^2_+ = 0 \qquad (8)$$

The condition of existence of the non-trivial Bloch function is represented in the determinant form:

$$\det\begin{pmatrix}\langle v^1,\mathcal{DN}_{11}v^1\rangle & \langle v^1,\mathcal{DN}_{12}v^2\rangle\\ \langle v^2,\mathcal{DN}_{21}v^1\rangle & \langle v^2,\mathcal{DN}_{22}v^2\rangle\end{pmatrix} = 0, \qquad (9)$$

where $\langle v^s, \mathcal{DN}_{st}v^t\rangle =$

$$\sum_{r=1}^{n}\sum_{sgn,sgn'}\frac{\langle v^s\frac{\partial\psi_r}{\partial n}\rangle_{\Gamma^s_{sgna}}\langle\frac{\partial\psi_r}{\partial n}, v^t\rangle_{\Gamma^t_{sgn'a}}}{\lambda-\lambda_r} + \langle v^s, Kv^t\rangle. \qquad (10)$$

Of course all above constructions and arguments concerning 1d slots of 2d periods are automatically transferred to the case of 2d slots of the 3d periods of a quasi-2d lattice in R_3. We leave an exact formulation and verification of the corresponding wording to the reader, but just use it next section of our paper.

3. DISPERSION EQUATION OF THE DOUBLE QUASI–2D LATTICE AND 2D LANDAU–ZENER EFFECT

We aim on the spectral analysis of a double periodic quasi-2d lattice with rhomboidal periods Ω^u, Ω^d playing the roles of basements of the upper and the lower cones of the two-storey joint period, see Fig. (3). Assume that first and the second storeys

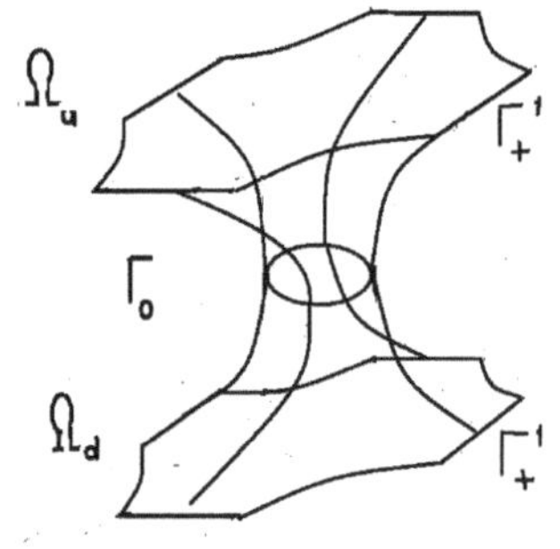

Fig. (3): Detail of the double square quasi-2d lattice with rhomboidal periods. The 2d slots are shown symbolically as 1d slots

.

are connected by the link constructed in a form of a double cone with the slot Γ_0 dividing the upper and lower cones and a tunneling boundary condition on it defined by a real antisymmetric matrix

$$\mathbf{B} : P_0^{\perp}\psi^u\Big|_{\Gamma_0} = 0,\ P_0^{\perp}\psi^d\Big|_{\Gamma_0} :$$

$$\begin{pmatrix} P_0 \frac{\partial \psi^u}{\partial n^u}\Big|_{\Gamma_0^u} \\ P_0 \frac{\partial \psi^d}{\partial n^d}\Big|_{\Gamma_0^d} \end{pmatrix} = \begin{pmatrix} 0 & -\beta \\ \beta & 0 \end{pmatrix} \begin{pmatrix} P_0 \psi^u\Big|_{\Gamma_0^u} \\ P_0 \psi^d\Big|_{\Gamma_0^d} \end{pmatrix}. \quad (11)$$

with the outward normals $n^u = -n^d$, and an orthogonal 1d projection $P_0 = e^0\rangle\langle e^0$ onto the open channels on the slot $\Gamma_0^{u,d}$. This tunneling boundary condition, with large β, emulates the potential barrier/well for the corresponding charge carriers (electrons/ holes, respectively) separating the upper and lower lattices, because implies $\psi^u \approx 0 \approx \psi^d$ if $\beta^{-1} \approx 0$. If the slots $\Gamma_{u,d}^s$ of the upper and lower periods are equipped with the matching boundary conditions on the contact with the neighboring periods, then the Schrödinger operator on the whole lattice, with a real, bounded and piecewise continuous periodic potential is selfadjoint, and the corresponding dispersion equation can be derived from the Bloch condition on a single period, via comparison of the boundary values

$$\vec{\Psi} = \left(\vec{\psi}_1^u, \vec{\psi}_2^u, \psi_0^u, \psi_0^d, \vec{\psi}_2^d, \vec{\psi}_1^d\right),$$

of the wave-functions on the slots $\Gamma_u^s, \Gamma_0, \Gamma_d^t$ of the upper and lower periods and the balance of the corresponding boundary currents with the tunneling boundary condition. Imposing the quasi-periodic boundary conditions on the slots $\Gamma_{sgn\,a}^s(u)$, $\Gamma_{sgn\,a}^s(d)$ and the tunneling boundary conditions on Γ_0^u, Γ_0^d, we obtain the linear system for the variables $\psi_+ = \left(\psi_{+a}^{u1}, \psi_{+a}^{2u}, \psi_0^u, \psi_0^d, \psi_{+a}^{u1}, \psi_{+a}^{u2}\right)$, similar to (9):

$$\langle v_u^1 \mathscr{D}\mathscr{N}_{11}^u v_u^1\rangle \psi_1^u + \langle v_u^1 \mathscr{D}\mathscr{N}_{12}^u v_u^2\rangle \psi_2^u + \langle v_u^1 \mathscr{D}\mathscr{N}_{10}^u\rangle \psi_0^u = 0,$$

$$\langle v_2^u \mathscr{D}\mathscr{N}_{21}^u v_u^1\rangle \psi_1^u + \langle v_u^2 \mathscr{D}\mathscr{N}_{22}^u v_u^2\rangle \psi_2^u + \langle v_u^2 \mathscr{D}\mathscr{N}_{20}^u\rangle \psi_0^u = 0,$$

$$\langle \mathscr{D}\mathscr{N}_{01}^u, v_u^1\rangle \psi_1^u + \langle \mathscr{D}\mathscr{N}_{02}^u, v_u^2\rangle \psi_2^u + \mathscr{D}\mathscr{N}_{00}^u \psi_0^u = -\beta \psi_0^d,$$

$$\langle \mathscr{D}\mathscr{N}_{01}^d, v_d^1\rangle \psi_1^d + \langle \mathscr{D}\mathscr{N}_{02}^d, v_d^2\rangle \psi_2^d + \mathscr{D}\mathscr{N}_{00}^d \psi_0^d = \beta \psi_0^u,$$

$$\langle v_d^2 \mathscr{D}\mathscr{N}_{21}^d v_d^1\rangle \psi_1^d + \langle v_d^2 \mathscr{D}\mathscr{N}_{22}^d v_d^2\rangle \psi_2^d + \langle v_d^2 \mathscr{D}\mathscr{N}_{20}^d\rangle \psi_0^d = 0,$$

$$\langle v_d^1 \mathscr{D}\mathscr{N}_{11}^d v_d^1\rangle \psi_1^d + \langle v_u^1 \mathscr{D}\mathscr{N}_{12}^d v_u^2\rangle \psi_2^d + \langle v_u^1 \mathscr{D}\mathscr{N}_{10}^d\rangle \psi_0^d = 0. \quad (12)$$

Existence of a non-trivial solution of this linear system is guaranteed by an appropriate determinant condition. Denote

$$\langle v_{u,d}^s \mathscr{D}\mathscr{N}_{11}^{u,d} v_{u,d}^s\rangle := d_{st}^{u,d},$$

$$\mathscr{D}\mathscr{N}^{u,d}(p) =:$$

$$= \begin{pmatrix} d_{11}^{u,d} & d_{12}^{u,d} & \langle v^1, \mathscr{D}\mathscr{N}_{10}^{u,d}\rangle \\ d_{21}^{u,d} & d_{22}^{u,d} & \langle v^2, \mathscr{D}\mathscr{N}_{20}^{u,d}\rangle \\ \langle \mathscr{D}\mathscr{N}_{01}^{u,d}, v_u^1\rangle & \langle \mathscr{D}\mathscr{N}_{02}^{u,d}, v_{u,d}^2\rangle & \mathscr{D}\mathscr{N}_{00}^{u,d} \end{pmatrix},$$

$$\mathscr{D}\mathscr{N}_T^{u,d}(p) =:$$

$$= \begin{pmatrix} \langle v_{u,d}^1 \mathscr{D}\mathscr{N}_{11}^{u,d} v_{u,d}^1\rangle & \langle v_{u,d}^1 \mathscr{D}\mathscr{N}_{12}^{u,d} v_{u,d}^2\rangle \\ \langle v_{u,d}^2 \mathscr{D}\mathscr{N}_{21}^{u,d} v_{u,d}^1\rangle & \langle v_{u,d}^2 \mathscr{D}\mathscr{N}_{22}^{u,d} v_{u,d}^2\rangle \end{pmatrix}. \quad (13)$$

Then the determinant condition, because of $v^s = v^s(p_s)$, gives the dispersion equation

$$\beta^{-2} \det \mathscr{D}\mathscr{N}^u \det \mathscr{D}\mathscr{N}^d + \det \mathscr{D}\mathscr{N}_T^u \det \mathscr{D}\mathscr{N}_T^d = 0. \quad (14)$$

In particular, if $\beta \to \infty$, the linear system splits into a pair of independent blocks, corresponding to the upper and lower period, with the dispersion equations

$$\det \mathscr{D}\mathscr{N}_T^u = 0 \text{ and } \det \mathscr{D}\mathscr{N}_T^d = 0$$

similar to ones we obtained in previous section. If β is large, then the intersection of terms

$$\det \mathscr{D}\mathscr{N}_T^u(p) \det \mathscr{D}\mathscr{N}_T^d(p) = 0$$

is transformed into a quasi-intersection. The transport properties for large β near to the intersection of the unperturbed terms (for $\beta = \infty$) are defined by the tensor of second derivatives of λ with respect to the quasi-momentum p. In particular, the mobility of the charge carriers at the quasi-intersection is defined by the tensor $m^{-1} = \left\{\frac{\partial \lambda}{\partial p_s \partial p_t}\right\}_{s,t=1}^{s,t=2}$ of the second derivatives of the dispersion function $\lambda(\vec{p})$. For a weak interaction of the layers, the maximal eigenvalue of the tensor (and hence the minimal effective mass and maximal mobility) is observed on the quasimomenta plane in the direction n orthogonal to

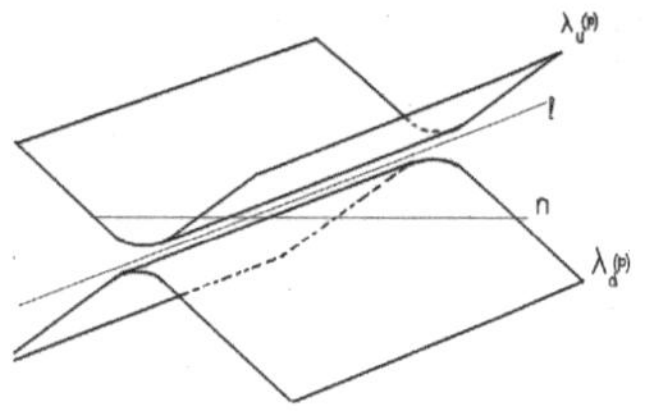

Fig. (**4**): Two-dimensional Landau-Zener effect.

the intersection *l* of the tangent planes of the dispersion surfaces of the upper and lower layers of the unperturbed double lattice. Depending on position of the Fermi level of the double lattice the charge carriers are either electrons or holes. Suggested in this paper analysis of the Bloch functions of 2d periodic lattices may serve a basement for calculation of the correlations of the wave-functions initially obtained in form of the Slater determinants, see the corresponding remark in [4] p. 293.

CONCLUSION

The physical approach and the mathematical technique suggested above for analysis of the resonance processes on the periodic systems of quantum dots are based on fitted "zero-range" solvable models, which has mathematical roots in the von Neumann operator extension theory, see [12]. Though the discovery of the theory was done by John von Neumann based on a deep physical motivation, physicists never used it, in original von Neumann form, before 1964, when the seminal paper [5] was published, where the direct connection between E. Fermi zero-range potential see [9] and von Neumann theory was established. Yet another 50 years were needed to notice that the zero-range model can be fitted, based on a special choice of the inner structure and another another 20 years were needed to see , that the fitted zero-range model may be used as a first approximation in the modified analytic perturbation procedure, see [1, 15]. Now we see some prospects of using of this mathematical technique for analysis of experimental results obtained in the studies of the edge channels in the ultra-narrow quantum wells as well as high mobility of charge carriers in self-assembled Silicon-based low-dimensional periodic structures. Specifically, this theoretical analysis allows to trace the influence of the quantum dots and single impurity centers, embedded in the edge channels, on the characteristics of the spin-dependent scattering, revealed by the quantum Hall effect and quantum spin Hall effect measurements as well as other features revealed by the electrically-detected electron paramagnetic resonance, see for instance [2, 3].

ACKNOWLEDGEMENT

The authors are grateful to Kyle Beloy and Anastasia Borschevsky for a crucial remark concerning the leading role of 2-electrons correlations in coupled clusters techniques and R. Bartlett for encouraging discussion of basic ideas and state of art of the coupled-clusters theory, as well as the reprint [4] provided.

REFERENCES

[1] N. Bagraev, A. Mikhailova, B. Pavlov, L. Prokhorov, and A. Yafyasov. Parameter regime of a resonance quantum switch. Phys Rev B 2005; 71: 1-16.

[2] Bagraev N et al. Superconducting properties of silicon nanostructures. Phys Tech Semicond 2009; 43: 1481-1495.

[3] Bagraev N, Gehlhoff W, Gets DS, Klyachkin LE. Kudryavtsev AA, Malyarenko AM, Mashkov VA, Romanov VV. EDEPR of impurity centers embedded in silicon microcavities. Physica B 2009; 404: 5140-5143.

[4] Bartlett RJ, Musial M. Coupled-cluster theory in quantum chemistry. Rev Mod Phys 2007; 79: 291-351.

[5] Berezin FA, Faddeev LD. A remark on Schrödinger equation with a singular potential. Soviet Math Dokl 1961; 2: 372-376.

[6] Brüning J, Martin G, Pavlov B. On calculation of Kirchhoff coefficients for Helmholtz Resonator. Russ J Math Phys 2009; 16: 188-207.

[7] Callaway J. Energy band theory. Academic Press, NY-London, 1964.

[8] Demkov YN, Kurasov PB, Ostrovski VN. Double-periodical in time and energy solvable system with two interacting set of states. J Physics A, Math and General 1995; 28: 4361-4380

[9] Fermi E. Sul motto dei neutroni nelle sostance idrogenate (in Italian), Ricerca Scientifica 1936; 7; 13.

[10] Fox C, Oleinik V, Pavlov B. A Dirichlet-to-Neumann approach to resonance gaps and bands of periodic networks. Contemp Math 2006; 412: 151-169.

[11] Harmer M, Pavlov B, Yafyasov A. Boundary condition at the junction. J Comput Electr 2007; 6: 153-157.

[12] von Neumann J. Mathematical foundations of quantum mechanics. Twelfth printing. Princeton Landmarks in Mathematics. Princeton Paperbacks. Princeton University Press, Princeton, NJ, 1996.

[13] Novoselov K, Geim A, Morozov S, Jiang D, Katsnelson I, Grigorieva I, Dubonos S. Two-dimensional gas of massless Dirac fermions in graphen. Nature 2005; 438: 197-200.

[14] Pavlov B. S-Matrix and Dirichlet-to-Neumann Operators. In: Encyclopedia of Scattering. ed. Pike R, Sabatier P, Academic Press, Harcourt Science and Tech. Company, 2001: 1678-1688.

[15] Pavlov B. A solvable model for scattering on a junction and a modified analytic perturbation procedure. Oper Theory Adv Appl 2009; 197: 281-335.

[16] Sylvester J, Uhlmann G. The Dirichlet to Neumann map and applications. In: Proceedings of the Conference Inverse problems in partial differential equations (Arcata,1989)". SIAM, Philadelphia, 1990: 101.

[17] Zener C. Non-adiabatic crossing of energy-levels. Proc R Soc A 1932; 137: 696.

[18] Ziman J. Electrons and phonons: the theory of transport phenomena in solids. Oxford: Clarendon Press, 1960.

[19] Ziman J, Mott N, Hirsch P. The Physics of Metals. London, Cambridge, 1969.

CHAPTER 2

Resonant Scattering by Open Periodic Waveguides

Stephen P. Shipman

Louisiana State University, Baton Rouge, Louisiana, USA.

Abstract: This chapter concerns the interaction between guided electromagnetic or acoustic modes of a penetrable periodic planar waveguide and plane waves originating from sources exterior to the waveguide. The interaction causes resonant enhancement of fields in the waveguide and anomalous transmission of energy across it. A guided mode is an eigenfunction of a member of the family of operators in the Floquet-Bloch decomposition of the periodic differential operator underlying the waveguide structure. The theory of existence or nonexistence of modes in ideal lossless waveguides is founded on variational principles. The mechanism for resonant scattering behavior is the dissolution of an embedded eigenvalue into the continuous spectrum, which corresponds to the destruction of a guided mode of a waveguide, upon perturbation of the wavevector or the material properties or geometry of the structure. Analytic perturbation of functions that unify the guided modes and the extended scattering states gives rise to asymptotic formulas for transmission anomalies.

1. INTRODUCTION

A planar structure acts both as a guide of electromagnetic or acoustic waves as well as a scatterer of waves that originate from sources exterior to it. An open waveguide is one that is in contact with the ambient space; the effect of this contact is twofold: (1) Fields produced by sources in the guide lose energy through radiation into the ambient space; and (2) fields originating in the ambient space interact with the guided modes of the slab. This Chapter is concerned with the latter, and, in particular, with resonant phenomena that result from this interaction.

Consider, for example, a periodically perforated film, an infinite wall of dielectric pillars, or the slab structure in Fig. (**1**). If a field strikes the structure, the coherent scattering by the periodic geometry can result in enhanced transmission or reflection of energy in narrow frequency intervals, causing the structure to act as a frequency-specific filter. In many cases, the explanation for this anomalous transmission is understood to be the resonant interaction between the incident field and the guided modes of the structure. There is a large body of literature dedicated to this phenomenon; we shall discuss some of it later on.

The fundamental mechanism of resonance can be understood as follows. A waveguide (closed or open) possesses a "dispersion relation" relating the frequency of a guided wave to its wavevector, which in the case of a slab waveguide is two-dimensional. The ambient space has its own dispersion relation, and, there, the wavevector is three-dimensional.

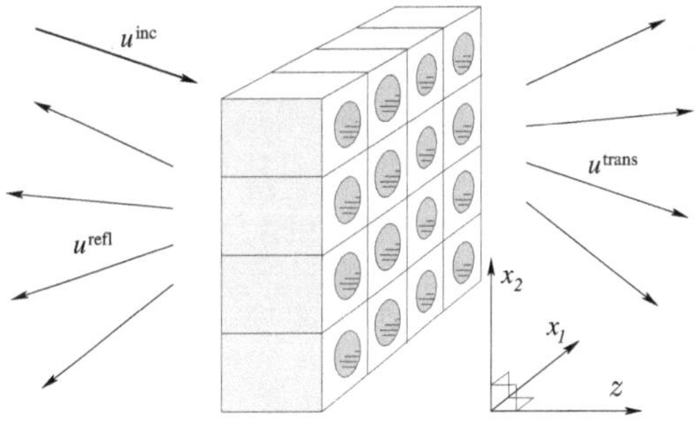

Fig. (**1**): A slab waveguide that is doubly periodic in $x = (x_1, x_2)$ and finite in z (sixteen periods are shown). It is in contact with a homogeneous ambient space.

This means that a two-dimensional wavevector parallel to the waveguide admits a discrete set of frequencies corresponding to guided modes but a continuum of frequencies corresponding to plane waves in the ambient space because of the additional spatial degree of freedom there. Resonance occurs when the frequency of a plane wave and the component of its wavevector parallel to the slab lie close to the dispersion relation for the waveguide itself. This is an expression of the idea of interaction between plane waves and guided modes.

The periodicity of the slab acts as a mechanism for this type of resonance. It causes incident plane waves to be diffracted into a finite number outward-propagating plane waves (Rayleigh-Bloch scattering), which are accompanied by slow surface waves that fall off exponentially with the distance from the slab (Rayleigh-Bloch surface waves). Periodicity causes the frequencies for (real) wavevectors of guided modes to be complex because of "coupling" to the Rayleigh-Bloch scattered waves. In physical terms, the mode is not truly guided but is "leaky". When the imaginary part of the frequency is very small, plane waves in the ambient medium that oscillate at the real part of the frequency scatter resonantly in the slab. We will analyze in detail the situation in which the slab structure admits a true guided mode (with real frequency) that is *isolated* in the sense that nearby frequencies of guided modes are leaky (with complex frequency). The frequency of such a guided mode can be conceived as an eigenvalue that is embedded in the continuous spectrum corresponding to a specific wavevector, and the dissolution of the eigenvalue upon perturbation of the wavevector or structural parameters corresponds to the destruction of the guided mode. This gives rise to anomalous transmission of plane-wave energy across the slab and resonant enhancement of the field in the structure. Some of these anomalies are seen in Figs. (**6**),(**7**),(**8**),(**9**) in Sec. 4.4.

The exposition of the material proceeds as follows.

§2. A tutorial on the effects of periodicity of an open slab waveguide; the emergence of Rayleigh-Bloch scattering and nonrobust guided modes. General context of scattering of fields by a periodic slab and formulation of the problem of scattering of plane waves through the Floquet transform.

§3. Variational formulation of the scattering problem for pseudo-periodic fields and guided modes. Existence and nonexistence of guided modes; dispersion relations for robust modes and embedded frequencies of nonrobust modes.

§4. The extension of the scattering problem to complex frequency and wavevector. Boundary-integral equations for scattering and guided modes and the role of the Calderón boundary-integral operators.

§5. Resonance. Discussion of some of the literature on enhanced transmission. The Fano resonance in quantum mechanics. Transmission anomalies near nonrobust guided modes of a periodic slab; formulas for the anomalies with error estimates; analysis of resonant amplitude enhancement.

Most of the analysis is presented for the case of scalar waves in a penetrable and lossless doubly periodic slab waveguide in three-dimensional space. The ideas and results are not specific to any particular geometry, such as layers or gratings. Moreover, they extend to the Maxwell system and slabs fabricated from perfect conductors or acoustically hard or soft components, as well as lattice models. The focus is on lossless materials, although much of the analysis is extensible to lossy materials.

Let us set the notation for scalar waves. The periodic slab structure is defined through two material coefficients, $\varepsilon(x,z)$ and $\mu(x,z)$, where

$$x \in \mathbb{R}^2, \quad z \in \mathbb{R},$$

that are doubly 2π-periodic in x between the parallel planes $z = z_-$ and $z = z_+$ and constant in the homogeneous ambient space outside the region between these planes. Both coefficients are positive measurable functions bounded from below and above:

$$\begin{aligned}
&\varepsilon(x+2\pi n,z) = \varepsilon(x,z),\ n \in \mathbb{Z}^2,\\
&\mu(x+2\pi n,z) = \mu(x,z),\ n \in \mathbb{Z}^2,\\
&0 < \varepsilon_- \le \varepsilon(x,z) \le \varepsilon_+, \quad 0 < \mu_- \le \mu(x,z) \le \mu_+,\\
&\varepsilon(x,z) = \varepsilon_0,\ \mu(x,z) = \mu_0 \ \text{ if } z \le z_- \text{ or } z \ge z_+.
\end{aligned}$$

The scalar wave equation in this open slab structure is

$$\varepsilon\frac{\partial^2 u}{\partial t^2} - \nabla\cdot\mu^{-1}\nabla u = 0. \tag{1}$$

The assumption of harmonic time dependence

$$u(x,z;t) = u(x,z)e^{-i\omega t}$$

results in the Helmholtz equation for the spatial factor $u(x,z)$,

$$\nabla\cdot\mu^{-1}\nabla u + \omega^2\varepsilon\, u = 0. \tag{2}$$

2. SCATTERING BY A PERIODIC SLAB

The purpose of this section is to expound the role that structural periodicity plays in the coupling of plane waves and guided modes of an open slab waveguide, as this mechanism will be central to the analysis of resonance later on. We then sketch how the problem of plane-wave scattering fits into the general context of time dynamics.

2.1 Effects of Periodicity

We begin in a nontechnical way by examining the simplest time-harmonic solutions of the wave equation in the presence of the slab. We shall progress from the simple case of scattering of plane waves by a slab with no genuine periodicity to scattering by a general lossless periodic slab in order to understand the concepts of Rayleigh-Bloch diffraction, robust and nonrobust (embedded) guided modes, cutoff frequencies, and the interaction between extended (scattering) states and guided modes.

Fig. (**2**)A shows a two-dimensional cross-section of a homogeneous slab, invariant in the x-variables. A plane wave w^{inc} with wavevector $\langle \kappa, \eta \rangle$, $\kappa = \langle \kappa_1, \kappa_2 \rangle$, satisfying the wave equation (1) is traveling to the right at an angle of $\arctan(|\kappa|/\eta)$ with the line perpendicular to the slab,

$$\begin{aligned} w^{\text{inc}}(x,z;t) &= \cos(\kappa \cdot x + \eta z - \omega t), \\ |\kappa|^2 + \eta^2 &= \varepsilon_0 \mu_0 \omega^2, \\ \omega > 0, &\quad \eta > 0, \end{aligned}$$

and strikes the slab from the left. Part of the energy is reflected back to the left and part is transmitted through to the right.

The incident field w^{inc} is the real part of $u^{\text{inc}}(x,z)e^{-i\omega t} = e^{i(\kappa\cdot x + \eta z)}e^{-i\omega t}$, and the spatial part u^{inc} satisfies the Helmholtz equation (2). Dropping the time factor, we can write the total field as $u(x,z) = e^{i\kappa \cdot x}\phi(z)$, where

$$\phi(z) = \begin{cases} e^{i\eta z} + a e^{-i\eta z}, & z < z_-, \\ c e^{i\nu z} + d e^{-i\nu z}, & z_- < z < z_+, \\ b e^{i\eta z}, & z_+ < z. \end{cases} \qquad (3)$$

The common x-factor is necessary for continuity, and, in the slab,

$$|\kappa|^2 + \nu^2 = \varepsilon_1 \mu_1 \omega^2. \qquad (4)$$

The complex reflection and transmission coefficients a and b, as well as c and d, are determined by the interface conditions at $z = z_-$ and $z = z_+$,

$$\begin{aligned} u(x, z_\pm - 0) &= u(x, z_\pm + 0), \\ \mu_0^{-1} \tfrac{\partial}{\partial z} u(x, z_- - 0) &= \mu_1^{-1} \tfrac{\partial}{\partial z} u(x, z_- + 0), \\ \mu_1^{-1} \tfrac{\partial}{\partial z} u(x, z_+ - 0) &= \mu_0^{-1} \tfrac{\partial}{\partial z} u(x, z_+ + 0). \end{aligned}$$

An analogous problem describes the scattering of a plane wave $u^{\text{inc}}(x,z) = e^{i(\kappa\cdot x - \eta z)}e^{-i\omega t}$ incident upon the slab from the right.

For each of these solutions,

$$|\kappa| < \omega\sqrt{\varepsilon_0 \mu_0}. \qquad (5)$$

This conical region in (κ, ω)-space is the *light cone* for the exterior medium (blue region in Fig. (**2**)A), and for each pair in this region, there is a two-parameter family of *scattering states* spanned by the solution with z-dependent factor (3) and its counterpart with incident field from the right.

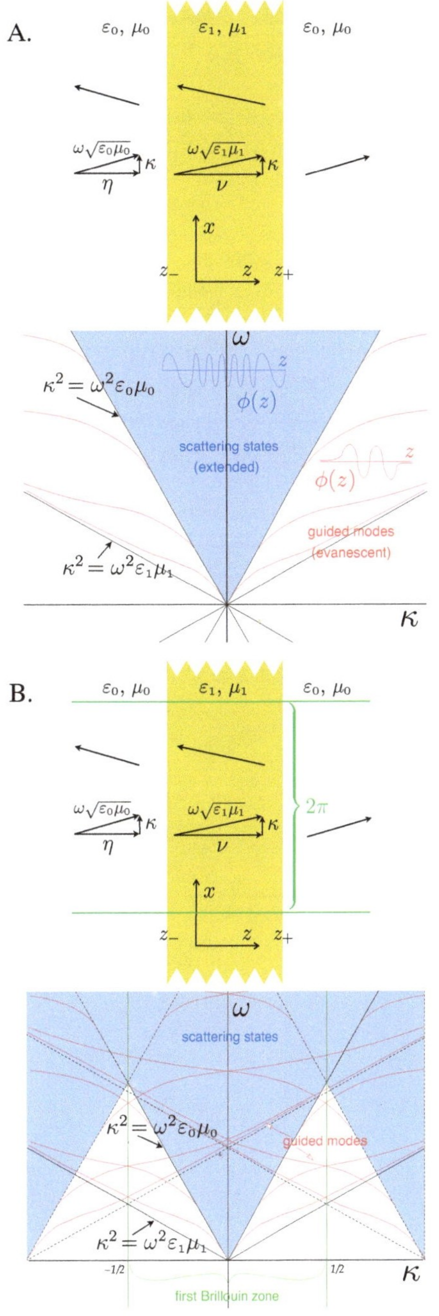

Fig. (**2**): **A.** A homogeneous 2D slab scatters a plane wave incident on the left. The arrows indicate the directions of the plane-wave components of the total field. The (κ, ω) for all scattering states lies in the blue light cone for the exterior medium. Dispersion relations (red) for guided modes lie between the interior and exterior light cones. The behavior of the z-factor in the separable fields $\phi(z)e^{i\kappa x}$ is indicated by the inset graphs. **B.** By imposing an artificial period of 2π in a uniform slab, the extended and guided Bloch (pseudo-periodic) states can be represented by their reduced wavenumber in the first Brillouin zone by shifting the wavenumber by an integer.

Assuming that $\varepsilon_0\mu_0 < \varepsilon_1\mu_1$, Fig. (2)A depicts three regions in (κ, ω)-space,

$$0 < \omega^2\varepsilon_0\mu_0 - |\kappa|^2 < \omega^2\varepsilon_1\mu_1 - |\kappa|^2,$$
$$\omega^2\varepsilon_0\mu_0 - |\kappa|^2 < 0 < \omega^2\varepsilon_1\mu_1 - |\kappa|^2,$$
$$\omega^2\varepsilon_0\mu_0 - |\kappa|^2 < \omega^2\varepsilon_1\mu_1 - |\kappa|^2 < 0.$$

The first region parameterizes the scattering states. It is the (κ, ω)-region in which $\phi(z)$ is oscillatory inside and outside the slab. In the third region, there are no scattering states because the pair (κ, ω) violates (5). In fact, $\phi(z)$ is exponential inside and outside the slab, and there are therefore no solutions that are uniformly bounded in magnitude over $z \in \mathbb{R}$, which we require for our Helmholtz fields (exponentially growing solutions do not play a direct role in the spectral decomposition).

In the second region, there are no scattering states either. But, although $\phi(z)$ is exponential outside the slab, it is still oscillatory inside. The requirement that a solution to the Helmholtz equation be bounded in magnitude leads to solutions of the form

$$\phi(z) = \begin{cases} e^{\gamma z}, & z < z_-, \\ ce^{i\nu z} + de^{-i\nu z}, & z_- < z < z_+, \\ be^{-\gamma z}, & z_+ < z, \end{cases} \tag{6}$$

where $\gamma > 0$. The solution $e^{i(\kappa\cdot x - \omega t)}\phi(z)$ is a *guided mode* of the slab, exponentially decaying as $|z| \to \infty$ and propagating along the slab in the direction of κ with phase velocity

$$v_{\text{ph}} = \omega/|\kappa|. \tag{7}$$

This velocity is less than that of waves in the ambient medium.

The condition that ϕ decay on both sides of the slab is satisfied only on certain relations between κ and ω that are branches of the *dispersion relation* for guided modes. There is one branch for each non-negative integer n, and one computes that they are expressed in the form

$$L\Theta(\kappa, \omega) - 2\Phi(\kappa, \omega) = n\pi, \; n \in \mathbb{N}, \tag{8}$$

in which

$$L = z_+ - z_-,$$
$$\Theta = (\omega^2\varepsilon_1\mu_1 - |\kappa|^2)^{1/2},$$
$$\alpha = \frac{\mu_1}{\mu_0}\frac{(|\kappa|^2 - \omega^2\varepsilon_0\mu_0)^{1/2}}{(\omega^2\varepsilon_1\mu_1 - |\kappa|^2)^{1/2}},$$
$$\Phi = \arctan\alpha.$$

These relations define a sequence of functions $W_n(\kappa)$ with domains $\mathscr{D}(W_n) = \mathbb{R}\setminus[-\kappa_n, \kappa_n]$ and images (ω_n, ∞), such that (8) is expressed as the union of the graphs

$$\omega = W_n(\kappa), \; \kappa \in \mathscr{D}(W_n), \; n \in \mathbb{N}.$$

These graphs are the *branches* of the dispersion relation for guided modes and are shown in red in Fig. (2)A. One can show that each function W_n is increasing, emanates tangentially from the light cone for the exterior medium at $(\pm\kappa_n, \omega_n)$, and is asymptotically tangent to the light cone for the interior medium as $|\kappa| \to \infty$. The points (κ_n, ω_n) are

$$(\kappa_n, \omega_n) = \frac{n\pi}{L\sqrt{\varepsilon_1\mu_1 - \varepsilon_0\mu_0}}(\sqrt{\varepsilon_0\mu_0}, 1).$$

Fig. (2)A shows that, for each value of κ, there is a continuous spectrum of frequencies $\omega \in (|\kappa|/\sqrt{\varepsilon_0\mu_0}, \infty)$ that admit scattering states and a finite number of guided mode frequencies given by $W_n(\kappa)$ for all n such that $\kappa_n < |\kappa|$. The frequency $|\kappa|/\sqrt{\varepsilon_0\mu_0}$ on the light cone is the *cutoff frequency* for the slab structure and wavevector κ.

We must keep in mind that the solutions we have constructed are purely monochromatic fields with infinite energy (but finite energy density), and, as such, are idealized fields that are not truly physically viable in isolation. The incident fields in the scattering states are thought of as originating from sources infinitely far away from the slab, and the guided modes are thought of as being excited by sources in or near the slab but infinitely far from an observer in the slab. Integral superpositions of monochromatic fields form finite-energy solutions of the wave equation; this spectral theory is discussed below.

In Fig. (3)C we see a two-dimensional depiction of a slab with a genuine periodicity of 2π. Solutions of the Helmholtz equation no longer have the separable form $u(x,z) = u_1(x)u_2(z)$. But since the slab is periodic and the incident field in the scattering problem exhibits only a phase difference of $2\pi\kappa\cdot n$ between the points (x,z) and $(x+2\pi n, z)$, we may take the point of view that, when the incident field is scattered by the slab, an observer at $(x+2\pi n, z)$ will experience a field that differs from that experienced by an observer at (x,z) only by a phase of $2\pi\kappa\cdot n$. This means that we are interested in *pseudo-periodic*, or *quasi-periodic*, solutions of the Helmholtz equation

$$u(x,z;\kappa) = e^{i\kappa\cdot x}\tilde{u}(x,z;\kappa), \tag{9}$$

in which $\tilde{u}(x,z;\kappa)$ is periodic in x. Such a solution u is a *Bloch wave* with *Bloch wavevector* κ, *Bloch*

C.

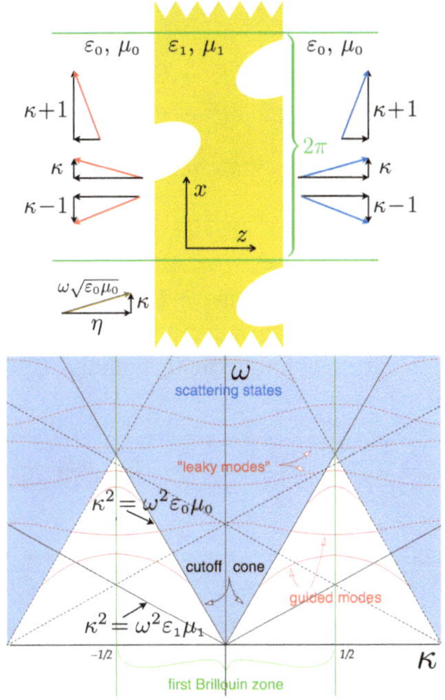

D.

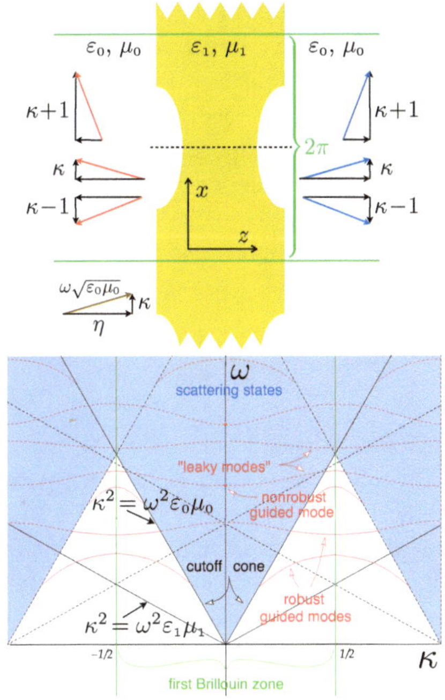

Fig. (**3**): **C.** The periodicity of the slab causes an incident plane wave (brown) to be scattered into a finite number of directions. The reflected diffractive orders are indicated in red and the transmitted in blue. Real dispersion relations split apart (cf. Fig. (**2**)B) at the crossing points (this diagram is qualitative and some of the branches are not drawn). The dotted curves indicate the real part of the frequency that has become complex due to the periodicity. **D.** Because of the symmetry of the structure about the dotted line, the frequency on the dispersion relation may be real for $\kappa = 0$. Such isolated real points on a complex dispersion relation correspond to nonrobust guided slab modes.

factor $e^{i\kappa\cdot x}$, and *Floquet multiplier* $e^{2\pi i\kappa\cdot n}$. The corresponding physical solution (with the (κ,ω)-dependence suppressed) is

$$\begin{aligned}\mathrm{Re}\,(\tilde{u}(x,z)e^{i(\kappa\cdot x-\omega t)}) \\ = |\tilde{u}(x,z)|\cos(\theta(x,z)+\kappa\cdot x-\omega t) \quad (10)\end{aligned}$$

$(\theta(x,z) = \arg\tilde{u}(x,z))$, which is a wave that is distorted in a periodic fashion in the x-variables by the slab structure.

Now, the periodic factor in a Bloch wave can be written as a Fourier series in the variable x parallel to the slab, with coefficients that depend on the transverse variable z,

$$\tilde{u}(x,z) = \sum_{m\in\mathbb{Z}^2} \phi_m(z)e^{im\cdot x}. \qquad (11)$$

Because u satisfies a homogeneous Helmholtz equation exterior to the slab, there, ϕ_m must have the form

$$\phi_m(z) = c_m^1\phi_m^1(z) + c_m^2\phi_m^2(z), \qquad (12)$$

in which $\phi^{1,2}$ are independent solutions of the ordinary differential equation $\phi_m'' + \eta_m^2\phi_m = 0$, where the numbers η_m are defined through

$$\eta_m^2 + |m+\kappa|^2 = \omega^2\varepsilon_0\mu_0. \qquad (13)$$

The functions $\phi_m^{\pm}$ are either oscillatory, linear, or exponential, depending on the numbers η_m. The Fourier components of u are known as the *spatial harmonics* or the *diffractive (or diffraction) orders* associated with the periodic structure. There are many references that expound these ideas, including C. Wilcox [88] and M. Nevière [62]. We distinguish between the following classes of spatial harmonics, for fixed parameters (κ,ω):

$$\begin{aligned} &m \in \mathscr{Z}_p \Leftrightarrow \eta_m^2 > 0,\ \eta_m > 0 && \text{(propagating)},\\ &m \in \mathscr{Z}_\ell \Leftrightarrow \eta_m^2 = 0,\ \eta_m = 0 && \text{(linear)},\\ &m \in \mathscr{Z}_e \Leftrightarrow \eta_m^2 < 0,\ -i\eta_m > 0 && \text{(evanescent)}. \end{aligned} \qquad (14)$$

The first class $\mathscr{Z}_p$ of oscillatory harmonics is finite, the class $\mathscr{Z}_\ell$ of linear harmonics is generically empty but always finite, and the class $\mathscr{Z}_e$ of exponential harmonics is infinite. The latter harmonics are called evanescent because of the requirement that the component that is exponentially growing as $|z| \to \infty$ vanish.

The general bounded pseudo-periodic solution of the Helmholtz equation can be interpreted as the field resulting from scattering of plane waves by the slab:

$$u(x,z) = \sum_{m\in\mathscr{Z}_p} a_m^{\text{inc}} e^{i\eta_m z} e^{i(m+\kappa)\cdot x} + \sum_{m\in\mathbb{Z}^2} a_m e^{-i\eta_m z} e^{i(m+\kappa)\cdot x} \quad (z<z_-), \quad (15)$$

$$u(x,z) = \sum_{m\in\mathscr{Z}_p} b_m^{\text{inc}} e^{-i\eta_m z} e^{i(m+\kappa)\cdot x} + \sum_{m\in\mathbb{Z}^2} b_m e^{i\eta_m z} e^{i(m+\kappa)\cdot x} \quad (z>z_+). \quad (16)$$

The sums over $\mathscr{Z}_p$ represent traveling source waves incident upon the slab from the left and right, generalizing the incident field to a sum of plane waves, each of whose Bloch wavevector differs from κ by an integer pair $m\in\mathscr{Z}_p$. The sums over $\mathbb{Z}^2$ represent the reflected and transmitted fields. The scattered, or diffracted, field is the difference $u(x,z)-u^{\text{inc}}(x,z)$, which is radiating. We say that a function u is *radiating* (or *outgoing*) if it is of the form (15,16) with $a_m^{\text{inc}}=0$ and $b_m^{\text{inc}}=0$ for all $m\in\mathscr{Z}_p$.

Condition 1 (Radiation) *A complex-valued function u defined on* $\mathbb{R}^3$ *is said to satisfy the* radiation condition *for the slab for the real pair* (κ,ω), *with* $\omega>0$, *if there exist a real number* z_0 *and complex coefficients* $\{c_m^{\pm}\}_{m\in\mathbb{Z}^2}$ *in* $\ell^2(\mathbb{Z}^2)$ *such that*

$$u(x,z) = \sum_{m\in\mathbb{Z}^2} c_m^{\pm} e^{\pm i\eta_m z} e^{i(m+\kappa)\cdot x} \quad \textit{for } \pm z > z_0.$$

Because the Bloch form $u(x,z)=e^{i\kappa\cdot x}\tilde{u}(x,z)$, is equivalent to the form $u(x,z)=e^{i(\kappa+m)\cdot x}\tilde{u}(x,z)$ by the change of periodic part $\tilde{u}(x,z)\mapsto\tilde{u}(x,z)e^{-im\cdot x}$, one can always assume that κ lies in the first *Brillouin zone*,

$$B=[-1/2,1/2)^2, \quad (17)$$

shown in Fig. (**3**)C for the two-dimensional case. It is then guaranteed that $|\kappa+m|\geq|\kappa|$ for all $m\in\mathbb{Z}^2$. This value of κ is called the *reduced wavevector* for u.

The harmonic pseudo-periodic scattering problem is stated below. Proof that it has a solution that is typically unique is achieved through a weak formulation of the problem, which we present later.

Problem 2 (Plane-wave scattering) *Given* $\omega>0$ *and* $\kappa\in B$, *find a function u on* $\mathbb{R}^3$ *that is doubly* 2π-*pseudo-periodic in x with Bloch wavevector* κ *and that satisfies the Helmholtz equation* (2) *and such that*

$$u(x,z)=u^{\text{inc}}(x,z)+u^{\text{sc}}(x,z),$$

in which

$$u^{\text{inc}}(x,z) = \sum_{m\in\mathscr{Z}_p} (a_m^{\text{inc}} e^{i\eta_m z} + b_m^{\text{inc}} e^{-i\eta_m z}) e^{i(m+\kappa)\cdot x}$$

and u^{sc} *satisfies the radiation Condition 1.*

A **guided mode** is a nontrivial solution to Problem 2 such that $a_m^{\text{inc}}=b_m^{\text{inc}}=0$ for all $m\in\mathscr{Z}_p$. As long as $\mathscr{Z}_\ell=\emptyset$, such solutions necessarily fall off exponentially with $|z|$ because of the conservation of energy law for Helmholtz fields with the expansions (15,16),

$$\sum_{m\in\mathscr{Z}_p} \eta_m(|a_m^{\text{inc}}|^2+|b_m^{\text{inc}}|^2) = \sum_{m\in\mathscr{Z}_p} \eta_m(|a_m|^2+|b_m|^2).$$

This can be proved by integration by parts applied to the Helmholtz equation (2) multiplied by $\bar{u}$.

Because of the periodicity of the slab, all the spatial harmonics for a given wavevector $\kappa\in B$ are coupled. They cannot exist in isolation; rather, in each Helmholtz field, all harmonics are generally present. A distinctive structure emerges from this observation: when a plane-wave source is scattered by a periodic slab, the energy of the scattered, or diffracted, field that propagates away from the slab is split into a finite number of distinct plane waves traveling at prescribed angles. These are known as *Rayleigh-Bloch waves*, and their angles α_m, depicted in Fig. (**3**)C are

$$\alpha_m = \arcsin\frac{|\kappa+m|}{\omega\sqrt{\varepsilon_0\mu_0}}, \quad m\in\mathscr{Z}_p. \quad (18)$$

The angles α_m depend only on κ, ω, the exterior material coefficients ε_0 and μ_0, and the periodicity of the slab (which we always normalize to 2π and therefore does not appear in the expression for α_m), and not on any other attributes of the structure. The evanescent diffractive orders do not carry energy away from the slab, but only along it. They are known as slow *Rayleigh-Bloch surface waves* because they have a phase velocity that is less than the wave speed of the ambient space. In the special case that $\eta_m=0$ for some $m\in\mathbb{Z}^2$, the corresponding linear harmonic is exactly grazing the slab; it carries no energy away from it yet is extended in the z-variable. If a pair (κ,ω) admits no propagating harmonics, then no incident field in (15,16) is available and there is no notion of scattering of plane waves originating from sources exterior to the slab. A nontrivial bounded pseudo-periodic solution of the Helmholtz equation is, in this case, always a guided mode, assuming $\mathscr{Z}_\ell=\emptyset$.

Because of the coupling of spatial harmonics due to the periodicity, we have seen that a pseudo-periodic solution of the Helmholtz equation is typically composed of all harmonics with wavevectors $\kappa+m$. It can be ascribed a unique Bloch wavevector κ in B, but, among the possible wavevectors $\kappa+m$, there is in general no preferred one. This is in contrast to the case of the x-invariant slab we considered first, in which the periodicity is absent. It is instructive to impose a periodicity artificially and view this case from the point of view of the Bloch theory to understand just how the transition from flat to periodic takes place. This is illustrated in Fig. (**2**)B, where all pairs (κ,ω) admitting scattering or guided states are shifted in κ by an integer into B. Thus all Bloch states are represented by a pair (κ,ω) in $B\times\mathbb{R}$.

By assigning a wavevector $\kappa\in B$ to each state of the flat (nonperiodic) slab, we view it as having a Fourier expansion in spatial harmonics for which all but one coefficient vanishes: indeed, by writing $\kappa=\bar{\kappa}+\bar{m}$, with $\bar{\kappa}\in B$ and $\bar{m}\in\mathbb{Z}^2$, and $\phi_{\bar{m}}(z)=\phi(z)$, a solution of the form $e^{i\kappa\cdot x}\phi(z)$ can be written as

$$\phi_{\bar{m}}(z)e^{(\bar{m}+\bar{\kappa})\cdot x},\ \bar{\kappa}\in B. \tag{19}$$

If the flat slab is perturbed periodically, we expect its Bloch states to be perturbed from their special form and typically attain all spatial harmonics. For those guided modes corresponding to (κ,ω) in the white region below the light cone, or below the cutoff frequency for $\kappa\in B$, all spatial harmonics are evanescent, and the guided mode should persist. We call these guide modes *robust*. Guided mode frequencies above the cutoff frequency (the blue region of Fig. (**2**)B) are embedded in the continuous spectrum of scattering states. If we expect a typical state of the periodically perturbed slab to have a full expansion in spatial harmonics, then a guided mode can typically no longer exist because of the presence of propagating harmonics. The destruction of a guided mode under this perturbation of the structure corresponds to the dissolution of an embedded eigenvalue into the continuous spectrum. This is made precise with the Floquet-Bloch theory, which we discuss soon.

The dissolution of the dispersion relations in the (κ,ω)-region of scattering states, as well as the typical splitting of crossings of dispersion relations is depicted in Fig. (**3**)C. In fact, the dispersion relation does not actually disappear, but the frequency becomes complex. The dotted curves correspond to the real part of a generalized dispersion relation, which we discuss in Sec. 3.2.2 and 4.4. Fields corresponding to points on the complex dispersion curve are *generalized guided modes*. Generalized modes whose frequency ω has a small but nonzero imaginary part form the theoretical basis of *leaky modes* or *quasi-guided modes*. Discussions of the topic of leaky modes as well as methods for computation of their dispersion relations can be found in [70, 84, 68, 30] and references therein.

Wave phenomena connected with periodicity of waveguide structures and applications are treated for a large class of structures in the nice review paper [16] of Elachi.

Even for periodic slabs, embedded guided mode frequencies can exist. The two-dimensional structure in Fig. (**3**)D, for example, is symmetric about a horizontal line, and we can therefore seek states for $\kappa=0$ that are symmetric or antisymmetric about this line. In the (κ,ω)-region of one propagating harmonic, $m=0$, this harmonic $e^{i\eta_0 z}$ is constant in x and therefore absent in antisymmetric states, making them exponentially decaying as $|z|\to\infty$. Because a perturbation of κ from 0 breaks the symmetry of the propagating harmonic, the guided mode is destroyed and ω takes on a nonzero imaginary part. We refer to such a guided mode as *nonrobust*. The destruction of a guided mode, this time due to a perturbation of κ, corresponds again to the dissolution of an embedded eigenvalue. The eigenvalue is characterized by an isolated pair (κ,ω) on the complex dispersion relation at which ω and κ are *real* and is illustrated in Fig. (**3**)D. Isolated true guided modes are discussed in [70] and [84]. We deal with their mathematical existence in Sec. 3.2.

It is the interaction between nonrobust guided slab modes and incoming plane waves that causes anomalous scattering behavior, including transmission anomalies, and this is the subject of Sec. 4.4. The analysis involves an analytic connection of generalized guided modes to scattering states and perturbation analysis in the vicinity of a nonrobust true guided mode pair (κ_0,ω_0), where κ_0 and ω_0 are both real.

2.2 Broader Context

Let us step back to the time-dependent wave equation (1) and consider its free solutions in $\mathbb{R}^3$ in the presence of the slab. To place the wave equation into the proper functional-analytic setting, we should write it as a first-order system by defining $v=\frac{\partial u}{\partial t}$:

$$\frac{\partial}{\partial t}\begin{bmatrix}u\\v\end{bmatrix}=\begin{bmatrix}0 & I\\ \varepsilon^{-1}\nabla\cdot\mu^{-1}\nabla & 0\end{bmatrix}\begin{bmatrix}u\\v\end{bmatrix}, \tag{20}$$

which may be supplemented by initial data $u(x,z;0)$ and $v(x,z;0)$. The appropriate functional spaces are the Hilbert spaces

$$\mathscr{H} = L^2(\mathbb{R}^3, \mathbb{C}, \varepsilon dV),$$
$$\mathscr{K} = L^2(\mathbb{R}^3, \mathbb{C}^3, \mu^{-1} dV),$$

with inner products

$$(u,w)_{\mathscr{H}} = \int_{\mathbb{R}^3} \varepsilon u\bar{w}\, dV,$$
$$(F,G)_{\mathscr{K}} = \int_{\mathbb{R}^3} \mu^{-1} F\cdot\bar{G}\, dV.$$

The symbol ∇ denotes the usual gradient operator in $\mathscr{H}$ (defined independently of the measure εdV). It is a closed operator, which takes values in $\mathscr{K}$ and whose domain is $\mathscr{D}(\nabla) = H^1(\mathbb{R}^3) \subset \mathscr{H}$, the Sobolev space of L^2 functions with weak L^2 derivatives. Similarly, the symbol $\nabla\cdot$ denotes the usual divergence operator in $\mathscr{K}$, with values in $\mathscr{H}$. With respect to the inner products in $\mathscr{H}$ and $\mathscr{K}$, one can verify that the adjoint of ∇ is $\nabla^\dagger = -(\varepsilon^{-1}\nabla\cdot)\mu^{-1}$.
The operator matrix in (20), which we denote by A, has a natural domain,

$$A = \begin{bmatrix} 0 & I \\ -\nabla^\dagger\nabla & 0 \end{bmatrix}, \quad \mathscr{D}(A) = \mathscr{D}(\nabla^\dagger\nabla) \oplus \mathscr{D}(\nabla),$$

and, thus defined, is an anti-self-adjoint operator in the Hilbert space

$$\mathscr{D}(\nabla) \oplus \mathscr{H}$$

with inner product

$$\left(\begin{bmatrix} u_1 \\ v_1 \end{bmatrix}, \begin{bmatrix} u_2 \\ v_2 \end{bmatrix} \right) = (\nabla u_1, \nabla u_2)_{\mathscr{K}} + (v_1, v_2)_{\mathscr{H}}.$$

Let E_λ be the standard spectral resolution of the identity in $\mathscr{H}$ for the operator $\nabla^\dagger\nabla$ (see [3], Ch. VI, for example). This means that, for each $\lambda \in \mathbb{R}$, E_λ is an orthogonal projection in $\mathscr{H}$; for each $u \in \mathscr{H}$, $(E_\lambda u, u)$ is a nondecreasing function of λ; and

$$\begin{aligned} u &= \int_0^\infty dE_\lambda u \qquad \forall u \in \mathscr{H}, \\ \nabla^\dagger\nabla u &= \int_0^\infty \lambda\, dE_\lambda u \quad \forall u \in \mathscr{D}(\nabla^\dagger\nabla). \end{aligned} \tag{21}$$

These integrals of $\mathscr{H}$-valued functions exist in the Lebesgue-Stieljes sense, and the lower limits are zero because the positivity of $\nabla^\dagger\nabla$ implies that $E_\lambda = 0$ if $\lambda < 0$.
Since A is anti-self-adjoint, it admits a spectral resolution of the identity in $\mathscr{D}(\nabla) \oplus \mathscr{H}$ with purely imaginary spectrum. For the wave equation, it is convenient to split the resolution into two projection-valued functions F_ω^+ and F_ω^- such that

$$\begin{aligned} \begin{bmatrix} u \\ v \end{bmatrix} &= \int_0^\infty dF_\omega^+ \begin{bmatrix} u \\ v \end{bmatrix} + \int_0^\infty dF_\omega^- \begin{bmatrix} u \\ v \end{bmatrix}, \\ A\begin{bmatrix} u \\ v \end{bmatrix} &= \int_0^\infty i\omega dF_\omega^+ \begin{bmatrix} u \\ v \end{bmatrix} + \int_0^\infty -i\omega dF_\omega^- \begin{bmatrix} u \\ v \end{bmatrix}. \end{aligned} \tag{22}$$

In order to obtain $F_\omega^\pm$ in terms of E_λ, we write A in terms of E_λ,

$$A\begin{bmatrix} u \\ v \end{bmatrix} = \int_0^\infty \begin{bmatrix} 0 & 1 \\ -\lambda & 0 \end{bmatrix} \begin{bmatrix} dE_\lambda u \\ dE_\lambda v \end{bmatrix}, \tag{23}$$

and use the spectral resolution of the matrix in this expression,

$$\begin{aligned} \begin{bmatrix} 1 & 0 \\ 0 & 1 \end{bmatrix} &= P_\omega^+ + P_\omega^-, \\ \begin{bmatrix} 0 & 1 \\ -\omega^2 & 0 \end{bmatrix} &= i\omega P_\omega^+ - i\omega P_\omega^-, \end{aligned} \tag{24}$$

in which the orthogonal projections $P^\pm$ are

$$P_\omega^+ = \frac{1}{2}\begin{bmatrix} 1 & (i\omega)^{-1} \\ i\omega & 1 \end{bmatrix}, \; P_\omega^- = \frac{1}{2}\begin{bmatrix} 1 & -(i\omega)^{-1} \\ -i\omega & 1 \end{bmatrix}.$$

By inserting (24) into (23), with $\lambda = \omega^2$ and $\omega \geq 0$, we deduce that the resolutions $F_\omega^\pm$ in (22) are given by

$$F_\omega^\pm \begin{bmatrix} u \\ v \end{bmatrix} = \int_0^{\omega^2} P_{\omega'}^\pm \begin{bmatrix} dE_{\omega'^2} u \\ dE_{\omega'^2} v \end{bmatrix}, \tag{25}$$

or, equivalently,

$$dF_\omega^\pm \begin{bmatrix} u \\ v \end{bmatrix} = P_\omega^\pm \begin{bmatrix} dE_{\omega^2} u \\ dE_{\omega^2} v \end{bmatrix}. \tag{26}$$

Because A is anti-self-adjoint, there exists a unique strongly continuous unitary group of operators in $\mathscr{D}(\nabla) \oplus \mathscr{H}$, denoted by e^{tA}, whose generator is A (see §VIII.4 of [73]). This means that

$$\tfrac{\partial}{\partial t} e^{tA} w = A e^{tA} w$$

for $w \in \mathscr{D}(A)$. The solution of the initial-value problem

$$\frac{\partial}{\partial t}\begin{bmatrix} u \\ v \end{bmatrix} = A\begin{bmatrix} u \\ v \end{bmatrix}, \quad \begin{bmatrix} u(0) \\ v(0) \end{bmatrix} = \begin{bmatrix} u_0 \\ v_0 \end{bmatrix},$$

in terms of the resolution (22) of A is

$$\begin{aligned} \begin{bmatrix} u(t) \\ v(t) \end{bmatrix} &= e^{tA}\begin{bmatrix} u_0 \\ v_0 \end{bmatrix} \\ &= \int_0^\infty e^{i\omega t} dF_\omega^+ \begin{bmatrix} u_0 \\ v_0 \end{bmatrix} + \int_0^\infty e^{-i\omega t} dF_\omega^- \begin{bmatrix} u_0 \\ v_0 \end{bmatrix} \end{aligned}$$

(the dependence on (x,z) is suppressed in $u(t)$).
Let us understand this solution more concretely in terms of generalized eigenfunctions. Since $F_\omega^\pm$ are expressed in terms of E_λ, we may focus our attention on $\nabla^\dagger\nabla = -\varepsilon^{-1}\nabla\cdot\mu^{-1}\nabla$. Suppose a pair $(u,v)^t \in \mathscr{D}(A)$ is expressed through integral superpositions

$$u(x,z) = \int_0^\infty w_1(x,z;\lambda)d\lambda,$$
$$v(x,z) = \int_0^\infty w_2(x,z;\lambda)d\lambda,$$

in which w_i satisfy the Helmholtz equation

$$-\varepsilon^{-1}\nabla\cdot\mu^{-1}\nabla w_i(x,z;\lambda) = \lambda w_i(x,z;\lambda)$$

in $\mathbb{R}^3$ but do not necessarily have finite L^2 norm. Then

$$\nabla^\dagger\nabla u(x,z) = \int_0^\infty \lambda w_1(x,z;\lambda)d\lambda.$$

The decomposition of $(w_1,w_2)^t$ according to $P_\omega^\pm$ for $\lambda = \omega^2$ $(\omega \geq 0)$ is

$$\begin{bmatrix} w_1 \\ w_2 \end{bmatrix} = \begin{bmatrix} u_+ \\ i\omega u_+ \end{bmatrix} + \begin{bmatrix} u_- \\ -i\omega u_- \end{bmatrix},$$

in which

$$u_\pm(x,z;\omega) = \tfrac{1}{2}\left(w_1(x,z;\omega^2) \pm (i\omega)^{-1}w_2(x,z;\omega^2)\right).$$

In the time dynamics, this represents a decomposition into harmonic fields oscillating with the time factors $e^{i\omega t}$ and $e^{-i\omega t}$:

$$\begin{bmatrix} u(x,z,t) \\ v(x,z,t) \end{bmatrix} = \int_0^\infty \left(e^{i\omega t}\begin{bmatrix} u_+(x,z;\omega) \\ i\omega\, u_+(x,z;\omega) \end{bmatrix} + e^{-i\omega t}\begin{bmatrix} u_-(x,z;\omega) \\ -i\omega\, u_-(x,z;\omega) \end{bmatrix} \right) d(\omega^2).$$

The functions $w_i(x,z;\lambda)$ in the generalized eigenfunction expansions include scattering states and guided modes for the slab, in other words, Helmholtz fields with the behavior (15,16). A rigorous development of a generalized Fourier transform in terms of generalized eigenfunctions that concretely realizes the spectral resolution E_λ through a unitary transformation is not given here. For specific related problems, the reader is referred to Theorem 4.1 of Goldstein [27] for infinite cylindrical scatterers, Theorem 8.5 of Wilcox [88] for diffraction gratings, and Theorem 3 of Groves [28] for an obstacle in a closed waveguide.
The treatment of scattering and resonance in this Chapter will remain in the frequency domain, that is, we study scattering of time-harmonic fields. The theory of scattering of general finite-energy disturbances for a diffraction grating is treated rigorously in [88]; other references on scattering theory include Lax and Phillips [50], Barut [5], Reed and Simon [74], and Newton [63].

The consideration of scattering of time-harmonic fields by a slab may begin with a field produced by a single harmonic monopole source off the guide, in other words, a fundamental solution of the Helmholtz equation, in the presence of the scatterer. One must determine from physical principles an appropriate radiation condition that makes the solution unique. Typically, this condition is obtained by means of the *principle of limiting absorption*, by which one begins with the unique finite-energy solution in a lossy ambient space and passes to the limit of vanishing loss (see [15, 87]). For scalar waves and a bounded obstacle in space, the condition is known as the "Sommerfeld radiation condition",

$$\lim_{r\to\infty} r\left(\tfrac{\partial u}{\partial r} - iku\right) = 0.$$

In the two-dimensional case, the factor of r is replaced by $\sqrt{r}$. For the Maxwell system, the same condition applies to all components of the electromagnetic field, and the additional conditions of Silver-Müller on the orthogonality of E, H, and the radial vector at the far field, are also satisfied [60, 61].
For unbounded scatterers, the Sommerfeld condition does not apply in general. This is because energy can be radiated not only into free space but also along the scatterer in the form of guided modes, which are excited by the evanescent waves emanating from the source. The contribution of the guided modes in addition to the scattering states appears in the radiating fundamental solution, which can then be used to derive the correct radiation condition for general sources. For the case of an impedance plane, the fundamental solution and radiation condition are treated rigorously by Nosich; see equations 20 and 24 in [65]. Equation 2 in the same reference gives the radiation condition for a closed waveguide, which is analogous to Condition 1. There is large body of literature on scattering by open waveguides and resonators, including discussions of radiation conditions; see, for example, [64, 77, 86].
Because our slab structure is invariant under the action of the group $2\pi\mathbb{Z}^2$ of transformations of $\mathbb{R}^3$ in the x-variables, the operator

$$S = \nabla^\dagger\nabla = -\varepsilon^{-1}\nabla\cdot\mu^{-1}\nabla \tag{27}$$

can be decomposed into components that act on pseudo-periodic functions. This is accomplished through the *Floquet transform*, by which a function on $\mathbb{R}^3$ is expressed as an integral superposition of pseudo-periodic functions. It is defined by

$$(\mathscr{F}u)(x,z;\kappa) = \sum_{n\in\mathbb{Z}^2} u(x+2\pi n,z)e^{-2\pi i\kappa\cdot n}.$$

For each κ, $\mathscr{F}u$ is doubly pseudo-periodic in x with fundamental domain $W\times\mathbb{R}$, $W=[0,2\pi)^2$, and Bloch wavevector κ. It is also periodic in κ with fundamental domain equal to the first Brillouin zone B. The function $u(x,z)$ is reconstructed from its Floquet transform through

$$u(x,z) = \int_B (\mathscr{F}u)(x,z;\kappa)d\kappa.$$

The Floquet transform is unitary in L^2,

$$\int_{\mathbb{R}^3} |u(x,z)|^2 dV = \int_{W\times\mathbb{R}}\int_B |(\mathscr{F}u)(x,z;\kappa)|^2 dA\,dV,$$

and it commutes with S in the sense that

$$(\mathscr{F}Su)(x,z;\kappa) = S_\kappa((\mathscr{F}u)(x,z;\kappa)),$$

where S_κ is essentially the restriction of S to pseudo-periodic functions and will be defined precisely in the next section. The solutions to the scattering Problem 2 are generalized eigenfunctions of S_κ. The family $\{S_\kappa\}_{\kappa\in B}$ provides an integral decomposition of S,

$$S = \int_B^{\oplus} S_\kappa d\kappa. \tag{28}$$

Direct integrals of operators such as this one are treated in §XIII.16 of [72], and more on application of the Floquet transform to waves in periodic structures can be found in [43] and references therein.

3. PLANE-WAVE SCATTERING AND GUIDED MODES

Because of the decomposition (28), solutions of the Helmholtz equation are integral superpositions of pseudo-periodic solutions, which we will investigate in the remainder of this Chapter. Specifically, we consider the problems of plane-wave scattering and guided modes, which are the inhomogeneous and homogeneous ($u^{\text{inc}}=0$) versions of the same Problem 2. The existence of a guided mode at (κ,ω) is equivalent to the nonuniqueness of solutions (for $\mathscr{Z}_\ell=\emptyset$).

3.1 Spectrum for the Slab

We may restrict analysis to one period in $\mathbb{R}^3$ of the functions ε and μ, which is the strip

$$\mathscr{S} = \{(x,z)\in\mathbb{R}^3 : x\in(0,2\pi)^2\}. \tag{29}$$

If ε or μ has a jump discontinuity along a surface Σ with normal vector n, the Helmholtz equation (2) must be supplemented by matching conditions

$$u \text{ and } \mu^{-1}\frac{\partial u}{\partial n} \text{ are continuous on } \Sigma. \tag{30}$$

By passing to a weak form of the equation, ε and μ are only required to be measurable, u is only required to have weak L^2 derivatives, and the matching conditions and pseudo-periodicity are automatically imposed. For this, we introduce the functional spaces[1]

$$\begin{aligned} H^1_{\kappa,\text{loc}}(\mathscr{S}) = \{u\in H^1_{\text{loc}}(\mathscr{S}) :\\ u(x_1,2\pi,z) = e^{2\pi i\kappa_2}u(x_1,0,z),\\ u(2\pi,x_2,z) = e^{2\pi i\kappa_1}u(0,x_2,z)\}, \end{aligned} \tag{31}$$

in which the boundary values of u are understood in the sense of the usual trace of H^1 functions, and

$$H^1_\kappa(\mathscr{S}) = H^1_{\kappa,\text{loc}}(\mathscr{S})\cap H^1(\mathscr{S}). \tag{32}$$

The pseudo-periodic scattering problem in weak form is as follows, where $C^\infty_{0,\kappa}(\mathscr{S})$ is the space of infinitely differentiable functions with compact support in the closure of $\mathscr{S}$ and that satisfy the same pseudo-periodic condition as the functions in $H^1_{\kappa,\text{loc}}(\mathscr{S})$.

Problem 3 (Scattering, weak form 1) *Find a function $u\in H^1_{\kappa,\text{loc}}(\mathscr{S})$ such that*

$$\int_{\mathscr{S}} (\mu^{-1}\nabla u\cdot\nabla\bar{v} - \omega^2\varepsilon u\bar{v})dV = 0 \;\; \forall v\in C^\infty_{0,\kappa}(\mathscr{S})$$

and

$$u(x,z) = u^{inc}(x,z) + u^{sc}(x,z),$$

where u^{inc} is a sum of plane waves as in Problem 2 and u^{sc} satisfies the radiation Condition 1.

It can be shown that this is equivalent to the strong form of the scattering Problem 2 if all functions involved are smooth.

[1] $H^1_{\text{loc}}(\mathscr{S})$ denotes the subspace of locally L^2 functions on $\mathscr{S}$ with locally L^2 weak gradients, and $H^1(\mathscr{S})$ denotes the Hilbert space of functions $u\in H^1_{\text{loc}}(\mathscr{S})$ for which the norm $\|u\|_{H^1} = (\int_{\mathscr{S}}(|u|^2+|\nabla u|^2))^{\frac{1}{2}}$ is finite.

A useful functional-analytic framework for Problem 3 is the Hilbert space $L^2(\mathscr{S},\varepsilon dV)$ of square-Lebesgue-integrable complex-valued functions in the strip with inner product

$$b(u,v) = \int_{\mathscr{S}} \varepsilon u\bar{v}\,dV \tag{33}$$

and the unbounded symmetric nonnegative quadratic form in $L^2(\mathscr{S},\varepsilon dV)$, with form domain $H^1_\kappa(\mathscr{S})$, defined by

$$a(u,v) := \int_{\mathscr{S}} \mu^{-1}\nabla u\cdot\nabla\bar{v}\,dV,\quad u,v\in H^1_\kappa(\mathscr{S}). \tag{34}$$

This form is closed, and the associated positive operator S_κ is given by[2]

$$S_\kappa u = -\varepsilon^{-1}\nabla\cdot\mu^{-1}\nabla u,\quad u\in\mathscr{D}(S_\kappa)\subset H^1_\kappa(\mathscr{S}). \tag{35}$$

Of course, there is a close relationship between the spectrum $\sigma(S_\kappa)$ of S_κ and the harmonic Bloch states of the slab. An eigenfunction of S_κ is a nontrivial function in $H^1_\kappa(\mathscr{S})$ that satisfies the Helmholtz equation; it can be extended to a pseudo-periodic solution in $\mathbb{R}^3$. By the general form (15,16), the coefficients of all propagating and linear harmonics must vanish, and the field is therefore a guided mode. The continuous spectrum corresponds to scattering states.

The spectrum of S_κ is characterized in part by the min-max principle (see [72], Ch. XIII, or[89], for example). The sequence $\{\lambda_j(\kappa)\}_{j=1}^\infty$ defined by

$$\lambda_j(\kappa) = \sup_{V^{j-1}<L^2(\mathscr{S})}\ \inf_{\substack{u\in(V^{j-1})^\perp\setminus\{0\}\\ u\in H^1_\kappa(\mathscr{S})}} \frac{a(u,u)}{b(u,u)}, \tag{36}$$

in which the supremum is taken over all $(j-1)$-dimensional subspaces, is nondecreasing, and converges to the infimum λ_- of the essential spectrum of S_κ. If $\lambda_n\neq\lambda_-$, then λ_n is the nth eigenvalue of S_κ, counting from the bottom and including multiplicity. In fact, as we expect, we will see that $\lambda_- = |\kappa|^2/(\varepsilon_0\mu_0)$, $\lambda_1 > |\kappa|^2/(\varepsilon_+\mu_+)$, and there are only finitely many eigenvalues below the essential spectrum.

Much of the analysis in sections 3.1 and 3.2 is adapted from the work of Bonnet-Bendhia and Starling [7]. We include proofs for completeness and consistency within the present framework.

[2]The operator $-\varepsilon^{-1}\nabla\cdot\mu^{-1} : \mathscr{D}(\varepsilon^{-1}\nabla\cdot\mu^{-1}) \subset L^2(\mathscr{S},\mathbb{C}^3,\mu^{-1}dV) \to L^2(\mathscr{S},\mathbb{C},\varepsilon dV)$ is the adjoint of $\nabla_0 : H^1_0(\mathscr{S})\subset L^2(\mathscr{S},\mathbb{C},\varepsilon dV)\to L^2(\mathscr{S},\mathbb{C}^3,\mu^{-1}dV)$, where $H^1_0(\mathscr{S})$ is the subspace of $H^1(\mathscr{S})$ with vanishing trace on $\partial\mathscr{S}$. ∇_0 is the restriction to $H^1_0(\mathscr{S})$ of the usual weak gradient operator ∇, and $\nabla\cdot$ is the usual weak divergence operator.

Theorem 4 (spectrum) *The essential spectrum of S_κ, for $\kappa\in B$, consists of all $\lambda=\omega^2$, where ω is above the light cone for the medium exterior to the slab; there are only finitely many eigenvalues below the essential spectrum. Precisely,*

1. $\sigma(S_\kappa)\subset[\frac{|\kappa|^2}{\varepsilon_+\mu_+},\infty)$;

2. $\sigma_{ess}(S_\kappa) = [\frac{|\kappa|^2}{\varepsilon_0\mu_0},\infty)$;

3. there are only finitely many eigenvalues λ_j strictly less than $\frac{|\kappa|^2}{\varepsilon_0\mu_0}$.

Proof. 1. Each $u\in H^1_\kappa(\mathscr{S})$ admits the Fourier series

$$u(x,z) = \sum_{m\in\mathbb{Z}^2} u_m(z)e^{i(m+\kappa)\cdot x}, \tag{37}$$

from which we obtain

$$\int_{\mathscr{S}}|u|^2 = 4\pi^2\int_{\mathbb{R}}\sum_{m\in\mathbb{Z}^2}|u_m(z)|^2dz,$$

$$\int_{\mathscr{S}}|\nabla_x u|^2 = 4\pi^2\int_{\mathbb{R}}\sum_{m\in\mathbb{Z}^2}|m+\kappa|^2|u_m(z)|^2dz.$$

Since $|m+\kappa|\geq|\kappa|$ for $m\in\mathbb{Z}^2$ and $\kappa\in B$, we obtain

$$\int_{\mathscr{S}}\mu^{-1}|\nabla u|^2 \geq \mu_+^{-1}\int_{\mathscr{S}}|\nabla_x u|^2 \geq \frac{|\kappa|^2}{\varepsilon_+\mu_+}\int_{\mathscr{S}}\varepsilon|u|^2.$$

In view of (36), part (1) follows.

2. Let $\psi(z)$ be a smooth function on $\mathbb{R}$ with compact support that vanishes in $[z_-,z_+]$. If $\lambda\geq|\kappa|^2/(\varepsilon_0\mu_0)$, then $\eta^2=\lambda\varepsilon_0\mu_0-|\kappa|^2\geq 0$, and one can verify that the functions

$$u^n(x,z) := n^{-\frac{1}{2}}\psi(z/n)e^{i(\eta z+\kappa\cdot x)},\quad n\in\mathbb{N}_+,$$

are in $\mathscr{D}(S_\kappa)$ and are bounded from below in $L^2(\mathscr{S},\varepsilon dV)$ uniformly in n and that

$$\varepsilon^{-1}\nabla\cdot\mu^{-1}\nabla u^n + \lambda u^n \to 0\quad \text{as } n\to\infty$$

in $L^2(\mathscr{S},\varepsilon dV)$. Conversely, for $\lambda\in\sigma_{ess}(S_\kappa)$, Weyl's criterion (Theorem 7.2 of [32]) provides a sequence u^n from $\mathscr{D}(S_\kappa)$ such that, in $L^2(\mathscr{S},\varepsilon dV)$, u^n has norm 1 and tends weakly to 0 and $(S_\kappa-\lambda)u^n\to 0$ strongly. These conditions imply

$$\int_{\mathscr{S}}\mu^{-1}|\nabla u^n|^2 = \int_{\mathscr{S}}\varepsilon\bar{u}^nS_\kappa u^n\to\lambda,$$

$$\int_{\mathscr{S}\setminus\Omega}\varepsilon|u^n|^2 - \int_{\mathscr{S}}\varepsilon|u^n|^2\to 0,$$

as $n\to\infty$, where Ω is defined below (41, Fig. (4)), and, as in part (1), we have

$$\int_{\mathscr{S}}\mu^{-1}|\nabla u^n|^2 \geq \mu_0^{-1}\int_{\mathscr{S}\setminus\Omega}|\nabla_x u^n|^2 \geq \frac{|\kappa|^2}{\varepsilon_0\mu_0}\int_{\mathscr{S}\setminus\Omega}\varepsilon|u^n|^2.$$

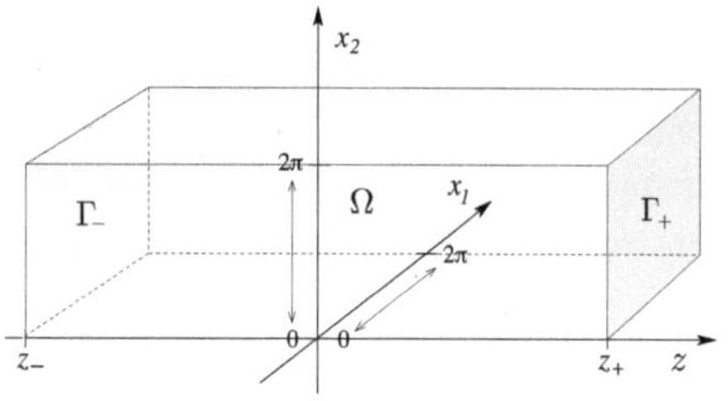

Fig. (**4**): The domain Ω, comprising one truncated period of the functions ε and μ.

We conclude that $\lambda \geq |\kappa|^2/(\varepsilon_0\mu_0)$.

3. We show below that $\lambda_j = \hat{\lambda}_j$ for $\lambda_j < |\kappa|^2/(\varepsilon_0\mu_0)$, where the $\hat{\lambda}_j$ arise from an equivalent formulation of the scattering problem in Ω and tend to infinity as $n \to \infty$. ■

The scattering Problem 3 can also be formulated in terms of the x-periodic part $\tilde{u}$ of the field $u(x,z) = \tilde{u}(x,z)e^{i\kappa\cdot x}$. The Helmholtz equation implies

$$(\nabla + i\kappa)\cdot\mu^{-1}(\nabla + i\kappa)\,\tilde{u} + \omega^2\varepsilon\,\tilde{u} = 0, \tag{38}$$

and its weak form is

$$\int_{\mathscr{S}} (\mu^{-1}(\nabla + i\kappa)\tilde{u}\cdot(\nabla - i\kappa)\bar{v} - \omega^2\varepsilon\,\tilde{u}\bar{v})dV = 0 \tag{39}$$

$\forall v \in C^\infty_{0,\mathrm{per}}(\mathscr{S})$ $(= C^\infty_{0,\kappa}(\mathscr{S})$ with $\kappa = 0)$, or, in expanded form,

$$\int_{\mathscr{S}} \mu^{-1}\big(\nabla\tilde{u}\cdot\nabla\bar{v} + i\kappa\cdot(\tilde{u}\nabla_x\bar{v} - \bar{v}\nabla_x\tilde{u}) + \\ + (|\kappa|^2 - \omega^2\varepsilon\mu)\tilde{u}\bar{v}\big)dV = 0. \tag{40}$$

The relevant quadratic form in $H^1_{\mathrm{per}}(\mathscr{S})$ is

$$a_\kappa(u,v) = \\ \int_{\mathscr{S}} \mu^{-1}\big(\nabla u\cdot\nabla\bar{v} + i\kappa\cdot(u\nabla_x\bar{v} - \bar{v}\nabla_x u) + |\kappa|^2 u\bar{v}\big)dV.$$

An equally important formulation of the scattering problem for a fixed pair (κ,ω) is obtained by truncating the strip $\mathscr{S}$ to a domain Ω finite length, outside of which $\varepsilon = \varepsilon_0$ and $\mu = \mu_0$ (Fig. (**4**)),

$$\Omega = \{(x,z) \in \mathscr{S} : z_- < z < z_+\}, \tag{41}$$
$$\Gamma_\pm = \mathscr{S} \cap \{(x,z_\pm) : x \in \mathbb{R}^2\},$$
$$\Gamma = \Gamma_- \cup \Gamma_+.$$

The normal vector n on Γ is taken to be directed out of Ω.

The radiation condition is enforced through a Dirichlet-to-Neumann operator $T = T^{\kappa,\omega}$ that characterizes radiating fields in the sense that, for pseudo-periodic Helmholtz fields u,

$$\partial_n u + Tu = 0 \text{ on } \Gamma \iff u \text{ is radiating.} \tag{42}$$

Technically, $T : H^{\frac{1}{2}}(\Gamma) \to H^{-\frac{1}{2}}(\Gamma)$ acts on traces on Γ of functions in $H^1_\kappa(\Omega)$ and is defined through the Fourier transform as follows. For any $f \in H^{\frac{1}{2}}(\Gamma)$, let $\hat{f}_m$ be the Fourier coefficients of $e^{-i\kappa\cdot x}f$; this is a pair of numbers $\hat{f}_m = (\hat{f}^-_m, \hat{f}^+_m)$, one giving the mth pseudoperiodic Fourier component of f on Γ_- and the other on Γ_+,

$$f(x,z_\pm) = \sum_{m\in\mathbb{Z}^2} \hat{f}^\pm_m e^{i(m+\kappa)\cdot x}. \tag{43}$$

Then T is defined by

$$T : H^{\frac{1}{2}}(\Gamma) \to H^{-\frac{1}{2}}(\Gamma), \quad (\widehat{Tf})_m = -i\eta_m\hat{f}_m. \tag{44}$$

The operator T has a nonnegative real part T_r and a nonpositive imaginary part T_i:

$$T = T_r + iT_i, \tag{45}$$

$$(\widehat{T_r f})_m = \begin{cases} -i\eta_m\hat{f}_m & \text{if } m \in \mathscr{Z}_e, \\ 0 & \text{otherwise.} \end{cases} \tag{46}$$

$$(\widehat{T_i f})_m = \begin{cases} -\eta_m\hat{f}_m & \text{if } m \in \mathscr{Z}_p, \\ 0 & \text{otherwise.} \end{cases} \tag{47}$$

For each pair (κ,ω), define the sesquilinear forms in $H^1_\kappa(\Omega)$ (we suppress the dependence on κ),

$$\hat{a}^\omega(u,v) = \int_\Omega \mu^{-1}\nabla u\cdot\nabla\bar{v} + \mu_0^{-1}\int_\Gamma (T^\omega u)\bar{v},$$
$$\hat{a}^\omega_r(u,v) = \int_\Omega \mu^{-1}\nabla u\cdot\nabla\bar{v} + \mu_0^{-1}\int_\Gamma (T^\omega_r u)\bar{v},$$
$$\hat{a}^\omega_i(u,v) = \mu_0^{-1}\int_\Gamma (T^\omega_i u)\bar{v},$$
$$\hat{b}(u,v) = \int_\Omega \varepsilon u\bar{v}.$$

We have $\hat{a}^\omega = \hat{a}^\omega_r + i\hat{a}^\omega_i$. Define also the bounded conjugate-linear functional f^ω_Γ on $H^1_\kappa(\Omega)$,

$$f^\omega_\Gamma(v) = \mu_0^{-1}\int_\Gamma (\partial_n + T^\omega)u^{\mathrm{inc}}\,\bar{v} \quad \forall v \in H^1_\kappa(\Omega).$$

The frequency dependence of T is exhibited because we will consider the problem for fixed κ, with ω^2 playing the role of an eigenvalue. Problem 3 is equivalent to the following one:

Problem 5 (Scattering, weak form 2) *Find a function $u \in H^1_\kappa(\Omega)$ such that*

$$\hat{a}^\omega(u,v) - \omega^2\hat{b}(u,v) = f^\omega_\Gamma(v) \;\; \forall v \in H^1_\kappa(\Omega). \tag{48}$$

The equivalence of Problems 2, 3, and 5 is in the sense of the following Proposition, whose proof we leave to the reader. The space $H^1_{\kappa,\mathrm{loc}}(\mathbb{R}^3)$ consists of those functions in $H^1_{\mathrm{loc}}(\mathbb{R}^3)$ that are pseudo-periodic with Bloch wavevector κ.

Proposition 6 (Equivalence of problems) *If $u \in H^1_{\kappa,\mathrm{loc}}(\mathbb{R}^3)$ solves Problem 2, then $u|_{\mathscr{S}}$ solves Problem 3 and $u|_\Omega$ solves Problem 5. If $u \in H^1_\kappa(\Omega)$ solves Problem 5, then there exists a unique extension $\tilde{u} \in H^1_{\kappa,\mathrm{loc}}(\mathbb{R}^3)$ of u that solves Problem 2.*

The scattering problem in the form of Problem 5 can be generalized to include harmonic pseudo-periodic source fields originating from sources interior to and exterior to Ω; these sources are realized by generalizing the right-hand side of (48) to an arbitrary element of the conjugate dual of $H^1_\kappa(\Omega)$.

Now, a guided mode of the slab for the pair (κ,ω) (with $\mathscr{Z}_\ell = \emptyset$) is the extension to $\mathbb{R}^3$ of a nontrivial solution to Problem 5 in the absence of a source field, that is, a function $u \in H^1_\kappa(\Omega)$ that satisfies the nonlinear eigenvalue problem

$$\hat{a}^\omega(u,v) - \omega^2\hat{b}(u,v) = 0 \quad \forall v \in H^1_\kappa(\Omega). \tag{49}$$

The form $\hat{a}^\omega$ is not real if ω lies above the cutoff frequency, or if $\omega^2 > |\kappa|^2/(\varepsilon_0\mu_0)$ because of the presence of propagating spatial harmonics for the pair (κ,ω). Because $\hat{a}^\omega_i$ is definite in sign, the condition (49) is equivalent to a pair of conditions; this is straightforward to prove.

Proposition 7 (Real eigenvalues) *If $\omega^2 \in \mathbb{R}$, then a function $u \in H^1_\kappa(\Omega)$ satisfies the homogeneous problem (49) if and only if it satisfies the equation*

$$\hat{a}^\omega_r(u,v) - i\hat{a}^\omega_i(u,v) - \omega^2\hat{b}(u,v) = 0 \;\; \forall v \in H^1_\kappa(\Omega) \tag{50}$$

and if and only if it satisfies the pair

$$\begin{aligned} &\hat{a}^\omega_r(u,v) - \omega^2\hat{b}(u,v) = 0 \quad \forall v \in H^1_\kappa(\Omega), \\ &(\widehat{u|_\Gamma})_m = 0 \quad \forall m \in \mathscr{Z}_p. \end{aligned} \tag{51}$$

The solutions u of the first condition in (51) are of the form (15,16), in which the normal derivatives of all the propagating harmonics vanish on Γ. The second condition requires that the propagating harmonics vanish altogether, leaving only those that are linear or evanescent. In the case that all η_m are nonzero, the field u is exponentially confined to the slab waveguide and is a true guided mode.

By means of the min-max principle applied to the real form $\hat{a}^\omega_r$, one obtains a sequence of frequencies $\{\omega_j\}$ that subsumes the frequencies of the guided modes. This is proved in the next theorem. We denote the set of square frequencies of the guided modes by $\hat{\lambda}_j$:

$$\{(\hat{\lambda}_j)^{\frac{1}{2}}\}_{j=1}^N \;:\; \text{guided-mode frequencies},$$

in which N may be ∞. Because of Proposition 6, the values $\hat{\lambda}_j$ that lie below $|\kappa|^2/(\varepsilon_0\mu_0)$ coincide with the eigenvalues λ_j:

$$\hat{\lambda}_j = \lambda_j \quad \text{for} \quad \lambda_j < |\kappa|^2/(\varepsilon_0\mu_0).$$

Theorem 8 (Guided-mode frequencies) *Given $\kappa \in B$, the frequencies ω for which the equation $\hat{a}^\omega_r(u,\cdot) - \omega^2\hat{b}(u,\cdot) = 0$ admits a nontrivial solution $u \in H^1_\kappa(\Omega)$ are the elements of a nondecreasing sequence $\{\omega_j\}_{j=1}^\infty$ of positive numbers that tends to ∞ and their additive inverses. The nonnegative frequencies for which the slab admits a guided mode with Bloch wavevector κ is a subset of this family that includes all ω_j less than $|\kappa|/\sqrt{\varepsilon_0\mu_0}$.*

Proof. For a fixed value of $\omega > 0$, the set of numbers α for which $\hat{a}^\omega_r(u,\cdot) - \alpha\hat{b}(u,\cdot) = 0$ admits a nontrivial solution is a nondecreasing nonnegative sequence $\alpha_j(\omega)$. This is seen as follows. Since $\hat{a}^\omega_r$ and $\hat{b}$ are bounded forms in $H^1_\kappa(\Omega)$, there exist linear operators A^ω_r and B from $H^1_\kappa(\Omega)$ to itself defined through

$$\begin{aligned} (A^\omega_r u, v) &= \hat{a}^\omega_r(u,v) + \hat{b}(u,v), \\ (Bu, v) &= \hat{b}(u,v). \end{aligned}$$

The operator A^ω_r is bijective with bounded inverse because $\hat{a}^\omega_r + \hat{b}$ is coercive:

$$\hat{a}^\omega_r(u,u) + \hat{b}(u,u) \geq$$

$$\mu_+^{-1}\int_\Omega |\nabla u|^2 + \varepsilon_-\int_\Omega |u|^2 \geq \min\{\mu_+^{-1}, \varepsilon_-\}\|u\|^2_{H^1_\kappa(\Omega)}.$$

The operator B is compact because of the compact embedding of $H^1_\kappa(\Omega)$ into $L^2(\Omega)$. Therefore the set of α that admit a nontrivial solution to $(A^\omega_r - (\alpha + 1)B)u = 0$ is a sequence converging to infinity. That the α_j must be positive is seen through their construction by the min-max principle,

$$\alpha_j(\omega) = \sup_{V^{j-1} < L^2(\Omega)} \; \inf_{\substack{u \in (V^{j-1})^\perp \setminus \{0\} \\ u \in H^1_\kappa(\Omega)}} \frac{\hat{a}^\omega_r(u,u)}{\hat{b}(u,u)}. \tag{52}$$

One shows that each α_j is a nonincreasing function of ω (details can be found in the proof of Theorem 3.3 of [7]). Therefore, for each $j = 1,2,\ldots$, there is exactly one number $\omega_j \geq 0$ such that $\alpha_j(\omega_j) = \omega_j^2$, and the sequence $\{\omega_j\}_{j=1}^\infty$ tends to infinity. These are the values of ω for which $\hat{a}^\omega_r(u,\cdot) - \omega^2\hat{b}(u,\cdot) = 0$ admits a nontrivial solution.

By the characterization (49) of guided modes and Proposition 7, the set of guided mode frequencies $\{(\hat{\lambda}_j)^{\frac{1}{2}}\}_{j=1}^N$ is a subset of $\{\omega_j\}_{j=1}^\infty$. If $\omega_j <$

$|\kappa|/\sqrt{\varepsilon_0\mu_0}$, then $\hat{a}^{\omega_j} = \hat{a}_r^{\omega_j}$, and, as (κ,ω_j) admits only evanescent harmonics, ω_j is a guided-mode frequency. ■

The existence of guided modes is treated in the next subsection. What we can say at this point is that, because the eigenvalues $\hat{\lambda}_j$ tend to infinity, their multiplicities must be finite. If a real pair (κ,ω) admits a guided mode, then any solution to the scattering problem is not unique because the addition of a guided mode results in another solution. For plane-wave sources there always exists a solution ([7]). This will turn out to be important in calculating the leading-order resonant amplitude enhancement in Sec. 5.5.

Theorem 9 (Existence of scattered fields)
Problem 5 has a solution, and the space of solutions is finite-dimensional.

Proof. Rewrite (48) in the following way:

$$\hat{a}^{\omega}(u,v)+\hat{b}(u,v)-(\omega^2+1)\hat{b}(u,v)=f_{\Gamma}^{\omega}(v). \quad (53)$$

As in the proof of Theorem 8, we may define the linear operators A^{ω} and C^{ω} from $H^1_{\kappa}(\Omega)$ into itself, as well as an element $\tilde{f}\in H^1_{\kappa}(\Omega)$ through

$$\begin{aligned}(A^{\omega}u,v) &= \hat{a}^{\omega}(u,v)+\hat{b}(u,v),\\ (C^{\omega}u,v) &= -(\omega^2+1)\hat{b}(u,v),\\ (\tilde{f},v) &= f(v).\end{aligned}$$

Now (53) takes the form

$$(A^{\omega}+C^{\omega})u=\tilde{f}. \quad (54)$$

The operator C^{ω} is compact, and A^{ω} is bijective with a bounded inverse because $\hat{a}^{\omega}+\hat{b}$ is coercive: As we have shown,

$$\mathrm{Re}\,(\hat{a}^{\omega}(u,u)+\hat{b}(u,u))\geq \min\{\mu_+^{-1},\varepsilon_-\}\|u\|^2_{H^1_{\kappa}(\Omega)}.$$

By the Fredholm alternative, (54) has a solution if and only if $(\tilde{f},v)=0$ for all $v\in \mathrm{Null}(A^{\omega}+C^{\omega})^{\dagger}$, that is, for all v that satisfy

$$(w,(A^{\omega}+C^{\omega})^{\dagger}v)=0\ \forall w\in H^1_{\kappa}(\Omega). \quad (55)$$

By using the relation

$$(w,(A^{\omega}+C^{\omega})^{\dagger}v)=\hat{a}_r^{\omega}(w,v)+i\hat{a}_i^{\omega}(w,v)-\omega^2\hat{b}(w,v),$$

(55) becomes

$$\hat{a}_r^{\omega}(v,w)-i\hat{a}_i^{\omega}(v,w)-\omega^2\hat{b}(v,w)=0$$

for all $w\in H^1_{\kappa}(\Omega)$. By Proposition 7, the propagating harmonics of such a function v vanish on Γ, and therefore, by the definitions of $\tilde{f}$ and f_{Γ}^{ω}, $(\tilde{f},v)$ vanishes.

The space of solutions is finite-dimensional because A^{ω} is invertible and C^{ω} is compact. ■

3.2 Guided Modes

We turn to proving (mathematical) existence of isolated and embedded guided-mode frequencies, as well as the the nonexistence of guided modes in certain structures for which $\varepsilon(x,z)\leq\varepsilon_0$ and $\mu(x,z)\leq\mu_0$. An example of the latter is an infinite homogeneous ceramic matrix in which a "slab" is created by a doubly periodic array of air holes.

At the end of this section, we will indicate a variety of related results in the literature, especially on guided modes on periodic surfaces and trapped modes in closed waveguides containing an obstacle.

3.2.1 Existence and Nonexistence

The following Theorem is adapted from Theorems 4.3 and 4.4 of [7]. It guarantees the existence of guided modes below the cutoff frequency for any $\kappa\in B$. If $\mu=\mu_0$, the converse of part (1) is also true. Let $\mathscr{N}(\kappa)$ be the number of eigenvalues λ_j less than $|\kappa|^2/(\varepsilon_0\mu_0)$.

Theorem 10 (Existence of guided modes)
1. If $\varepsilon\mu>\varepsilon_0\mu_0$ on a set of positive measure and

$$\int_{\mathscr{S}}\left(\frac{\varepsilon}{\varepsilon_0}-\frac{\mu_0}{\mu}\right)dV\geq 0, \quad (56)$$

then, for all $\kappa\in B\setminus\{0\}$, $\mathscr{N}(\kappa)\geq 1$, that is, there exists a guided mode at a frequency below $|\kappa|/\sqrt{\varepsilon_0\mu_0}$.
2. Let K be an open set in S, and let $\{\beta_j\}_{j=1}^{\infty}$ be the spectrum of the Dirichlet Laplacian in K ($-\nabla^2$ with $u=0$ on ∂K). If $\kappa\in B\setminus\{0\}$ and $\varepsilon>\varepsilon_*$, $\mu>\mu_*$ on K, with $\varepsilon_*\mu_*>\beta_j\frac{\varepsilon_0\mu_0}{|\kappa|^2}$, then $\mathscr{N}(\kappa)\geq j$, that is, there are at least j independent guided modes with Bloch wavevector κ and frequency below $|\kappa|/\sqrt{\varepsilon_0\mu_0}$.

Proof. We adapt the arguments of [7].

1. Set $H_0=\max\{|z_-|,|z_+|\}$, and define, for $H>H_0$,

$$u^H(x,z)=\begin{cases}1 & \text{if } |z|<H,\\ \frac{2H-|z|}{H} & \text{if } H<|z|<2H,\\ 0 & \text{if } |z|>2H,\end{cases}$$

on $\mathscr{S}$. By the assumption that $\varepsilon\mu>\varepsilon_0\mu_0$ on a set of positive measure, there exists a real-valued test function $w\in C^{\infty}_{0,\mathrm{per}}(\mathscr{S})$ with support in $|z|<H_0$ such that

$$\int_{\mathscr{S}}\left(\frac{\mu_0}{\mu}-\frac{\varepsilon}{\varepsilon_0}\right)w\,dV<0. \quad (57)$$

From the definition of a_{κ}, if u is real-valued,

$$a_{\kappa}(u,u)=\int\mu^{-1}[|\nabla u|^2+|\kappa|^2|u|^2].$$

Let α be an arbitrary real number, and define

$$w_\alpha^H = u^H + \alpha w, \quad (58)$$

which is real-valued on $\mathscr{S}$. Thus

$$\frac{a_\kappa(w_\alpha^H, w_\alpha^H)}{\int \varepsilon (w_\alpha^H)^2} - \frac{|\kappa|^2}{\varepsilon_0\mu_0} = \frac{1}{\int \varepsilon (w_\alpha^H)^2}\left[\frac{8\pi^2}{\mu_0 H} + \alpha^2 \int \mu^{-1}|\nabla w|^2 + + \frac{|\kappa|^2}{\mu_0}\int \left(\frac{\mu_0}{\mu} - \frac{\varepsilon}{\varepsilon_0}\right)(1 + 2\alpha w + \alpha^2 w^2)\right].$$

In the case of strict inequality in (56), this expression is negative for $\alpha = 0$ and sufficiently large H. In the case of equality, the expression is negative for sufficiently small nonzero α and sufficiently large H.

2. If $u \in H^1_\kappa(\mathscr{S})$ has support in the closure of K, then

$$\frac{\int \mu^{-1}|\nabla u|^2}{\int \varepsilon |u|^2} \le \frac{1}{\varepsilon_* \mu_*}\frac{\int_K |\nabla u|^2}{\int_K |u|^2}.$$

Since each function $u \in H^1_0(K)$ is extensible to a function in $H^1_\kappa(\Omega)$, this inequality, together with the min-max principle for $\{\lambda_j\}$ and $\{\beta_j\}$ imply

$$\lambda_j(\kappa) \le \frac{1}{\varepsilon_* \mu_*}\beta_j.$$

As long as $\varepsilon_* \mu_* > \beta_j \frac{\varepsilon_0\mu_0}{|\kappa|^2}$, we obtain $\lambda_j(\kappa) \le \frac{|\kappa|^2}{\varepsilon_0\mu_0}$. ∎

Theorem 11 states that there are continuous dispersion relations for robust guided modes. The condition $\varepsilon\mu \ge \varepsilon_0\mu_0$ can be eliminated if $\mu = \mu_0$.

Theorem 11 (Dispersion relations)
1. The eigenvalues $\lambda_j(\kappa)$ are continuous functions of $\kappa \in B$.
2. If $\varepsilon\mu \ge \varepsilon_0\mu_0$, then, for each unit vector $\hat{\kappa} \in \mathbb{R}^2$, the functions $\lambda_j(s\hat{\kappa}) - \frac{s^2}{\varepsilon_0\mu_0}$ are nonincreasing in s for $s\hat{\kappa} \in B$, and therefore $\mathscr{N}(s\hat{\kappa})$ is nondecreasing.

Proof. The proof follows [7]. For part (1), assume that $\int_{\mathscr{S}} |u|^2 = 1$, and, for arbitrary κ^1 and κ^2 in B, consider the difference

$$a_{\kappa^1}(u,u) - a_{\kappa^2}(u,u) = 2(\kappa^1 - \kappa^2)\cdot \int_{\mathscr{S}} \mu^{-1} \mathrm{Im}\, \bar{u} \nabla_x u \, dV + + (|\kappa^1|^2 - |\kappa^2|^2)\int_{\mathscr{S}} \mu^{-1}|u|^2 dV,$$

which yields

$$|a_{\kappa^1}(u,u) - a_{\kappa^2}(u,u)| \le |\kappa^1 - \kappa^2|\frac{2}{\mu_-}\left[\left(\textstyle\int_{\mathscr{S}} |\nabla_x u|^2\right)^{\frac{1}{2}} + \max(|\kappa^1|, |\kappa^2|)\right]. \quad (59)$$

In order to replace the L^2-norm of $\nabla_x u$ with an expression involving a_{κ^2}, we use Young's inequality

$$|2\mathrm{Im}\,(\kappa \cdot \nabla u)\bar{u}| \le \tfrac{1}{2}|\nabla_x u|^2 + 2|\kappa|^2|u|^2$$

to obtain the coercivity estimate

$$a_\kappa(u,u) + |\kappa|^2 \int_{\mathscr{S}} \mu^{-1}|u|^2 \ge \int_{\mathscr{S}} \mu^{-1}(|\partial_z u|^2 + \tfrac{1}{2}|\nabla_x u|^2).$$

Using this in the estimate (59) yields

$$a_{\kappa^1}(u,u) \le a_{\kappa^2}(u,u) + |\kappa^1 - \kappa^2|\tfrac{2}{\mu_-} \times \left[\left(2\mu_+ (a_{\kappa^2}(u,u) + \tfrac{|\kappa^2|^2}{\mu_-})\right)^{\frac{1}{2}} + \max(|\kappa^1|, |\kappa^2|)\right].$$

This implies the inequality

$$\lambda_j(\kappa^1) \le \lambda_j(\kappa^2) + |\kappa^1 - \kappa^2|\tfrac{2}{\mu_-} \times \left[\left(2\mu_+ (\lambda_j(\kappa^2) + \tfrac{|\kappa^2|^2}{\mu_-})\right)^{\frac{1}{2}} + \max(|\kappa^1|, |\kappa^2|)\right].$$

Using this and its analog with κ^1 and κ^2 interchanged, we obtain

$$\limsup_{\kappa^2 \to \kappa^1} \lambda_j(\kappa^2) \le \lambda_j(\kappa^1) \le \liminf_{\kappa^2 \to \kappa^1} \lambda_j(\kappa^2),$$

which implies continuity of λ_j at κ^1.
To prove part (2), assume again that $\int_{\mathscr{S}} |u|^2 = 1$ and observe that, since $\lambda_j \le \frac{|\kappa|^2}{\varepsilon_0\mu_0}$ for $\kappa \in B$,

$$\lambda_j(\kappa) - \frac{|\kappa|^2}{\varepsilon_0\mu_0} = \sup\inf\min\left(\frac{a_\kappa(u,u) - |\kappa|^2 \varepsilon_0^{-1}\mu_0^{-1}\int \varepsilon |u|^2 dV}{\int \varepsilon |u|^2 dV}, 0\right). \quad (60)$$

One computes that

$$a_\kappa(u,u) - \frac{|\kappa|^2}{\varepsilon_0\mu_0}\int_{\mathscr{S}} \varepsilon |u|^2 dV = a|\kappa|^2 + b\cdot\kappa + c,$$

where

$$a = \int_{\mathscr{S}} \mu^{-1}(1 - \tfrac{\varepsilon\mu}{\varepsilon_0\mu_0})|u|^2 dV \le 0,$$
$$c = \int_{\mathscr{S}} \mu^{-1}|\nabla u|^2 \ge 0,$$

and this implies that

$$\min\left(a_\kappa(u,u) - |\kappa|^2\varepsilon_0^{-1}\mu_0^{-1}\int\varepsilon|u|^2dV, 0\right),$$

with $\kappa = s\hat{\kappa}$ is nonincreasing as a function of s as long as $s\hat{\kappa} \in B$. This, in turn, implies that

$$\lambda_j(s\hat{\kappa}) - \frac{s^2}{\varepsilon_0\mu_0} \tag{61}$$

is nonincreasing. ■

As we have discussed, real dispersion relations for guided modes in a (κ, ω) region of $\mathbb{R}^3$ that admits at least one propagating spatial harmonic typically do not exist. Instead, the imaginary part of the frequency corresponding to κ vanishes only for isolated real values of κ, giving rise to isolated real pairs in (κ, ω)-space that admit guided modes. At these isolated pairs, the frequency, wavevector, and structure are in such a relation that all propagating harmonics of the solutions u of the first equation in (51) vanish, that is, the second condition in (51) is satisfied. This situation occurs, for example, if the the structure has symmetry about a plane transverse to the waveguide and $\kappa_1 = 0$ or $\kappa_2 = 0$; this is treated in [82].

Theorem 12 (Embedded eigenvalues)
If $\kappa = (0, \kappa_2)$ (resp. $\kappa = (\kappa_1, 0)$) is in the interior of B, then there exist functions $\varepsilon(x,z)$ and $\mu(x,z)$ that are symmetric about the x_2z-plane (resp. the x_1z-plane) that admit a guided-mode frequency above the cutoff frequency $|\kappa|/\sqrt{\varepsilon_0\mu_0}$.

Proof. We give a sketch of the proof. If $\varepsilon(x)$ and $\mu(x)$ are symmetric about the x_2z-plane and $\kappa = (0, \kappa_2)$, then the symmetric part $H_\kappa^{1\,\text{sym}}(\Omega)$ and antisymmetric part $H_\kappa^{1\,\text{ant}}(\Omega)$ of $H_\kappa^1(\Omega)$ with respect to the x_2z-plane are orthogonal with respect to $\hat{a}^\omega$ and $\hat{b}^\omega$. Thus a function that is antisymmetric about the x_2z-plane and satisfies $\hat{a}^\omega(u,v) - \omega^2\hat{b}(u,v) = 0$ for all $v \in H_\kappa^{1\,\text{ant}}(\Omega)$ also satisfies the equation for all $v \in H_\kappa^1(\Omega)$. The frequencies of antisymmetric modes are therefore a subset of a sequence $\{\omega_j^{ant}\}_{j=1}^\infty$ defined in analogy to the ω_j in Theorem 8 by restricting the evaluation of the forms in the Rayleigh quotient (52) to functions in $H_\kappa^{1\,\text{ant}}(\Omega)$. By making ε_+ and/or μ_+ large enough, at least one of the ω_j^{ant} can be adjusted so that

$$0 < \varepsilon_0\mu_0(\omega_j^{ant})^2 - \kappa_2^2 < 1$$

and $\mathscr{Z}_\ell = \emptyset$. In this regime, there is at least one propagating spatial harmonic and all of them have $m = (0, m_2)$. Since all harmonics with $\kappa_1 = m_1 = 0$ are symmetric about the x_2z-plane and the corresponding field that satisfies $\hat{a}_r^\omega(u,v) - (\omega_j^{ant})^2\hat{b}(u,v) = 0$ for all $v \in H_\kappa^1(\Omega)$ is antisymmetric, the coefficients of these harmonics for u must vanish and u is therefore a guided mode. ■

The existence of these antisymmetric guided modes with zero Bloch wavenumber in the variable of structural symmetry can also be proved using a harmonic Lippmann-Schwinger equation [80]. Embedded eigenvalues at nonzero wavenumber are constructed in the two-dimensional case for periodic slabs with metal inclusions [7], and analogous results are obtained for a two-dimensional lattice model [69].

Theorem 13 (Nonexistence of guided modes)
Let ω and κ be real, and let one of the following conditions be satisfied:

1. In Ω, $\varepsilon_- < \varepsilon(x,z) \le \varepsilon_0$ and $\mu_- < \mu(x,z) \le \mu_0$, and

$$L(\omega^2\varepsilon_0\mu_0 - |\kappa|^2)^{1/2} < \pi \text{ if } \mathscr{Z}_p \neq \emptyset; \tag{62}$$

2. There is a real number z_0 such that $\varepsilon(x, z_0 + z)$, $\varepsilon(x, z_0 - z)$, $\mu(x, z_0 + z)$, and $\mu(x, z_0 - z)$ are nondecreasing functions of z for all $x \in \mathbb{R}^2$.

Then there exists no function $u \in H_\kappa^1(\Omega)$ for which (49) holds. This means that the periodic slab structure admits no guided modes at the pair (κ, ω).

Condition (62) imposes no restriction on the width of the slab if there are no propagating spatial harmonics for the parameters ω, κ, ε_0, and μ_0. Otherwise, the restriction is interpreted as demanding that the width be less than half the wavelength in the direction perpendicular to the slab of the spatial harmonic whose propagation direction is closest to the normal ($m = 0$), and, in fact, less than half that of all propagating spatial harmonics. Thus the restriction becomes more severe as the frequency increases. Whether this or some weaker restriction is necessary for the prohibition of guided modes is an open question.

Proof. Following Shipman and Volkov [82], we take a different approach to making the form $a(u,v)$ real. Instead of eliminating its imaginary part, we restrict the form to a subspace $X \subset H_\kappa^1(\Omega)$ on which the imaginary part vanishes,

$$u \in X \Leftrightarrow \int_{\Gamma_\pm} u(x,z)e^{-i(m+\kappa)x}dA = 0 \;\; \forall m \notin \mathscr{Z}_e.$$

In other words, the nonevanescent spatial harmonics must vanish on Γ_- and Γ_+, a condition we know must be satisfied for guided modes. The form a is closed on its domain X, and $a(u,u) \geq 0$ for $u \in X$. Suppose that $u \in X$ satisfies

$$\hat{a}^{\omega}(u,v) - \omega^2 \hat{b}(u,v) = 0 \tag{63}$$

for all $v \in X$. In order that u be extensible to a solution of the Helmholtz equation in the strip $\mathscr{S}$, it must satisfy (63) for all $v \in H^1_\kappa(\Omega)$, or (49). (This amounts to a finite number of additional constraints, given by taking, say, $v = ze^{i(m+\kappa)x}$ and $v = (L-z)e^{i(m+\kappa)x}$ for each $m \notin \mathscr{Z}_e$, and is equivalent to setting the nonevanescent spatial harmonics of $\partial_n u$ equal to zero on $\Gamma_\pm$ when the normal derivative exists.)

Let us fix ω and κ and turn to the problem of finding the numbers γ for which there exists a nontrivial function $u \in X$ such that

$$\hat{a}^{\omega}(u,v) - \gamma \hat{b}(u,v) = 0 \;\; \forall v \in X. \tag{64}$$

These nonnegative numbers $\gamma = \gamma_j^{\kappa,\omega}(\varepsilon,\mu)$ are given by the min-max expression (52), but with the infimum taken over $u \in X$. For the special choice of material coefficients $\varepsilon \equiv \varepsilon_0$ and $\mu \equiv \mu_0$, when the slab ceases to exist, they can be calculated explicitly. Finding solutions to (64) with these constant coefficients amounts to solving

$$\Delta u + \gamma \varepsilon_0 \mu_0 u = 0 \tag{65}$$

in Ω with pseudo-periodic conditions on the planes parallel to the z-axis and appropriate boundary conditions on the sides $\Gamma_\pm$. As $\gamma\varepsilon_0\mu_0$ is constant, it suffices to seek separable solutions

$$u(x,z) = e^{i(m+\kappa)x} u_1(z), \tag{66}$$

where u_1 is a combination of $e^{\pm i\nu z}$ with

$$|m+\kappa|^2 + \nu^2 = \gamma \varepsilon_0 \mu_0. \tag{67}$$

For $m \in \mathscr{Z}_p \cup \mathscr{Z}_\ell$, $u \in X$ requires that $u_1(0) = u_1(L) = 0$, so that

$$u(x,z) = e^{i(m+\kappa)x} \sin(\nu z), \;\; \nu L = \pi j, \tag{68}$$

where $j \in \mathbb{N}_+$. Thus, we have a set of eigenvalues γ_{mj} of (64) given by

$$\gamma_{mj} = \tfrac{1}{\varepsilon_0\mu_0}\left(|m+\kappa|^2 + \left(\tfrac{\pi j}{L}\right)^2\right), \; m \notin \mathscr{Z}_e, \; j \in \mathbb{N}_+.$$

For $m \in \mathscr{Z}_e$, because of the term of $\hat{a}^{\omega}$ containing the Dirichlet-to-Neumann map T^{ω}, we have $u_1'(0) = \alpha_m u_1(0)$ and $u_1'(L) = -\alpha_m u_1(L)$, with $\alpha_m = -i\eta_m > 0$, and these conditions require that u_1 be oscillatory, or $\nu^2 > 0$ in (67). But $\eta_m^2 < 0$ in the equation $|m+\kappa|^2 + \eta_m^2 = \omega^2\varepsilon_0\mu_0$, and we infer that $\gamma > \omega^2$.

From this we see that the lowest eigenvalue $\gamma_1^{\omega,\kappa}(\varepsilon_0,\mu_0)$ is greater than ω^2 if and only if either $\mathscr{Z}_p \cup \mathscr{Z}_\ell = \emptyset$ or $\mathscr{Z}_p \cup \mathscr{Z}_\ell \neq \emptyset$ and $\gamma_{01} > \omega^2$. This is equivalent to the condition

$$L(\omega^2\varepsilon_0\mu_0 - |\kappa|^2)^{1/2} < \pi \text{ if } \mathscr{Z}_p \cup \mathscr{Z}_\ell \neq \emptyset. \tag{69}$$

Now since $0 < \varepsilon(x,z) \leq \varepsilon_0$ and $0 < \mu(x,z) \leq \mu_0$ in Ω, the quotient $\hat{a}^{\omega}(u,u)/\hat{b}(u,u)$ with coefficients (ε,μ) is greater than or equal to the quotient with coefficients (ε_0,μ_0) so that

$$\gamma_1^{\kappa,\omega}(\varepsilon,\mu) \geq \gamma_1^{\kappa,\omega}(\varepsilon_0,\mu_0) > \omega^2.$$

Thus (63) is never satisfied for all $v \in X$, and therefore neither is (49) satisfied for any $u \in H^1_\kappa(\Omega)$. Finally, we observe that, if $\mathscr{Z}_p = \emptyset$ and $\mathscr{Z}_\ell \neq \emptyset$, then $\omega^2\varepsilon_0\mu_0 - |\kappa|^2 = 0$, so (69) can be replaced by (62).

To prove the Theorem subject to condition (2), we follow Theorem 3.5 of [7]. It is convenient to take $z_- < z_0 < z_+$ with $z_0 = 0$. For pseudo-periodic u, we begin with the identity

$$\int_\Omega z\tfrac{\partial u}{\partial z}\nabla\cdot\mu^{-1}\nabla\bar{u} + \int_\Omega \mu^{-1}\nabla\bar{u}\cdot\left(\tfrac{\partial u}{\partial z}e_z + z\tfrac{\partial}{\partial z}\nabla u\right) = \mu_0^{-1}\int_\Gamma z\tfrac{\partial u}{\partial z}\tfrac{\partial \bar{u}}{\partial n},$$

($e_z = \langle 0,0,1\rangle$) use the Helmholtz equation in the first term on the left, and add the complex conjugate of the equation to obtain

$$-\omega^2\int_\Omega \varepsilon z\tfrac{\partial}{\partial z}|u|^2 + \int_\Omega \mu^{-1}\left(2\left|\tfrac{\partial u}{\partial z}\right|^2 + z\tfrac{\partial}{\partial z}|\nabla u|^2\right) = 2\mu_0^{-1}\int_\Gamma |z|\left|\tfrac{\partial u}{\partial z}\right|^2.$$

Then use integration by parts in z on the two terms with the symbol $z\frac{\partial}{\partial z}$ and replace $\int \mu^{-1}|\nabla u|^2$ using

$$\int_\Omega\left(-\mu^{-1}|\nabla u|^2 + \omega^2\varepsilon|u|^2\right) = \mu_0^{-1}\int_\Gamma \bar{u}\,T_r u$$

to obtain

$$\int_\Omega\left[2\mu^{-1}\left|\tfrac{\partial u}{\partial z}\right|^2 + \omega^2 z|u|^2\tfrac{\partial\varepsilon}{\partial z} - z|\nabla u|^2\tfrac{\partial\mu^{-1}}{\partial z}\right] + \mu_0^{-1}\int_\Gamma \bar{u}\,T_r u = \mu_0^{-1}\int_\Gamma |z|\left[\omega^2\mu_0\varepsilon_0|u|^2 + \left|\tfrac{\partial u}{\partial z}\right|^2 - |\nabla_x u|^2\right]. \tag{70}$$

By condition (2) of the Theorem (with $z_0 = 0$), $z\frac{\partial\varepsilon}{\partial z}$ and $-z\frac{\partial\mu^{-1}}{\partial z}$ are nonnegative, so the left-hand side is nonnegative. If we assume that u is a guided mode and write and u in its expansion in spatial harmonics, we see that the integral on the right-hand side is a sum over $m \in \mathscr{Z}_e$ of terms that are multiples of

$$\omega^2 \mu_0 \varepsilon_0 + |\eta_m|^2 - |m+\kappa|^2. \qquad (71)$$

But for $m \in \mathscr{Z}_e$, $\eta_m^2 < 0$, and by the definition of η_m, (71) vanishes. Therefore the right-hand side of (70) vanishes and we conclude that u vanishes identically in Ω. ∎

3.2.2 Modes in Related Problems

It is worth mentioning a few problems in wave propagation that are closely related to guided modes in open slab scatterers.

The problem of waves guided by periodically corrugated planar surfaces (diffraction gratings) is in many ways no different from ours. There is no transmission of energy across the surface, but the concepts of Rayleigh-Bloch scattering and the mathematical techniques for their analysis are essentially the same. Existence of scalar surface waves is treated by Linton and McIver [48] using variational techniques, by Grikurov, *et. al.,* [26] using the "augmented scattering matrix", as well as others [12, 35, 56, 66, 88]. Guided modes for the Maxwell equations in metal strip gratings over a substrate, in which the strips or the space between the strips is very thin, are established by Ammari, *et. al.* [1, 2].

One can consider a complementary structure to our slab system in which the ambient space is replaced by a photonic crystal and the slab is replaced by a homogeneous material. In fact, one of the primary attractions of photonic crystals is that they can be used to guide electromagnetic fields along paths carved out of the crystal at frequencies whose propagation is prohibited in the bulk. Guided modes of planar defects in two-dimensional photonic crystals are discussed by Ammari and Santosa [4], and the existence of modes in linear defects in three-dimensional photonic crystals is proved by Kuchment and Ong [37].

There is a large body of literature concerning the closely related problem in which closed waveguides with an obstacle placed inside admit trapped modes whose energy is concentrated at the obstacle. Much of this literature is presented in the context of water or sound waves by Evans, Linton, McIver, Ursell, and others. Trapped modes for water waves in a channel with a free surface were shown to exist by Ursell [85]. There, the governing equation is the Laplace equation, which admits no propagating spatial harmonics in the channel. The method of matched spatial harmonics and the residue calculus (see [55]) can be used to construct trapped acoustic (Helmholtz) modes [21, 18]. A variety of methods are demonstrated in [49]. Infinite sequences of trapped mode frequencies are treated in [47], and results on the dependence of the number of trapped modes on the structure are given in [14, 38].

Of particular interest are trapped modes whose frequencies are embedded in the continuous spectrum for a waveguide with an obstacle. Several techniques have been used to construct them, such as multipole expansions [10], variational formulations [19], generalized eigenfunction expansions [28], boundary integrals [57], construction of obstacles from trial trapped-mode solutions [56], and mode-matching and residue calculus [20]. In the latter, one can observe how the presence of a propagating spatial harmonic imposes an additional condition that is not present for trapped modes at frequencies below the cutoff. This condition corresponds to the second condition in the characterization (51).

4. COMPLEX EXTENSION AND BOUNDARY INTEGRALS

Rigorous analysis of resonance near nonrobust guided modes requires a formulation of the scattering problem for complex (κ, ω) and the complex dispersion relation for generalized guided modes. This is achieved by "reducing" the problem to an auxiliary one that is posed on a bounded domain. The method we expound in this section is that of boundary integrals.

Another approach, which we shall not expound, is to allow (κ, ω) to be complex in the weak formulation 5 and thereby extend the resolvent of the operator S_κ. The poles in the closed lower half ω-plane ($\text{Im}\,\omega \leq 0$) of this extension for real κ are the branches of the dispersion relation discussed in Sec. , on which a real value of ω corresponds to a true guided mode (exponentially decaying in space). This approach is taken by Lenoir, *et. al.*, for acoustic scattering by a rigid bounded obstacle. The poles are in the open lower half plane, $\text{Im}\,\omega < 0$, which expresses the fact that a bounded scatterer can support no bound (exponentially localized) acoustic modes.

Regardless of the method of reduction, the resulting auxiliary problem is an equation

$$A(\kappa, \omega)\psi(\kappa, \omega) = \phi(\kappa, \omega),$$

in which A is an operator that is jointly analytic in (κ, ω) and acts in a suitable function space that is independent of (κ,ω). The function ϕ is determined by the source field and the solution ψ contains data about the total field (source plus scattered) that is sufficient for reconstructing the total physical field in space. A generalized guided mode is represented by a solution to the homogeneous equation

$$A(\kappa,\omega)\psi(\kappa,\omega)=0,$$

and the locus of (κ,ω) pairs in $\mathbb{C}^3$ (or $\mathbb{C}^2$ in the two-dimensional case) for which a nontrivial solution exists is the complex multi-branched dispersion relation.

If the open waveguide is constructed from homogeneous components with smooth interfaces, the auxiliary problem is naturally formulated in terms of boundary integrals. The functions ϕ and ψ are traces of the source and total fields on the interfaces, and the operator A is composed of layer potentials. For simplicity, we shall suppose that our structure is composed of a homogeneous and isotropic medium occupying a region $\tilde{D}$ that is bounded in the z variable and periodic in the x variables and that has a smooth boundary and outwardly directed normal vector $n(\mathbf{r})$ for $\mathbf{r}\in\partial\tilde{D}$. Let $D=\tilde{D}\cap\mathscr{S}$ be one period of the structure, set $D^c=\mathscr{S}\setminus\bar{D}$, and denote by ∂D the part of the boundary of $\tilde{D}$ that lies in $\mathscr{S}$. Denote the dielectric and magnetic constants in D by ε_1 and μ_1 and those in the ambient medium by ε_0 and μ_0.

We will first describe the extension of the problems of scattering and guided modes to complex (κ,ω) and introduce the outgoing pseudo-periodic Green function for the Helmholtz equation. The Green function underlies the boundary-integral equations from which the auxiliary problem we described above is constructed. The associated Calderón boundary-integral projectors allow for an elegant and organized approach.

4.1 Complex Extension

The extension to the complex domain is not only important for the mathematical analysis of resonant behavior at real (κ,ω); the fields themselves for complex (κ,ω) have physical significance. As we have already understood, true guided slab modes are nonzero pseudo-periodic solutions to the Helmholtz equation with real (κ,ω) that fall of exponentially away from the slab. If we keep κ real but now allow the frequency ω of a generalized guided mode to have a nonzero imaginary part, it turns out (Theorem 15) that this imaginary part must be negative. This means that the mode decays in time but grows exponentially with distance away from the slab; this is made clear through the analytic continuation of the spatial harmonics as we discuss presently. The physical interpretation of these modes as leaky modes must be treated with care, and we refer the reader to the literature, for example [70, 84, 30]. If, on the other hand, ω remains real while the κ of a generalized mode attains an imaginary part, the mode is a harmonic field that is attenuated due to radiation losses as it travels along the slab; see [68] and the chapter by Tausch of this book.

In the definition (14) of the numbers η_m,

$$\eta_m=\left[\varepsilon_0\mu_0\omega^2-(m_1+\kappa_1)^2-(m_2+\kappa_2)^2\right]^{\frac{1}{2}},$$

the choice of square root was made to give the correct radiation Condition 1. For each $m\in\mathbb{Z}^2$, the branch cut for the square root can be taken to be the negative imaginary axis. When ω decreases through a real value at which η_m vanishes, we say that the mth Rayleigh diffractive order is cut off, as this spatial harmonic passes from propagating to evanescent. For ω just above the cutoff frequency, this harmonic is at a grazing angle with the slab and gives rise to an anomaly in the transmission coefficient known as the Wood anomaly (see [84, 53], for example). In the regime of leaky modes in which $\kappa\in\mathbb{R}^2$ and ω has a small imaginary part, η_m must jump from one branch of the square root function to another as the real part of ω passes a cutoff value. We will not treat this important case but focus on anomalies that are the result of nonrobust guided modes at real (κ,ω) at which all η_m are nonzero; this type of resonance is also discussed in [84]. We will show how these anomalies generalize the Fano resonance derived originally in the context of quantum mechanics.

If $(\kappa,\omega)\in\mathbb{R}^3$ and $\mathscr{Z}_\ell$ is empty, that is, for all $m\in\mathbb{Z}^2$,

$$\varepsilon_0\mu_0\omega^2-|m+\kappa|^2\neq 0,$$

then the numbers η_m can be extended analytically in a complex neighborhood of (κ,ω) in $\mathbb{C}^3$. If ω attains a small negative imaginary part, the outgoing propagating harmonics become exponentially growing as $|z|\to\infty$ (recall that we take $\operatorname{Re}\omega>0$) and the incoming harmonics become decaying; the reverse occurs if ω attains a small positive imaginary part. The evanescent harmonics remain evanescent under small perturbations of ω and κ. The radiation Condition 1 extends in this neighborhood to a *generalized outgoing condition.*

Condition 14 (Outgoing) *A complex-valued function u defined on $\mathbb{R}^3$ is said to satisfy the* generalized outgoing condition *for the slab for the complex pair (κ,ω), with $\text{Re}\,\omega>0$, if there exist a real number z_0 and complex coefficients $\{c_m^\pm\}_{m\in\mathbb{Z}^2}$ in $\ell^2(\mathbb{Z}^2)$ such that*

$$u(x,z)=\sum_{m\in\mathbb{Z}^2} c_m^\pm e^{\pm i\eta_m z} e^{i(m+\kappa)\cdot x} \quad \text{for } \pm z>z_0.$$

An analogous condition holds for the two-dimensional Helmholtz equation.

The outgoing condition is extended to electromagnetic fields by requiring that each component of the electric and magnetic fields satisfy Condition 14. Additional constraints are imposed by the Maxwell system: the E and H fields of each propagating harmonic are perpendicular to each other and to the propagation direction ([34] §7.1).

The outgoing pseudo-periodic Green function $G(\mathbf{r})$ ($\mathbf{r}=(x,z)$) for the Helmholtz equation in a homogeneous medium with coefficients (ε_0,μ_0) is a function that satisfies the equation in $\mathbb{R}^3$ except on a two-dimensional periodic array of source points $(2\pi n,0)$, whose strengths differ by a phase determined by the Bloch wavevector κ:

$$(\nabla^2+\omega^2\varepsilon_0\mu_0)G(\mathbf{r})=-\sum_{n\in\mathbb{Z}^2}\delta(x-2\pi n,z)e^{i\kappa\cdot x}.$$

Its representation in spatial Fourier harmonics is

$$G(\mathbf{r})=-\frac{1}{8\pi^2}\sum_{m\in\mathbb{Z}^2}\frac{1}{i\eta_m}e^{i\eta_m|z|}e^{i(m+\kappa)\cdot x}. \tag{72}$$

For $z\neq 0$, the convergence is exponential in m and one can see that G satisfies the Helmholtz equation; G also satisfies the outgoing Condition 14. A proof of the two-dimensional analog is given in [79]. The two-dimensional Green function looks the same except that the sum is taken over $\mathbb{Z}$ and $8\pi^2$ is replaced by 4π.

The scattering Problem 2 can be generalized by means of the outgoing condition, as can the definition of a guided mode. A *generalized guide mode* is a function that satisfies the Helmholtz equation as well as the outgoing condition. This means that it exists in the absence of generalized source fields ($u^{\text{inc}}=0$). If $\text{Im}\,\kappa=\text{Im}\,\omega=0$, the propagating harmonics necessarily vanish and the field falls off exponentially with distance from the slab. (Note that the outgoing condition encompasses exponentially decaying fields.)

The following theorem from plays a crucial role in the analysis of resonant transmission anomalies in Sec. 4.4. It asserts that generalized guided modes can exist only for $\text{Im}\,\omega\le 0$. (We always take $\text{Re}\,\omega>0$.) We prove it here in the case of the Helmholtz equation; it extends generally to other harmonic wave equations, continuous and discrete.

Theorem 15 (Generalized modes) *Suppose that (κ,ω) is such that $\mathscr{Z}_\ell=\emptyset$ and that u is pseudo-periodic with real wavevector κ and satisfies the Helmholtz equation and the generalized outgoing Condition 14. Then $\text{Im}\,\omega\le 0$. In addition, $u\to 0$ as $|z|\to\infty$ if and only if ω is real.*

Proof. For this proof, let $z_0=-z_-=z_+>0$. The Helmholtz equation and integration by parts gives

$$\begin{aligned}0&=\int_\Omega(\nabla\cdot\mu^{-1}\nabla u+\omega^2\varepsilon u)\,\bar{u}\\&=\int_\Omega(-\mu^{-1}|\nabla u|^2+\omega^2\varepsilon|u|^2)+\int_\Gamma\mu_0^{-1}(\partial_n u)\bar{u}.\end{aligned}$$

Using the outgoing condition, one computes that, for sufficiently large $z_0>0$,

$$\int_\Gamma\mu_0^{-1}(\partial_n u)\bar{u}=\frac{4\pi^2}{\mu_0}\sum_{m\in\mathbb{Z}^2}i\eta_m(|c_m^-|^2+|c_m^+|^2)e^{-2\text{Im}\,\eta_m|z_0|}.$$

These equations yield

$$-\text{Im}\,(\omega^2)\int_\Omega\varepsilon|u|^2=\frac{4\pi^2}{\mu_0}\sum_{m\in\mathbb{Z}^2}\text{Re}\,\eta_m(|c_m^-|^2+|c_m^+|^2)e^{-2\text{Im}\,\eta_m|z_0|}. \tag{73}$$

Let κ be real and ω be in a neighborhood of a point on the real axis for which all numbers η_m are analytic functions of ω. If $\text{Im}\,\omega>0$ (with our convention that $\text{Re}\,\omega>0$), then $\text{Im}\,\eta_m>0$ for all $m\in\mathbb{Z}^2$ and we obtain a contradition by letting z_0 tend to ∞. If $\text{Im}\,\omega=0$, then, for all $m\in\mathscr{Z}_p$, $\eta_m>0$ and thus $c_m^\pm=0$. Therefore, u is exponentially decaying as $|z|\to\infty$. Conversely, if $u\to 0$ as $|z|\to\infty$, then, because $\text{Im}\,\omega\le 0$, all harmonics with $m\in\mathscr{Z}_p$ are exponentially growing in $|z|$ and therefore $c_m^\pm=0$ for all $m\in\mathscr{Z}_p$. The rest of the terms decay exponentially; hence letting $z_0\to\infty$ in (73) shows that $\text{Im}\,\omega^2$ and therefore also $\text{Im}\,\omega$ vanishes. ■

4.2 The Helmholtz Equation

We seek a solution of the Helmholtz equation in $\mathbb{R}^3$ such that

$$\begin{aligned}\nabla^2 u+\omega^2\alpha_1 u&=f \quad \text{in } D,\\ \nabla^2 u+\omega^2\alpha_0 u&=f \quad \text{in } D^c,\end{aligned} \tag{74}$$

in which $\alpha_k = \varepsilon_k\mu_k$, $k = 1,2$, subject to the interface conditions

$$\left.\begin{aligned} u(\mathbf{r}-0n(\mathbf{r})) &= u(\mathbf{r}+0n(\mathbf{r})), \\ \mu_1^{-1}\partial_n u(\mathbf{r}-0n(\mathbf{r})) &= \mu_0^{-1}\partial_n u(\mathbf{r}+0n(\mathbf{r})) \end{aligned}\right\} \mathbf{r}\in\partial D \tag{75}$$

(the notation ± 0 indicates limits as $h \to 0^{\pm}$) and the pseudo-periodicity condition for wavevector κ. We may restrict analysis to the strip $\mathscr{S}$.

We will show how a solution u to the generalized scattering problem can be elegantly decomposed into source and scattered fields in the interior of D and in the exterior of D separately,

$$\begin{aligned} u|_D &= u^{\text{int}} = u^{\text{int}}_{\text{so}} + u^{\text{int}}_{\text{sc}}, \\ u|_{D^c} &= u^{\text{ext}} = u^{\text{ext}}_{\text{so}} + u^{\text{ext}}_{\text{sc}}. \end{aligned}$$

The scattered field should satisfy the homogeneous equation, and the source field is produced by the sources represented by f:

$$\begin{aligned} (\nabla^2+\omega^2\alpha_1)u^{\text{int}}_{\text{so}} &= f|_D, \\ (\nabla^2+\omega^2\alpha_1)u^{\text{int}}_{\text{sc}} &= 0, \end{aligned}$$

$$\begin{aligned} (\nabla^2+\omega^2\alpha_0)u^{\text{ext}}_{\text{so}} &= f|_{D^c}, \\ (\nabla^2+\omega^2\alpha_0)u^{\text{ext}}_{\text{sc}} &= 0, \end{aligned}$$

so that the equations (74) are satisfied.

We will see that unique determination of u is generically guaranteed by the following additional conditions, which complete the formulation of the physical problem.

1. $u^{\text{int}}_{\text{so}}$ is taken to be the restriction to D of the outgoing pseudo-periodic field $U^{\text{int}}_{\text{so}}$ in $\mathscr{S}$ satisfying

$$(\nabla^2+\omega^2\alpha_1)U^{\text{int}}_{\text{so}} = f\chi_D \quad \text{in } \mathscr{S},$$

where χ_D is the characteristic function of D. This field is given by

$$U^{\text{int}}_{\text{so}}(\mathbf{r}) = -\int_D G(\mathbf{r}-\mathbf{r}')f(\mathbf{r}')\,dV.$$

2. $u^{\text{ext}}_{\text{so}}$ is taken to be the restriction to D^c of a pseudo-periodic field $U^{\text{ext}}_{\text{so}}$ in $\mathscr{S}$ satisfying

$$(\nabla^2+\omega^2\alpha_0)U^{\text{ext}}_{\text{so}} = f\chi_{D^c} \quad \text{in } \mathscr{S}.$$

Such a field $U^{\text{ext}}_{\text{so}}$ is not unique, as it could be modified by a field emanating from sources at infinity, such as a plane wave U^{∞}_{so} satisfying $(\nabla^2 + \omega^2\alpha_0)U^{\infty}_{\text{so}} = 0$ in $\mathscr{S}$.

3. $u^{\text{ext}}_{\text{sc}}$ satisfies the generalized outgoing condition.

4. The interface conditions (75) must hold on ∂D.

The scattering problem can be reduced to integral equations on the boundary of D that involve the interface data of the total field u as unknown variables and the interface data of the source fields $u^{\text{int}}_{\text{so}}$ and $u^{\text{ext}}_{\text{so}}$ as the term of inhomogeneity. The interface data of a field u consists of the limits of u and $\mu^{-1}\partial_n u$ to ∂D, which are taken from the interior or exterior of D. We refer to this data collectively as the interior or exterior trace of u.

$$\psi_{\text{int}}(\mathbf{r}) = \begin{bmatrix} u^{\text{int}}(\mathbf{r}) \\ \mu_1^{-1}\partial_n u^{\text{int}}(\mathbf{r}) \end{bmatrix}, \quad \psi_{\text{ext}} = \begin{bmatrix} u^{\text{ext}}(\mathbf{r}) \\ \mu_0^{-1}\partial_n u^{\text{ext}}(\mathbf{r}) \end{bmatrix},$$

for $\mathbf{r}\in\partial D$. The traces are naturally considered in the space $H^{\frac{1}{2}}(\partial D)\oplus H^{-\frac{1}{2}}(\partial D)$.

We shall omit the details of the proper function spaces for the fields and their traces as well as technical aspects of the proofs and focus on the mathematical structure and how a boundary-integral formulation of the scattering problem arises from it. A through rigorous treatment of the theory, including the ensuing theory of layer potentials for the Helmholtz equation with applications to acoustic scattering is available in Costabel and Stephan [11] or Nédélec [61] §3.1, as well as Colton and Kress [9], Kress [39], and many other works. The first two works make explicit reference to the Calderón boundary-integral projectors, which we use below in the formulation of the auxiliary problem for scattering by a slab. Our presentation essentially follows that presented for the two-dimensional case by Shipman and Venakides [79].

The development of a boundary-integral formulation begins with the boundary-integral representations of Helmholtz fields in D and in D^c. Let us first take ε and μ to be constant over all of $\mathbb{R}^3$, as if the structure were not present, but retain the knowledge of the domain D. We will see how an arbitrary pair of functions (ξ,η) on ∂D can be decomposed uniquely into the sum of the boundary data of an interior Helmholtz field and the boundary data of an exterior outgoing Helmholtz field, both with the same coefficients ε and μ.

If u satisfies $(\nabla^2+\omega^2\alpha)u = 0$ with $\alpha = \varepsilon\mu$ in D, then u can be reconstructed from its boundary data on ∂D,

$$\begin{bmatrix} u(\mathbf{r}) \\ \mu^{-1}\partial_n u(\mathbf{r}) \end{bmatrix}, \quad \mathbf{r}\in\partial D, \tag{76}$$

through

$$u(\mathbf{r}) = \int_{\partial D}\left[-\frac{\partial G(\mathbf{r}-\mathbf{r}')}{\partial n_{\mathbf{r}'}}u(\mathbf{r}') + \right. \\ \left. +\mu G(\mathbf{r}-\mathbf{r}')\mu^{-1}\frac{\partial u}{\partial n}(\mathbf{r}')\right]ds_{\mathbf{r}'}, \quad \mathbf{r}\in D. \tag{77}$$

If u satisfies $(\nabla^2+\omega^2\alpha)u=0$ in D^c and the generalized outgoing condition, then

$$u(\mathbf{r})=\int_{\partial D}\left[\frac{\partial G(\mathbf{r}-\mathbf{r}')}{\partial n_{\mathbf{r}'}}u(\mathbf{r}')+ -\mu G(\mathbf{r}-\mathbf{r}')\mu^{-1}\frac{\partial u}{\partial n}(\mathbf{r}')\right]ds_{\mathbf{r}'},\quad \mathbf{r}\in D^c. \quad (78)$$

Both equations are proved by using the divergence theorem (or Green's identities) in the truncated period Ω of the slab structure. The contributions from the sides of $\mathscr{S}$ that are parallel to the z-axis vanish because of the pseudo-periodicity of u and G, and the contributions from the perpendicular sides $\Gamma_\pm$ vanish because of the outgoing condition satisfied by both u and G.
These representations show one way in which u is generated by a combination of single- and double-layer potentials. It is natural to extend these formulas to allow an arbitrary pair (ξ,η) of functions on ∂D in place of the trace of u and to define

$$u^{\text{int}}(\mathbf{r})=\int_{\partial D}\left[-\frac{\partial G(\mathbf{r}-\mathbf{r}')}{\partial n_{\mathbf{r}'}}\xi(\mathbf{r}')+ +\mu G(\mathbf{r}-\mathbf{r}')\eta(\mathbf{r}')\right]ds_{\mathbf{r}'},\quad \mathbf{r}\in D, \quad (79)$$

$$u^{\text{ext}}(\mathbf{r})=\int_{\partial D}\left[\frac{\partial G(\mathbf{r}-\mathbf{r}')}{\partial n_{\mathbf{r}'}}\xi(\mathbf{r}')+ -\mu G(\mathbf{r}-\mathbf{r}')\eta(\mathbf{r}')\right]ds_{\mathbf{r}'},\quad \mathbf{r}\in D^c. \quad (80)$$

The function u^{int} satisfies $(\nabla^2+\omega^2\alpha)u=0$ in D, and the function u^{ext} satisfies the same equation in D^c plus the outgoing condition because the Green function G does. Both fields have traces in $H^{\frac{1}{2}}(\partial D)\oplus H^{-\frac{1}{2}}(\partial D)$.
Taking the limits of u and $\partial_n u$ as $\mathbf{r}\to\partial D$ leads to the following representations. The analysis of the singularity in the integral is subtle and is based on the Sokhotski-Plemelj formulas.

$$u^{\text{int}}(\mathbf{r})=\tfrac{1}{2}\xi(\mathbf{r})+ \int_{\partial D}\left[-\frac{\partial G(\mathbf{r}-\mathbf{r}')}{\partial n_{\mathbf{r}'}}\xi(\mathbf{r}')+\mu G(\mathbf{r}-\mathbf{r}')\eta(\mathbf{r}')\right]ds_{\mathbf{r}'},$$

$$\mu^{-1}\partial_n u^{\text{int}}(\mathbf{r})=\tfrac{1}{2}\eta(\mathbf{r})+ \int_{\partial D}\left[-\mu^{-1}\frac{\partial^2 G(\mathbf{r}-\mathbf{r}')}{\partial n_{\mathbf{r}}\partial n_{\mathbf{r}'}}\xi(\mathbf{r}')+\frac{\partial G(\mathbf{r}-\mathbf{r}')}{\partial n_{\mathbf{r}}}\eta(\mathbf{r}')\right]ds_{\mathbf{r}'}, \quad (81)$$

$$u^{\text{ext}}(\mathbf{r})=\tfrac{1}{2}\xi(\mathbf{r})+ \int_{\partial D}\left[\frac{\partial G(\mathbf{r}-\mathbf{r}')}{\partial n_{\mathbf{r}'}}\xi(\mathbf{r}')-\mu G(\mathbf{r}-\mathbf{r}')\eta(\mathbf{r}')\right]ds_{\mathbf{r}'},$$

$$\mu^{-1}\partial_n u^{\text{ext}}(\mathbf{r})=\tfrac{1}{2}\eta(\mathbf{r})+ \int_{\partial D}\left[\mu^{-1}\frac{\partial^2 G(\mathbf{r}-\mathbf{r}')}{\partial n_{\mathbf{r}}\partial n_{\mathbf{r}'}}\xi(\mathbf{r}')-\frac{\partial G(\mathbf{r}-\mathbf{r}')}{\partial n_{\mathbf{r}}}\eta(\mathbf{r}')\right]ds_{\mathbf{r}'}. \quad (82)$$

These formulas are composed of four operators giving the values and normal derivatives on ∂D of the single- and double-layer potentials,

$$(S\eta)(\mathbf{r})=\int_{\partial D}G(\mathbf{r}-\mathbf{r}')\eta(\mathbf{r}')ds_{\mathbf{r}'},$$

$$(K\xi)(\mathbf{r})=\int_{\partial D}\frac{\partial G(\mathbf{r}-\mathbf{r}')}{\partial n_{\mathbf{r}'}}\xi(\mathbf{r}')ds_{\mathbf{r}'},$$

$$(K'\eta)(\mathbf{r})=\int_{\partial D}\frac{\partial G(\mathbf{r}-\mathbf{r}')}{\partial n_{\mathbf{r}}}\xi(\mathbf{r}')ds_{\mathbf{r}'},$$

$$(T\xi)(\mathbf{r})=\int_{\partial D}\frac{\partial^2 G(\mathbf{r}-\mathbf{r}')}{\partial n_{\mathbf{r}}\partial n_{\mathbf{r}'}}\xi(\mathbf{r}')ds_{\mathbf{r}'}.$$

The singular integrals, especially T, must be treated carefully (see [11]). These operators are such that the following matrix operator A is bounded in $H^{\frac{1}{2}}(\partial D)\oplus H^{-\frac{1}{2}}(\partial D)$,

$$A=\begin{bmatrix} K & -\mu S \\ \mu^{-1}T & -K' \end{bmatrix}.$$

Let I be the identity operator, and define the operators

$$P_{\text{int}}=\tfrac{1}{2}I-A,\quad P_{\text{ext}}=\tfrac{1}{2}I+A.$$

The equations (81) and (82) are expressed in terms of P_{int} and P_{ext} as

$$\begin{bmatrix} u^{\text{int}} \\ \mu^{-1}\partial_n u^{\text{int}} \end{bmatrix}=P_{\text{int}}\begin{bmatrix} \xi \\ \eta \end{bmatrix},\quad \begin{bmatrix} u^{\text{ext}} \\ \mu^{-1}\partial_n u^{\text{ext}} \end{bmatrix}=P_{\text{ext}}\begin{bmatrix} \xi \\ \eta \end{bmatrix}.$$

Now, because of the integral representations (77) and (78), we also have

$$\begin{bmatrix} u^{\text{int}} \\ \mu^{-1}\partial_n u^{\text{int}} \end{bmatrix}=P_{\text{int}}\begin{bmatrix} u^{\text{int}} \\ \mu^{-1}\partial_n u^{\text{int}} \end{bmatrix},$$

$$\begin{bmatrix} u^{\text{ext}} \\ \mu^{-1}\partial_n u^{\text{ext}} \end{bmatrix}=P_{\text{ext}}\begin{bmatrix} u^{\text{ext}} \\ \mu^{-1}\partial_n u^{\text{ext}} \end{bmatrix}.$$

This shows that P_{int} and P_{ext} are projection operators, and, by their definition, they are complementary. These are the *Calderón projectors* for the Helmholtz equation in the doubly pseudo-periodic setting. They depend on the parameters ω, κ, and α. In summary, we have

1. $P_{\text{int}}+P_{\text{ext}}=I$,
2. $P_{\text{int}}^2=P_{\text{int}}$ and $P_{\text{ext}}^2=P_{\text{ext}}$,

3. The range of P_{int} is the subspace consisting of traces of free κ-pseudo-periodic Helmholtz fields in D.

4. The range of P_{ext} is the subspace consisting of traces of free κ-pseudo-periodic Helmholtz fields in D^c that satisfy the outgoing condition.

Let us return to the interior and exterior constants $\alpha_1 = \varepsilon_1\mu_1$ and $\alpha_0 = \varepsilon_0\mu_0$ and the scattering problem and put

$$\bar{\varepsilon} = \tfrac{1}{2}(\varepsilon_0 + \varepsilon_1), \quad \bar{\mu} = \tfrac{1}{2}(\mu_0 + \mu_1).$$

Consider the traces of all the fields involved:

$$\psi = \begin{bmatrix} u \\ \mu^{-1}\partial_n u \end{bmatrix},$$

$$\phi_{\text{so}}^{\text{int}} = \begin{bmatrix} u_{\text{so}}^{\text{int}} \\ \mu_1^{-1}\partial_n u_{\text{so}}^{\text{int}} \end{bmatrix}, \quad \phi_{\text{sc}}^{\text{int}} = \begin{bmatrix} u_{\text{sc}}^{\text{int}} \\ \mu_1^{-1}\partial_n u_{\text{sc}}^{\text{int}} \end{bmatrix}, \tag{83}$$

$$\phi_{\text{so}}^{\text{ext}} = \begin{bmatrix} u_{\text{so}}^{\text{ext}} \\ \mu_0^{-1}\partial_n u_{\text{so}}^{\text{ext}} \end{bmatrix}, \quad \phi_{\text{sc}}^{\text{ext}} = \begin{bmatrix} u_{\text{sc}}^{\text{ext}} \\ \mu_0^{-1}\partial_n u_{\text{sc}}^{\text{ext}} \end{bmatrix}.$$

Because of the requirement of continuity of the boundary data,

$$\begin{aligned} \psi &= \phi_{\text{so}}^{\text{int}} + \phi_{\text{sc}}^{\text{int}}, \\ \psi &= \phi_{\text{so}}^{\text{ext}} + \phi_{\text{sc}}^{\text{ext}}. \end{aligned} \tag{84}$$

Recall that the interior scattered field $u_{\text{sc}}^{\text{int}}$ is a free pseudo-periodic Helmholtz field in D with constant α_1 and that the interior source field $u_{\text{so}}^{\text{int}}$, by our definition, extends to a free outgoing Helmholtz field $U_{\text{so}}^{\text{int}}$ in D^c with the same constant α_1. This means that $\phi_{\text{sc}}^{\text{int}}$ is in the nullspace of P_{ext}^1 and $\phi_{\text{so}}^{\text{int}}$ is in the range (the superscript refers to the value α_1). Similarly, the exterior scattered field $u_{\text{sc}}^{\text{ext}}$ is a free pseudo-periodic outgoing Helmholtz field in D^c with constant α_0 and the exterior source field $u_{\text{so}}^{\text{ext}}$ extends to a free Helmholtz field in D also with α_0; thus $\phi_{\text{sc}}^{\text{ext}}$ is in the nullspace of P_{int}^0 and $\phi_{\text{so}}^{\text{ext}}$ is in the range. This all means that we can project the trace of the field we seek onto traces of the known interior and exterior source fields that produce it:

$$\begin{aligned} P_{\text{ext}}^1\psi &= \phi_{\text{so}}^{\text{int}}, \\ P_{\text{int}}^0\psi &= \phi_{\text{so}}^{\text{ext}}. \end{aligned} \tag{85}$$

For the problem of scattering of a plane wave $u^{\text{inc}}(x,z) = e^{i(\bar{m}+\kappa)x}e^{i\eta_{\bar{m}}z}$, the source traces are

$$\phi_{\text{so}}^{\text{ext}} = \begin{bmatrix} 1 \\ i\mu_0^{-1}(\bar{m}+\kappa, \eta_{\bar{m}})\cdot n \end{bmatrix} u^{\text{inc}}(x,z), \quad \phi_{\text{so}}^{\text{int}} = \begin{bmatrix} 0 \\ 0 \end{bmatrix}.$$

The Calderón projectors in the equations (85) have a second-order derivative of G in the integral kernel with leading singularities differing by a multiplicative constant. To cause these to cancel in a linear combination, we can multiply them by the matrices

$$\Lambda_k = \begin{bmatrix} 1 & 0 \\ 0 & \mu_k/\bar{\mu} \end{bmatrix}, \tag{86}$$

for $k = 0, 1$, to obtain

$$[\Lambda_1 P_{\text{ext}}^1 + \Lambda_0 P_{\text{int}}^0]\psi = \Lambda_1\phi_{\text{so}}^{\text{int}} + \Lambda_0\phi_{\text{so}}^{\text{ext}}. \tag{87}$$

With the notation

$$\eta(\mathbf{r}) = \mu_1^{-1}\frac{\partial u^{\text{int}}}{\partial n}(\mathbf{r}) = \mu_0^{-1}\frac{\partial u^{\text{ext}}}{\partial n}(\mathbf{r}), \quad r \in \partial D,$$

the resulting system of boundary-integral equations for $[u(\mathbf{r}), \eta(\mathbf{r})]^t$ is

$$\begin{aligned} u(\mathbf{r}) + \int_{\partial D}\Big[\frac{\partial(G_1 - G_0)(\mathbf{r}-\mathbf{r}')}{\partial n_{\mathbf{r}'}}u(\mathbf{r}') + \\ -(\mu_1 G_1 - \mu_0 G_0)(\mathbf{r}-\mathbf{r}')\,\eta(\mathbf{r}')\Big]ds_{\mathbf{r}'} \\ = u_{\text{so}}^{\text{int}}(\mathbf{r}) + u_{\text{so}}^{\text{ext}}(\mathbf{r}), \end{aligned} \tag{88}$$

$$\begin{aligned} \eta(\mathbf{r}) + \bar{\mu}^{-1}\int_{\partial D}\Big[\frac{\partial^2(G_1 - G_0)(\mathbf{r}-\mathbf{r}')}{\partial n_{\mathbf{r}}\partial n_{\mathbf{r}'}}u(\mathbf{r}') + \\ -(\mu_1 G_1 - \mu_0 G_0)(\mathbf{r}-\mathbf{r}')\,\eta(\mathbf{r}')\Big]ds_{\mathbf{r}'} \\ = \bar{\mu}^{-1}\frac{\partial}{\partial n}(u_{\text{so}}^{\text{int}}(\mathbf{r}) + u_{\text{so}}^{\text{ext}}(\mathbf{r})). \end{aligned} \tag{89}$$

The system in which we are interested is the pair (85), for the trace of the scattered field is recovered from the decomposition (84) and this or the total field is then used to determine the scattered field in D and D^c by means of boundary-integral representation formulas (77,78). Any solution of the pair (85) is also a solution to the combination (87). What we must now determine is if a solution to the combination is also a solution of the pair. To ascertain this, we observe that (87) is equivalent to

$$\left.\begin{aligned} \Lambda_1 P_{\text{ext}}^1\psi &= \Lambda_1\phi_{\text{so}}^{\text{int}} + f \\ \Lambda_0 P_{\text{int}}^0\psi &= \Lambda_0\phi_{\text{so}}^{\text{ext}} - f \end{aligned}\right\}\ \text{for some } f, \tag{90}$$

or, alternatively,

$$\left.\begin{aligned} P_{\text{ext}}^1\psi &= \phi_{\text{so}}^{\text{int}} + \Lambda_1^{-1}f \\ P_{\text{int}}^0\psi &= \phi_{\text{so}}^{\text{ext}} - \Lambda_0^{-1}f \end{aligned}\right\}\ \text{for some } f. \tag{91}$$

Such a function pair $f = [f_1, f_2]^t$ is characterized by the property that $\Lambda_1^{-1}f$ is in the range of the projection P_{ext}^1 (because $\phi_{\text{so}}^{\text{int}}$ is) and $\Lambda_0^{-1}f$ is in the range of P_{int}^0 (because $\phi_{\text{so}}^{\text{ext}}$ is), or, equivalently,

$$P_{\text{int}}^1\Lambda_1^{-1}f = 0, \quad P_{\text{ext}}^0\Lambda_0^{-1}f = 0. \tag{92}$$

Now, one can calculate that the Calderón projectors for the *reciprocal* coefficients

$$\begin{aligned} \varepsilon^0 &= \frac{2\varepsilon_1\mu_1}{\mu_0+\mu_1}, & \mu^0 &= \frac{\mu_0+\mu_1}{2}, \\ \varepsilon^1 &= \frac{2\varepsilon_0\mu_0}{\mu_0+\mu_1}, & \mu^1 &= \frac{\mu_0+\mu_1}{2}, \end{aligned} \tag{93}$$

are related to those of the original coefficients by conjugation by $\Lambda_{0,1}$. In particular, if we distinguish the projectors for the reciprocal coefficients by a bar,

$$\bar{P}^0_{\text{int}} = \Lambda_1 P^1_{\text{int}} \Lambda_1^{-1} \tag{94}$$

is the interior Calderón projector for the coefficients (ε^0, μ^0) and

$$\bar{P}^1_{\text{ext}} = \Lambda_0 P^0_{\text{ext}} \Lambda_0^{-1} \tag{95}$$

is the exterior projector for the coefficients (ε^1, μ^1). Then, because of (92), we obtain the pair

$$\begin{aligned} \bar{P}^0_{\text{int}} f &= 0, \\ \bar{P}^1_{\text{ext}} f &= 0. \end{aligned} \qquad \text{(reciprocal system)} \tag{96}$$

A function f that satisfies this pair is simultaneously the trace of an exterior pseudo-periodic outgoing Helmholtz field with constants (ε^0, μ^0) and the trace of an interior Helmholtz field with constants (ε^1, μ^1). If $f \neq 0$, this field corresponds to a generalized guided mode of a *reciprocal structure* characterized by these new constants. Since

$$\begin{aligned} \alpha_0 &= \varepsilon_0\mu_0 = \varepsilon^1\mu^1, \\ \alpha_1 &= \varepsilon_1\mu_1 = \varepsilon^0\mu^0, \end{aligned}$$

the mode satisfies the Helmholtz equation with the interior and exterior valus of α switched relative to those of the original structure, but the multiplicative jump in the normal derivative is replaced by continuity because $\mu^0 = \mu^1$.

If no guided mode exists in the reciprocal structure, that is, if the pair (96) admits only the trivial solution, then the combination (87) is equivalent to the pair (85). In other words, uniqueness of the solution of the reciprocal scattering problem, in which the ambient medium is characterized by (ε^0, μ^0) and periodic structure are characterized by (ε^1, μ^1), implies equivalence of the original scattering Problem 2 and the boundary integral equations (88,89).

The essential results can be summarized in the following theorem.

Theorem 16 *If the (reciprocal) periodic structure with coefficients*

$$\begin{aligned} \varepsilon &= \varepsilon^1,\ \mu = \mu^1 && \text{in } D, \\ \varepsilon &= \varepsilon^0,\ \mu = \mu^0 && \text{in } D^c, \end{aligned}$$

defined in terms of given constants (ε_1, μ_1) and (ε_0, μ_0) by (93) does not admit a free pseudo-periodic outgoing Helmholtz field in the absence of a source (a generalized guided mode), then the boundary-integral system (88,89) is equivalent to the scattering Problem 2 in the (original) structure with

$$\begin{aligned} \varepsilon &= \varepsilon_1,\ \mu = \mu_1 && \text{in } D, \\ \varepsilon &= \varepsilon_0,\ \mu = \mu_0 && \text{in } D^c. \end{aligned}$$

More specifically, with ϕ^{int}_{so} and ϕ^{ext}_{so} defined as in (83)*, the scattered field u^{sc} of the scattering problem with source fields u^{int}_{so} and u^{ext}_{so} is obtained by using the solution to (88,89),*

$$\begin{bmatrix} u(\mathbf{r}) \\ \eta(\mathbf{r}) \end{bmatrix} = \begin{bmatrix} u(\mathbf{r}) \\ \mu^{-1}\partial_n u(\mathbf{r}) \end{bmatrix}, \quad \mathbf{r} \in \partial D,$$

in the representation formulas (77) *and* (78).

The reciprocal relations have the property that, if R is defined by

$$R(\varepsilon_0, \mu_0; \varepsilon_1, \mu_1) = (\varepsilon^0, \mu^0; \varepsilon^1, \mu^1),$$

subject to the relations (93), then

$$R^2(\varepsilon_0, \mu_0; \varepsilon_1, \mu_1) = (\varepsilon_0 \tfrac{\mu_0}{\bar{\mu}}, \bar{\mu}; \varepsilon_1 \tfrac{\mu_1}{\bar{\mu}}, \bar{\mu})$$

and

$$R^3 = R^1;$$

and the image of R is those sets of coefficients for which $\mu_0 = \mu_1$.

4.3 Two-Dimensional Reduction

In its two-dimensional form, the Helmholtz equation describes a variety of harmonic waves, including acoustic waves in structures that are invariant in one direction, water waves in certain regimes, and polarized electromagnetic waves. If the periodicity of the slab degenerates to invariance in, say, the x_2-direction and we assume that the electromagnetic waves are also invariant in the x_2-direction, then the Maxwell system decouples into two polarizations.

In the E-polarized case, the E field is directed out of the x_1z-plane and H lies in the plane. If we denote by u this out-of-plane component, then $H = (i\omega\mu)^{-1}\langle -u_z, 0, u_{x_1}\rangle$. The Maxwell system implies the Helmholtz equation for u:

$$\nabla \cdot \mu^{-1}\nabla u + \omega^2 \varepsilon u = 0, \tag{97}$$

which, considered in the distributional sense, implies continuity of u and $\mu^{-1}\partial_n u$ at ∂D. The foregoing analysis of the boundary-integral equations is

valid in two dimensions if the Green functions are replaced with their two-dimensional analogues.
In the H-polarized case, the H field is directed out of the x_1z-plane and E lies in the plane. If u is this out-of-plane component, then $E = -(i\omega\varepsilon)^{-1}\langle -u_z, 0, u_{x_1}\rangle$. The equation in distributional form is

$$\nabla\cdot\varepsilon^{-1}\nabla u+\omega^2\mu u=0, \tag{98}$$

which implies continuity of u and $\varepsilon^{-1}\partial_n u$ at ∂D. In this case, the results on the boundary-integral formulation must be modified by interchanging the roles of ε and μ.
Let us examine the case of nonmagnetic materials, or $\mu\equiv 1$. In the E-polarized case, we wish to solve the scattering problem with

$$\begin{aligned}\varepsilon=\varepsilon_1,\ \mu=1 \quad &\text{in } D,\\ \varepsilon=\varepsilon_0,\ \mu=1 \quad &\text{in } D^c.\end{aligned}$$

The system of boundary-integral equations, with $\eta=\partial_n u$ continuous on ∂D, is

$$\begin{aligned}u(\mathbf{r})+\int_{\partial D}\Big[&\frac{\partial(G_1-G_0)(\mathbf{r}-\mathbf{r}')}{\partial n_{\mathbf{r}'}}u(\mathbf{r}')+\\ &-(G_1-G_0)(\mathbf{r}-\mathbf{r}')\,\eta(\mathbf{r}')\Big]ds_{\mathbf{r}'}\\ &=u^{\text{int}}_{\text{so}}(\mathbf{r})+u^{\text{ext}}_{\text{so}}(\mathbf{r}),\end{aligned} \tag{99}$$

$$\begin{aligned}\eta(\mathbf{r})+\int_{\partial D}\Big[&\frac{\partial^2(G_1-G_0)(\mathbf{r}-\mathbf{r}')}{\partial n_{\mathbf{r}}\partial n_{\mathbf{r}'}}u(\mathbf{r}')+\\ &-(G_1-G_0)(\mathbf{r}-\mathbf{r}')\,\eta(\mathbf{r}')\Big]ds_{\mathbf{r}'}\\ &=\frac{\partial}{\partial n}(u^{\text{int}}_{\text{so}}(\mathbf{r})+u^{\text{ext}}_{\text{so}}(\mathbf{r})).\end{aligned} \tag{100}$$

The reciprocal problem of Theorem 16 for E-polarization has

$$\begin{aligned}\varepsilon=\varepsilon_0,\ \mu=1 \quad &\text{in } D,\\ \varepsilon=\varepsilon_1,\ \mu=1 \quad &\text{in } D^c.\end{aligned}$$

In the H-polarized case, we wish to solve the scattering problem with the following replacement in the forgoing analysis:

$$\begin{aligned}\varepsilon\mapsto 1,\ \mu\mapsto\varepsilon_1 \quad &\text{in } D,\\ \varepsilon\mapsto 1,\ \mu\mapsto\varepsilon_0 \quad &\text{in } D^c.\end{aligned}$$

The system of boundary-integral equations in the H-polarization case, with $\eta=\varepsilon^{-1}\partial_n u$ continuous on ∂D, is

$$\begin{aligned}u(\mathbf{r})+\int_{\partial D}\Big[&\frac{\partial(G_1-G_0)(\mathbf{r}-\mathbf{r}')}{\partial n_{\mathbf{r}'}}u(\mathbf{r}')+\\ &-(\varepsilon_1 G_1-\varepsilon_0 G_0)(\mathbf{r}-\mathbf{r}')\,\eta(\mathbf{r}')\Big]ds_{\mathbf{r}'}\\ &=u^{\text{int}}_{\text{so}}(\mathbf{r})+u^{\text{ext}}_{\text{so}}(\mathbf{r}),\end{aligned} \tag{101}$$

$$\begin{aligned}\eta(\mathbf{r})+\bar{\varepsilon}^{-1}\int_{\partial D}\Big[&\frac{\partial^2(G_1-G_0)(\mathbf{r}-\mathbf{r}')}{\partial n_{\mathbf{r}}\partial n_{\mathbf{r}'}}u(\mathbf{r}')+\\ &-(\varepsilon_1 G_1-\varepsilon_0 G_0)(\mathbf{r}-\mathbf{r}')\,\eta(\mathbf{r}')\Big]ds_{\mathbf{r}'}\\ &=\bar{\varepsilon}^{-1}\frac{\partial}{\partial n}(u^{\text{int}}_{\text{so}}(\mathbf{r})+u^{\text{ext}}_{\text{so}}(\mathbf{r})).\end{aligned} \tag{102}$$

The reciprocal problem has

$$\begin{aligned}\varepsilon=\varepsilon_0\bar{\varepsilon}^{-1},\ \mu=\bar{\varepsilon} \quad &\text{in } D,\\ \varepsilon=\varepsilon_1\bar{\varepsilon}^{-1},\ \mu=\bar{\varepsilon} \quad &\text{in } D^c.\end{aligned}$$

But Helmholtz fields in a structure with these coefficients coincide with fields in a structure with ε replaced with $c\varepsilon$ and μ replaced with $c^{-1}\mu$, with c a constant. By taking $c=\bar{\varepsilon}$, the reciprocal problem is seen to be identical to that for the E-polarized case. In summary, we have obtained Theorem 4.4 of [79].

Theorem 17 *If the (reciprocal) two-dimensional structure with dielectric constants*

$$\varepsilon=\varepsilon_1 \text{ in } D^c,\quad \varepsilon=\varepsilon_0 \text{ in } D,$$

and magnetic constant $\mu=1$ admits no nontrivial E-polarized field in the absence of a source, (generalized guided mode) then the systems (99,100) and (101,102) are equivalent to the E-polarized and H-polarized scattering problems, respectively, with $\mu=1$.

We have seen already that the condition in this theorem is satisfied when ω and κ are real, $\varepsilon_1>\varepsilon_0$, $\mu=1$, and an additional condition from Theorem 13 holds.

4.4 The Harmonic Maxwell System

For the harmonic Maxwell system, rigorous treatment of the technical aspects of the boundary-integral operators can be found, for example, in Müller [60] and [61], and the Calderón projectors are treated for bounded objects in $\mathbb{R}^3$ in [61] §5.5.
The development of the boundary-integral equations and the reciprocal problem for the Maxwell system parallels that of the Helmholtz equation. We present the framework, leaving the technical details to the references cited above.
We seek a pseudo-periodic solution of the harmonic Maxwell system in $\mathbb{R}^3$,

$$\begin{aligned}\nabla\times H+i\omega\varepsilon E&=J_1,\\ \nabla\times E-i\omega\mu H&=J_2,\end{aligned}$$

subject to continuity of the tangential traces of E and H on ∂D. The sources J_1 and J_2 are electric and magnetic currents. If we put

$$v = \begin{bmatrix} H \\ E \end{bmatrix}, \quad f = \begin{bmatrix} J_1 \\ J_2 \end{bmatrix},$$

$$L = \begin{bmatrix} \nabla\times & i\omega\varepsilon \\ -i\omega\mu & \nabla\times \end{bmatrix}, \quad L_k = \begin{bmatrix} \nabla\times & i\omega\varepsilon_k \\ -i\omega\mu_k & \nabla\times \end{bmatrix},$$

for $k = 1,2$, the Maxwell system is written compactly as

$$Lv = f. \tag{103}$$

A solution v can be decomposed into source and scattered fields in the interior of D and in the exterior of D,

$$\begin{aligned} v|_D &= v^{\text{int}} = v^{\text{int}}_{\text{so}} + v^{\text{int}}_{\text{sc}}, \\ v|_{D^c} &= v^{\text{ext}} = v^{\text{ext}}_{\text{so}} + v^{\text{ext}}_{\text{sc}}. \end{aligned} \tag{104}$$

The scattered field should satisfy the homogeneous Maxwell system,

$$\begin{cases} L_1 v^{\text{int}}_{\text{so}} = f|_D \\ L_1 v^{\text{int}}_{\text{sc}} = 0 \end{cases} \Longrightarrow L_1 v^{\text{int}} = f|_D.$$

$$\begin{cases} L_0 v^{\text{ext}}_{\text{so}} = f|_{D^c} \\ L_0 v^{\text{ext}}_{\text{sc}} = 0 \end{cases} \Longrightarrow L_0 v^{\text{ext}} = f|_{D^c}.$$

The conditions that generically determine a unique solution v are the following:

1. $v^{\text{int}}_{\text{so}}$ is taken to be the restriction to D of the outgoing pseudo-periodic field $V^{\text{int}}_{\text{so}}$ satisfying

$$L_1 V^{\text{int}}_{\text{so}} = f\chi_D \quad \text{in } \mathscr{S}.$$

2. $v^{\text{ext}}_{\text{so}}$ is taken to be the restriction to D^c of a field $V^{\text{ext}}_{\text{so}}$ satisfying

$$L_0 V^{\text{ext}}_{\text{so}} = f\chi_{D^c} \quad \text{in } \mathscr{S}.$$

Such a field is not unique, as it could be modified by fields emanating from sources at infinity, such as plane waves V^{∞}_{so} satisfying $L_0 V^{\infty}_{\text{so}} = 0$ in $\mathscr{S}$.

3. $v^{\text{ext}}_{\text{sc}}$ satisfies the generalized outgoing condition.

4. The tangential traces of the fields v^{int} and v^{ext} must match on ∂D.

The integral representation formulas for a Maxwell field $[H,E]^t$ involve the tangential traces of H and E, or the electric current j and magnetic current m,

$$\begin{aligned} j(\mathbf{r}) &= -n(\mathbf{r}) \times H(\mathbf{r}) \\ m(\mathbf{r}) &= n(\mathbf{r}) \times E(\mathbf{r}) \end{aligned} \quad (\mathbf{r} \in \partial D). \tag{105}$$

If $[H,E]^t$ is an outgoing pseudo-periodic Maxwell field in D^c with constant coefficients ε and μ,

$$\begin{aligned} E(\mathbf{r}) = &-i\omega\mu \int_{\partial D} G(\mathbf{r}-\mathbf{r}')j(\mathbf{r}')dS(\mathbf{r}') + \\ &+ \frac{1}{i\omega\varepsilon}\nabla \int_{\partial D} G(\mathbf{r}-\mathbf{r}')\text{div}_{\partial D} j(\mathbf{r}')dS(\mathbf{r}') + \\ &+ \nabla\times \int_{\partial D} G(\mathbf{r}-\mathbf{r}')m(\mathbf{r}')dS(\mathbf{r}'), \end{aligned}$$

$$\begin{aligned} H(\mathbf{r}) = &-i\omega\varepsilon \int_{\partial D} G(\mathbf{r}-\mathbf{r}')m(\mathbf{r}')dS(\mathbf{r}') + \\ &+ \frac{1}{i\omega\mu}\nabla \int_{\partial D} G(\mathbf{r}-\mathbf{r}')\text{div}_{\partial D} m(\mathbf{r}')dS(\mathbf{r}') + \\ &- \nabla\times \int_{\partial D} G(\mathbf{r}-\mathbf{r}')j(\mathbf{r}')dS(\mathbf{r}'), \end{aligned} \tag{106}$$

and the integral representation for interior fields has an additional factor of -1 in each term of the right-hand sides (see [6, 61]).

The boundary-integral formulation of the scattering problem involves the tangential traces $[j(\mathbf{r}), m(\mathbf{r})]^t$ of the total field $v = [H,E]^t$ as unknown variables and the tangential traces of the source fields $v^{\text{int}}_{\text{so}}$ and $v^{\text{ext}}_{\text{so}}$ in the term of inhomogeneity,

$$\begin{aligned} q_1(\mathbf{r}) &= \tfrac{\mu_1}{\bar{\mu}}(-n \times H^{\text{int}}_{\text{so}}) + \tfrac{\mu_0}{\bar{\mu}}(-n \times H^{\text{ext}}_{\text{so}}), \\ q_2(\mathbf{r}) &= \tfrac{\varepsilon_1}{\bar{\varepsilon}}(n \times E^{\text{int}}_{\text{so}}) + \tfrac{\varepsilon_0}{\bar{\varepsilon}}(n \times E^{\text{ext}}_{\text{so}}). \end{aligned} \quad (\mathbf{r} \in \partial D)$$

The analogue of the system (88,89) is the pair

$$\begin{aligned} j(\mathbf{r}) &- \frac{1}{\bar{\mu}} \int_{\partial D} n(\mathbf{r}) \times [j(\mathbf{r}') \times \nabla(\mu_1 G_1 - \mu_0 G_0)] dS(\mathbf{r}') \\ &- \frac{1}{i\omega\bar{\mu}} \int_{\partial D} [n(\mathbf{r}) \times m(\mathbf{r}')][\omega^2\varepsilon_1\mu_1 G_1 - \omega^2\varepsilon_0\mu_0 G_0]|dS(\mathbf{r}') \\ &- \frac{1}{i\omega\bar{\mu}} \int_{\partial D} n(\mathbf{r}) \times [(m(\mathbf{r}')\cdot\nabla)\nabla(G_1 - G_0)] dS(\mathbf{r}') \\ &= q_1(\mathbf{r}), \end{aligned} \tag{107}$$

$$\begin{aligned} m(\mathbf{r}) &- \frac{1}{\bar{\varepsilon}} \int_{\partial D} n(\mathbf{r}) \times [m(\mathbf{r}') \times \nabla(\varepsilon_1 G_1 - \varepsilon_0 G_0)] dS(\mathbf{r}') \\ &+ \frac{1}{i\omega\bar{\varepsilon}} \int_{\partial D} [n(\mathbf{r}) \times j(\mathbf{r}')][\omega^2\varepsilon_1\mu_1 G_1 - \omega^2\varepsilon_0\mu_0 G_0] dS(\mathbf{r}') \\ &+ \frac{1}{i\omega\bar{\varepsilon}} \int_{\partial D} n(\mathbf{r}) \times [(j(\mathbf{r}')\cdot\nabla)\nabla(G_1 - G_0)] dS(\mathbf{r}') \\ &= q_2(\mathbf{r}), \end{aligned} \tag{108}$$

in which the functions G_0 and G_1 are evaluated at $\mathbf{r}-\mathbf{r}'$.

To derive these equations, we begin as before by letting ε and μ be constant over all of $\mathbb{R}^3$. We will see how arbitrary pairs of tangential fields (j,m) on ∂D can be decomposed uniquely into the sum of the tangential trace of an interior Maxwell field and the tangential trace of an exterior outgoing Maxwell field, both with Bloch wavevector κ. We shall omit the details of the proper functional spaces for the fields and their traces as well as technical aspects of the proofs; the reader can find this material in [61], Ch. 5, as well as [6, 60]. Instead, we present the analogue of the structure developed above for the Helmholz equation.

Define the boundary-integral operator

$$\mathscr{A}\begin{bmatrix} j \\ m \end{bmatrix} = \begin{bmatrix} A \\ A' \end{bmatrix}\begin{bmatrix} j \\ m \end{bmatrix}, \tag{109}$$

in which

$$\begin{aligned} A\begin{bmatrix} j \\ m \end{bmatrix}(\mathbf{r}) &= \int_{\partial D} n(\mathbf{r}) \times [j(\mathbf{r}') \times \nabla_{\mathbf{r}'} G(\mathbf{r}-\mathbf{r}')] dS(\mathbf{r}') \\ &- \frac{1}{i\omega\mu}\int_{\partial D} [n(\mathbf{r}) \times m(\mathbf{r}')]\omega^2 \varepsilon\mu G(\mathbf{r}-\mathbf{r}') dS(\mathbf{r}') \\ &- \frac{1}{i\omega\mu}\int_{\partial D} n(\mathbf{r}) \times [(m(\mathbf{r}')\cdot\nabla_{\mathbf{r}'})\nabla_{\mathbf{r}'} G(\mathbf{r}-\mathbf{r}')] dS(\mathbf{r}'), \end{aligned}$$

$$\begin{aligned} A'\begin{bmatrix} j \\ m \end{bmatrix}(\mathbf{r}) &= \int_{\partial D} n(\mathbf{r}) \times [m(\mathbf{r}') \times \nabla_{\mathbf{r}'} G(\mathbf{r}-\mathbf{r}')] dS(\mathbf{r}') \\ &+ \frac{1}{i\omega\varepsilon}\int_{\partial D} [n(\mathbf{r}) \times j(\mathbf{r}')]\omega^2 \varepsilon\mu G(\mathbf{r}-\mathbf{r}') dS(\mathbf{r}') \\ &+ \frac{1}{i\omega\varepsilon}\int_{\partial D} n(\mathbf{r}) \times [(j(\mathbf{r}')\cdot\nabla_{\mathbf{r}'})\nabla_{\mathbf{r}'} G(\mathbf{r}-\mathbf{r}')] dS(\mathbf{r}'). \end{aligned}$$

Let $\mathscr{I}$ be the identity operator $\mathscr{I}[j,m]^t = [j,m]^t$. The Calderón projectors for the Maxwell system are

$$\mathscr{P}_{\text{int}} = \tfrac{1}{2}\mathscr{I} - \mathscr{A}; \quad \mathscr{P}_{\text{ext}} = \tfrac{1}{2}\mathscr{I} + \mathscr{A}. \tag{110}$$

The following statements can be proved; the first is trivial, and the others are nontrivial and involve the representation formulas (106) and the Sokhotski-Plemelj formulas.

1. $\mathscr{P}_{\text{int}} + \mathscr{P}_{\text{ext}} = \mathscr{I}$,
2. $\mathscr{P}_{\text{int}}^2 = \mathscr{P}_{\text{int}}$ and $\mathscr{P}_{\text{ext}}^2 = \mathscr{P}_{\text{ext}}$,
3. The range of $\mathscr{P}_{\text{int}}$ is the space of tangential traces $\begin{bmatrix} -n\times H \\ n\times E \end{bmatrix}$ of pseudo-periodic interior free Maxwell fields $\begin{bmatrix} H \\ E \end{bmatrix}$.
4. The range of $\mathscr{P}_{\text{ext}}$ is the space of tangential traces $\begin{bmatrix} -n\times H \\ n\times E \end{bmatrix}$ of pseudo-periodic exterior free Maxwell fields $\begin{bmatrix} H \\ E \end{bmatrix}$ that satisfy the outgoing condition.

Returning to the scattering problem with interior and exterior material constants (ε_1,μ_1) and (ε_0,μ_0), consider the traces of all the fields involved:

$$\begin{aligned} &\psi = \begin{bmatrix} -n\times H \\ n\times E \end{bmatrix}, \\ &\phi_{\text{so}}^{\text{int}} = \begin{bmatrix} -n\times H_{\text{so}}^{\text{int}} \\ n\times E_{\text{so}}^{\text{int}} \end{bmatrix}, \quad \phi_{\text{sc}}^{\text{int}} = \begin{bmatrix} -n\times H_{\text{sc}}^{\text{int}} \\ n\times E_{\text{sc}}^{\text{int}} \end{bmatrix}, \\ &\phi_{\text{so}}^{\text{ext}} = \begin{bmatrix} -n\times H_{\text{so}}^{\text{ext}} \\ n\times E_{\text{so}}^{\text{ext}} \end{bmatrix}, \quad \phi_{\text{sc}}^{\text{ext}} = \begin{bmatrix} -n\times H_{\text{sc}}^{\text{ext}} \\ n\times E_{\text{sc}}^{\text{ext}} \end{bmatrix}. \end{aligned} \tag{111}$$

Because of the requirement of continuity of the tangential electric and magnetic fields, we have

$$\begin{aligned} \psi &= \phi_{\text{so}}^{\text{int}} + \phi_{\text{sc}}^{\text{int}}, \\ \psi &= \phi_{\text{so}}^{\text{ext}} + \phi_{\text{sc}}^{\text{ext}}. \end{aligned} \tag{112}$$

In analogy with the Helmholtz case, the interior scattered field $v_{\text{sc}}^{\text{int}} = [H_{\text{sc}}^{\text{int}}, E_{\text{sc}}^{\text{int}}]^t$ satisfies the homogeneous Maxwell system in D and the interior source field extends to a field in D^c that satisfies the homogeneous Maxwell system with (ε_1,μ_1) and the outgoing condition. This means that $\phi_{\text{sc}}^{\text{int}}$ is in the nullspace of $\mathscr{P}_{\text{ext}}^1$ and $\phi_{\text{so}}^{\text{int}}$ is in the range. Similarly, the exterior scattered field $v_{\text{sc}}^{\text{ext}} = [H_{\text{sc}}^{\text{ext}}, E_{\text{sc}}^{\text{ext}}]^t$ satisfies the homogeneous Maxwell system in D^c and the interior source field extends to a field in D that satisfies the homogeneous Maxwell system with (ε_0,μ_0); thus $\phi_{\text{sc}}^{\text{ext}}$ is in the nullspace of $\mathscr{P}_{\text{int}}^0$ and $\phi_{\text{so}}^{\text{ext}}$ is in the range. By projecting to the source fields, we obtain

$$\begin{aligned} \mathscr{P}_{\text{ext}}^1 \psi &= \phi_{\text{so}}^{\text{int}}, \\ \mathscr{P}_{\text{int}}^0 \psi &= \phi_{\text{so}}^{\text{ext}}. \end{aligned} \tag{113}$$

For the problem of scattering of plane waves, the source traces are

$$\phi_{\text{so}}^{\text{ext}} = \begin{bmatrix} j_{\text{inc}} \\ m_{\text{inc}} \end{bmatrix}, \quad \phi_{\text{so}}^{\text{int}} = \begin{bmatrix} 0 \\ 0 \end{bmatrix}$$

where $\phi_{\text{so}}^{\text{ext}}$ is the tangential trace of a plane electromagnetic wave.

To cause the leading-order singularity in the Calderón projectors to cancel in a linear combination, we multiply them by the matrices

$$\Lambda_k = \begin{bmatrix} \mu_k/\bar{\mu} & 0 \\ 0 & \varepsilon_k/\bar{\varepsilon} \end{bmatrix},$$

for $k = 0, 1$,

$$\left[\Lambda_1 \mathscr{P}_{\text{ext}}^1 + \Lambda_0 \mathscr{P}_{\text{int}}^0\right]\psi = \Lambda_1 \phi_{\text{so}}^{\text{int}} + \Lambda_0 \phi_{\text{so}}^{\text{ext}}. \tag{114}$$

This equation is the boundary-integral system (107,108) for $j(\mathbf{r})$ and $m(\mathbf{r})$.

We are interested in solving the pair (113); the solution of the scattering problem is obtained from ψ through the decomposition (112) and the boundary-integral representation formulas (106). Any solution of the pair (113) is also a solution to the combination (114), and we must determine if a solution to the combination of the equations is a solution of the pair. As before, we find that (114) is equivalent to

$$\left.\begin{array}{l}\mathscr{P}^1_{\text{ext}}\psi = \phi^{\text{int}}_{\text{so}} + \Lambda_1^{-1} f \\ \mathscr{P}^0_{\text{int}}\psi = \phi^{\text{ext}}_{\text{so}} - \Lambda_0^{-1} f\end{array}\right\} \text{ for some } f. \tag{115}$$

The reciprocal coefficients for the Maxwell system are

$$\begin{array}{lll}\varepsilon^0 = \mu_1 \dfrac{\varepsilon_0+\varepsilon_1}{\mu_0+\mu_1}, & \mu^0 = \varepsilon_1 \dfrac{\mu_0+\mu_1}{\varepsilon_0+\varepsilon_1} & \text{in } D, \\ \varepsilon^1 = \mu_0 \dfrac{\varepsilon_0+\varepsilon_1}{\mu_0+\mu_1}, & \mu^1 = \varepsilon_0 \dfrac{\mu_0+\mu_1}{\varepsilon_0+\varepsilon_1} & \text{in } D^c.\end{array} \tag{116}$$

and one can check that

$$\bar{\mathscr{P}}^0_{\text{int}} = \Lambda_1 \mathscr{P}^1_{\text{int}} \Lambda_1^{-1}$$

is the interior Calderón projector for the constants (ε^0, μ^0) and that

$$\bar{\mathscr{P}}^1_{\text{ext}} = \Lambda_0 \mathscr{P}^0_{\text{ext}} \Lambda_0^{-1} \tag{117}$$

is the exterior projector for the coefficients (ε^1, μ^1). The pair (115) implies that $\mathscr{P}^1_{\text{int}}\Lambda_1^{-1} f = 0$ and $\mathscr{P}^0_{\text{ext}}\Lambda_0^{-1} f = 0$, or

$$\left.\begin{array}{l}\bar{\mathscr{P}}^0_{\text{int}} f = 0, \\ \bar{\mathscr{P}}^1_{\text{ext}} f = 0.\end{array}\right. \quad \text{(reciprocal system)} \tag{118}$$

A function f that satisfies this pair is simultaneously the trace of an exterior outgoing Maxwell field with constants (ε^0, μ^0) and the trace of an interior Maxwell field with constants (ε^1, μ^1). The extension of this field to $\mathbb{R}^3$ is a generalized guided mode of the reciprocal structure characterized by these new constants. Notice that

$$\varepsilon_0\mu_0 = \varepsilon^1\mu^1, \quad \varepsilon_1\mu_1 = \varepsilon^0\mu^0. \tag{119}$$

As in the case of the Helmholtz equation, uniqueness of the solution of the reciprocal scattering problem implies equivalence of the original scattering problem and the boundary integral equations (107,108). This result is summarized in the following theorem.

Theorem 18 *If the (reciprocal) structure with coefficients*

$$\begin{array}{ll}\varepsilon = \varepsilon^1,\ \mu = \mu^1 & \textit{in } D, \\ \varepsilon = \varepsilon^0,\ \mu = \mu^0 & \textit{in } D^c,\end{array}$$

defined in terms of given constants (ε_1, μ_1) *and* (ε_0, μ_0) *by* (116) *does not admit a free Maxwell field in the absence of a source (a generalized guided mode), then the boundary-integral system (107,108) is equivalent to the scattering problem for the harmonic Maxwell system in the (original) structure with*

$$\begin{array}{ll}\varepsilon = \varepsilon_1,\ \mu = \mu_1 & \textit{in } D, \\ \varepsilon = \varepsilon_0,\ \mu = \mu_0 & \textit{in } D^c.\end{array}$$

More specifically, with ϕ^{int}_{so} *and* ϕ^{ext}_{so} *defined as in* (111), *the scattered field* $v_{sc} = [H_{sc}, E_{sc}]$ *with source fields* $v^{int}_{so} = [H^{int}_{so}, E^{int}_{so}]$ *and* $v^{ext}_{so} = [H^{ext}_{so}, E^{ext}_{so}]$ *is obtained by using the solution to (107,108),*

$$\begin{bmatrix} j(\mathbf{r}) \\ m(\mathbf{r}) \end{bmatrix} = \begin{bmatrix} -n(\mathbf{r}) \times H(\mathbf{r}) \\ n(\mathbf{r}) \times E(\mathbf{r}) \end{bmatrix}, \quad \mathbf{r} \in \partial D \tag{120}$$

in the representation formula (106) *for* $\mathbf{r} \in D$ *and its analog for* $\mathbf{r} \in D^c$.

For the Maxwell system, the reciprocal relations (116) satisfy

$$R^2 = I.$$

This means that every structure is the reciprocal of its own reciprocal.

Theorems 16 and 18 lead to a condition for the existence of guided modes in periodic slab structures. The following theorem can be strengthened somewhat by the condition $R^2(\varepsilon_0, \mu_0, \varepsilon_1, \mu_1) = (\varepsilon_0, \mu_0, \varepsilon_1, \mu_1)$, which holds for the Helmholtz equation with $\mu_0 = \mu_1$ and for the Maxwell system. It is given in [79] for polarized electromagnetic fields in two-dimensional nonmagnetic structures.

Theorem 19 (Existence of guided modes)

1. If the boundary-integral system (88,89) with zero source field, or equivalently,

$$(\Lambda_1 P^1_{ext} + \Lambda_0 P^0_{int})\psi = 0 \tag{121}$$

has a nontrivial solution, then the periodic structure with interior constants (ε_1, μ_1) *and exterior constants* (ε_0, μ_0) *or the reciprocal structure with constants given by the relations* (93) *admits a generalized guided mode of the Helmholtz equation (at the frequency and wavevector appearing in the projectors).*

2. If the boundary-integral system (107,108) with $q_1 = q_2 = 0$, *or equivalently,*

$$(\Lambda_1 \mathscr{P}^1_{ext} + \Lambda_0 \mathscr{P}^0_{int})\psi = 0 \tag{122}$$

has a nontrivial solution, then the periodic structure with interior constants (ε_1, μ_1) *and exterior constants* (ε_0, μ_0) *or the reciprocal structure with constants given by the relations* (116) *admits a generalized guided mode of the Maxwell equation.*

5. RESONANCE

Let us recapitulate what we have learned about resonant interaction of guided modes with plane waves and make clear the nature of the resonance that we wish to investigate.

We have seen that, if a slab admits a true guided mode at a real pair (κ_0, ω_0) in a (κ, ω)-regime for which at least one spatial harmonic is propagating, the frequency ω_0 of the guided mode is embedded in the spectrum of the pseudo-periodic Helmholtz or Maxwell operator in $\mathscr{S}$ for the Bloch wavevector κ. Restricted to the strip, the guided mode is a finite-energy eigenfunction. An embedded eigenvalue is typically nonrobust with respect to perturbation of (real) κ from κ_0 or perturbation of the geometry or material coefficients of the waveguide. The dissolution of the embedded eigenvalue coincides with the frequency's attaining an imaginary part as (κ, ω) remains on the complex dispersion relation for generalized guided modes. As we have discussed, the corresponding generalized modes with small imaginary part are leaky: they interact with the propagating spatial harmonics and therefore cannot persist as true guided modes. This resonant interaction lies behind the phenomenon of transmission anomalies and the enhancement of field intensity within the waveguide when the guide is illuminated by a plane wave. We will focus on perturbations of κ.

Physically speaking, true guided modes are idealized entities. They exist in a mathematical sense, as exact solutions to the Helmholtz or Maxwell equations in the absence of any external sources, oscillating with undiminished intensity for all time in an infinite waveguide. Every physical structure, in contrast, is subject to thermal and radiation losses due to material and fabrication limitations, and thus all guided modes in the laboratory or in nature must be initiated and sustained by a source of energy.

It is often useful to take the point of view that resonance in physical systems is the result of the proximity of the system to a idealized one that admits mathematically a guided mode or bound (finite-energy) state. The energy of the idealized bound state is embedded within a continuous spectrum corresponding to extended states. As the system is perturbed from the idealized one, the bound and extended states become coupled. The eigenvalue dissolves into the continuous spectrum and the bound state is destroyed, that is, the perturbed system possesses no finite-energy state. Instead, extended states near the bound-state frequency are sharply modified by the perturbation, and the perturbed system exhibits behavior that we call resonant.

In the context of quantum mechanics, this type of resonance is often called Feshbach resonance. Let us briefly discuss this setting. The autoionizing (Auger) states of the Helium atom provide the simplest example, which is treated in detail by Reed and Simon in §XXII.6 [72]. The idealized system in this case is described by the Hamiltonian of two *uncoupled* electrons in the presence of the (fixed) potential created by the positively charged Helium nucleus. The energy associated with a bound state in which both electrons are excited (above the ground state) is an eigenvalue of the idealized Hamiltonian that is embedded in the continuous spectrum corresponding to the extended states. When this idealized Hamiltonian is perturbed through a Coulomb coupling between the electrons, the eigenvalue disappears, and, instead of possessing a bound state, the physical system exhibits sharply modified states that are extended in the variable of one of the electrons; in other words, one of the electrons breaks free from the atom, causing it to ionize.

This Coulomb interaction between the electrons gives rise to anomalies in the graph of absorption *vs.* energy of the atom that are close to the energies of the idealized bound state of two excited electrons. An approximate formula for the anomalies is derived in [72] by continuation of the resolvent of the perturbed operator into the lower half complex energy plane. The graph of the function is the familiar Lorentzian, or Breit-Wigner, resonance shape

$$f(E) = \text{const.} \frac{1}{(E - E_r)^2 + (\Gamma/2)^2}, \tag{123}$$

in which E_r is a resonant frequency. The width Γ of the resonance at half the maximum height is given by the Fermi golden rule; a similar principle, as we shall show, arises for the double-spiked anomaly observed in the transmission of classical waves across periodic slabs.

The term "Fano resonance" is commonly applied to this double-spiked anomaly, which is observed in many resonant systems, including the Auger states of the noble gases (as Helium). The name originates from the work of Ugo Fano [22], in which he derived a formula for the anomaly with a parameter q that controls the relative size of the peak and dip as a function of energy,

$$f_q(\mathrm{e}) = \text{const.} \frac{|q + \mathrm{e}|^2}{1 + \mathrm{e}^2}, \tag{124}$$

in which e represents the energy normalized to a characteristic width Γ of the anomaly,

$$\mathrm{e} = \frac{E - E_{\text{res}}}{\Gamma/2}. \tag{125}$$

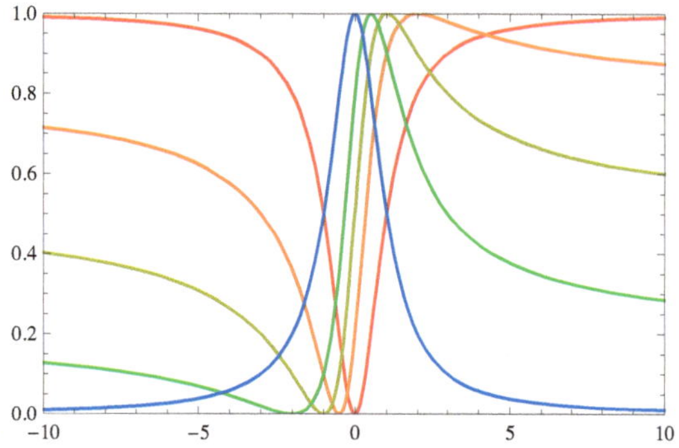

Fig. (**5**): The Fano resonance (124), with $q = 0, \frac{1}{2}, 1, 2, \infty$, normalized to a maximum height of 1.

As q ranges over real numbers from infinity to zero, the graph morphs from a Lorentzian $1/(1+e^2)$ to an inverted Lorentzian $e^2/(1+e^2)$, as illustrated in Fig. (**5**),

Our discussions treat resonance as a phenomenon observed in the frequency domain. The coupling of a bound state to extended states also has interesting and important implications for the time dynamics of the system. Soffer and Weinstein [83] approach the time-dependent resonant theory in quantum systems by treating the perturbed Schrödinger equation as a dynamical system under very general hypotheses. They demonstrate the Fermi golden rule for decay of transient fields and derive intermediate- and long-time behavior of the coupled system. We will restrict our analysis to the frequency domain.

5.1 Fano Resonance

Fano's model is the simplest one that describes the linear coupling of a bound state to extended states under a perturbation of a system. One begins with a self-adjoint operator H_0 that admits an eigenvalue corresponding to a bound (finite-energy) state embedded within a continuous spectrum corresponding to extended states. As the operator is modified by a self-adjoint perturbation W that couples the bound and extended states of H_0, the eigenvalue dissolves into the continuous spectrum. The operator $H = H_0 + W$ possesses no finite-energy state, and extended states with energy near that of the bound state of H_0 are sharply modified.

A careful discussion of the derivation and application of the Fano formula is provided by Rau [71], whose close collaboration with Fano provides important insight into the subtleties of the physical context.

In $\mathbb{C} \times L^2(\mathbb{R})$, define the "unperturbed" self-adjoint operator H_0 by its explicit spectral representation,

$$H_0 : \begin{bmatrix} a \\ b(E) \end{bmatrix} \mapsto \begin{bmatrix} E_0\, a \\ E\, b(E) \end{bmatrix},$$

in which E_0 is real. The domain of H_0 is

$$\mathscr{D}(H_0) = \{[a, b(E)]^t : b(E), Eb(E) \in L^2(\mathbb{R})\},$$

and its spectrum is $\sigma(H_0) = \mathbb{R}$. The vector $[1,0]^t$ is a proper eigenfunction with embedded eigenvalue E_0, and the rest of the spectrum is absolutely continuous. For $\hat{E} \in \sigma(H_0)$, $[0, \delta(E-\hat{E})]^t$ are generalized eigenfunctions.

Define the perturbation

$$W : \begin{bmatrix} a \\ b(E) \end{bmatrix} \mapsto \begin{bmatrix} V_0 a + \int V^*(E')b(E')dE' \\ V(E)a + \int V_1(E,E')b(E')dE' \end{bmatrix},$$

in which V_0 and V_1 are real valued. If we assume that V and V_1 satisfy

$$V(E)(E+i)^{-1} \in L^2(\mathbb{R}),$$
$$V_1(E,E')(E'+i)^{-1} \in L^2(\mathbb{R}^2),$$

then W is symmetric on $\mathscr{D}(H_0)$ and, since $W(H_0 + i)^{-1}$ is of Hilbert-Schmidt class, W is compact relative to H_0. Therefore $H_0 + W$ is self-adjoint with domain $\mathscr{D}(H_0 + W) = \mathscr{D}(H_0)$ and the essential spectrum is unchanged,

$$\sigma_{\text{ess}}(H_0 + W) = \sigma_{\text{ess}}(H_0) = \mathbb{R}.$$

This is an instance of a Theorem of Weyl; see Corollary 2 in §XIII.4 of [72].

Consider the generalized eigenvalue problem

$$H \begin{bmatrix} a \\ b(E) \end{bmatrix} = \hat{E} \begin{bmatrix} a \\ b(E) \end{bmatrix},$$

which is equivalent to the following equations:

$$\begin{aligned} &(\hat{E} - E)b(E) - \int V_1(E,E')b(E')dE' = V(E)a, \\ &\int V^*(E')b(E')dE' = (\hat{E} - E_0 - V_0)a. \end{aligned} \tag{126}$$

We allow b to be a distribution (which depends on $\hat{E}$), with at most a δ-singularity at $E = \hat{E}$,

$$b(E) = \tilde{b}(E) + c\,\delta(E - \hat{E}),$$

in which c is a constant to be determined and $\tilde{b}$ is a function. Both c and $\tilde{b}$ depend on $\hat{E}$. The system (126) becomes

$$\begin{aligned} &(\hat{E} - E)\tilde{b}(E) - \int V_1(E,E')\tilde{b}(E')dE' \\ &\qquad = cV_1(E,\hat{E}) + aV(E), \\ &\int V^*(E')\tilde{b}(E')dE' = a(\hat{E} - E_0 - V_0) - cV^*(\hat{E}). \end{aligned} \tag{127}$$

Fano's calculation. Fano's result in Sec. 2 of [22] is obtained if we set $V_1 = 0$ (he also has $V_0 = 0$). System (127) reduces to

$$(\hat{E}-E)\tilde{b}(E) = aV(E),$$
$$\int V^*(E')\tilde{b}(E')\,dE' + cV^*(\hat{E}) = a(\hat{E}-E_0-V_0).$$

For the solution, we obtain

$$\begin{bmatrix} a \\ b(E) \end{bmatrix} = \begin{bmatrix} a \\ a\dfrac{V(E)}{\hat{E}-E} + c\delta(E-\hat{E}) \end{bmatrix}, \tag{128}$$

(the function $V(E)/(\hat{E}-E)$ is understood in the sense of distributions as a principal value) subject to the relation between a and c,

$$cV^*(\hat{E}) = a(\hat{E}-\mathring{E}(\hat{E})), \tag{129}$$

in which the shifted resonant frequency $\mathring{E}(\hat{E})$ is

$$\mathring{E}(\hat{E}) = E_0 + V_0 + F(\hat{E}),$$
$$F(\hat{E}) = \text{P.V.}\int \frac{|V(E')|^2}{\hat{E}-E'}\,dE'. \tag{130}$$

For nonresonant $\hat{E}$, it is enlightening to set $c = C(\hat{E}-\mathring{E}(\hat{E}))$ with C a fixed arbitrary constant independent of $\hat{E}$ and write

$$\begin{bmatrix} a \\ b(E) \end{bmatrix} = C\begin{bmatrix} V^*(\hat{E}) \\ V^*(\hat{E})\dfrac{V(E)}{\hat{E}-E} + (\hat{E}-\mathring{E}(\hat{E}))\delta(E-\hat{E}) \end{bmatrix}$$
$$\text{if } \hat{E} \neq \mathring{E}(\hat{E}). \tag{131}$$

Then, as $\hat{E}-\mathring{E}(\hat{E})$ vanishes and $V^*(\hat{E}) \neq 0$ when $\hat{E}-\mathring{E}(\hat{E}) = 0$, the δ part of the generalized eigenfunction vanishes, and we obtain

$$\begin{bmatrix} a \\ b(E) \end{bmatrix} = a\begin{bmatrix} 1 \\ \dfrac{V(E)}{\hat{E}-E} \end{bmatrix}, \quad \hat{E} = \mathring{E}(\hat{E}) \text{ and } V(\hat{E}) \neq 0.$$

When both $\hat{E}-\mathring{E}(\hat{E}) = 0$ and $V^*(\hat{E}) = 0$, the generalized eigenspace for $\hat{E}$ is two-dimensional:

$$\begin{bmatrix} a \\ b(E) \end{bmatrix} = \begin{bmatrix} a \\ a\dfrac{V(E)}{\hat{E}-E} + c\delta(E-\hat{E}) \end{bmatrix}$$
$$\text{if } \hat{E} = \mathring{E}(\hat{E}) \text{ and } V(\hat{E}) = 0.$$

Let us now assume, as Fano does, that $V(E) \neq 0$ for all E and fix a generalized eigenvector, that is, choose C in (131). By defining the real-valued functions of real $\hat{E}$

$$z(\hat{E}) = \frac{\hat{E}-\mathring{E}(\hat{E})}{|V(\hat{E})|^2} \tag{132}$$

and

$$\Delta(\hat{E}) = \operatorname{arccot}(-z(\hat{E})/\pi), \tag{133}$$

an appropriately scaled eigenvector is written conveniently as

$$a_{\hat{E}} = \frac{\sin\Delta(\hat{E})}{\pi V(\hat{E})},$$
$$b_{\hat{E}}(E) = \frac{\sin\Delta(\hat{E})}{\pi V(\hat{E})}\frac{V(E)}{\hat{E}-E} - \cos\Delta(\hat{E})\delta(E-\hat{E}). \tag{134}$$

With this definition, $[a_{\hat{E}}, b_{\hat{E}}(E)]^t$ coincides with the generalized eigenvectors $\pm[0, \delta(E-\hat{E})]^t$ of H_0 as $\hat{E} \to \pm\infty$.

Fano considers the concrete situation in which the vector $[1,0]^t$ corresponds to a bound state $\varphi(x)$ of H_0 in $\mathbb{R}^3$, exponentially decaying as $|x| = r \to \infty$ ($x \in \mathbb{R}^3$), and the generalized eigenfunctions correspond to extended states $\psi_E(x)$ of H_0 that have oscillatory far-field behavior:

$$\psi_E(x) \to \sin(k(E)r + \alpha(E)), \quad (r \to \infty).$$

Through a generalized Fourier transform, an arbitrary state represented by $[a, b(E)]^t$ can be expressed as a superposition of the bound state and the extended states:

$$\begin{bmatrix} a \\ b(E) \end{bmatrix} \rightsquigarrow a\,\varphi(x) + \int b(E)\psi_E(x)\,dE. \tag{135}$$

To determine the far-field behavior of the spatial state $\Psi_{\hat{E}}(x)$ corresponding to the generalized eigenfunction (134) of the perturbed operator $H = H_0 + W$, we must compute

$$\Psi_{\hat{E}}(x) \to \int b_{\hat{E}}(E)\sin\left(k(E)r + \alpha(E)\right)dE, \ (r \to \infty).$$

In the limit as $r \to \infty$, the δ part of the integral contributes $-\cos\Delta(\hat{E})\sin(k(\hat{E})r + \alpha(\hat{E}))$ and the principal-value part contributes $-\sin\Delta(\hat{E})\cos(k(\hat{E})r + \alpha(\hat{E}))$. In the case that $k(\hat{E}) = m\hat{E}$ (space is homogeneous far from the origin), the latter is shown as follows.

Let us assume $V \in L^2(\mathbb{R})$, and define

$$U(\hat{E}) = \text{P.V.}\int \frac{V(E)}{\hat{E}-E}\sin(mEr + \alpha(E))dE,$$
$$f_\pm(E) = V(E)e^{\pm i\alpha(E)};$$

then

$$U(\hat{E}) = \frac{1}{2i}\int \frac{1}{\hat{E}-E}\left(f_+(E)e^{imEr} - f_-(E)e^{-imEr}\right)dE.$$

Now go over to the Fourier variable and use the representation of the Hilbert transform there (*e.g.* [13] §2.3),

$$g(\hat{E}) = -\frac{1}{\pi i}\int \frac{1}{\hat{E}-E} f(E)dE, \iff \mathscr{F}g(\xi) = \mathrm{sgn}(\xi)\mathscr{F}f(\xi),$$

to obtain

$$\mathscr{F}U(\xi) = -\tfrac{\pi}{2}\mathrm{sgn}(\xi)\left[\mathscr{F}f_+(\xi - \tfrac{mr}{2\pi}) - \mathscr{F}f_-(\xi + \tfrac{mr}{2\pi})\right],$$

which tends to

$$-\tfrac{\pi}{2}\left[\mathscr{F}f_+(\xi - \tfrac{mr}{2\pi}) + \mathscr{F}f_-(\xi + \tfrac{mr}{2\pi})\right]$$

in $L^2(\mathbb{R})$ as $r \to \infty$. Thus,

$$\left\| -\tfrac{\pi}{2}\left[f_+(\hat{E})e^{im\hat{E}r} + f_-(\hat{E})e^{-im\hat{E}r}\right] - U(\hat{E})\right\| \to 0$$

in $L^2(\mathbb{R})$ as $r \to \infty$, and the term in brackets is $2V(\hat{E})\cos(m\hat{E}r + \alpha(\hat{E}))$, as desired.
The result is that

$$\begin{aligned}\Psi_{\hat{E}}(x) \to &-\left[\sin\Delta(\hat{E})\cos(k(\hat{E})r + \alpha(\hat{E})) + \right.\\ &\left. + \cos\Delta(\hat{E})\sin(k(\hat{E})r + \alpha(\hat{E}))\right]\\ = &-\sin[k(\hat{E})r + \alpha(\hat{E}) + \Delta(\hat{E})] \quad (r \to \infty).\end{aligned}$$

As $\hat{E}$ traverses the real line, $z(\hat{E})$ passes from $-\infty$, through zero near the shifted resonance frequency $\mathring{E}(\hat{E})$ (assuming this frequency is unique), to ∞. This means that $\Delta(\hat{E})$ runs from 0 to π as $\hat{E}$ traverses $\mathbb{R}$, and therefore, the spatial asymptotic behavior of $\Psi_{\hat{E}}$ is as $-\psi_{\hat{E}}$ for $\hat{E}$ large and negative and as $\psi_{\hat{E}}$ for $\hat{E}$ large and positive. Thus the extended states are modified sharply near the resonant frequency but remain unaltered far from resonance.
Fano was interested more particularly in the way observable properties are modified by the resonance. Let T be a linear functional represented by a smooth function in the spectral variable E so that T can be applied to generalized eigenfunctions[3]. Evaluation at $x \in \mathbb{R}^3$ described above is an example. We are interested in comparing the effect of the perturbation W on the values that $|T|^2$ takes on generalized eigenfunctions, that is, we wish to compare $|T(\Psi_{\hat{E}})|^2$ to $|T(\psi_{\hat{E}})|^2$. By inserting the solution $[a_{\hat{E}}, b_{\hat{E}}(E)]^t$(134) into the general Fourier integral (135) and defining

$$\Phi_{\hat{E}} = \varphi + \mathrm{P.V.}\int \frac{V(E)}{\hat{E}-E}\psi_E\, dE,$$

we find that $T(\Psi_{\hat{E}})$ is related to $T(\psi_{\hat{E}})$ through

$$T(\Psi_{\hat{E}}) = \sin\Delta(\hat{E})\frac{T(\Phi_{\hat{E}})}{\pi V(\hat{E})} - \cos\Delta(\hat{E})T(\psi_{\hat{E}}).$$

In terms of Fano's reduced energy variable

$$\mathrm{e} = -\cot\Delta(\hat{E}) = \frac{\hat{E} - \mathring{E}(\hat{E})}{\pi|V(\hat{E})|^2} = \frac{\hat{E} - \mathring{E}(\hat{E})}{\Gamma_{\hat{E}}/2},$$

the spectral width of the resonance

$$\Gamma_{\hat{E}} = 2\pi|V(\hat{E})|^2,$$

and the shape parameter

$$q = \frac{T(\Phi_{\hat{E}})}{\pi V(\hat{E})T(\psi_{\hat{E}})},$$

the ratio we seek is

$$\frac{|T(\Psi_{\hat{E}})|^2}{|T(\psi_{\hat{E}})|^2} = \frac{|q+\mathrm{e}|^2}{1+\mathrm{e}^2}. \tag{136}$$

The idea now is to use e instead of $\hat{E}$ a vicinity of the resonant frequency $\hat{E}^*$ for which $\hat{E}^* = \mathring{E}(\hat{E}^*)$ and hold q fixed. Such an approximation is justified if $V(\hat{E}) \ll 1$ near $\hat{E}^*$, allowing q and Γ to be approximated by their values at $\hat{E}^*$ for values of $\hat{E}$ with $|\hat{E} - \hat{E}^*|$ on the order of a few times $\Gamma_{\hat{E}^*}$. This is the situation in which the coupling of continuum to bound states is weak near $\hat{E}^*$ and the width Γ of the resonance is therefore narrow.
If one linearly interpolates of the graph of the Fano anomaly through the resonance, the result is flat, whereas experimental data show a nonzero "background" slope. Fano treats this discrepancy for the Helium atom by fitting a modified formula with a nonzero slope to the data. He also extends his treatment to systems with multiple continua or multiple bound states. It has become common in the literature on resonance in classical and quantum systems to fit a Fano formula to experimental data.

5.2 Transmission Resonance

Examples of anomalous transmission of energy through slab structures and related resonant systems abound in the literature. Since the late 1990s,

[3]Fano [22] considers the conjugate-linear functional $(\Psi_E|T|i)$, which is the "matrix element of a suitable transition operator T between an initial state i and the state Ψ_E", and whose square modulus is "the probability of excitation of the stationary state Ψ_E."

there has appeared a vast amount of literature on enhanced transmission of electromagnetic waves through periodic structures, in particular, optical transmission of through metallic sheets with sub-wavelength arrays of holes or dimples. In the case of metal structures, it is generally understood that the transmission is enhanced or inhibited because of coupling of incident plane waves with the surface plasmons of the structure. In fact, there has been quite a lively discussion and controversy surrounding the mechanism. We cannot adequately represent the scope of the recent literature here, but we will presently indicate some of the basic issues and provide some references.

The theory that we present in this Chapter is a part of the story. While one does not expect the ensuing analysis to extend to all cases of anomalous transmission, it is reasonable to expect that it can be adapted to those situations in which one can identify an idealized (lossless) structure that is somehow close to the resonant one. One important phenomenon that does not fall into the setting of the dissolution of an embedded eigenvalue is the Wood anomaly that occurs at cutoff frequencies of the spatial harmonics; this phenomenon also plays a role in enhanced transmission, and is treated mathematically in [53]. The peaks observed in the transmission of plane-wave energy across metal slabs with thin slits or in Fabry-Perot resonance (the mirror effect of organized reflection from the walls of a slab) [40, 76] are also not generally connected with guided modes.

We will analyze in detail the particular case of perturbation of the Bloch wavenumber in scalar (acoustic or polarized electromagnetic) waves in lossless two-dimensional slabs, which encompasses those composed of lossless penetrable materials as well as perfect conductors and acoustically hard or soft surfaces. A numerical example is shown in Fig. (**6**). If the source field is taken to be incident upon the slab from the left,

$$u^{\text{inc}}(x,z) = \sum_{m \in \mathscr{Z}_p} a_m^{\text{inc}} e^{i\eta_m z} e^{i(m+\kappa)\cdot x},$$

with all $b_m^{\text{inc}} = 0$ in (16), the transmitted time-averaged energy flux of one period of the scattered field through a plane parallel to the slab is given by

$$\mathscr{E}^{\text{trans}} = \operatorname{Im} \int_{\Gamma_+} \mu_0^{-1} \bar{u} \partial_n u = \mu_0^{-1} \sum_{m \in \mathscr{Z}_p} \eta_m |b_m|^2.$$

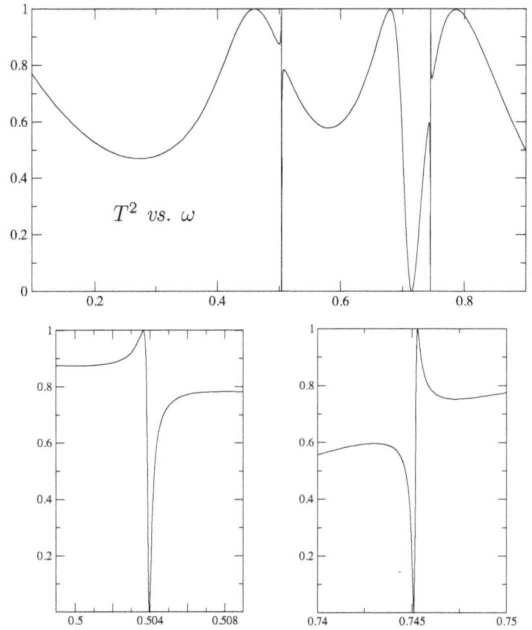

Fig. (**6**): Numerical computation of the transmitted energy T^2 as a function of reduced frequency ω, with $\kappa = 0.02$. The scatterer is a single infinite 2π-periodic row of infinitely tall rods with radius $\pi/2$, $\varepsilon_1 = 10$, and $\mu_1 = 1$. In the exterior medium, $\varepsilon_0 = \mu_0 = 1$. There are guided modes at the (κ, ω) pairs $(0, \sim 0.4039)$ and $(0, \sim 0.7452)$. In this frequency range, there is exactly one propagating spatial harmonic.

The corresponding energy flux $\mathscr{E}^{\text{inc}}$ for the incident field is obtained by using a_m^{inc} in place of b_m in the rightmost expression, and we define

$$T^2 = \frac{\mathscr{E}^{\text{trans}}}{\mathscr{E}^{\text{inc}}}.$$

The extension of the analysis of transmission resonance to doubly periodic structures and the full Maxwell system as well as to perturbations of the geometry and material properties, including small losses, of the structure will certainly lead to new interesting features and formulas. Several categories of transmission anomalies at normal incidence that arise from the introduction of a channel after every six rows of an otherwise perfect square-lattice photonic crystal slab are reported in [33], and it appears that the sharpest of them, which coincide with very high field amplitude enhancement in the slab, are a result of the dissolution of a guided mode.

In [17], Ebbesen, *et. al.*, reported the extraordinary transmission of light through metal sheets with a periodic array of holes whose spacing is smaller than the wavelength of the incident light. Since then, there has been much literature by many authors expounding the role of plasmons and the Wood anomaly in the theory of enhanced transmission, [8, 24, 29, 25, 41, 42, 44, 45, 51, 46, 58, 59], and the role of evanescent fields [52]. Many investigations involve coupled-mode analysis, in which the slab/ambient-space system is modeled by the prominent features of the waves it supports: the incoming waves, the surface plasmon polaritons, and the modes of the holes in a metal sheet, [46, 70]. Electric circuit models have also been successfully used to compute transmission anomalies [59]. A review of literature on this subject is given in [23].

Our analysis of transmission anomalies is based on an analytic connection of the scattering states of a slab to the generalized guided modes and analytic perturbation about an isolated real point (κ_0, ω_0) on the complex dispersion relation for generalized guided modes. This is accomplished through the auxiliary problem

$$A(\kappa,\omega)\psi(\kappa,\omega) = \phi(\kappa,\omega)$$

described at the beginning of Sec. 3.2.2. The operators $A(\kappa,\omega)$ act in a common Banach space $\mathscr{H}$. In the case of acoustic or electromagnetic scattering by a homogeneous periodic slab with a smooth boundary, the operator A is the boundary-integral operator derived in Sec. 4.2 and 4.4,

$$A = \Lambda_1 P^1_{\text{ext}} + \Lambda_0 P^0_{\text{int}}.$$

It can be shown that this operator is of the form $I + C$, where C is compact. For the two-dimensional scalar case, this is shown in [79]; the operator is posed in $\mathscr{H} = H^s(\partial D) \oplus H^{s-1}(\partial D)$, where s is most naturally taken to be $1/2$.

The following exposition follows that of [81] and [67] for two-dimensional problems in which the scatterer is periodic in one direction so that κ is scalar, and where the guided mode is simple. We shan't treat the case of guided modes of multiplicity greater than one, which are associated with multiple anomalies that emanate from the modes' characteristic (κ,ω) pair.

Consider a nonrobust guided slab mode corresponding to the simple eigenvalue 0 of $A(\kappa_0,\omega_0)$. This means that there is a neighborhood $U \subset \mathbb{C}^2$ of (κ_0,ω_0) and a simple closed curve C encircling the origin in the complex λ plane such that, for all $(\kappa,\omega) \in U$, $A(\kappa,\omega)$ has a unique, simple eigenvalue $\tilde{\ell}(\kappa,\omega)$ contained in the region bounded by C and that $(\kappa,\omega) = (\kappa_0,\omega_0)$ is the unique point in $U \cap \mathbb{R}^2$ satisfying $\tilde{\ell}(\kappa,\omega) = 0$. The relation

$$\tilde{\ell}(\kappa,\omega) = 0 \quad \text{(dispersion relation)}$$

is a dispersion relation in U for generalized guided modes, and (κ_0,ω_0) represents an isolated point on the relation corresponding to a true guided mode, which falls off exponentially with distance from the slab.

By projecting onto the eigenspace corresponding to $\tilde{\ell}(\kappa,\omega)$, one is able to split the source and scattered fields into "resonant" and "nonresonant" components. Specifically, the spectral projection

$$P_1(\kappa,\omega) = \frac{1}{2\pi i}\oint_C (\lambda I - A(\kappa,\omega))^{-1} d\lambda,$$

in which C is a sufficiently small circle about 0 in the complex λ-plane, is jointly analytic in (κ,ω) at (κ_0,ω_0) (*i.e.*, in a neighborhood U of (κ_0,ω_0)). P_1 is a rank-one projection-valued function of (κ,ω) that commutes with A and whose image is the eigenspace of $A(\kappa,\omega)$ with eigenvalue $\tilde{\ell}(\kappa,\omega)$. An analytic eigenvector $\hat{\psi}(\kappa,\omega)$ is obtained by fixing an eigenvector $\hat{\psi}_0$ of $A(\kappa_0,\omega_0)$ and setting

$$\hat{\psi}(\kappa,\omega) = P_1(\kappa,\omega)\hat{\psi}_0.$$

To see that $\tilde{\ell}(\kappa,\omega)$ is analytic, observe that

$$P_1(\kappa_0,\omega_0)\hat{\psi}(\kappa,\omega) = \beta(\kappa,\omega)\hat{\psi}_0,$$

where $\beta(\kappa_0,\omega_0) = 1$ and $\beta(\kappa,\omega)$ is analytic because $\hat{\psi}(\kappa,\omega)$ is, and that

$$P_1(\kappa_0,\omega_0)A(\kappa,\omega)\hat{\psi}(\kappa,\omega) = \tilde{\ell}(\kappa,\omega)\beta(\kappa,\omega)\hat{\psi}_0$$

is also analytic. Theory of spectral projections and analytic perturbation of eigenvalues can be found in many classic references, including Ch. 7 §3 of Kato [36], §XII.2 of [72], §5.6 of Hille and Phillips [31], Ch. XI of Riesz and Sz.-Nagy [75], as well as Steinberg [78].

The projection P_1 together with its complement $P_2 = I - P_1$ form a partial spectral resolution of the identity on $\mathscr{H}$:

$$I = P_1 + P_2,$$
$$A = \tilde{\ell}P_1 + AP_2.$$

Now write A as

$$A = (\tilde{\ell} - 1)P_1 + \tilde{A},$$

where $\tilde{A} = P_1 + AP_2$, and observe that $\tilde{A}$ is analytic with analytic inverse in a neighborhood of (κ_0,ω_0)

and commutes with P_1 and P_2. One obtains the relation

$$AP_2\tilde{A}^{-1}P_2 = P_2. \tag{137}$$

An analytic connection between scattering states and generalized guided modes is made as follows. Choose an analytic vector $\phi(\kappa,\omega)$ that represents a source field (such as the incident plane wave we take below) and split it into its resonant and nonresonant parts

$$\phi = P_1\phi + P_2\phi = \alpha\hat{\psi} + \phi_2.$$

The multiple α is analytic at (κ_0,ω_0). Indeed, since $\hat{\psi}$ and $\alpha\hat{\psi}$ are analytic, so are $P_1(\kappa_0,\omega_0)\hat{\psi} = \beta\hat{\psi}(\kappa_0,\omega_0)$ and $P_1(\kappa_0,\omega_0)\alpha\hat{\psi} = \alpha\beta\hat{\psi}(\kappa_0,\omega_0)$ and therefore also the functions α and $\alpha\beta$. Since $\beta(\kappa_0,\omega_0) = 1$, α is analytic at (κ_0,ω_0).
For (κ,ω) near the dispersion relation $\tilde{\ell}(\kappa,\omega) = 0$, the "resonant part" $\alpha\hat{\psi}$ of the source field will induce a proportionally high field, and we therefore normalized the source by a constant multiple (to be specified soon) $\ell = c\tilde{\ell}$ of the eigenvalue. The equation $A\psi = \ell\phi$ has a solution $\psi(\kappa,\omega)$ that is analytic at (κ_0,ω_0), namely

$$\psi = c\alpha\hat{\psi} + \ell\tilde{A}^{-1}\phi_2,$$

where we have used (137) and $\phi_2 = P_2\phi_2$. This is seen clearly in the matrix form,

$$A\psi = \begin{bmatrix} \tilde{\ell} & 0 \\ 0 & AP_2 \end{bmatrix} \begin{bmatrix} c\alpha\hat{\psi} \\ \ell\tilde{A}^{-1}\phi_2 \end{bmatrix} = \begin{bmatrix} \ell\alpha\hat{\psi} \\ \ell\phi_2 \end{bmatrix} = \ell\phi.$$

It is the analytic field $\psi(\kappa,\omega)$ that connects scattering states with guided modes. If $\tilde{\ell}(\kappa,\omega) = 0$, $\psi(\kappa,\omega)$ represents a generalized guided mode, and otherwise, it represents a scattering state.
Let us return to the concrete situation in which the source field ϕ represents the incident plane waves $e^{i(\kappa\cdot x+\eta_0 z)}$. In the case of the Helmholtz equation or the Maxwell system, the integral representation formulas (77,78) or (106), together with the form (72) of the Green function, show that the coefficients of the propagating spatial harmonics of the reflected and transmitted fields are analytic functions of (κ,ω) at (κ_0,ω_0). We shall consider the regime in which there is exactly one propagating harmonic and therefore a single complex reflected coefficient $a(\kappa,\omega)$ and a single transmitted coefficient $b(\kappa,\omega)$. Because of the conservation of energy relation $|\ell|^2 = |a|^2 + |b|^2$ for real pairs (κ,ω), we deduce the important condition that ℓ, *a, and b have a common root at* (κ_0,ω_0). We continue to assume that $\mathrm{Re}\,\omega_0 > 0$, so that $\mathrm{Im}\,\omega \le 0$ whenever $\ell(\kappa,\omega) = 0$ for real κ near κ_0 (Theorem 15), as well as the generic condition that $\partial\ell/\partial\omega$, $\partial a/\partial\omega$, and $\partial b/\partial\omega$ do not vanish at (κ_0,ω_0). The analysis of transmission anomalies is based on the following conditions alone:

$$\ell(\kappa_0,\omega_0) = a(\kappa_0,\omega_0) = b(\kappa_0,\omega_0) = 0 \quad \text{with } (\kappa_0,\omega_0) \in \mathbb{R}^2, \tag{138}$$

$$|\ell(\kappa,\omega)|^2 = |a(\kappa,\omega)|^2 + |b(\kappa,\omega)|^2 \quad \forall(\kappa,\omega) \in \mathbb{R}^2, \tag{139}$$

$$\ell(\kappa,\omega) = 0 \text{ with } \kappa \in \mathbb{R} \implies \mathrm{Im}\,\omega \le 0, \tag{140}$$

$$\tfrac{\partial\ell}{\partial\omega} \ne 0,\ \tfrac{\partial a}{\partial\omega} \ne 0,\ \tfrac{\partial b}{\partial\omega} \ne 0, \text{ at } (\kappa_0,\omega_0). \tag{141}$$

The analytic perturbation theory of these three analytic functions about the true guided mode pair (κ_0,ω_0) is facilitated by the Weierstraß Preparation Theorem ([54] §16), which, because of conditions (138) and (141) provides the following forms in the variables $\tilde{\omega} = \omega - \omega_0$ and $\tilde{\kappa} = \kappa - \kappa_0$:

$$\begin{aligned}
\ell &= \left[\tilde{\omega} + \ell_1\tilde{\kappa} + \ell_2\tilde{\kappa}^2 + O(|\tilde{\kappa}|^3)\right]\left[1 + O(|\tilde{\kappa}|+|\tilde{\omega}|)\right],\\
a &= \left[\tilde{\omega} + r_1\tilde{\kappa} + r_2\tilde{\kappa}^2 + O(|\tilde{\kappa}|^3)\right]\left[r_0 e^{i\theta_1} + O(|\tilde{\kappa}|+|\tilde{\omega}|)\right],\\
b &= \left[\tilde{\omega} + t_1\tilde{\kappa} + t_2\tilde{\kappa}^2 + O(|\tilde{\kappa}|^3)\right]\left[t_0 e^{i\theta_2} + O(|\tilde{\kappa}|+|\tilde{\omega}|)\right].
\end{aligned} \tag{142}$$

All series are convergent for $(\tilde{\kappa},\tilde{\omega})$ in a neighborhood of $(0,0)$. The constant c has been chosen so that the leading coefficient in the second factor for ℓ is unity, and both r_0 and t_0 are positive. These forms guarantee an explicit analytic dispersion relation $\ell(\kappa,\omega) = 0$ near (κ_0,ω_0),

$$\omega = W(\kappa) := \omega_0 - \ell_1(\kappa-\kappa_0) - \ell_2(\kappa-\kappa_0)^2 + \ldots$$

and similar explicit expressions for the zero sets of a and b.

Theorem 20 *The following relations hold among the coefficients in the forms* (142)*:*

1. $r_0 > 0,\ t_0 > 0,\ r_0^2 + t_0^2 = 1,$
2. $\ell_1 = r_1 = t_1 \in \mathbb{R},$
3. $Im\,\ell_2 \ge 0,$
4. $\ell_2 \in \mathbb{R} \iff r_2 = t_2 \in \mathbb{R} \iff \ell_2 = r_2 = t_2 \in \mathbb{R}.$

If r_2 *and* t_2 *are real and* ℓ_2 *is imaginary, then*

5. $r_2 t_2 = -|\ell_2|^2.$

If $\ell_1 = 0$*, then*

6. $Re\,\ell_2 = r_0^2 Re\,r_2 + t_0^2 Re\,t_2,$
7. $|\ell_2|^2 = r_0^2|r_2|^2 + t_0^2|t_2|^2.$

Proof. The positivity of r_0 and t_0 is a consequence of the Weierstraß Preparation Theorem, and the third relation in (1.) follows from (139). Because of condition (140), ℓ_1 must be real and $\mathrm{Im}\,\ell_2$ positive.

To prove (2.), let $(\tilde{\kappa},\tilde{\omega})$ tend to $(0,0)$ along the set $\{(\tilde{\kappa},\tilde{\omega}) : \tilde{\omega}+\ell_1\tilde{\kappa}=0\} \subset \mathbb{R}^2$. The forms (142) imply

$$\begin{aligned}
&|\ell|^2 = O(\tilde{\kappa}^4),\\
&|a|^2 = r_0^2(|\mathrm{Re}\, r_1-\ell_1|^2+|\mathrm{Im}\, r_1|^2)\tilde{\kappa}^2+O(|\tilde{\kappa}|^3),\\
&|b|^2 = t_0^2(|\mathrm{Re}\, t_1-\ell_1|^2+|\mathrm{Im}\, t_1|^2)\tilde{\kappa}^2+O(|\tilde{\kappa}|^3),\\
&\qquad (\tilde{\kappa},\tilde{\omega})\to(0,0),\ \tilde{\omega}+\ell_1\tilde{\kappa}=0.
\end{aligned}$$

The balance of these powers in equation (139) implies $\ell_1=r_1=t_1$.
Now assume that r_2 and t_2 are real and let $(\tilde{\kappa},\tilde{\omega})$ tend to $(0,0)$ along the set $\{(\tilde{\kappa},\tilde{\omega}) : \tilde{\omega}+\ell_1\tilde{\kappa}+t_2\tilde{\kappa}^2=0\} \subset \mathbb{R}^2$ to obtain

$$\begin{aligned}
&|\ell|^2 = (|\mathrm{Re}\,\ell_2-t_2|^2+|\mathrm{Im}\,\ell_2|^2)\tilde{\kappa}^4+O(|\tilde{\kappa}|^5),\\
&|a|^2 = r_0^2(r_2-t_2)^2\kappa^4+O(|\tilde{\kappa}|^5),\\
&|b|^2 = O(\tilde{\kappa}^6),\\
&\qquad (\tilde{\kappa},\tilde{\omega})\to(0,0),\ \tilde{\omega}+\ell_1\tilde{\kappa}+t_2\tilde{\kappa}^2=0,
\end{aligned}$$

and then along the set $\{(\tilde{\kappa},\tilde{\omega}) : \tilde{\omega}+\ell_1\tilde{\kappa}+r_2\tilde{\kappa}^2=0\} \subset \mathbb{R}^2$ to obtain

$$\begin{aligned}
&|\ell|^2 = (|\mathrm{Re}\,\ell_2-r_2|^2+|\mathrm{Im}\,\ell_2|^2)\tilde{\kappa}^4+O(|\tilde{\kappa}|^5),\\
&|a|^2 = O(\tilde{\kappa}^6),\\
&|b|^2 = t_0^2(t_2-r_2)^2\kappa^4+O(|\tilde{\kappa}|^5),\\
&\qquad (\tilde{\kappa},\tilde{\omega})\to(0,0),\ \tilde{\omega}+\ell_1\tilde{\kappa}+r_2\tilde{\kappa}^2=0,
\end{aligned}$$

Using (139) again, each of these asymptotic regimes yields a relation among the coefficients, and the sum of them gives

$$(r_2-t_2)^2 = (\mathrm{Re}\,\ell_2-r_2)^2+(\mathrm{Re}\,\ell_2-t_2)^2+2|\mathrm{Im}\,\ell_2|^2 \quad (\mathrm{Im}\, r_2=\mathrm{Im}\, t_2=0). \tag{143}$$

If ℓ_2 is imaginary, this relation simplifies to (5.). If $r_2=t_2$, then it implies $\ell_2=r_2=t_2$, which proves part of (4.).
To complete the proof of (4.), assume that $\ell_2\in\mathbb{R}$ and let $(\tilde{\kappa},\tilde{\omega})$ tend to $(0,0)$ along the set $\{(\tilde{\kappa},\tilde{\omega}) : \tilde{\omega}+\ell_1\tilde{\kappa}+\ell_2\tilde{\kappa}^2=0\} \subset \mathbb{R}^2$ to obtain

$$\begin{aligned}
&|\ell|^2 = O(\tilde{\kappa}^6),\\
&|a|^2 = r_0^2(|\mathrm{Re}\, r_2-\ell_2|^2+|\mathrm{Im}\, r_2|^2)\tilde{\kappa}^4+O(|\tilde{\kappa}|^5),\\
&|b|^2 = t_0^2(|\mathrm{Re}\, t_2-\ell_2|^2+|\mathrm{Im}\, t_2|^2)\tilde{\kappa}^4+O(|\tilde{\kappa}|^5),\\
&\qquad (\tilde{\kappa},\tilde{\omega})\to(0,0),\ \tilde{\omega}+\ell_1\tilde{\kappa}+\ell_2\tilde{\kappa}^2=0.
\end{aligned}$$

It follows from balancing powers in (139) that $\ell_2=r_2=t_2$.
To prove the last two relations, we compute $|\ell|^2$, $|a|^2$, and $|b|^2$ using the forms (142):

$$\begin{aligned}
\ell\bar{\ell} = \big[&\tilde{\omega}^2+\ell_1^2\tilde{\kappa}^2+2\ell_1\tilde{\omega}\tilde{\kappa}+2\mathrm{Re}\,\ell_2\tilde{\omega}\tilde{\kappa}^2+\\
&+2\ell_1\mathrm{Re}\,\ell_2\tilde{\kappa}^3+(2\ell_1\mathrm{Re}\,\ell_3+|\ell_2|^2)\tilde{\kappa}^4+\\
&+\ldots\big]\,[1+O(|\tilde{\kappa}|+|\tilde{\omega}|)],
\end{aligned}$$

$$\begin{aligned}
a\bar{a} = \big[&\tilde{\omega}^2+|r_1|^2\tilde{\kappa}^2+2r_1\tilde{\omega}\tilde{\kappa}+2\mathrm{Re}\, r_2\tilde{\omega}\tilde{\kappa}^2+\\
&+2r_1\mathrm{Re}\, r_2\tilde{\kappa}^3+(2r_1\mathrm{Re}\, r_3+|r_2|^2)\tilde{\kappa}^4+\\
&+\ldots\big]\,[r_0^2+O(|\tilde{\kappa}|+|\tilde{\omega}|)],
\end{aligned}$$

and there is a similar expression for $|b|^2$. Provided $\ell_1=0$, the $\tilde{\kappa}^4$ and $\tilde{\omega}\tilde{\kappa}^2$ terms simplify to

$$\begin{aligned}
&\tilde{\kappa}^4 \text{ term:} && |\ell_2|^2 = r_0^2|r_2|^2+t_0^2|t_2|^2,\\
&\tilde{\omega}\tilde{\kappa}^2 \text{ term:} && \mathrm{Re}\,\ell_2 = r_0^2\mathrm{Re}\, r_2+t_0^2\mathrm{Re}\, t_2.
\end{aligned}$$

■

The condition $\ell_1=r_1=t_1$ leads to useful expressions for the zero sets of ℓ, a, and b near (κ_0,ω_0):

$$\begin{aligned}
&\ell(\kappa,\omega)=0 \iff\\
&\qquad \omega=\omega_0-\ell_1(\kappa-\kappa_0)-\ell_2(\kappa-\kappa_0)^2-\ldots,\\
&a(\kappa,\omega)=0 \iff\\
&\qquad \omega=\omega_0-\ell_1(\kappa-\kappa_0)-r_2(\kappa-\kappa_0)^2-\ldots,\\
&b(\kappa,\omega)=0 \iff\\
&\qquad \omega=\omega_0-\ell_1(\kappa-\kappa_0)-t_2(\kappa-\kappa_0)^2-\ldots.
\end{aligned} \tag{144}$$

The statement that the guided mode at (κ_0,ω_0) is nonrobust with respect to perturbations of κ is implied by the generic inequality $\mathrm{Im}\,\ell_2>0$. While we are interested in calculating transmission anomalies near nonrobust guided modes, the ensuing analysis does not exclude the case that ℓ_2 is real.
Figs. (**7**) and (**8**) show resonant anomalies in the transmission coefficient

$$T(\kappa,\omega) = \left|\frac{b(\kappa,\omega)}{\ell(\kappa,\omega)}\right|, \tag{145}$$

which is the square root of the time-averaged energy flux transmitted across one period the slab relative to that of the incident wave.
Using the expansions (142), approximate formulas for ℓ, a, and b give an analytic formula for T^2 to any desired degree of accuracy. In order to capture the essential features of the anomaly, one must include the second-order terms in (144). Assuming $\mathrm{Im}\,\ell_2>0$, the resulting approximation has an error of first order in κ, as stated in Theorem 21 below.
Four of the graphs in Figs. (**7**) and (**8**) exhibit peaks and dips that reach 100% and 0%. This occurs when all of the coefficients r_n and t_n are real and the zero set of each of $a(\kappa,\omega)$ and $b(\kappa,\omega)$ in $\mathbb{C}^2$ intersects $\mathbb{R}^2$ in a curve (rather than just a point) described by the real function given in (144). The zero set of a in $\mathbb{R}^2$ describes the frequency ω as a function of κ at which 100% transmission is achieved, whereas the zero set of b describes the frequencies of 0% transmission.

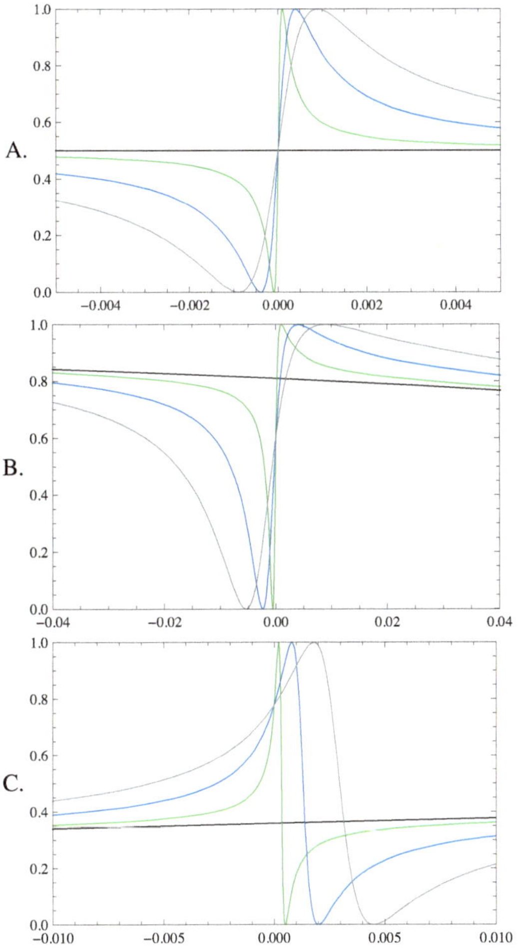

Fig. (**7**): Transmission anomaly (T^2 *vs.* $\tilde{\omega}$) using the approximate formulas in Theorem 21 with $\ell_1 = 0$ and various values of the parameters $(t_0, \ell_1, t_2, r_2, \zeta)$.

A. $(2^{-\frac{1}{2}}, 0, 1, -1, 0)$; $\tilde{\kappa} = 0.0, 0.01, 0.02, 0.03$.

B. $(0.9, 0, 1.5, -2.5, 3)$; $\tilde{\kappa} = 0.0, 0.02, 0.04, 0.06$.

C. $(0.6, 0, -5, -2, -4)$; $\tilde{\kappa} = 0.0, 0.01, 0.02, 0.03$.

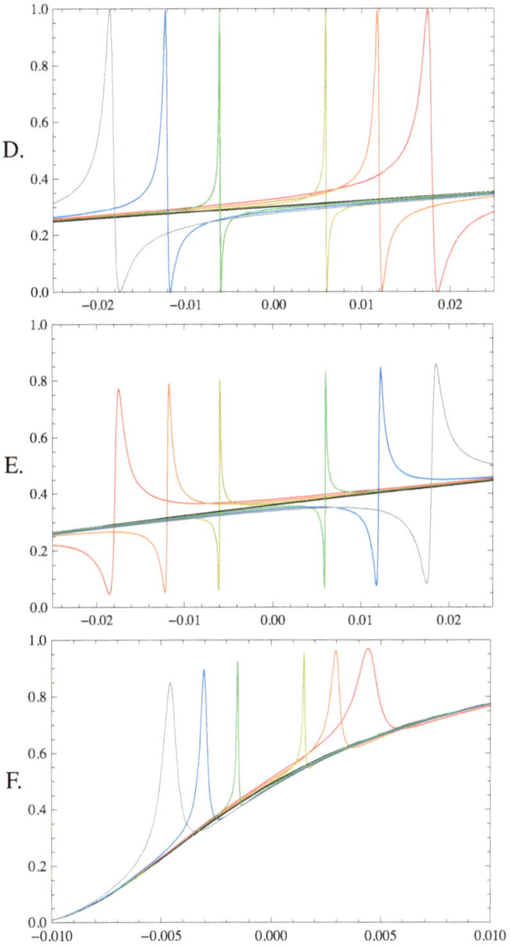

Fig. (**8**): Transmission anomaly (T^2 *vs.* $\tilde{\omega}$) using the approximate formulas in Theorem 21 with $\ell_1 \neq 0$ and various values of the parameters $(t_0, \ell_1, t_2, r_2, \zeta)$; $\tilde{\kappa} =$ -0.009, -0.006, -0.003, 0.0, 0.003, 0.006, 0.009.

D. $(0.55, 2, -7, 7, -5)$

E. $(0.6, -2, 5+4i, -5+4i, -8)$

F. $(0.7, 0.5, -1-8i, 1-2i, -90)$

The graphs exhibit noteworthy features that are established rigorously by the expansions. Assuming the coefficients of a and b are real, we observe the following:

1. For $\kappa = \kappa_0$, the structure supports a guided mode at frequency ω_0, yet there is no anomaly present in the graph of T *vs.* ω.

2. As κ is perturbed from κ_0, a guided mode is no longer supported, and a peak and a dip in the graph of T *vs.* ω emanate from the frequency ω_0 of the destroyed mode. To order $O((\kappa-\kappa_0)^2)$, both the peak and dip occur at $\omega \approx \omega_0 - \ell_1(\kappa-\kappa_0)$. Thus, if $\ell_1 \neq 0$, the spike passes through ω_0 at a speed of $-\ell_1$ as κ passes through κ_0 and for each $\kappa \neq \kappa_0$, both the peak and the dip are on the same side of ω_0.

3. To order $O((\kappa-\kappa_0)^3)$, the frequencies of the peak and the dip differ from one another by $(t_2 - r_2)(\kappa-\kappa_0)^2$. As long as $r_2 \neq t_2$, this implies that they appear in the same order, regardless of whether they are to the left or to the right of or straddling ω_0.

We will show presently that T^2 can approximated to order $O(|\tilde{\kappa}| + \tilde{\omega}^2)$ by an expression involving seven real parameters. For real (κ, ω),

$$|\ell(\kappa,\omega)| = \left[\left|\tilde{\omega} + \ell_1\tilde{\kappa} + \ell_2\tilde{\kappa}^2\right| + O(|\tilde{\kappa}|^3)\right] \times$$
$$\times \left[1 + c_1\tilde{\omega} + c_2\tilde{\kappa} + O(\tilde{\kappa}^2 + \tilde{\omega}^2)\right],$$
$$|a(\kappa,\omega)| = \left[r_0\left|\tilde{\omega} + \ell_1\tilde{\kappa} + r_2\tilde{\kappa}^2\right| + O(|\tilde{\kappa}|^3)\right] \times$$
$$\times \left[1 + a_1\tilde{\omega} + a_2\tilde{\kappa} + O(\tilde{\kappa}^2 + \tilde{\omega}^2)\right],$$
$$|b(\kappa,\omega)| = \left[t_0\left|\tilde{\omega} + \ell_1\tilde{\kappa} + t_2\tilde{\kappa}^2\right| + O(|\tilde{\kappa}|^3)\right] \times$$
$$\times \left[1 + b_1\tilde{\omega} + b_2\tilde{\kappa} + O(\tilde{\kappa}^2 + \tilde{\omega}^2)\right],$$

in which the coefficients of the second factors are real. The $\tilde{\kappa}^2$-terms in the first factors have been retained because, as we shall see, they are necessary for ensuring an error of $O(|\tilde{\kappa}|)$ in the ratio, which is possible if we assume that $\operatorname{Im}\ell_2 \neq 0$.
Now, we have

$$\left|\frac{a}{b}\right| = \frac{r_0|\tilde{\omega} + \ell_1\tilde{\kappa} + r_2\tilde{\kappa}^2| + O(|\tilde{\kappa}|^3)}{t_0|\tilde{\omega} + \ell_1\tilde{\kappa} + t_2\tilde{\kappa}^2| + O(|\tilde{\kappa}|^3)} \times$$
$$\times \left(1 + \zeta\tilde{\omega} + O(|\tilde{\kappa}| + \tilde{\omega}^2)\right), \quad (146)$$

where $\zeta = a_1 - b_1 \in \mathbb{R}$, and by means of this ratio, T^2 expressed as

$$T^2 = \frac{|b|^2}{|b|^2 + |a|^2} = \frac{1}{1 + |a/b|^2}. \quad (147)$$

More terms in both factors of the expansions of $|b|$ and $|a|$ would be needed in order to obtain an approximation of order $(\tilde{\kappa}^2 + \tilde{\omega}^2)$.

The validity of Theorem 20 relies on the assumption that $\operatorname{Im}\ell_2 \neq 0$, which, by part 4 of Theorem 20, is equivalent to $r_2 \neq t_2$ in the case that these are both real. The case that $t_2 = r_2 = \ell_2 \in \mathbb{R}$ is a singular situation in which there is nearly a real dispersion relation for guided modes in the presence of a propagating harmonic; the formula degenerates and exhibits no anomaly. In this case, higher orders in the expansions of a, b, and ℓ should be taken, up to the first coefficient of ℓ with a nonzero imaginary part, in order to capture the anomaly. A true real dispersion relation exists in the highly degenerate case that all coefficients ℓ_n vanish, in which case the guided mode at (κ_0, ω_0) is in fact robust with respect to perturbations of κ and no anomaly is present. This situation occurs, for example, if the structure is not genuinely periodic, as illustrated in Fig. (**2**)B.
As we have discussed, the coefficient ℓ_1 controls the position of the transmission anomaly as a function of $\tilde{\kappa}$. If $\ell_1 \neq 0$, it is the derivative of a sort of dispersion relation for leaky modes, giving the real part of the frequency as a function of real κ.
The parameters in the formulas of the theorem have the following significance.

1. If $\ell_1 \neq 0$, then $\lim_{(\kappa,\omega)\to(\kappa_0,\omega_0)} T(\kappa,\omega) = t_0$. Thus, $T(\kappa,\omega)$ is continuous at (κ_0,ω_0).
2. If $\ell_1 = 0$, then $\lim_{\omega\to\omega_0} T(\kappa_0,\omega) = t_0$ and

$$\lim_{\kappa\to\kappa_0} T(\kappa,\omega_0) = \frac{t_0|t_2|}{(t_0^2|t_2|^2 + r_0^2|r_2|^2)^{\frac{1}{2}}}. \quad (148)$$

The different limits can be seen in Fig. (**7**).

3. $\dfrac{\partial}{\partial\omega}T(\kappa_0,\omega)|_{\omega_0} = -2\zeta\dfrac{r_0^2 t_0^2}{(r_0^2 + t_0^2)^2}.$

It is also possible to derive an approximate formula for the phase of the transmitted field, which undergoes sharp variation near the resonant pair (κ_0, ω_0).

Theorem 21 *Given that ℓ, a, and b have a common root at $(\kappa_0,\omega_0) \in \mathbb{R}^2$; that their partial derivatives with respect to ω do not vanish at (κ_0,ω_0); and that $\operatorname{Im}\ell_2 \neq 0$ in the form* (142)*, the following approximations hold.*

$$T^2(\kappa,\omega) = \frac{1}{1 + D_0^2} + O(|\tilde{\kappa}| + \tilde{\omega}^2)$$
$$= \frac{E_0^2}{E_0^2 + 1} + O(|\tilde{\kappa}| + \tilde{\omega}^2)$$
$$= \frac{t_0^2|\tilde{\omega} + \ell_1\tilde{\kappa} + t_2\tilde{\kappa}^2|^2}{|\tilde{\omega} + \ell_1\tilde{\kappa} + \ell_2\tilde{\kappa}^2|^2}(1 + c_1\tilde{\omega})^2 + O(|\tilde{\kappa}| + \tilde{\omega}^2),$$

as $(\tilde{\kappa},\tilde{\omega}) \to (0,0)$ *in* $\mathbb{R}^2$*, where* D_0 *and* E_0 *are defined by*

$$D_0 = \frac{r_0|\tilde{\omega}+\ell_1\tilde{\kappa}+r_2\tilde{\kappa}^2|}{t_0|\tilde{\omega}+\ell_1\tilde{\kappa}+t_2\tilde{\kappa}^2|}(1+\zeta\tilde{\omega}),$$

$$E_0 = \frac{t_0|\tilde{\omega}+\ell_1\tilde{\kappa}+t_2\tilde{\kappa}^2|}{r_0|\tilde{\omega}+\ell_1\tilde{\kappa}+r_2\tilde{\kappa}^2|}(1-\zeta\tilde{\omega}).$$

Proof. The treatment of the asymptotics as $(\tilde{\kappa},\tilde{\omega}) \to (0,0)$ is subtle. What allows obtention of the $O(|\kappa|)$ part of the estimate is the assumption that $\operatorname{Im}\ell_2 \neq 0$. To prove the first approximation (the second and third are handled similarly), let us define A and B to be the numerator and denominator of the first factor in (146) and C to be the second factor. Their approximations that appear in D_0 are

$$\begin{aligned}&A_0 = r_0|\tilde{\omega}+\ell_1\tilde{\kappa}+r_2\tilde{\kappa}^2|,\ A = A_0+O(|\tilde{\kappa}|^3),\\ &B_0 = t_0|\tilde{\omega}+\ell_1\tilde{\kappa}+t_2\tilde{\kappa}^2|,\ B = B_0+O(|\tilde{\kappa}|^3),\\ &C_0 = 1+\zeta\tilde{\omega},\quad C = C_0+O(|\tilde{\kappa}|+\tilde{\omega}^2).\end{aligned}$$

With these definitions, we have

$$\begin{aligned}T^2 &= \frac{B^2}{B^2+A^2C^2},\\ \frac{1}{1+D_0^2} &= \frac{B_0^2}{B_0^2+A_0^2C_0^2}.\end{aligned} \tag{149}$$

The crucial inequality is the lower bound

$$\begin{aligned}&A_0^2+B_0^2 \geq \tfrac{1}{4}(A_0+B_0)^2 \geq\\ &\tfrac{1}{4}\min(t_0^2,r_0^2)(|\tilde{\omega}+\ell_1\tilde{\kappa}+r_2\tilde{\kappa}^2|+|\tilde{\omega}+\ell_1\tilde{\kappa}+t_2\tilde{\kappa}^2|)^2\\ &\geq m_0^2\tilde{\kappa}^4,\end{aligned} \tag{150}$$

in which

$$m_0 = \tfrac{1}{4}\min(t_0,r_0)\left[|\operatorname{Re}(r_2-t_2)|+|\operatorname{Im}r_2|+|\operatorname{Im}t_2|\right].$$

By the assumption that $\operatorname{Im}\ell_2 \neq 0$ and part (4) of Theorem 20, m_0 is strictly positive.

In what follows, the symbol $O(|\tilde{\kappa}|^n)$ is used in place of any function that is "big-oh" of $|\tilde{\kappa}|^n$ as $(\tilde{\kappa},\tilde{\omega}) \to (0,0)$, that is any function that is bounded in magnitude by a constant multiple by $|\tilde{\kappa}|^n$ for sufficiently small $(\tilde{\kappa},\tilde{\omega})$.

Since C_0^2 and C^2 as well as their reciprocals are $O(1)$, it follows that

$$\begin{aligned}&A_0^2 = (B_0^2+A_0^2C_0^2)O(1),\\ &B_0+A_0C^2 = (A_0+B_0)O(1) = (A_0^2+B_0^2)^{\frac{1}{2}}O(1),\\ &A_0^2+B_0^2 = (B_0^2+A_0^2C^2)O(1).\end{aligned} \tag{151}$$

We compare the denominators in (149):

$$\begin{aligned}B^2+A^2C^2 = B_0^2+A_0^2C_0^2+A_0^2O(|\tilde{\kappa}|+\tilde{\omega}^2)+\\ +(B_0+A_0C^2)f_1(\tilde{\kappa})+f_2(\tilde{\kappa}),\end{aligned} \tag{152}$$

in which $f_1(\tilde{\kappa}) = O(|\tilde{\kappa}|^3)$ and $f_2(\tilde{\kappa}) = O(\tilde{\kappa}^6)$. The lower bound (150) gives

$$\begin{aligned}f_1(\tilde{\kappa}) = O(|\tilde{\kappa}|^3) = m_0|(t_2-r_2)\tilde{\kappa}^2|O(|\tilde{\kappa}|) =\\ = (A_0^2+B_0^2)^{\frac{1}{2}}O(|\tilde{\kappa}|).\end{aligned} \tag{153}$$

This, together with the second and third equations in (151), gives

$$\begin{aligned}&(B_0+A_0C^2)f_1(\tilde{\kappa}) =\\ &= (A_0^2+B_0^2)O(|\tilde{\kappa}|) = (B_0^2+A_0^2C_0^2)O(|\tilde{\kappa}|).\end{aligned} \tag{154}$$

The term $f_2(\tilde{\kappa})$ is estimated similarly:

$$\begin{aligned}&f_2(\tilde{\kappa}) = O(\tilde{\kappa}^6) = m_0^2|(t_2-r_2)\tilde{\kappa}^2|^2O(\tilde{\kappa}^2) =\\ &= (A_0^2+B_0^2)O(\tilde{\kappa}^2) = (B_0^2+A_0^2C_0^2)O(\tilde{\kappa}^2).\end{aligned} \tag{155}$$

This, together with (154) and the first equation in (151), give

$$B^2+A^2C^2 = (B_0^2+A_0^2C_0^2)(1+O(|\tilde{\kappa}|+\tilde{\omega}^2)). \tag{156}$$

Next, we compare B^2 and B_0^2:

$$B^2 = B_0^2+A_0g_2(\tilde{\kappa})+g_2(\tilde{\kappa}), \tag{157}$$

where $g_1(\tilde{\kappa}) = O(|\tilde{\kappa}|^3)$ and $g_2(\tilde{\kappa}) = O(\tilde{\kappa}^6)$. In a similar fashion, we obtain

$$\begin{aligned}&A_0g_1(\tilde{\kappa}) = A_0(A_0^2+B_0^2)^{\frac{1}{2}}O(|\tilde{\kappa}|) =\\ &= (A_0^2+B_0^2)O(|\tilde{\kappa}|) = (B_0^2+A_0^2C_0^2)O(|\tilde{\kappa}|),\end{aligned} \tag{158}$$

$$g_2(\tilde{\kappa}) = (B_0^2+A_0^2C_0^2)O(|\tilde{\kappa}|^2). \tag{159}$$

This gives

$$B^2 = B_0^2+(B_0^2+A_0^2C_0^2)O(|\tilde{\kappa}|) \tag{160}$$

Finally, using (156) and (160), we obtain the result of the theorem:

$$\begin{aligned}&(B^2+A^2C^2)^{-1}B^2 = (B_0^2+A_0^2C_0^2)^{-1}\times\\ &\times(1+O(|\tilde{\kappa}|+\tilde{\omega}^2))\left[B_0^2+(B_0^2+A_0^2C_0^2)O(|\tilde{\kappa}|)\right]\\ &= \left[(B_0^2+A_0^2C_0^2)^{-1}B_0^2+O(|\tilde{\kappa}|)\right](1+O(|\tilde{\kappa}|+\tilde{\omega}^2))\\ &= (B_0^2+A_0^2C_0^2)^{-1}B_0^2+O(|\tilde{\kappa}|+\tilde{\omega}^2).\end{aligned}$$

■

5.3 Relation to Fano Resonance

The approximations of T^2 given in Theorem 21 generalize the Fano resonance (124) when viewed as functions of $\tilde{\omega}$. The seven real parameters reduce to two if the following conditions are satisfied (Fig. (7)A):

1. $\ell_1 = 0$ (the anomaly remains at about ω_0);
2. r_2 and t_2 are real (the extremal values of the anomaly are 0 and 1);
3. $\mathrm{Re}\,\ell_2 = 0$ (the dispersion relation is purely imaginary to leading order in $\tilde{\kappa}^2$);
4. $\zeta = 0$ (the background transmission is flat).

Under these conditions, the connection between T^2 and the Fano resonance is made through

$$\begin{aligned} \Gamma &= 2\tilde{\kappa}^2 \mathrm{Im}\,\ell_2, \\ q &= \frac{t_2}{\mathrm{Im}\,\ell_2}, \\ \mathrm{e} &= \frac{\tilde{\omega}}{\tilde{\kappa}^2\,\mathrm{Im}\,\ell_2}. \end{aligned} \qquad (161)$$

The relation between the width of the resonance and the imaginary part of ℓ_2 should be compared with the formulation of Fermi's golden rule by Reed and Simon at the end of §12.6 of [72].
With only the first three conditions, there arises a description of the parameter Γ in terms of r_2 and t_2. The relations (5) and (6) in Theorem 20 imply

$$\begin{aligned} t_2 &= \pm \mathrm{Im}\,\ell_2 \frac{r_0}{t_0}, \\ r_2 &= \mp \mathrm{Im}\,\ell_2 \frac{t_0}{r_0}. \end{aligned}$$

In particular, t_2 and r_2 are of opposite sign. This leads to

$$\Gamma = 2\tilde{\kappa}^2 \mathrm{Im}\,\ell_2 = 2\tilde{\kappa}^2 \frac{t_0}{r_0}|t_2| = 2\tilde{\kappa}^2 \frac{r_0}{t_0}|r_2|.$$

5.4 Structural Perturbations and Bifurcation of Anomalies

As Figs. (7) and (8) illustrate, if $\ell_1 \neq 0$, the entire anomaly (peak and dip) is always to one side of ω_0 because it travels with speed $-\ell_1$ as a function of κ near κ_0 and widens proportionally to $\tilde{\kappa}^2$. But when $\ell_1 = 0$, the peak and dip may straddle the resonant frequency ω_0. For symmetric structures with $\kappa_0 = 0$, the dispersion relation (ω *vs.* κ) is necessarily even in κ, and we obtain $\ell_1 = 0$. In this case, the non-robust mode at κ_0 is a standing wave exponentially confined to the slab. These two types of behavior

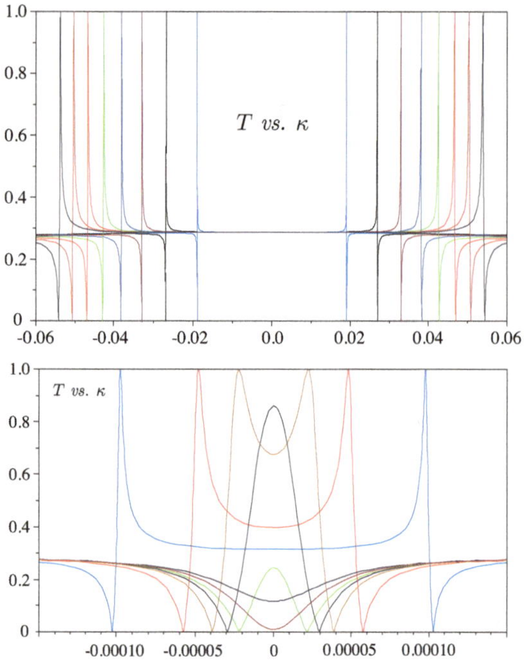

Fig. (**9**): The bifurcation discussed in point (2) of Sec. 5.4. **Top:** At the critical value $\gamma = \gamma_0$ for which there is a single guided mode at $(\kappa_0(\gamma_0), \omega_0(\gamma_0)) = (0, \sim 0.977886)$, T is graphed as a function of κ for values of ω that run from $\omega_0 - 0.001$ to $\omega_0 - 0.008$ in increments of -0.001 as the colors run from blue to violet. **Bottom:** For a value of γ after bifurcation, the transmission is graphed as a function of κ for several fixed values of ω greater than the guided mode frequency $\omega_0(\gamma)$. The values of κ lie between the guided mode wavenumbers $\pm\kappa_0(\gamma)$.

can be connected through a variation of the structure where the guided mode pair (κ_0, ω_0) is a function of a structural parameter γ. There are two scenarios.

1. One can choose a parameter γ that controls the asymmetry of the slab, with $\gamma = 0$ corresponding to a symmetric one. An example of this is a two-dimensional slab whose period consists of an elliptical rod that, for $\gamma = 0$, has the z-axis (perpendicular to the slab) as one of its principle axes, and where γ is an angle of rotation of the rod. The wavenumber κ_0 of the guided mode is an odd function of γ, as is ℓ_1, which passes from negative to positive as γ passes through zero.

2. One can define γ so that, for $\gamma < 0$ there are no guided mode pairs (κ_0, ω_0) in an open set $U \subset \mathbb{R}^2$ but that, for $\gamma = 0$, there is a single pair that bifurcates into two for $\gamma > 0$. The simplest model for which such a bifurcation can be constructed is a discrete one, in which the open waveguide is modeled

by a one-dimensional lattice that is coupled to a uniform two-dimensional lattice, where the masses and spring constants of the waveguide as well as the coupling have period two. Any such structure is symmetric, so if the real pair (κ_0, ω_0) admits a guided mode, so does the pair $(-\kappa_0, \omega_0)$. In [69], Ptitsyna uses one of the two coupling constants as the structural parameter γ and analyzes this type of bifurcation in detail. The bifurcation is most clearly seen if the anomaly is graphed as function of κ for different fixed values of ω; it is shown in Fig. (**9**).

5.5 Amplitude Enhancement

When resonant scattering occurs due to a perturbation that results in the destruction of a guided mode, the transmission anomalies that we have analyzed are accompanied by the phenomenon of enhancement of the field amplitude in the waveguide.

We have seen that the transmission coefficient exhibits no anomaly as a function of real ω at the wavenumber of the nonrobust guided mode $\kappa = \kappa_0$, whereas a very sharp anomaly appears near the frequency of the guided mode for arbitrarily small deviations $\tilde{\kappa} = \kappa - \kappa_0$. In numerical simulations of scattering of scalar plane waves by two-dimensional periodic dielectric waveguides, we observe a corresponding phenomenon in the field in the waveguide: There is no significant field amplitude enhancement near the guided mode frequency for fixed wavenumber $\kappa = \kappa_0$, whereas very high enhancement is observed for small nonzero $\tilde{\kappa}$, for ω in the vicinity of the transmission anomaly.

Determination of the field amplitude enhancement as a function of κ involves a subtlety of the source field ϕ that did not play a role in the derivation of the transmission anomaly, namely, that the resonant part in the decomposition of the trace of a plane-wave source field vanishes at the guided-mode pair (κ_0, ω_0). This is seen as follows. Recall the decomposition of the trace $\phi(\kappa, \omega)$ of an analytic source field

$$\phi(\kappa, \omega) = \alpha(\kappa, \omega)\hat{\psi}(\kappa, \omega) + \phi_2(\kappa, \omega). \quad (162)$$

We have shown that the multiple α is analytic at (κ_0, ω_0). If ϕ is the trace of a plane-wave source field for the Helmholtz equation,

$$u^{\text{inc}}(x,z) = e^{i(\tilde{m}+\kappa)x}e^{i\eta_{\tilde{m}}z}, \quad (163)$$

with $\eta_{\tilde{m}} > 0$ at (κ_0, ω_0), then $\phi(\kappa, \omega)$ is analytic at (κ_0, ω_0). We will prove that $\alpha(\kappa_0, \omega_0) = 0$. Because of Theorem 9, there is a solution to the scattering problem at (κ_0, ω_0) (although a solution is not unique). Let ψ be the trace of a solution, and let its partial spectral decomposition be

$$\psi = \gamma\hat{\psi} + \psi_2, \quad (164)$$

in which γ is a complex constant. Using the decompositions $A = \tilde{\ell}P_1 + AP_2$, (162), and (164) in the equation

$$A(\kappa_0, \omega_0)\psi = \phi(\kappa_0, \omega_0),$$

we obtain

$$\gamma\tilde{\ell}(\kappa_0, \omega_0)\hat{\psi} + A(\kappa_0, \omega_0)\psi_2 = \alpha(\kappa_0, \omega_0)\hat{\psi} + \phi_2(\kappa_0, \omega_0).$$

Since $\tilde{\ell}(\kappa_0, \omega_0) = 0$, we obtain $\alpha(\kappa_0, \omega_0) = 0$ and therefore an expansion

$$\alpha(\kappa, \omega) = \alpha_1\tilde{\kappa} + \alpha_2\tilde{\omega} + \dots. \quad (165)$$

The enhancement of the field amplitude should be manifest in the ratio

$$\left|\frac{\psi}{\ell\phi}\right| = \frac{|c\alpha\hat{\psi} + \ell\tilde{A}^{-1}\phi_2|}{|\ell\alpha\hat{\psi} + \ell\phi_2|}.$$

Since the nonresonant component of ψ as well as the source are of order ℓ, any meaningful enhancement may be measured by the ratio

$$\mathscr{A} = \frac{|c\alpha|}{|\ell|} = \frac{|\alpha|}{|\tilde{\ell}|} = \frac{|\alpha_1\tilde{\kappa} + \alpha_2\tilde{\omega}| + O(\tilde{\kappa}^2 + \tilde{\omega}^2)}{|\tilde{\omega} + \ell_1\tilde{\kappa} + \ell_2\tilde{\kappa}^2| + O(|\tilde{\kappa}|^3)}(|c| + O(|\tilde{\kappa}| + |\tilde{\omega}|), \quad (166)$$

as $(\kappa, \omega \to 0)$.

The numerical observations we mentioned above, namely the absence of field enhancement at the wavenumber of the guided mode and enhancement inversely proportional to $\tilde{\kappa}$ in the vicinity of the anomaly for small $\tilde{\kappa} \neq 0$, are borne out through analysis of $\mathscr{A}$.

By setting $\tilde{\kappa} = 0$ in (166), we obtain

$$\mathscr{A} = |c\alpha_2| + O(|\omega|), \quad \tilde{\kappa} = 0, \tilde{\omega} \to 0,$$

which demonstrates the absence of field enhancement at $\tilde{\kappa} = 0$. Assuming that $\operatorname{Im}\ell_2 \neq 0$, for nonzero $\tilde{\kappa}$, the denominator in (166) reaches its minimal value to order $O(|\tilde{\kappa}|^3)$ when $\tilde{\omega} = -\ell_1\tilde{\kappa} - \operatorname{Re}\ell_2\tilde{\kappa}^2$. Part 6 of Theorem 20 shows that, if $\ell_1 = 0$, this frequency is between the peak and the dip of the transmission coefficient. Under this relation between $\tilde{\omega}$ and $\tilde{\kappa}$, (166) yields

$$\mathscr{A} = \frac{1}{|\tilde{\kappa}|}\frac{|\alpha_1 - \alpha_2\ell_1|}{|\mathrm{Im}\,\ell_2|}(|c| + O(|\tilde{\kappa}|),$$

$$\tilde{\omega} + \ell_1\tilde{\kappa} + \mathrm{Re}\,\ell_2\tilde{\kappa}^2 = 0\,, \tilde{\kappa} \to 0.$$

The leading order behavior of $1/|\tilde{\kappa}|$ is confirmed by numerical data in the two-dimensional Helmholtz case [81] and for a two-dimensional lattice model [69].

ACKNOWLEDGMENT

This work was supported by National Science Foundation grant DMS-0807325. Jessica Dowd produced the graphs in Figs. (**5**), (**7**), and (**8**), and Natalia Ptitsyna produced the graphs in Fig. (**9**).

REFERENCES

[1] Ammari H, Béreux N, Nédélec J-C. Resonant frequencies for a narrow strip grating. Math Meth Appl Sci 1999; 22: 1121-1152.
[2] Ammari H, Béreux N, Nédélec J-C. Resonances for Maxwell's equations in a periodic structure. Japan J Indust Appl Math 2000; 17: 149-198.
[3] Akhiezer NI, Glazman IM. Theory of linear operators in Hilbert space. Dover, 1993.
[4] Ammari H, Santosa F. Guided waves in a photonic bandgap structure with a line defect. SIAM J Appl Math 2004; 64: 2018-2033.
[5] Barut AO. The theory of the scattering matrix. Macmillan, 1967
[6] Barnes A. Electromagnetic scattering by three-dimensional periodic structures. PhD thesis, Duke University, 2003.
[7] Bonnet-Bendhia A-S, Starling F. Guided waves by electromagnetic gratings and nonuniqueness examples for the diffraction problem. Math Meth Appl Sci 1994; 17: 305-338.
[8] Barnes WL, Murray WA, Dintinger J, Devaux E, Ebbesen TW. Surface plasmon polaritons and their role in the enhanced transmission of light through periodic arays of subwavelength holes in a metal film. Phys Rev Lett 2004; 92: 107401.
[9] Colton D, Kress R. Integral equation methods in scattering theory. Krieger, 1992.
[10] Callan M, Linton CM, Evans DV. Trapped modes in two-dimensional waveguides. J Fluid Mech 1991; 229: 51-64.
[11] Costabel M, Stephan E. A direct boundary integral equation method for transmission problems. J Math Anal Appl 1985; 106: 367-413.
[12] Chandler-Wilde SN, Monk P. Existence, uniqueness, and variational methods for scattering by unbounded rough surfaces. SIAM J Math Anal 2005; 37: 598-618.
[13] Dym H, McKean HP. Fourier series and integrals. Probability and Mathematical Statistics. Academic Press, 1972.
[14] Edward J. Trapped modes for periodic structures in waveguides. Math Meth Appl Sci 2003; 27: 91-99.
[15] Eidus DM. The principle of limiting absorption. Amer Math Soc Transl 1965; 47:157-191.
[16] Elachi C. Waves in active and passive periodic structures: a review. Proc IEEE 1976; 64: 1666-1698.
[17] Ebbesen TW, Lezec HJ, Ghaemi HF, Thio T, Wolff PA. Extraordinary optical transmission through sub-wavelength hole arrays. Letters to Nature 1998; 391: 667-669.
[18] Evans DV, Linton CM, Ursel Fl. Trapped mode frequencies embedded in the continuous spectrum. Q J Mech Appl Math 1993; 46: 253-274.
[19] Evans DV, Levitin M, Vassiliev D. Existence theorems for trapped modes. J Fluid Mech 1994; 261: 21-31.
[20] Evans DV, Porter R. On the existence of embedded surface waves along periodic arrays of parallel plates. Q J Mech Appl Math 2002; 55: 481-494.
[21] Evans, DV. Trapped acoustic modes. IMA J Appl Math 1992; 49: 45-60.
[22] Fano U. Effects of configuration interaction on intensities and phase shifts. Phys Rev 1961; 124: 1866-1878.
[23] García de Abajo FJ. Light scattering by particle and hole arrays. Rev Mod Phys 2007; 79: 1267-1290.
[24] Genet C, Ebbesen TW. Light in tiny holes. Nature 2007; 445: 39-46.
[25] Grupp DE, Lezec HJ, Ebbesen TW, Pellerin KM, Thio T. Crucial role of metal surface in enhanced transmission through subwavelength apertures. Appl Phys Lett 2000; 77: 1569-1571.
[26] Grikurov VE, Lyalinov MA, Neittaanmäki, Plamenevskii BA. On surface waves in diffraction gratings. Math Meth Appl Sci 2000; 23: 1513-1535.
[27] Goldstein CI. Eigenfunction expansions associated with the Laplacian for certain domains with infinite boundaries. Transactions of the American Mathematical Society 135, 1969.
[28] Groves MD. Examples of embedded eigenvalues for problems in acoustic waveguides. Math Method Appl Sci 1998; 21: 479-488.
[29] Ghaemi HF, Thio T, Grupp DE, Ebbesen TW, Lezec HJ. Surface plasmons enhance optical transmission through subwavelength holes. Phys Rev B 1998; 58: 6779-6782.
[30] Haus HA, Miller DAB. Attenuation of cutoff modes and leaky modes of dielectric slab structures. IEEE J Quantum Elect 1986; 22: 310-318.
[31] Hille E, Phillips RS. Functional analysis and semi-groups. AMS Colloquium Pub. 31, rev. ed. 1981.
[32] Hislop PD, Sigal IM. Introduction to spectral theory with applications to Schrödinger operators. Applied Mathematical Sciences 113. Springer-Verlag, 1996.
[33] Haider MA, Shipman SP, Venakides S. Boundary-integral calculations of two-dimensional electromagnetic scattering in infinite photonic crystal slabs: channel defects and resonances. SIAM J Appl Math 2002; 62: 2129-2148.
[34] Jackson JD. Classical electrodynamics. John Wiley & Sons, Inc., 3rd ed. 1999.
[35] Jones DS. The eigenvalues of $\delta u + \lambda u = 0$ when the boundary conditions are given on semi-infinite domains. Proc Camb Philos Soc 1953; 49: 668-684.
[36] Kato T. Perturbation theory for linear operators. Springer-Verlag, 1980.
[37] Kuchment P, Ong BS. On guided waves in photonic crystal waveguides. Waves in Periodic and Random Media. Contemp Math AMS 2004; 339: 105-115.
[38] Khallaf NSA, Parnovski L, Vassiliev D. Trapped modes in a waveguide with a long obstacle. J Fluid Mech 2000; 403: 251-261.
[39] Kress R. Linear integral equations. Applied Mathematical Sciences 82. Springer-Verlag, 2nd ed. 1999.
[40] Kriegsmann GA. Complete transmission through a two-dimensional diffraction grating. SIAM J Appl Math 2004, 65: 24-42.
[41] Kim TJ, Thio T, Ebbesen TW, Grupp DE, Lezec HJ. Control of optical transmission through metals perforated with subwavelength hole arrays. Opt Lett 1999, 24: 256-258.
[42] Krishnan A, Thio T, Kim TJ, Lezec HJ, Ebbesen TW, Wolff PA, Pendry J, Martin-Moreno L, Garcia-Vidal FJ. Evanescently coupled resonance in surface plasmon enhanced transmission. Opt Commun 2001; 2000: 1-7.
[43] Kuchment P. The mathematics of photonic crystals. in: Mathematical modeling in optical science. Frontiers in Applied Mathematics 22, chapter 7, pp 207-272, SIAM, 2001.
[44] Kukhlevsky SV. Enhanced transmission of light through subwavelength nanoapertures by far-field multiple-beam interference. Phys Rev A 2008; 78: 023826-1-6.
[45] Kukhlevsky SV. Interference-induced enhancement of intensity and energy of a quantum optical field by a subwavelength array of coherent light sources. Appl Phys B 2008; 93: 145-150.
[46] Liu H, Lalanne P. Microscopic theory of the extraordinary optical transmission. Letters to Nature 2008; 452: 728-731.
[47] Linton CM, McIver M. Trapped modes in cylindrical waveguides. Q J Mech Appl Math 1998; 51: 389-412.
[48] Linton CM, McIver M. The existence of rayleigh–bloch surface waves. J Fluid Mech 2002; 470: 85-90.
[49] Linton CM, McIver M, McIver P, Ratcliffe K, Zhang J. Trapped modes for off-centre structures in guides. Wave Motion 2002; 36: 67-85.
[50] Lax PD, Phillips RS. Scattering theory. Academic Press, 1967.
[51] Lalanne P, Rodier JC, Hugonin JP. Surface plasmons of metallic surfaces perforated by nanohole arrays. J Opt A: Pure Appl Opt 2005; 7: 422-426.

[52] Lezec H,Thio T. Diffracted evanescent wave model for enhanced and suppressed optical transmission through subwavelength hole arrays. Opt Express 2004; 12: 3629-3651.

[53] Linton CM, Thompson I. Resonant effects in scattering by periodic arrays. Wave Motion 2007; 44: 165-175.

[54] Markushevich AJ. Theory of functions of a complex variable 2. Prentice Hall, 1965.

[55] Mittra R, Lee SW. Analytical techniques in the theory of guided waves. Macmillan, 1971.

[56] McIver P, Linton CM, McIver M. Construction of trapped modes for wave guides and diffraction gratings. Proc R Soc Lond A 1998; 454: 2593-2616.

[57] McIver M, Linton CM, McIver P, Zhang J. Embedded trapped modes for obstacles in two-dimensional waveguides. Q J Mech Appl Math 2001; 54: 273-293.

[58] Martín-Moreno L, García-Vidal FJ, Lezec HJ, Pellerin KM, Thio T, Pendry JB, Ebbesen TW. Theory of extraordinary optical transmission through subwavelength hole arrays. Phys Rev Lett 2001; 86: 1114-1117.

[59] Medina F, Mesa F, Marqués R. Extraordinary transmission through arrays of electrically small holes from a circuit theory perspective. IEEE Trans Microwave Theory Tech 2008; 56: 3108âĂŞ3120.

[60] Müller C. Foundations of the mathematical theory of electromagnetic waves. Springer-Verlag, 1969.

[61] Nédélec J-C. Acoustic and electromagnetic equations: integral representations for harmonic problems. Applied Mathematical Sciences 144. Springer-Verlag, New York, 2000.

[62] Nevière M. The homogeneous problem, in electromagnetic theory of gratings. Chapter 5, pp 123-157. Springer-Verlag, Berlin, 1980.

[63] Newton RG. Scattering theory of waves and particles. Springer-Verlag, 2nd ed. 1982.

[64] Nosich AI. Radiation conditions for open waveguides. Sov Phys Dokl 1987; 32: 720-722.

[65] Nosich AI. Radiation conditions, limiting absorption principle, and general relations in open waveguide scattering. J Electromagnet Wave 1994; 8: 329-353.

[66] Porter R, Evans DV. Embedded Rayleigh–Bloch surface waves along periodic rectangular arrays. Wave Motion 2008; 43: 29-50, 2005.

[67] Ptitsyna N, Shipman SP, Venakides S. Fano resonance of waves in periodic slabs. pp 73-78. MMET, IEEE, 2008.

[68] Peng ST, Tamir T, Bertoni HL. Theory of periodic dielect waveguides. IEEE Trans Microwave Theory Techn 1975; 23: 123-133.

[69] Ptitsyna N. A discrete model of guided modes and anomalous scattering in periodic structures. PhD thesis, Louisiana State University, 2009.

[70] Paddon P, Young JF. Two-dimensional vector-coupled-mode theory for textured planar waveguides. Phys Rev B 2000; 61: 2090-2101.

[71] Rau ARP. Perspectives on the Fano resonance formula. Phys Scr 2004; 69: C10-C13.

[72] Reed M, Simon B. Methods of mathematical physics IV: analysis of operators. Academic Press, 1980.

[73] Reed M, Simon B. Methods of mathematical physics I: functional analysis. Academic Press, 1980.

[74] Reed M, Simon B. Methods of mathematical physics III: scattering theory. Academic Press, 1980.

[75] Riesz F, Sz.-Nagy B. Functional analysis. Frederick Ungar Publishing Co, 1955.

[76] Shen JT, Catrysse PB, Fan S. Mechanism for designing metallic metamaterials with a high index of refraction. Phys Rev Lett 2005; 94: 197401.

[77] Shestopalov VP, Shestopalov YV. Spectral theory and excitation of open structures. IEE Electromagnetic Waves Series 42. The Institution of Electrical Engineers, 1996.

[78] Steinberg S. Meromorphic families of compact operators. Arch Rational Mech Anal 1968; 31: 372-379.

[79] Shipman SP, Venakides S. Resonance and bound states in photonic crystal slabs. SIAM J Appl Math 2003; 64: 322-342.

[80] Shipman SP, Volkov D. Existence of guided modes in periodic slabs. In: Discrete and continuous dynamical systems. Supplemental Volume, pp. 784-791. Fourth International Conference on Dynamical Systems and Differential Equations, 2004.

[81] Shipman SP, Venakides S. Resonant transmission near non-robust periodic slab modes. Phys Rev E 2005; 71: 026611-1-10.

[82] Shipman SP, Volkov D. Guided modes in periodic slabs: existence and nonexistence. SIAM J Appl Math 2007; 67: 687-713.

[83] Soffer A, Weinstein MI. Time-dependent resonance theory. GAFA, Geom Funct Anal 1998; 8: 1086-1128.

[84] Tikhodeev SG, Yablonskii AL, Muljarov EA, Gippius NA, Ishihara T. Quasiguided modes and optical properties of photonic crystal slabs. Phys Rev B 2002; 66: 045102-1-17.

[85] Ursell F. Trapping modes in the theory of surface waves. Proc Camb Phil Soc 1951; 47: 347-358.

[86] Veremei NB, Nosich AI. Electrodynamic modeling of open resonators with diffraction gratings. Radiophys Quant Electr 1989; 32: 213-219.

[87] Weder R. Spectral and scattering theory for wave propagation in perturbed stratified media. Applied Mathematical Sciences 87. Springer-Verlag, 1991.

[88] Wilcox CH. Scattering Theory for Diffraction Gratings. Applied Mathematical Sciences 46. Springer-Verlag, 1984.

[89] Weinstein A, Stenger W. Methods of Intermediate Problems for Eigenvalues. Mathematics in Science and Engineering 89. Academic Press, 1972.

CHAPTER 3

Mathematical and Numerical Techniques for Open Periodic Waveguides

Johannes Tausch

Department of Mathematics, Southern Methodist University, Dallas, TX 75275, U.S.A.

Abstract: The propagation of electromagnetic waves in dielectric slab waveguides with periodic corrugations is described by the spectrum of the Helmholtz operator on an infinite strip with quasiperiodic boundary conditions. This chapter reviews the basic properties of this spectrum, which typically consists of guided modes, radiation modes and leaky modes. A great deal of attention will be devoted to planar waveguides which share some of the important features of the periodic case. To compute the eigenmodes and the associated propagation constants numerically, one usually truncates the domain that contains the grating and imposes certain radiation conditions on the artificial boundary. An alternative to this approach is to decompose the infinite strip into a rectangle, which contains the grating, and two semi-infinite domains. The guided and leaky modes can be computed by matching the Dirichlet-to-Neumann operator on the interfaces of these three domains. The discretized eigenvalue problem is nonlinear because of the appearance of the propagation constant in the artificial boundary condition. We will discuss how such problems can be solved by numerical continuation. In this approach, one starts with an approximating planar waveguide and then follows the solutions by a continuous transition to the multilayer periodic structure. The chapter is concluded with a brief description of how the perfectly matched layer can be used to compute the guided modes of a waveguide.

1. INTRODUCTION

Periodic media are of considerable interest because there are certain directions and frequencies in which waves cannot propagate. This phenomenon is known as a spectral band gap and occurs because of cancellation due to coherent scattering. In electromagnetics, materials in which the refractive index varies periodically are called photonic crystals. They can be classified as one- two or three dimensional, depending on the number of space directions in which the refractive index varies.

A classical example of a one-dimensional photonic crystal is the Bragg mirror which is an infinite stack of alternating dielectric layers. Lord Rayleigh was able to demonstrate that the mirror exhibits band gaps [39]. Two dimensional photonic crystals consist of a square lattice of infinite dielectric rods. In this case it usually suffices to consider polarized waves which reduces the Maxwell problem to the scalar Helmholtz equation in the transverse plane. An example of a three dimensional crystal is the so-called woodpile where parallel dielectric logs are stacked in alternating directions [24, 43]. For an excellent introduction to photonic crystals and their use in controlling the propagation of light we refer the reader to the recent book [28].

The modes of a periodic structure are the nontrivial solutions of the Maxwell equations without source terms. The appropriate mathematical tool in this context is Floquet Theory which addresses differential equations with periodic coefficients. This topic has been surveyed in [31] and [32]. The type of solutions that are possible in periodic media is described by Floquet's Theorem, which states that the modes are the product of periodic function and a plane wave. Thus the problem of finding the modes can be reduced to computing the eigenfunctions of the Maxwell operator on the periodic cell with quasi periodic boundary conditions.

In the earlier mentioned examples of photonic crystals the periodic cell is finite and the resulting differential operator has a discrete spectrum. The eigenvalues are the frequencies that are allowed to propagate and depend on the wave number of the plane wave. This dependence is usually referred to as the dispersion relation. In order to describe the propagation characteristics of a given periodic structure, the eigenvalue problem must be discretized and solved for all wave vectors in the Brillouin zone. The frequencies that are not an eigenvalue for any wave vector constant make up the band gaps. There are numerous papers that describe this process and its analysis. Representative examples are [4, 16, 17, 29].

Another important class of photonic structures are layered materials. The best known example is the dielectric slab waveguide, which consists of a high index core sandwiched between two infinite lower index materials. Such a structure can support a finite number of guided waves that propagate in a direction parallel to the core. It is a standard textbook exercise to derive the dispersion relation of such a structure [34, 52]. The situation can be complicated by increasing the number of layers and by adding a periodic perturbation, for instance, a grating. Open and periodic waveguides can be found in many integrated-optics devices, such as semiconductor lasers, waveguide couplers, and leaky-wave antennas.

Gratings have been studied for a long time, mainly in the context of scattering [38, 51]. In a grating the refractive index varies in two directions, but is periodic only in one direction. While Floquet's Theorem is still applicable to describe the spectrum of the Maxwell operator in such a geometry, there are considerable differences to photonic crystals. The main difficulty is the fact that the periodic cell is an unbounded region, namely an infinite strip in the transverse direction of propagation. Thus the spectrum of the Maxwell operator typically consists of a finite number of discrete eigenvalues and a continuum of eigenmodes. Leaky modes are a third type of physically relevant eigensolutions. They are unbounded solutions and result from constructive scattering of a guided mode into the semi-infinite regions.

The goal of this chapter is to review some mathematical and numerical methods for planar and periodic dielectric waveguides. The history of this problem goes back to the engineering analysis of open waveguides in the 1960s (see the references in [37]). The first mathematical existence studies are much more recent, and appear to be limited to guided modes and the continuous spectrum [7]. The first numerical methods appear in the first engineering papers, see, again [37] and the review paper [25]. The following decades have seen vast improvements of discretization methods and iterative methods for solving the nonlinear eigenvalue problem.

After fixing the mathematical notations in Section we will first discuss planar non-periodic waveguides in . We devote a considerable amount of space for this topic as it is important for the understanding of the more complicated periodic case. Moreover, because of numerical instabilities, the computation of the modes is much more difficult than one might have expected. Section 3.5 reviews some analytical results about the existence of the continuous and discrete spectrum of periodic waveguides. The following Sections 4.4 and 5.1 describe several discretization methods for the operator equations.

In numerical computations the infinite periodic cell must be truncated and suitable conditions must be imposed on the artificial boundary. Since the frequency goes into the boundary condition the resulting eigenvalue value problem is non-linear. A popular way to solve this type of problem is the continuation method. In Section 6.1 we describe how the modes of a periodic structure can be computed by following the solutions of an averaged planar structure.

In the recent years, the perfectly matched layer (PML) has become a very popular method to truncate infinite domains in scattering problems. This methodology has also been applied to periodic structures and leads to very different computational issues. This will be mentioned briefly in the concluding Section 6.1.

2. PRELIMINARIES

Throughout this article, we assume that the materials in which the electromagnetic fields exist are linear, non-dispersive and isotropic. In this case, the relationship between the electric displacement and the electric field is $\mathbf{D} = \varepsilon \mathbf{E}$ and between the magnetic flux density and field is $\mathbf{B} = \mu \mathbf{H}$. Here, the electric permittivity ε and the magnetic permeability μ are scalars that may be functions of the position. Although this assumption is only an approximation, it is sufficiently accurate in many application of interest. In photonics applications the materials are usually non-magnetic, hence

$$\mu = \mu_0, \quad \varepsilon = \varepsilon_0 \varepsilon_r \quad \text{and} \quad n = \sqrt{\varepsilon_r}.$$

Here, ε_0 and μ_0 are the permittivity and permeability of free-space, ε_r is the relative permittivity and n is the refractive index.

In most applications the electromagnetic fields are time-harmonic, i.e., the time dependence is $\exp(-i\omega t)$, where ω is the frequency. Furthermore, there are no currents and charges. Under these assumptions, Ampère's and Faraday's law are

$$\begin{aligned} \nabla \times \mathbf{H} &= -i\omega\varepsilon\mathbf{E}. && (1) \\ \nabla \times \mathbf{E} &= i\omega\mu\mathbf{H}. && (2) \end{aligned}$$

In photonics problems one is usually interested in the type of solutions that can occur in a structure with a given refractive index. Mathematically, this is an eigenvalue problem where ω is the eigenvalue and $\mathbf{E},\mathbf{H}$ are the non-trivial eigenfunctions. Upon taking the divergence of (1) and (2) it becomes immediately clear that

$$\begin{aligned} \nabla\cdot(\varepsilon\mathbf{E}) &= 0, &(3)\\ \nabla\cdot(\mu\mathbf{H}) &= 0. &(4)\end{aligned}$$

must hold. Equations (1-4) are the source-free time-harmonic Maxwell's equations which appear in this form in many textbooks. At first glance, equations (3) and (4) are superfluous since they follow directly from (1) and (2). Nevertheless, they must be satisfied explicitly to avoid difficulties with the infinitely dimensional nullspace of the curl operator.
In this article we limit our attention to structures that are invariant in the y-direction. From basic electromagnetic theory it is well known that in this case the eigensolutions of the Maxwell equations come in two polarizations:

- Transverse Electric (TE) waves. The electric field is parallel to the y-axis $\mathbf{E}(x,y,z) = u(x,z)\,\mathbf{e}_y$, where the scalar function $u(x,z)$ satisfies the Helmholtz equation

  $$\Delta u + n^2k^2u = 0. \qquad (5)$$

  where $k = \omega\sqrt{\varepsilon_0\mu_0}$ is the wave number, which is the unknown eigenvalue. From Ampère's law it follows that the magnetic field is given by

  $$\mathbf{H}(x,y,z) = \frac{i}{\mu\omega}\Big(u_z(x,z)\mathbf{e}_x - u_x(x,z)\mathbf{e}_z\Big).$$

- Transverse Electric (TM) waves. The magnetic field is parallel to the y-axis $\mathbf{H}(x,y,z) = u(x,z)\,\mathbf{e}_y$, where the scalar function $u(x,z)$ satisfies the Helmholtz equation

  $$\nabla\cdot(\frac{1}{n^2}\nabla u) + k^2u = 0. \qquad (6)$$

  The electric field is given by

  $$\mathbf{E}(x,y,z) = \frac{1}{i\varepsilon\omega}\Big(u_z(x,z)\mathbf{e}_x - u_x(x,z)\mathbf{e}_z\Big),$$

  which is a consequence of Faraday's law.

In the literature there is an apparent disagreement about which plane is transverse. Here, transverse refers to the plane perpendicular to the direction of the periodic variation, i.e., the xy-plane. This is the convention found in, e.g. [37]. Another possibility is to denote the plane perpendicular to the grooves as the transverse plane. In this case, the terms TE and TM must be switched. This is the choice of. e.g [7].
For both polarization the problem is considerably simpler than the Maxwell problem, since only a scalar equation in two spatial variables must be solved. Moreover, the nullspaces of the Helmholtz operators are finite-dimensional, hence there is no need to incorporate the conditions (3) and (4).
The equations for the TE and TM polarizations can be unified into one, by writing

$$\mathscr{L}u := -\nabla\cdot p\nabla u = k^2qu \qquad (7)$$

where

$$p(x,z) = \begin{cases} 1, & \text{TE case,}\\ n^{-2}(x,z), & \text{TM case,}\end{cases}$$

and

$$q(x,z) = \begin{cases} n^2(x,z), & \text{TE case,}\\ 1, & \text{TM case.}\end{cases}$$

A typical grating structure can consist of several, but finitely many stratified layers and a layer that contains a periodic grating. The refractive index is a periodic function in z

$$n(x,z) = n(x,z+\Lambda)$$

where Λ is the grating period. With the exception of the grating layer, the index in each layer is a piecewise constant function of x only, and constant outside the finite interval $x \in [0,w]$. Usually, the two semi-infinite intervals are referred to as the superstate and substrate and the region in between is called the stack. In many applications the refractive indices in the two semi-infinite layers are the same, in others, they are different. We will see that the latter case complicates matters somewhat. We orient the direction of the x-axis such that the "negative" semi-infinite layer contains the smaller refractive index. A typical geometry is shown in Figure (**1**).
For a piecewise constant index, (7) is understood in the distributional sense. This implies that on the interface of two layers the solution is continuous

$$u^+ = u^-. \qquad (8)$$

The conditions for the derivative in the normal direction $\mathbf{n}$ of the interface depend on the polarization

$$\begin{aligned} \frac{\partial u^+}{\partial\mathbf{n}} &= \frac{\partial u^-}{\partial\mathbf{n}}, && \text{TE case,}\\ \frac{1}{n_+^2}\frac{\partial u^+}{\partial\mathbf{n}} &= \frac{1}{n_-^2}\frac{\partial u^-}{\partial\mathbf{n}}, && \text{TM case.}\end{aligned}$$

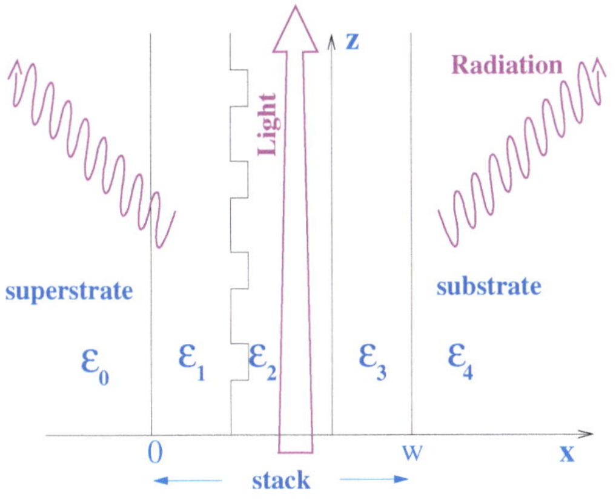

Fig. (**1**): Typical waveguide with grating layer.

3. PLANAR WAVEGUIDES

In a planar waveguide there is no grating layer, thus the refractive index is a function of x only

$$n(x,z) = n(x),$$

where $n(x)$ is a piecewise constant function with a finite number of discontinuities inside the interval $[0,w]$. While planar waveguides lack some of the interesting features of periodic waveguides, they are still helpful for the understanding the spectral properties of open structures: as in the periodic case, planar waveguides can have a set of discrete eigenmodes as well as a continuous spectrum. In addition, both types of waveguides have so-called leaky modes, which are characterized by complex propagation constants.

The periodic waveguide may be regarded as a perturbation of the planar waveguide, where inside the grating region the refractive index is replaced by its average value

$$n_g^2 = \frac{1}{w_g\Lambda}\int_0^{\Lambda}\int_{x_-}^{x_+} n^2(x,z)\,dxdz. \tag{9}$$

Here, w_g is the length of the interval $[x_-,x_+]$ on the x-axis that is occupied by the grating region. This concept is useful for computing the modes of a periodic structure by continuation from the averaged planar structure. In such an approach the modes of the structure with index

$$n_\nu^2(x,z) = \nu n^2(x,z) + (1-\nu)n_g^2 \tag{10}$$

are followed in the interval $\nu \in [0,1]$. This will be discussed in more detail in Section 6.1.

Since the structure is invariant in z-direction, the solutions of (7) have the form

$$u(x,z) = \exp(i\beta z)\Phi(x)$$

hence substitution into (7) leads to

$$\Big(p(x)\phi'(x)\Big)' + k^2 q(x)\phi(x) = \beta^2 p(x)\phi(x), \tag{11}$$

Since $x \in \mathbb{R}$ this is a singular Sturm-Liouville problem, where the propagation constant β is the unknown eigenvalue, and k is a fixed parameter. Of course, one could also revert the roles of β and k and would obtain a similar problem.

A material consisting of $J+1$ layers is completely described by $J+1$ refractive indices $n_0,\ldots,n_J$ and the positions $x_1,\ldots x_J$ of the J interfaces, c.f. Figure (**2**). The layer with index n_0 is the cover and layer n_J is the substrate. The other layers are the interior layers. To simplify notations we shift the origin to the first interface, i.e., $x_1 = 0$ and write $x_J = w$.

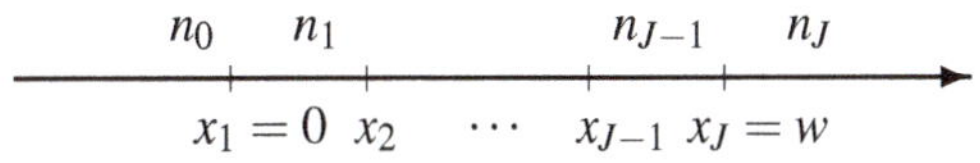

Fig. (**2**): The geometry parameters

In each layer the functions p and q are constant, therefore the solution for both polarizations is a linear combination of left and right going harmonics

$$\phi(x) \in \operatorname{span}\{\exp(\pm i\alpha_j x)\},\ x_j \le x \le x_{j+1}.$$

Here, the transverse wave numbers are

$$\alpha_j = \sqrt{k^2 n_j^2 - \beta^2}, \tag{12}$$

where we adopt the convention that the square root function has a branch cut on the negative real axis. Thus for a real value of β the propagation constant is either positive real or positive imaginary.

It is convenient to write the solution in the j-th layer in terms of the solution of the left endpoint of the layer

$$\begin{aligned}\phi(x) &= \cos(\alpha_j(x-x_j)\phi(x_j) \\ &\quad + \frac{\sin(\alpha_j(x-x_j))}{\alpha_j}\phi'(x_j).\end{aligned} \tag{13}$$

The state vector

$$\vec{\Phi}(x) = \begin{bmatrix}\phi(x)\\ p(x)\phi'(x)\end{bmatrix}$$

is continuous across layer interfaces. Using (13) it follows that the state vectors at the interfaces x_j and x_{j+1} transforms according to the formula

$$\vec{\Phi}(x_{j+1}) = T_j\vec{\Phi}(x_j)$$

where T_j is the matrix

$$T_j = \begin{bmatrix} \cos(\alpha_j w_j) & \sin(\alpha_j w_j)/\tilde{\alpha}_j \\ -\sin(\tilde{\alpha}_j w_j)\tilde{\alpha}_j & \cos(\alpha_j w_j) \end{bmatrix},$$

and

$$\tilde{\alpha}_j = p_j \alpha_j$$

where p_j denotes the value of the function $p(x)$ in the jth layer. The translation matrix is always invertible because the determinant of T_j is unity. The translation from 0 to w is simply the product

$$T = T_{J-1} \cdot \ldots \cdot T_1$$

The matrix T is useful to construct solutions of (11). We will demonstrate this now.

3.1 Continuous Spectrum

The value of α_0 is real if $\beta^2 \leq n_0^2 k^2$. Thus the left and right going harmonics $\exp(\pm i\alpha_0 x)$ are two linearly independent bounded solutions in the left semi-infinite layer $x \leq 0$. Using translation operators, both harmonics can be extended to a solution of (11) on the whole real axis. In the right semi-infinite layer the solution has the form of (13) for $j = J$.

Recall that the orientation of the x-axis is such that $n_0 \leq n_J$, hence the value of α_J is real for the given range of β^2 and thus the two solutions are bounded and are usually called two-sided radiation modes. If $0 < \beta^2 \leq n_0^2 k^2$ the modes propagate in z direction, if $\beta^2 < 0$ the modes are evanescent, that is, they decay in z.

Now consider a value of β^2 in the interval $n_0^2 k^2 < \beta^2 \leq n_J^2 k^2$. Here the value of α_0 is imaginary whereas the value of α_J is real. Hence there is only one linearly independent bounded mode in the left semi-infinite layer, namely $\exp(-i\alpha_0 x)$. Translation through the layers will result in a bounded solution on the whole real axis. These solutions radiate only in the right semi-infinite layer and are therefore called substrate modes.

3.2 Discrete Spectrum

So far, we have found bounded solutions for every value of β^2, thus the interval $\beta^2 \leq n_J^2 k^2$ is the continuous spectrum. The situation is different when $\beta^2 > n_J^2 k^2$, because then the transverse propagation constants in both semi-infinite layers are purely imaginary. In this case there is only one bounded exponential on either side. In general, the bounded solution in the left layer will translate to a combination of the bounded and the unbounded exponential on the right side, unless the coefficient of the unbounded exponential vanishes. If that happens, the mode is exponentially decaying in both semi-infinite layers and is called a guided mode.

In the following we derive a condition to characterize such a mode. Suppose that the mode is normalized to satisfy $\phi(0) = 1$ then

$$\phi(x) = \exp(-i\alpha_0 x) \quad \text{for} \quad x \leq 0$$

and

$$\phi(x) = \phi(x_J)\exp(i\alpha_J(x - x_J)) \quad \text{for} \quad x \geq x_J$$

In each layer of the stack a guided mode is again of the form (13). Using translation operators it follows that the state vectors at the points $x = 0$ and $x = w$ must satisfy the relationship

$$T \begin{bmatrix} 1 \\ -i\tilde{\alpha}_0 \end{bmatrix} = \phi(w) \begin{bmatrix} 1 \\ i\tilde{\alpha}_J \end{bmatrix}$$

where $\tilde{\alpha}_j = p_j \alpha_j$ and p_0 and p_J are the values of $p(x)$ in the semi-infinite layers. Note that the variables T, $\tilde{\alpha}_0$ and $\tilde{\alpha}_J$ are functions of β. Simple algebra shows that the above relationship is equivalent to finding the roots of the characteristic function

$$F(\beta) = t_{12}\tilde{\alpha}_0\tilde{\alpha}_J - t_{21} + i\Big(t_{11}\tilde{\alpha}_J + t_{22}\tilde{\alpha}_0\Big) = 0, \quad (14)$$

where the t_{ij} are the coefficients of the matrix T.

Since a guided mode has exponential decay it is a function $L^2(\mathbb{R})$. By multiplying (11) with ϕ and integrating by parts it follows that

$$-\int_{\mathbb{R}} p(\phi')^2 + k^2 \int_{\mathbb{R}} q\phi^2 = \beta^2 \int_{\mathbb{R}} p\phi^2$$

from which it can be concluded that

$$\left(\frac{\beta}{k}\right)^2 \leq \frac{\int_{\mathbb{R}} q\phi^2}{\int_{\mathbb{R}} p\phi^2}.$$

For both polarizations the right hand side can be estimated by

$$n_M^2 := \max_{x \in \mathbb{R}} n^2(x).$$

Hence there are guided modes only in the interval $n_J^2 k^2 \leq \beta^2 \leq n_M^2 k^2$. We will see in Section 3.3 that their number is finite. The types of solutions to (11) are summarized in Figure (3).

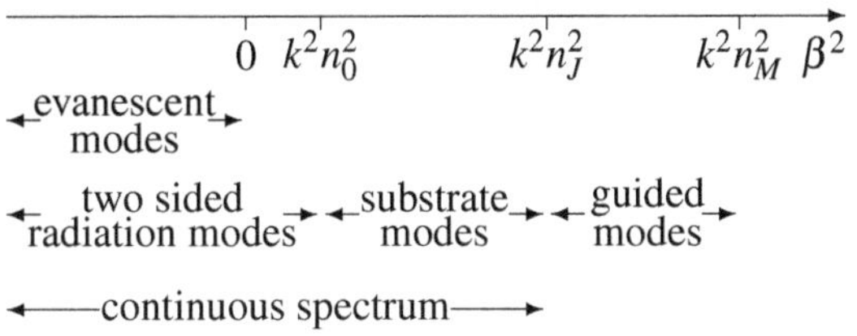

Fig. (**3**): Classification of the spectrum of a planar waveguide.

3.3 Leaky Modes

In addition to real solutions in the interval $n_J^2k^2 < \beta^2 < n_M^2k^2$ the characteristic equation (14) can also have complex roots. A complex value of β implies that the transverse wave numbers α_0 and α_J are in the complex plane and thus the branch choice of the square root in (12) becomes an important issue. The characteristic function is four-valued depending on the signs of the square root. Note that the sign choice of the interior α_j's is irrelevant because the translation operators T_j are even functions of α_j.

The solutions that have physical significance are those that radiate energy away from the stack and are known as leaky modes, see [13]. Although leaky waves have infinite energy they are physically significant and have been verified experimentally in finite regions of the waveguide [49]. The leaky modes form a discrete set of expansion functions in the stack and can represent field solutions in this region [35]. Leaky-wave analysis has the advantage that in the representation of a field the superposition integral of radiation modes is replaced by a discrete sum of leaky modes. In practice, only a few modes are necessary to obtain good approximations. This type of analysis has been applied in several photonics applications, especially for waveguide transitions. See, e.g. [33] and the references cited therein.

An elegant way to deal with the four-valuedness of the characteristic function is to make the change of variables

$$z = \frac{1}{2i}\left(\sqrt{k^2n_0^2-\beta^2}+\sqrt{k^2n_J^2-\beta^2}\right). \tag{15}$$

which was suggested in the paper [41]. Simple algebra shows that

$$\alpha_0 = i\left(z+\frac{\delta^2}{4z}\right) \tag{16}$$

$$\alpha_J = i\left(z-\frac{\delta^2}{4z}\right) \tag{17}$$

where

$$\delta^2 = k^2(n_J^2-n_0^2). \tag{18}$$

Because of $\alpha_j^2 = k^2(n_j^2-n_0^2)+\alpha_0^2$ all transverse wave numbers are analytic functions of z when $z \neq 0$. Furthermore, the coefficients of the translation matrices are analytic functions of α_j and therefore they are also analytic in $z \neq 0$. Thus the characteristic function is an analytic function of z in $\mathbb{C}\setminus\{0\}$ and has essential singularities in $z=0$ and $z=\infty$. If $n_0 = n_J$ then $\delta = 0$ and the transformation is simply

$$\alpha_0 = \alpha_J = iz$$

and the characteristic function is an entire function with an essential singularity at $z=\infty$.

Transformation (15) maps the interval $\beta^2 \in [n_J^2k^2, n_M^2k^2]$ to a finite interval on the positive real z-axis that does not contain the origin when $\delta \neq 0$. Since the function is analytic, there is at most a finite number of roots in this interval.

Another result about the z-roots that can be obtained from the Great Picard Theorem of complex analysis. Because of the essential singularities it follows that characteristic function either has no root at all, or an infinite number of roots that accumulate at infinity and, in addition, if $\delta \neq 0$, accumulate in the origin. If the frequency k is large enough, and $n_M > \max\{n_0, n_J\}$ it well-known that there are guided modes and hence there is an infinite number of roots. The spectrum in this situation is illustrated in Figure (**4**). The situation depicted here appears to be the generic case. However, the author of this chapter is not aware of results that guarantee existence of roots under more general conditions.

So far, we have considered the eigenvalue β for a given value of the frequency k. In the analysis of a waveguide one is usually interested how the l-th eigenvalue changes with the frequency k. The function $\beta_l(k)$ is called the dispersion relation and describes the way in which the propagation speed and attenuation varies with the frequency of the mode.

As the frequency increases, the number of guided modes increases. The smallest frequency for which a given mode is guided is called the cut-off frequency for that mode. This condition is characterized by a bifurcation of two dispersion curves in the complex plane [22].

We use the guide

$$\begin{aligned} n_0 &= 1.0, \\ n_1 &= n_4 = 1.66, \\ n_2 &= 1.53, \\ n_3 &= 1.60, \end{aligned}$$

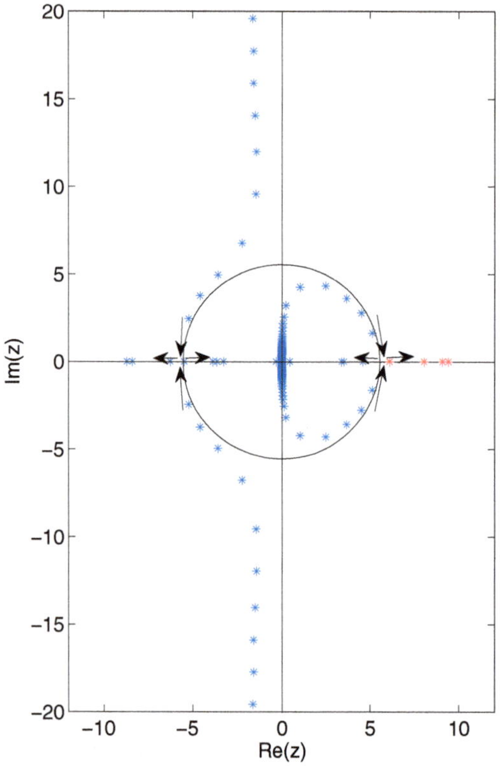

Fig. (**4**): Roots of the characteristic equation in the complex z-plane for $k = 9.929$. Guided modes are colored in red. The circle has radius $\delta/2$. Arrows indicate the motion of the roots with increasing frequency.

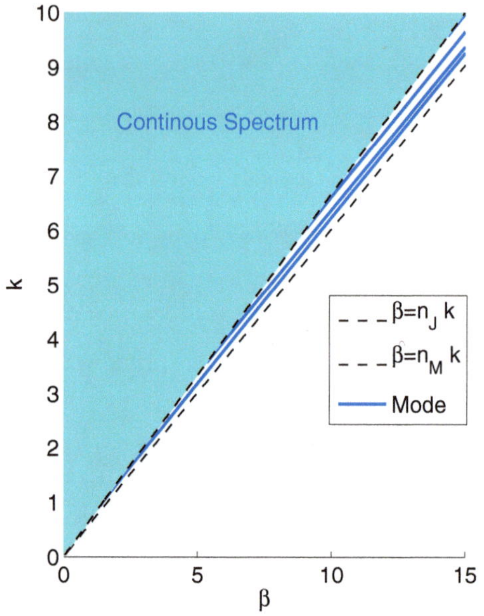

Fig. (**5**): Guided modes as a function of the frequency for the planar waveguide in Figure (**4**).

$$\begin{aligned} n_5 &= 1.66, \\ n_6 &= 1.5, \\ w_j &= 0.5, \quad 1 \le j \le 5, \end{aligned}$$

to illustrate the dependence of the eigenmodes on the frequency. This structure appears frequently in the literature [13]. The real dispersion curves are shown in (**5**). The dispersion of the leaky modes are best visualized in the z-plane. As k increases, the modes emanate from the origin in complex conjugate pairs until they bifurcate into two real solutions at $z = \delta/2$. Only the solution $z > \delta/2$ represents a guided solution. The other solution is non-physical as it increases exponentially in one layer. At the same time, there are complex conjugate pairs of solutions that move near the imaginary axis towards the point $z = -\delta/2$. Here the pairs bifurcate into another set of non-physical real solutions. The motion of the roots in the complex planes is indicated by arrows in Figure (**4**).

3.4 Variational Formulation

The previous discussion was based on deriving a characteristic equation of (11) using translation operators. A different way to treat the problem of finding the guided and leaky modes of a planar multilayer waveguide is the variational approach. Unlike the characteristic function method, this approach generalizes to arbitrary index profiles as long as the index varies only in a finite interval. Furthermore, we will show in the following section that numerical methods based on a variational formulation are more stable than algorithms that find the roots of a characteristic function.

Since guided and leaky modes are defined by only one complex exponential in the semi-infinite layers they solve (11) in the interval $[0, w]$ with Robin-type boundary conditions at the endpoints. Thus

$$\begin{aligned} \left(p\phi'\right)' + k^2 q\phi &= \beta^2 p\phi, && (19) \\ p_1\phi'(0) + ip_0\alpha_0\phi(0) &= 0, && (20) \\ p_{J-1}\phi'(w) - ip_J\alpha_J\phi(w) &= 0. && (21) \end{aligned}$$

Since the transverse wave numbers α_0 and α_J depend in a nonlinear way on the eigenvalue β^2 the system (19-21) does not constitute a regular Sturm-Liouville problem.

The actual nature of this problem becomes more apparent if the unknown β^2 is replaced by the z-variable in (15). It was recently discovered [44] that with this change of variables the variational form of (19-21) transforms into a quartic eigenvalue problem.

To derive this result we multiply (19) by a test function ψ and integrate by parts. This leads to

$$\begin{aligned}(\psi', p\phi') - k^2(\psi, q\phi) + \beta^2(\psi, p\phi) \\ -p_{J-1}\bar{\psi}(w)\phi'(w) + p_1\bar{\psi}(0)\phi'(0) = 0.\end{aligned}$$

Here, $(\cdot,\cdot)$ denotes the usual complex $L^2[0,w]$-inner product. Incorporating the boundary conditions (20) and (21) leads to

$$\begin{aligned}(\psi', p\phi') - k^2(\psi, q\phi) + \beta^2(\psi, p\phi) \\ -i\alpha_J\bar{\psi}(w)\phi(w) - i\alpha_0\bar{\psi}(0)\phi(0) = 0. \quad (22)\end{aligned}$$

The problem at hand is to find β and $\phi \neq 0$ such that the above equation is satisfied. Since α_0, α_J depend on β (22) is a nonlinear eigenvalue problem in β.

To obtain a formulation in the z variable, note that it follows from (12), (16) and (17) that

$$\begin{aligned}\beta^2 &= \frac{k^2}{2}(n_0^2 + n_J^2) - \frac{1}{2}(\alpha_0^2 + \alpha_J^2) \\ &= \frac{k^2}{2}(n_0^2 + n_J^2) + \left(z^2 + \frac{\delta^4}{16z^2}\right).\end{aligned}$$

Thus (22) is transformed to

$$\begin{aligned}&(\psi', p\phi') - k^2(\psi, q\phi) + \frac{k^2}{2}(n_0^2 + n_J^2)(\psi, p\phi) \\ &+\left(z - \frac{\delta^2}{4z}\right)\bar{\psi}(w)\phi(w) + \left(z + \frac{\delta^2}{4z}\right)\bar{\psi}(0)\phi(0) \\ &+\left(z^2 + \frac{\delta^4}{16z^2}\right)(\psi, p\phi) = 0.\end{aligned}$$

We introduce the bilinear forms

$$\begin{aligned}a(\psi,\phi) &= (\psi', p\phi') - k^2(\psi, q\phi) \\ &\quad + \frac{k^2}{2}(n_0^2 + n_J^2)(\psi, p\phi), \\ a^{\pm}(\psi,\phi) &= \bar{\psi}(0)\phi(0) \pm \bar{\psi}(w)\phi(w).\end{aligned}$$

After multiplying with z^2 and sorting out equal powers of z we obtain

$$\begin{aligned}&\frac{\delta^4}{16}(\psi, p\phi) - z\frac{\delta^2}{4}a^-(\psi,\phi) + z^2 a(\psi,\phi) \\ &+z^3 a^+(\psi,\phi) + z^4(\psi, p\phi) = 0, \quad (23)\end{aligned}$$

which is a quartic eigenvalue problem. In the case $\delta = 0$ the above simplifies to the quadratic eigenvalue problem

$$a(\psi,\phi) + za^+(\psi,\phi) + z^2(\psi,\phi) = 0. \quad (24)$$

3.5 Computation of Modes

The most obvious approach to computing the guided and leaky modes of a planar waveguide is to find the roots of the characteristic equation (14) with Newton's method. Because of the square root in (12) this function has two branch cuts in the complex plane. If one of the roots is near a branch cut, then iterates tend to jump across branch cuts without converging. To avoid such difficulties one can use the change of variables (15).

In applications one is usually interested in how many guided modes a given structure has or which the dominant leaky modes are. Therefore it is important to have a numerical method which is capable to find *all* roots at least in a specified range. Employing Newton's method with different initial guesses is not a very satisfactory approach as it would be hard decide when to stop searching for new roots. A better approach is to use the argument principle of complex analysis, which relates the number of roots in a domain to a contour integral over its boundary. Delves and Lyness [15] propose a method which is based on this principle and guarantees all roots of an analytic function in a given region of the complex plane. Variations of this idea with application to dielectric waveguides are discussed in [2, 3, 12] and [42].

Another approach to obtain all roots in a given range is to use a continuation method [1]. Petracek and Singh were the first to apply this technique for planar dielectric waveguides [30]. Their idea is to truncate the structure in the semi-infinite layers and to impose homogeneous Dirichlet boundary condition at the endpoints. The truncated problem is a regular Sturm-Liouville problem which can be formulated as a standard eigenvalue problem. Its solutions are the initial points that are followed by a in a smooth transition from the closed to the open structure. A more detailed description of continuation methods will be given later in Section 6.1 in the context of periodic waveguides.

Methods that compute the roots of (14) are capable of finding the propagation constants of certain structures, but they have their limitations because the exponential scaling of the characteristic function. This property causes significant cancellation errors when evaluating the function in floating point arithmetic.

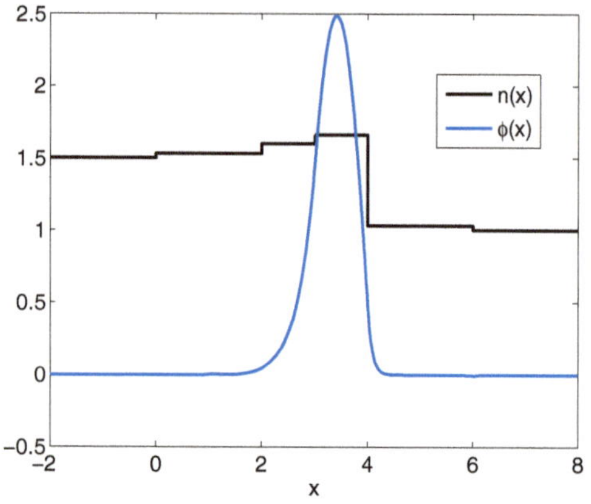

Fig. (6): Refractive index and dominant guided mode for $k = 9$ and $\beta/k = 1.63986$.

To illustrate how this happens consider the waveguide whose refractive index and dominant guided mode is shown in Figure (6). When $k = 9$ the mode has propagation constant $\beta/k \approx 1.64$. Hence the transverse wave number α_4 is purely imaginary in layer 4. The translation operator T_4 has eigenvalues $\exp(\pm i\alpha_4 w_4)$, and the eigenvectors $\Psi_\pm$ are state vectors that are either magnified or reduced when translated across this layer. The state vector of the eigenmode is a linear combination of $\Psi_\pm$, but since the mode decays in layer 4, the weight of Ψ_- must be very small. However, in the presence of floating point errors, this weight can be increased with the result that the function value of the numerically computed characteristic function is significantly different from zero. This effect becomes more noticeable when the width of the layer or the frequency is increased. This is illustrated in Figure (7) which shows the numerically computed characteristic function in the interval of β/k where the guided modes occur. For $k = 5$ the roots are clearly visible, but when $k = 7$ or $k = 9$, the roots disappear in numerical noise.

A different approach that does not rely on finding the roots of the characteristic function is the finite element method. Here, one begins with a partition of the interval $[0, w]$

$$0 = \xi_0 < \xi_1 < \ldots < \xi_N = w$$

where the set of node points $\{\xi_j\}$ contains the layer interfaces. The mesh width is $h = \max_j(\xi_{j+1} - \xi_j)$, and the finite element space S_h consists of functions that are piecewise polynomial on the partition and globally continuous. The finite element method seeks the solution of the variational formulation in

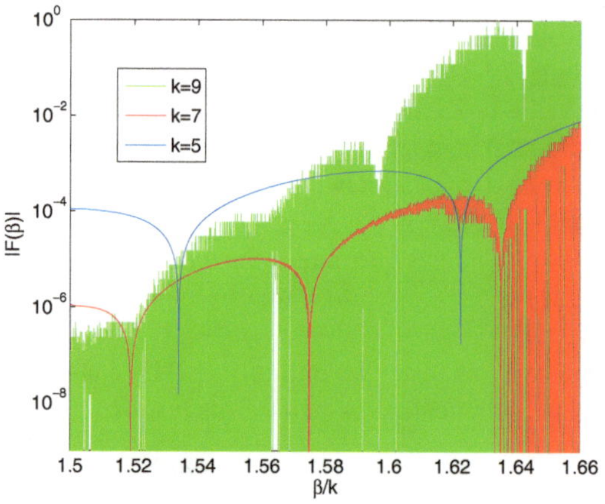

Fig. (7): Modulus of the characteristic function computed in double precision arithmetic of the structure with index profile shown in Figure (6). Three frequencies, TE polarization.

the space S_h. Since this space has finite dimension, the problem is reduced to a matrix problem.

If we work with the variational form in the β-variable (22), the resulting task is to find the values of β such that

$$A(\beta) \quad \text{is singular.} \tag{25}$$

This is a highly non-linear eigenvalue problem and was considered in a slightly more general setting in [50]. This approach avoids the numerical instabilities associated with the evaluation of the characteristic function, but does not guarantee that a complete set of solutions is found, since the solutions depend on the quality of the initial guess.

The alternative is to work with the variational form in the z-variable(23), in which case one obtains the quartic eigenvalue problem: Find z such that

$$A_0 + zA_1 + z^2A_2 + z^3A_3 + z^4A_4 \quad \text{is singular.}$$

When the finite elements are the piecewise linear hat functions, the matrices A_0, A_2 and A_4 are tridiagonal, and

$$A_1 = \frac{\delta^2}{4}\text{diag}[1,0,\ldots,0,-1],$$
$$A_3 = \frac{\delta^2}{4}\text{diag}[1,0,\ldots,0,1].$$

There are several numerical methods to treat polynomial eigenvalue problems, they are reviewed in [48]. The standard approach is to solve the equivalent generalized eigenvalue problem: Find z such that

$$\begin{bmatrix} & -I & & \\ & & -I & \\ & & & -I \\ A_0 & A_1 & A_2 & A_3 \end{bmatrix} + z \begin{bmatrix} I & & & \\ & I & & \\ & & I & \\ & & & A_4 \end{bmatrix}$$

is singular.

Thus a generalized eigenvalue problem of size $4N$ must be solved. For the latter the direct method, based on the QZ-factorization can be used. However, since the companion matrix is sparse, it is preferable to use Arnoldi methods, which are implemented in ARPACK.

The advantage of solving the quartic eigenvalue problem is that all dominant eigenvalues can be found by solving one eigenvalue problem. Iterative methods for (25) can only provide one eigenvalue at a time. Thus the numerical effort to compute several modes can be much higher and there is no guarantee that no modes have been missed.

4. PERIODIC WAVEGUIDES

We now turn our attention to waveguides that constitute a periodic perturbation of a multilayer planar structure. The discrete spectrum of the periodic structure is the perturbation of the spectrum of the planar structure.

However, because of coherent scattering, a guided mode can be changed into a leaky mode as the frequency is increased. Such a mode radiates energy away from the grating region into free space. This effect is equivalent to the occurrence of stop-bands in the dispersion relation and is characterized by a complex propagation constant.

In this section we only give a minimal description of the spectral properties for the periodic case to make our presentation self consistent. For more detail and mathematical rigour we refer to Stephen Shipman's chapter in the same book [40].

4.1 Floquet's Theorem

The modes of a grating structure such as the one shown in Figure (**1**) are the solutions of the eigenvalue problem (7), which is posed in the xz-plane. The refractive index has discrete translational symmetry in z-direction, that is, it remains unchanged under translations of an integer multiple of the grating period Λ. This allows us to consider the eigenvalue problem on the fundamental domain of the symmetry, which is one period of the grating.

The mathematical result behind this reduction is known as Floquet's theorem, which states that the eigenfunctions are quasi-periodic in the z-direction. This result can be seen by applying the Floquet transform, which, for a z-periodic medium is given by

$$u_\beta(x,z) := \mathscr{U}u(x,z,\beta) = \sum_{l\in\mathbb{Z}} u(x,z-\Lambda l)\exp(i\beta\Lambda l)$$

The transformed variable β is the quasimomentum, or propagation constant. It is obvious from the definition that the inverse transform is

$$u(x,z) = \frac{\Lambda}{2\pi}\int_{-\pi/\Lambda}^{\pi/\Lambda} u_\beta(x,z)d\beta.$$

Furthermore,

$$\begin{aligned} u_\beta(x,z+\Lambda) &= \exp(i\beta\Lambda)u_\beta(x,z), && (26)\\ u_{\beta+2\pi/\Lambda}(x,z) &= u_\beta(x,z). && (27) \end{aligned}$$

Moreover, the Floquet transform commutes with the operator $\mathscr{L}$ defined in (7), hence it satisfies the differential equation

$$\begin{aligned} \mathscr{L}u_\beta &= k^2 q u_\beta \quad \text{in } \Omega_\Lambda && (28)\\ u_\beta\big|_{\Gamma_\Lambda} &= e^{i\beta\Lambda}\, u_\beta\big|_{\Gamma_0} && (29)\\ \left.\frac{\partial u_\beta}{\partial z}\right|_{\Gamma_\Lambda} &= e^{i\beta\Lambda}\left.\frac{\partial u_\beta}{\partial z}\right|_{\Gamma_0} && (30) \end{aligned}$$

Here $\Omega_\Lambda = \mathbb{R}\times[0,\Lambda]$ is an infinite strip of width Λ and $\Gamma_0 = \mathbb{R}\times\{0\}$ and $\Gamma_\Lambda = \mathbb{R}\times\{\Lambda\}$ are its boundaries. In the following we will drop the subscript β when it is clear that a solution of (28-30) is meant.

Because of (27) we only have to consider propagation constants in the so-called Brillouin zone

$$-\frac{\pi}{\Lambda} < \mathrm{Re}\beta \leq \frac{\pi}{\Lambda}. \qquad (31)$$

Since the fundamental domain of the symmetry is infinite, it is possible that both the propagation constant or the frequency are complex. Thus the dispersion relation $k(\beta)$ of a mode is a complex function. Strictly speaking, such solutions are not part of the spectrum of the operator $\mathscr{L}$ as they are unbounded in the xz-plane. However, they are of great physical interest, as they represent waves that leak out energy away from the grating region as they travel in z-direction.

There are two equivalent viewpoints to characterize the spectrum of $\mathscr{L}$.

1. Fix the propagation constant $\beta \in \mathbb{R}$ and find the possibly complex values of k such that (28-30) has non-trivial solutions.

2. Fix the wavenumber $k \in \mathbb{R}$ and find the corresponding values of β, which again can be real or complex.

For theoretical purposes it more convenient to work with the first alternative as this viewpoint leads to a more standard eigenvalue formulation. Since problem (28-30) is posed in an infinite domain, the spectrum for a given β usually consists of a discrete and a continuous part. The real discrete eigensolutions are the guided modes of the structure, whereas the continuous spectrum consist of solutions of a scattering problem with an incoming quasi-periodic plane wave. The eigenvalue problem for the guided modes may also have an infinite number of discrete complex solutions.
The second viewpoint is often preferred in the engineering literature as the solutions with real k and complex β have the physical interpretation of a leaky mode, provided that certain radiation conditions are satisfied.

4.2 DtN Maps

For both theoretical and numerical purposes it is necessary to reduce the eigenvalue problem on the infinite strip to an equivalent problem on a finite domain. To that end consider two x-coordinates outside the grating region $x_\pm$ and define the subregions

$$\begin{aligned} \Omega_- &= (-\infty, x_-) \times (0, \Lambda), \\ \Omega_0 &= (x_-, x_+) \times (0, \Lambda), \\ \Omega_+ &= (x_+, \infty) \times (0, \Lambda) \end{aligned}$$

and the interfaces

$$\Gamma_\pm = \{x_\pm\} \times [0, \Lambda].$$

In the exterior regions $\Omega_\pm$ the refractive index is a piecewise constant function of x and hence it is possible to write down the solution and the Dirichlet-to-Neumann (DtN) operator in closed form.
For numerical purposes it desirable to make Ω_0 as small as possible, which means that $x_\pm$ are the endpoints of the grating region. However, to keep the discussion simple, we begin by setting $x_- = x_0 = 0$ and $x_+ = x_J = w$, so that the exterior regions have a constant refractive index, $n_- = n_0$ and $n_+ = n_J$, respectively. The modifications for allowing uniform layers in $\Omega_\pm$ are discussed later.
Because of quasiperiodicity, the Fourier expansion of a solution in an exterior region $\Omega_\pm$ is

$$u(x,z) = \sum_{l \in \mathbb{Z}} u_l^\pm \exp\left(\pm i \alpha_l^\pm x + i \beta_l z \right), \qquad (32)$$

where

$$\begin{aligned} \beta_l &= \frac{2\pi l}{\Lambda} + \beta, &(33) \\ \alpha_l^\pm &= \left(n_\pm^2 k^2 - \beta_l^2 \right)^{\frac{1}{2}} &(34) \end{aligned}$$

and $n_\pm$ is the refractive index in $\Omega_\pm$. The coefficients in (32) are

$$u_l^\pm = \frac{1}{\Lambda} \int_0^\Lambda u(x_\pm, z) \exp\left(-i\beta_l z \right) dz.$$

For the uniqueness in the expansion (32) the appropriate branch choice of the square root in (34) must be determined. When β is real, the standard definition of the square root, with $\sqrt{-1} = i$ and the branch cut on the negative real axis, ensures that solutions are either outgoing or exponentially decaying in the exterior regions. The latter case happens for an infinite number of l's.
To determine the branch choice for complex β one stipulates that the transverse wavenumber depends continuously on β [23]. This is based on the assumption that a guided mode of a planar waveguide undergoes small changes when a small periodic perturbation is introduced.
To clarify this concept consider the function

$$\zeta \mapsto \left(\kappa^2 - \zeta^2 \right)^{\frac{1}{2}}$$

for some $\kappa > 0$ in the complex plane. This function has two branch cuts at $(-\infty, -\kappa)$ and (κ, ∞). An analytic continuation of the function can be defined on a two-sheeted Riemann surface, with the top sheet characterized by the positive square root and the bottom sheet by the negative square root. Along the branch cut (κ, ∞) the first quadrant of the top sheet and the forth quadrant of the bottom sheet are connected. Furthermore, the forth quadrant of the top and first quadrant of the bottom are connected. The connection along the other branch cut is similar.
The DtN-map is an infinite-valued function because each $\alpha_l^\pm$ possesses one such Riemann surface. The appropriate sheet in the presence of a periodic perturbation is the neighborhood of the real axis on the top sheet, because this the choice for a real β in the planar case. This is illustrated in Figure (**8**).
Since in (33) the imaginary part of β_l is the same for all $l \in \mathbb{Z}$, all terms in expansion (32) are either exponentially increasing or decreasing in z-direction. If we pick a decreasing mode ($\operatorname{Im}\beta > 0$), then it can be seen from Figure (**8**) that the positive square root must be selected when $\operatorname{Re}\beta_l < \kappa$ and the negative square root must be selected when $\operatorname{Re}\beta_l > \kappa$.

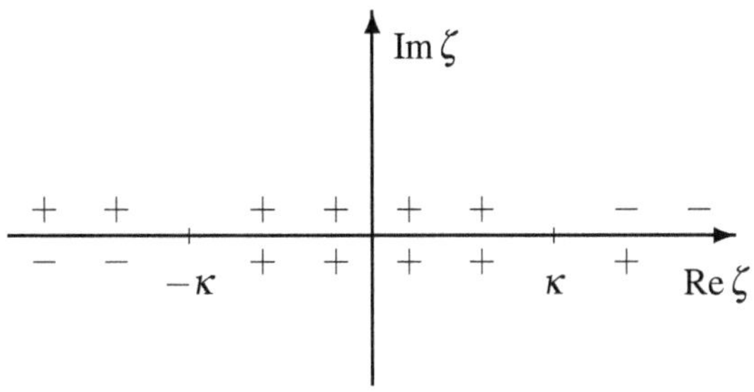

Fig. (**8**): The neighborhood of the top real axis on the Riemann surface for $\zeta \mapsto (\kappa^2 - \zeta^2)^{\frac{1}{2}}$. Plus signs indicate the top sheet, minus signs the bottom sheet.

Thus the solution $u(x,z)$ in the exterior regions is composed of three types of waves, depending on the behavior in the semi-infinite regions:

I: $\operatorname{Re}\beta_l \in (-\infty, 0) \Rightarrow$ decreasing, outgoing,
II: $\operatorname{Re}\beta_l \in (0, n_\pm k) \Rightarrow$ increasing, outgoing,
III: $\operatorname{Re}\beta_l \in (n_\pm k, \infty) \Rightarrow$ decreasing, incoming.

Case I and III occurs for an infinite number of $l \in \mathbb{Z}$, whereas Case II only finitely many times or not at all in one or both of the exterior regions.
With the appropriate branch choice clarified, the exterior DtN map is uniquely defined for any function on the interfaces $\Gamma_\pm$ by

$$u(z) = \sum_{l\in\mathbb{Z}} u_l^\pm \exp(i\beta_l z) \Rightarrow$$
$$T_\pm^e(\beta,k)u(z) = \sum_{l\in\mathbb{Z}} i\alpha_l^\pm u_l^\pm \exp(i\beta_l z). \quad (35)$$

We now address the modifications when there are several uniform layers in the exterior regions. We limit ourselves to Ω_+ as the treatment of Ω_- completely analogous. The expansion of u and its derivative is for $(x,z) \in \Omega_+$

$$\begin{aligned} u(x,z) &= \sum_{l\in\mathbb{Z}} u_l^\pm \phi_l(x)\exp(i\beta_l z), \\ \frac{\partial u}{\partial x}(x,z) &= \sum_{l\in\mathbb{Z}} u_l^\pm \phi_l'(x)\exp(i\beta_l z) \end{aligned}$$

where the functions ϕ_l satisfy

$$\begin{aligned} (p\phi_l')' + (qk^2 - p\beta^2)\phi_l &= 0, \quad \text{in } (x^+,\infty) \\ \phi_l(x_+) &= 1. \end{aligned}$$

Just as in the planar case, the boundary value problem above can be solved using translation operators. The state vector at $x = x^+$ is given by

$$\begin{bmatrix} \phi_l(x_+) \\ p(x_+)\phi_l'(x_+) \end{bmatrix} = T_{l+}^{-1} \begin{bmatrix} 1 \\ i\tilde{\alpha}_{J,l} \end{bmatrix}$$

where T_{l+} is the product of translation operators from x_+ to x_J. Then the coefficients in DtN operator of (35) is given by

$$\alpha_l^+ = \frac{\phi_l'(x_+)}{\phi_l(x_+)}. \quad (36)$$

In Section we have seen that the evaluation of translation operators can suffer from numerical instabilities. In this case it is better to solve the two-point boundary value problem

$$\begin{aligned} (p\phi_l')' + (qk^2 - p\beta^2)\phi_l &= 0, \quad \text{in } (x^+, x_J), \\ \phi_l(x_+) &= 1, \\ \phi_l'(x_J) - i\tilde{\alpha}_{J,l}\phi_l(x_J) &= 0 \end{aligned}$$

using a finite-element discretization.

4.3 The Spectrum of the Periodic Case

Just like we did for planar waveguides, we begin with the continuous spectrum of (28-30). These modes consist of scattering solutions of plane wave approaching the grating region from one of the semi-infinite layers. To simplify notations, we only consider a plane wave approaching from Ω_+, in which case the incoming field is

$$u^{\text{inc}}(x,z) = \exp(-i\alpha x + i\beta z). \quad (37)$$

Here, k and β are real and the transverse wavenumber is given by

$$\alpha = \sqrt{n_J^2 k^2 - \beta^2}.$$

Because of the Floquet theorem, we can limit ourselves to values of β in the Brillouin zone. Furthermore, by the incoming wave condition α must be non-negative, thus the restrictions

$$|\beta| \le n_J k, \qquad |\beta| \le \frac{\pi}{\Lambda} \quad (38)$$

must hold for any incoming field.
In Ω_+, the total field is the superposition of the incoming and the reflected field where both fields are quasi-periodic. The reflected field u^{ref} solves the Helmholtz equation with the outgoing radiation condition enforced by the branch choice in (34). On the

interface Γ_+, the total field u satisfies

$$\begin{aligned} p^i \frac{\partial u^i}{\partial n} &= p^e \left\{ \frac{\partial u^{\text{inc}}}{\partial n} + \frac{\partial u^{\text{ref}}}{\partial n} \right\} \\ &= p^e \left\{ \frac{\partial u^{\text{inc}}}{\partial n} + T_+^e(\beta,k)\left(u^i - u^{\text{inc}}\right) \right\} \end{aligned}$$

Here, the superscript i and e denote a limit approached from the interior or the exterior of Ω_0. With the incoming field given by (37), the last condition simplifies to

$$p^i \frac{\partial u^i}{\partial n} - p^e T_+^e(\beta,k)u(z) = 2p^e i\alpha \exp(i\beta z). \quad (39)$$

In Ω_-, the total field is the transmitted wave. By continuity of u and $p\frac{\partial u}{\partial n}$ it follows that on Γ_-

$$p^i \frac{\partial u^i}{\partial n} - p^e T_-^e(\beta,k)u^i = 0 \quad (40)$$

holds. Thus the total field in Ω_0 must satisfy the boundary value problem

$$\begin{aligned} \mathscr{L}u &= k^2 qu \quad \text{in } \Omega_0, & (41) \\ u|_{\Gamma_\Lambda} &= e^{i\beta\Lambda}\, u|_{\Gamma_0}, & (42) \\ \left.\frac{\partial u}{\partial z}\right|_{\Gamma_\Lambda} &= e^{i\beta\Lambda} \left.\frac{\partial u}{\partial z}\right|_{\Gamma_0}, & (43) \\ p^i \left.\frac{\partial u}{\partial n}\right|_{\Gamma_\pm} &= p^e T^\pm(\beta,k)\, u|_{\Gamma_\pm} + f_\pm. & (44) \end{aligned}$$

Here $f_\pm$ denote the right hand sides in (39) and (40). Once a solution of (41-44) has been found, it can be extended to a solution of (28-30) by expansion (32). The existence of solutions to the scattering problem, and therefore the nature of the continuous spectrum of (28-30) is linked to the solvability of (41-44). These questions can be settled by formulating the partial differential equation in variational form.
Before we do that we recall the following spaces of quasiperiodic functions.

1. $C_\beta^\infty(\mathbb{R}^2)$ is the space of C^∞-functions which satisfy (26) and vanish for large $|z|$.
2. $C_\beta^\infty(\Omega_0)$ consists of restrictions of functions in $C_\beta^\infty(\mathbb{R}^2)$ to Ω_0.
3. $H_\beta^1(\Omega_0)$ is the closure of $C_\beta^\infty(\Omega_0)$ in the H^1-norm.
4. On the interfaces $\Gamma_\pm$ the spaces $H_\beta^s(\Gamma_\pm)$ are defined by the Fourier transform

$$H_\beta^s(\Gamma_\pm) = \left\{ \sum_{l\in\mathbb{Z}} v_l e^{i\beta_l z} : \sum_{l\in\mathbb{Z}} \left(1+\beta_l^2\right)^s |v_l|^2 < \infty \right\}$$

The space $H_\beta^{\frac{1}{2}}(\Gamma_\pm)$ consists exactly of the traces of all functions of $H_\beta^1(\Omega_0)$ on $\Gamma_\pm$. Furthermore, the DtN operators $T_\pm^e : H_\beta^{\frac{1}{2}}(\Gamma_\pm) \to H_\beta^{-\frac{1}{2}}(\Gamma_\pm)$ are continuous.
The variational form is easily obtained by multiplying (41-44) with $v \in H_\beta^1(\Omega_0)$ and applying the divergence theorem. Note that the contribution of the boundaries Γ_0 and Γ_Λ cancel because of the quasi-periodicity condition.
The resulting problem is to find k and $u \in H_\beta^1(\Omega_0)$ such that

$$a(\beta,k;v,u) = (v,f) \quad (45)$$

holds for all $v \in H_\beta^1(\Omega_0)$. Here, $\beta \in \mathbb{R}$ is given and

$$\begin{aligned} a(\beta,k;v,u) &= \int_{\Omega_0} p\nabla\bar{v}\cdot\nabla u - k^2 \int_{\Omega_0} q\bar{v}u \\ &\quad - \int_{\Gamma_\pm} p\bar{v}T_\pm^e(\beta,k)u. \end{aligned}$$

The following existence and uniqueness results have been established in [7]. For more related results see also [5, 20, 19].

1. Problem (45) has at least one solution $u \in H_\beta^1(\Omega_0)$ and the solution set is a finite dimensional affine subspace.
2. If the solution of (45) is non-unique then k is called a singular frequency. This condition is equivalent to the existence of a non-trivial solution of the homogeneous equation. For such a solution the transverse wave numbers $\alpha_l^\pm$ in (32) are either zero or positive imaginary.
3. The singular frequencies form at most a countable sequence with no accumulation point.

By the first result, there is a nontrivial solution of (28-30) corresponding to any plane wave approaching the grating from Ω_+. Likewise, there is a solution for any plane wave approaching from Ω_-. Thus the continuous spectrum is the region

$$k \geq \frac{|\beta|}{\max(n_0,n_J)} = \frac{|\beta|}{n_J} \quad (46)$$

which is called light cone.
If $kn_0 \leq \beta \leq kn_J$ then the transverse wave number α_0^+ is real whereas all α_l^- are purely imaginary. Modes in this interval are substrate radiation modes. Below the light cone $0 \leq \beta \leq n_0 k$ there is no incident field, because it would violate condition (37). By the second and third result, it is possible that this range contains a finite number of guided modes.

Furthermore, the stated existence results do not exclude the possibility of further real frequencies above the light cone, although this appears to be an exceptional case. The reason is that (45) is nonhermitian and nonlinear because of the appearance of the DtN operators.

The derivation of variational formulation (45) is based on the assumption that β is real. In order to find leaky modes we must allow for complex β, in which case the boundary terms on Γ_0 and Γ_Λ do not cancel. Thus the variational formulation is

$$\tilde{a}(\beta,k;v,u) = 0 \tag{47}$$

where

$$\tilde{a}(\beta,k;v,u) = a(\beta,k;v,u) - e^{-2\mathrm{Im}\beta\Lambda}\int_{\Gamma_0} p\bar{v}\frac{\partial u}{\partial n}.$$

A leaky mode is characterized by a complex β and a real k for which (47) has a nontrivial solution.

4.4 A Simple Example

We illustrate the spectral properties of the simple grating structure with grating period $\Lambda = 1$, which is shown in Figure (**9**).

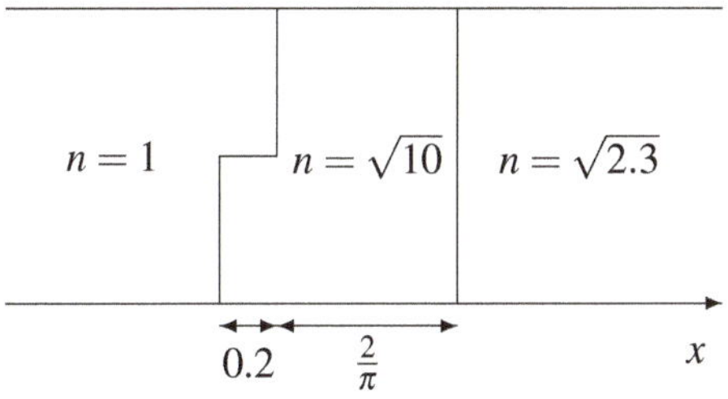

Fig. (**9**): Geometry of the waveguide. $\Lambda = 1$

Before we present the dispersion curves for this structure we consider the dispersion of the approximating planar structure, where the grating layer has been replaced by a uniform layer with the averaged refractive index (9). The dispersion curves of this planarized structure can of course be obtained with the techniques of Section . On the other hand, the results of the previous section apply as well since the planarized structure is also invariant with respect to translations of length Λ along the z-axis.

Suppose now that ϕ is an eigenfunction of

$$(p\phi')' + (qk^2q - p\beta_l^2)\phi = 0.$$

Then $u(x,z) = \exp(i\beta_l z)\phi(x)$ satisfies $\mathscr{L}u = k^2 q$ and has β-periodic boundary conditions on Γ_0 and Γ_Λ.

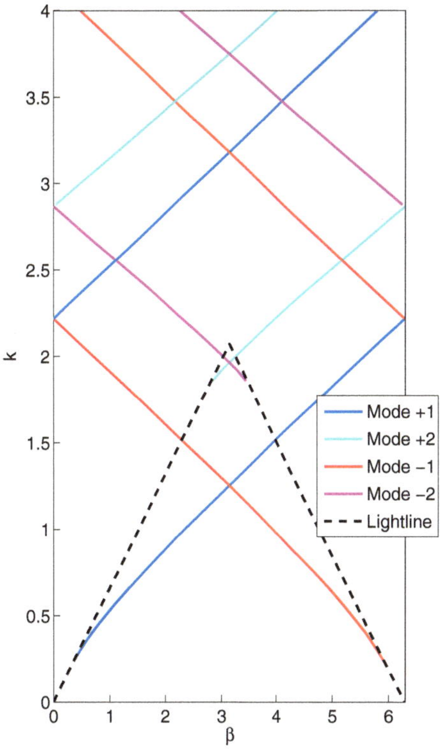

Fig. (**10**): k-β plot for the averaged planar waveguide. TE polarization

The converse holds as well. We see that if $k = k(\beta)$ is a dispersion of the planarized structure then $k(\beta_l)$, $l \in \mathbb{Z}$, are all eigenvalues of (28- 30). Thus the k-β plot is obtained by periodically wrapping the dispersion curves into the in the Brillouin zone.

Figure (**10**) shows the resulting picture. Here we choose the interval $[0, 2\pi/\Lambda]$ in order to clarify how modes in the periodic structure will couple. Note that the planarized structure is an example where there is an infinite number of guided modes inside the light cone.

Now we return to the periodic structure, which may be regarded as a perturbation of the planar structure.

Figure (**11**) shows the dispersion curves for the periodic structure. To keep the information given in this plot manageable, we display only modes ± 1 in blue and red and ± 2 in cyan and magenta. Modes ± 1 start out as guided modes until they couple at $\beta = \pi$, the first Bragg condition. After that, they become guided again, until they leave the light cone. The two modes couple again at $\beta = 2\pi$, the second Bragg.

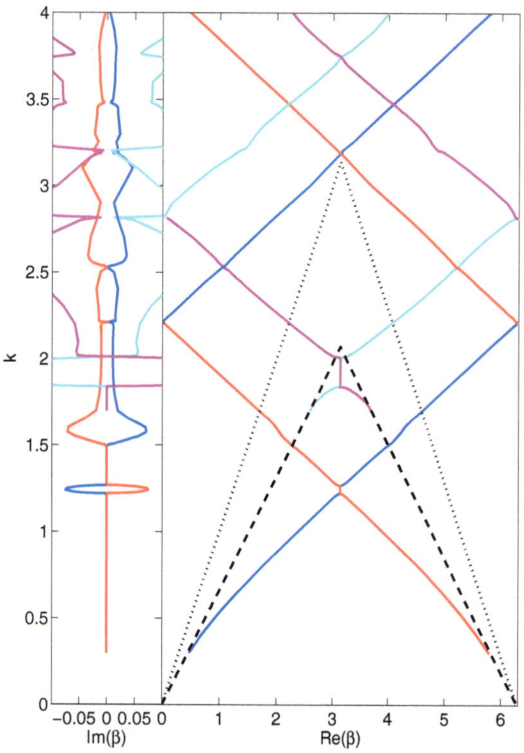

Fig. (**11**): k-β plot for the waveguide shown in Figure (**9**).

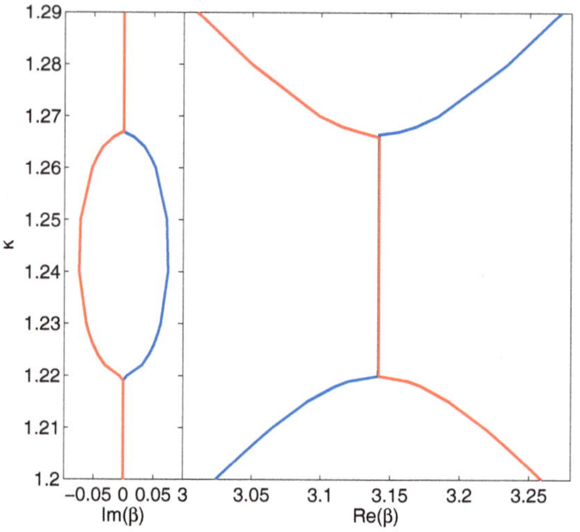

Fig. (**12**): Enlargement of Figure (**11**) near the first Bragg of the first mode.

The behavior of the dispersion relations near the first and second Bragg conditions is shown in Figures (**12**) and (**13**). Note that the second Bragg is an isolated point where the mode is guided.

We now illustrate the field profiles $\mathrm{Re}(u(\cdot,0))$ for a few selected modes. Figure (**14**) shows three examples of profiles of the dominant mode inside the light cone, inside the region of substrate modes and inside the region of two-sided modes. The former is a guided mode, the other two radiate into the substrate or into both exterior regions. Note that that latter two modes are leaky, hence their magnitude increases exponentially away from the grating structure. Since the rate of increase is small this is hard to see on the plots. Figure (**15**) shows a similar plot for the second mode. The plots shown in this section have been computed numerically by matching the interior and exterior DtN operators. This method will be described later in Section 5.1.

5. DISCRETIZATION METHODS

We now turn to numerical techniques for finding the modes of an open periodic waveguide. Most of these methods are based on some discretization of (41-44) and the truncation of the series expansion of the exterior DtN operator. The result is usually the the nonlinear eigenvalue problem: Find combinations of (β,k) such that

$$G(\beta,k) \in \mathbb{C}^{N\times N} \tag{48}$$

is singular. Here, N is the number of degrees of freedom in the discretization. It is not our goal here to give a complete account of all the methods that have appeared in the past, instead, we only mention a few representative examples and refer the reader to the original papers.

1. Finite elements. The papers [5, 19] address the solution of the scattering problem (45) with known k and β, but the methodology can also be applied to solve the eigenvalue problem.
2. Boundary elements. The papers [21, 8] consider a boundary integral formulation of the partial differential equation in Ω_0 that is coupled with the exterior DtN operators. The advantage of boundary elements is that there are unknowns only on the interfaces of the dielectric materials. Thus the resulting eigenvalue problem is much smaller than with finite elements.
3. The eigenvalue method is based on expanding the grating regions into the eigenfunctions of the differential operator. This suggests itself if the grating geometry is simple enough such

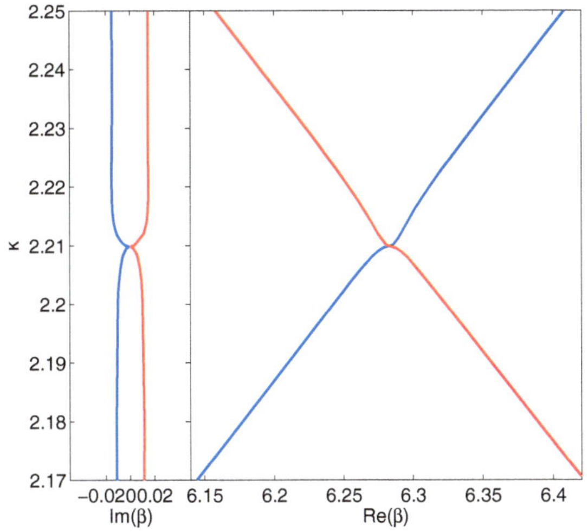

Fig. (**13**): Enlargement of Figure (**11**) near the second Bragg of the first mode.

that the eigenfunctions are known, and was first presented by Peng, Tamir and Bertoni [37] and is probably the oldest rigorous numerical method. A more recent review is [53]. In the following section we give a brief description of the basic ideas.

5.1 Eigenvalue Method

Probably the oldest convergent method for analyzing periodic waveguide structures and solving for the Floquet-Bloch modes numerically was developed by Peng et al. [37] and later extended in [11]. It should be mentioned that there are earlier investigations of open periodic waveguides using Floquet Theory are which are based on perturbation methods [27] or coupled-mode theory [14, 34]. Since these techniques are approximate in nature they can only provide a quantitative idea of the propagation characteristics near the first and second Bragg resonances. These methods can be accurate when the grating region is small, but fail to predict, for instance, a saturation effect of the attenuation of wave propagation when the grating width is increased, see Fig. 8 of [37].

The derivation of the eigenvalue method begins with the observation that the solution to (7) is quasi-periodic, which allows the following Fourier series expansion in the z-variable

$$u(x,z) = \sum_{l \in \mathbb{Z}} \psi_l(x) \exp\left(i\beta_l z\right). \tag{49}$$

The Fourier coefficient $\psi_l(x)$ is called the lth space

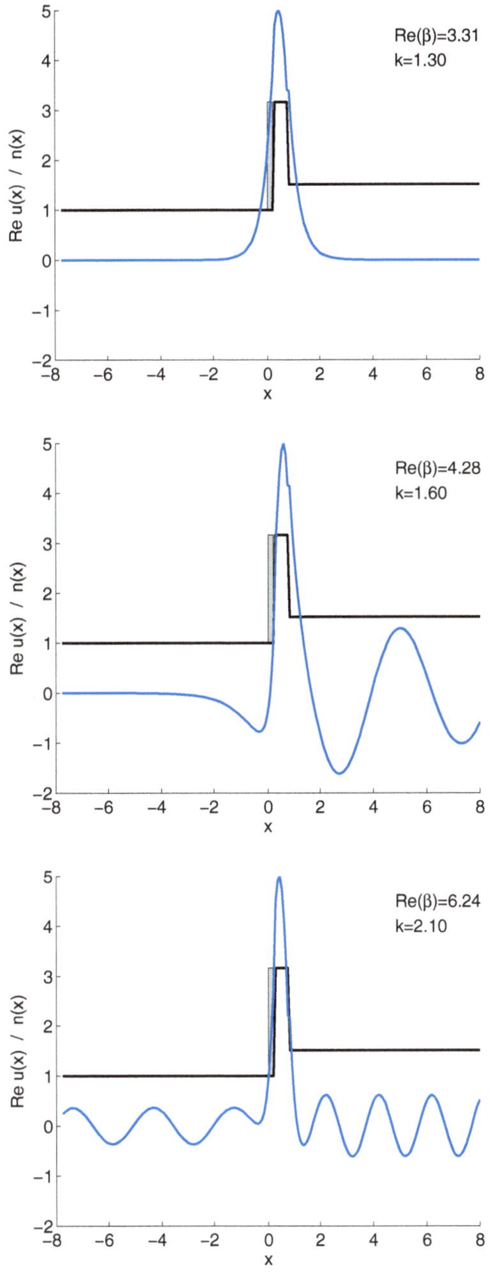

Fig. (**14**): Plot of the refractive index and field profiles of a guided, substrate radiation and two sided radiation mode for the first (dominant) eigenvalue. The location of the modes in the k-β-plane can be seen in Figure (**13**). The grating region is shaded.

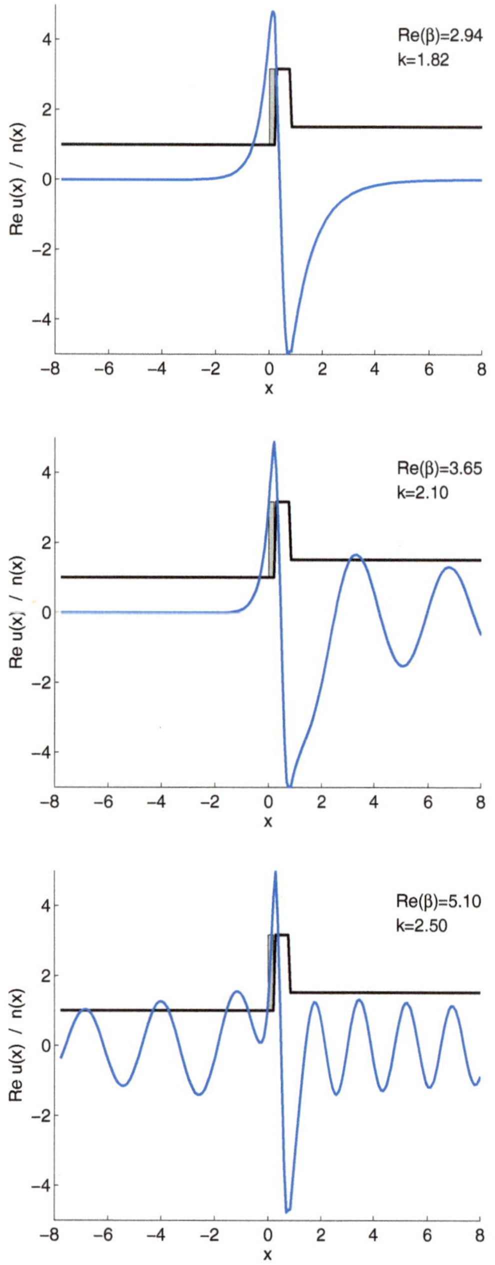

Fig. (**15**): Similar plot as Figure (**14**) for the second eigenvalue.

harmonic.

To simplify notations we limit the discussion to TE-modes (5), the extension to the TM case is straight forward. The technique limits the form of the grating to a rectangular tooth profile, hence the waveguide consists of stratified layers, where the refractive index is a function of x only, and the grating region, where the refractive index is a function of z only. To allow arbitrary shaped profiles, the grating layer must be partitioned into multiple layers with rectangular geometry, see Figure (**16**), but we will consider only the case of one grating layer.

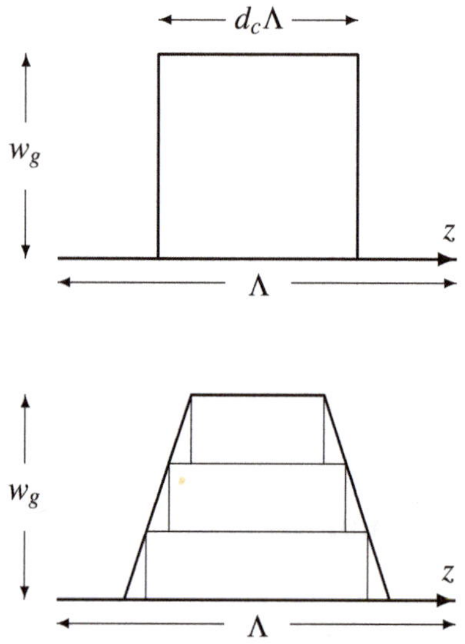

Fig. (**16**): Top:Rectangular tooth shape in the grating region. The tooth height is w_g the tooth width is $d_c\Lambda$. Bottom: Approximation of an arbitrary profile by multiple rectangular regions.

In the grating layer the refractive index is a periodic function of z and may be expanded in a Fourier series

$$n^2(z) = \sum_{l\in\mathbb{Z}} \kappa_l \exp\left(i\frac{2\pi}{\Lambda} lz\right) \tag{50}$$

and

$$\kappa_l = \begin{cases} d_c n_t^2 + (1-d_c)n_f^2, & l = 0, \\ (n_t^2 - n_f^2)\frac{\sin(l\pi d_c)}{\pi l}, & l \neq 0. \end{cases}$$

Here, n_t, n_f are the indices inside and outside the tooth and d_c is the fraction of the grating period occupied by the tooth, which is commonly referred to as the duty cycle.

Substitution of (49) into (5) leads to a system of constant coefficient second-order ordinary differential equations for the space harmonics

$$\psi_l''(x) - \beta_l^2 \psi_l(x) + k^2 \sum_{m \in \mathbb{Z}} \kappa_{l-m} \psi_m(x) = 0,$$

for $l \in \mathbb{Z}$. In matrix notation, the system appears in the form

$$\frac{d^2}{dx^2}\vec{\psi} - P\vec{\psi} = \vec{0} \tag{51}$$

where $\vec{\psi} = \vec{\psi}(x)$ is an infinite vector with coefficients ψ_n and P is an infinite matrix with constant coefficients

$$P_{lm} = -\beta_l^2 \delta_{lm} + k^2 \kappa_{l-m}.$$

The boundary conditions at the endpoints of the grating layer will be incorporated later. At this point we consider a general solution, which may be assumed to be of the form

$$\vec{\psi}(x) = \vec{c}\exp(i\rho x),$$

where $\vec{c}$ is a constant vector. Substituting this form into (51) shows that ρ and $\vec{c}$ are an eigenvalue and eigenvector of

$$P\vec{c} = \rho\vec{c}.$$

Suppose that $\{\vec{c}_m\}$ is a complete set of eigenvectors with eigenvalues ρ_n, then $\vec{\psi}$ has the representation

$$\vec{\psi} = \sum_m \vec{c}_m \left[\exp(i\rho_m x)v_m^+ + \exp(-i\rho_m x)v_m^-\right]$$

where the $v_m^{\pm}$ are coefficients that will be determined next. From (49) it follows that the solution in the grating region is

$$\begin{aligned} u(x,z) = &\sum_{l,m} \exp(i\beta_l z)c_{lm} \\ &\times \left[\exp(i\rho_m x)v_m^+ + \exp(-i\rho_m x)v_m^-\right] \end{aligned}$$

where c_{lm} are the coefficients of the eigenvector $\vec{c}_m$. Outside the grating the solution has the expansion (32), thus there is another set of coefficients, $u_l^{\pm}$, that determine the solution in $\Omega_{\pm}$. These coefficients provide the degrees of freedom needed to match the function and the derivative on the interfaces $\Gamma_{\pm}$ in order to obtain a solution in the whole strip Ω_Λ. This leads to the linear system

$$\begin{aligned} S_{++}\vec{v^+} + S_{+-}\vec{v^-} &= u^+ \\ S_{-+}\vec{v^+} + S_{--}\vec{v^-} &= u^- \\ RS_{++}\vec{v^+} - R^{-1}S_{+-}\vec{v^-} &= D_+u^+ \\ RS_{-+}\vec{v^+} - R^{-1}S_{+-}S_{--}\vec{v^-} &= D_-u^- \end{aligned}$$

where the matrices are defined as

$$\begin{aligned} [S_{\pm,\pm}]_{lm} &= c_{lm}\exp(i \pm \rho_m x_{\pm}) \\ R_{lm} &= \delta_{lm}\rho_m \\ [D_{\pm}]_{lm} &= \delta_{lm}\alpha_l^{\pm} \end{aligned}$$

In the above system the $\vec{u}^{\pm}$ coefficients can be eliminated easily. The result is that $[\vec{v^+}, \vec{v^-}]^T$ must be in the nullspace of the matrix

$$G = \begin{bmatrix} (R-D_+)S_{++} & (R-D_+)S_{+-} \\ (R-D_+)S_{-+} & (R-D_+)S_{--} \end{bmatrix}.$$

Since the coefficients of G depend on k and β, a mode is characterized by a combination of these parameters such that G has a nontrivial nullspace. This is a nonlinear eigenvalue problem, which can be solved with the methods described in Section 6.1.

6. Matching the Interior and Exterior DtN Operators

In the numerical methods discussed above the size of the matrix is primarily determined by the number of degrees of freedom used for the discretization of the interior domain. There are additional unknowns from the discretization of the interfaces $\Gamma_{\pm}$ which is small. This section discusses a technique which reduces the problem to an eigenvalue problem on the interfaces [46]. Thus the number of unknowns is significantly reduced which will allow the use of dense matrix techniques for the solution of the nonlinear eigenvalue problem.

To that end, consider the interior problem with Dirichlet boundary conditions

$$\begin{aligned} \mathscr{L}u - k^2 qu &= 0 \quad \text{in } \Omega_0 && (52) \\ u\big|_{\Gamma_\Lambda} &= e^{i\beta\Lambda}\, u\big|_{\Gamma_\Lambda} && (53) \\ \left.\frac{\partial u}{\partial z}\right|_{\Gamma_\Lambda} &= e^{i\beta\Lambda}\left.\frac{\partial u}{\partial z}\right|_{\Gamma_0} && (54) \\ u\big|_{\Gamma_\pm} &= \varphi_\pm && (55) \end{aligned}$$

If β and k are given then the problem is uniquely solvable with the exception of a number of discrete frequencies, called interior resonances. If k is not a resonance, then the mapping from the Dirichlet conditions on the interfaces to the normal derivative of the solution on the interface is well defined. This is the interior DtN operator

$$T^i(\beta,k)\begin{bmatrix} \varphi_+ \\ \varphi_- \end{bmatrix} = \begin{bmatrix} \frac{\partial u^+}{\partial n} \\ \frac{\partial u^-}{\partial n} \end{bmatrix}$$

Now suppose that (β,k) is such that (41-44) has a nontrivial solution. Denote its restriction to the interfaces by φ, then

$$\left[p^iT^i(\beta,k)-p^eT^e(\beta,k)\right]\varphi=0.$$

To simplify the following notations will set

$$T(\beta,k)=p^iT^i(\beta,k)-p^eT^e(\beta,k). \tag{56}$$

On the other hand, if (56) holds for some $\varphi\neq 0$, then φ can be extended to Ω_0 by solving (52-55) and to $\Omega_\pm$ by equation (32). Thus the eigenvalue problem is reduced to finding (β,k) such that (56) holds for some $\varphi\neq 0$.

The DtN operator is only defined if (β,k) is not an interior resonance. This is not really a problem, because the singular frequencies are different from the interior resonances, and therefore the interior DtN maps are well defined in a neighborhood of the solutions of (56). However, it is possible to modify the approach by matching the Robin-to-Robin (RtR) maps instead of DtN maps. The interior problem with Robin boundary conditions is well posed hence the RtR operator is well defined. This idea has been used in the context of computing artificial boundary conditions in, for instance [18].

6.1 Discretization of the DtN Operator

Equation (56) is discretized using a variational approach. To that end, we introduce the functions

$$e_l^\pm(z)=\begin{cases}\exp\left(i\beta_l z\right), & \text{on } \Gamma_\pm,\\ 0 & \text{on } \Gamma_\mp,\end{cases} \tag{57}$$

for $l\in\mathbb{Z}$, and the $4p+2$-dimensional space

$$S_p:=\text{span}_{|l|\leq p}\{e_l^\mp\}.$$

The discretized version of (56) is to find (β,k) and $0\neq\varphi\in S_p$ such that

$$(\psi,T(\beta,k)\varphi)=0$$

holds for all $\phi\in S_p$. In the basis (57) the variational form reduces to a nonlinear eigenvalue problem: Find (β,k) such that $T_p(\beta,k)$ is singular, where T_p is a 2×2-block matrix, whose coefficients are

$$[T_{\pm,\pm}(\beta,k)]_{l,l'}=\left(e_l^\mp,T(\beta,k)e_{l'}^\mp\right).$$

Since this is a small matrix, direct methods are suitable for the numerical solution. Before we discuss that, we describe how the matrix coefficients can be computed.

From its definition in (35) it follows that the exterior DtN operator contributes the diagonal factor $i\alpha_l\pm$ (36). In general, the interior part does not have a closed form and must be computed numerically. To obtain the matrix coefficients of the interior DtN operator the interior problem (52-55) must be solved $2p+1$ times with Dirichlet data $\phi=e_l^\mp$, $|l'|\leq p$. One such solution gives one column of $T_{\pm,\pm}$ which consists of the l-th Fourier coefficient of the normal derivative. There are several alternatives to solve the interior problem. The finite element approach was considered in [46]. An alternative is the boundary element approach, which was considered [47].

7. SOLUTION OF THE EIGENVALUE PROBLEM

We have seen that after discretization the problem of finding the modes of (7) reduces to finding the values of $\beta=\beta(k)$ such that the matrix $G(\beta,k)$ is singular. Of course, one could also compute $k_m(\beta)$, but if we are interested in leaky modes the first formulation is more natural. In the case of the planar waveguide the nonlinear eigenvalue problem can be formulated as a quartic problem, but for the periodic structure this is no longer possible.

Suppose now we want to compute a value for β such that $G(\beta,k)$ is singular for a given k. As k is fixed in the following discussion, we will omit the k-dependence and simply write $G(\beta)$. Since working with the determinant is problematic for computations, see [45], a popular way to solve this problem is by converting the matrix equation into an equivalent system of nonlinear equations. This can be accomplished, for instance, by the system

$$\mathbf{F}(\mathbf{x},\beta):=\left[\begin{array}{c}G(\beta)\mathbf{x}\\ \mathbf{b}^*\mathbf{x}-1\end{array}\right]=\mathbf{0}, \tag{58}$$

which has $N+1$ equations and unknowns. The additional unknown $\mathbf{x}\in\mathbb{C}^N$ is the eigenvector. In (58) the vector $\mathbf{b}\in\mathbb{C}^N$ must be chosen such that it is not in the orthogonal complement of the eigenspace of G. Since this space is unknown it is impossible to give a vector that will always work, but in practice a randomly selected $\mathbf{b}$ will be sufficient. Note that it is not a good idea to replace $\mathbf{b}$ by $\mathbf{x}$ because the function $\mathbf{x}\to\mathbf{x}^*\mathbf{x}$ is not analytic.

The system can be solved by Newton's method, which consists of solving a linear system with the Jacobian

$$F'(\mathbf{x},\beta)=\left[\begin{array}{cc}G(\beta)\mathbf{x} & G'(\beta)\mathbf{x}\\ \mathbf{b}^* & 0\end{array}\right]$$

in each step. A few remarks are in order

1. The Jacobian involves computing $G'(\beta)$, the derivative of $G(\beta)$ with respect to β. For the discretizations of the interior DtN operator it is usually easy to find this derivative analytically. For the exterior problem one can derive translation operators for the for the derivative of the state vector with respect to β, in a similar fashion as for the statevector itself. This is described in [46].

2. The Newton iteration must be performed on the Riemann surface discussed in Section 4.2. Thus every sheet of $\alpha_l^{\pm}$ in (34) must be changed, whenever a branch cut is crossed, see [47].

3. The Jacobian at the solution is invertible only if the dimension of the eigenspace is one. If two dispersion relations cross this assumption is no longer satisfied and the convergence of Newton's method applied to (58) will be slower.

If the interior problem is discretized with a finite-element type approach then the matrices G and G' are large but sparse which can be effectively exploited when solving system (58). On the other hand, if G is the matrix that arises from matching the interior and exterior DtN maps, as described in Section 5.1 then G is dense, but very small. In this setting the cost of computing the matrices G and G' dominates over the cost of a matrix inversion or diagonalization. It is preferable to solve the system with the Matrix-Newton method [46] instead of converting it into a nonlinear system. The idea of this iteration to find the correction σ such that the linearization at the current iterate β

$$G(\beta+\sigma) \approx G_0 + \sigma G_1 \tag{59}$$

is singular. Here, $G_0 = G(\beta)$ and $G_1 = G'(\beta)$. The next iterate is the smallest generalized eigenvalue (in modulus). This can be realized numerically by computing the largest eigenvalue λ_M of $G_0^{-1}G_1$ and setting $\sigma = -1/\lambda_M$. Then the next iterate is given by $\beta \to \beta + \sigma$. The method has several advantages over solving (58). It only requires an initial guess for β but not for $\mathbf{x}$, furthermore, the convergence does not deteriorate near crossing dispersion relations.

Regardless of which solver is used, it is essential to have a good initial guess to start the iteration. Since there are many solutions of the nonlinear eigenproblem in the complex plane, the iteration often converges to an undesired root. Especially near points where two modes couple, the region of attraction to one of the solutions is usually very small.

A good choice for the initial guess of β is the solution of a 'nearby' planar structure, which was already mentioned at the beginning of Section . If the grating is weak and the root is sufficiently far from coupling modes this initial guess suffices to obtain convergence. However, for strong gratings and in the neighborhood of Bragg points the convergence regions are small and the choice of an initial guess is more delicate.

In this case it is preferable to use a continuation method. This is a numerical technique where one follows the solution path $u(v)$ of an underdetermined nonlinear system $F(u,v) = 0$, with $F : \mathbb{C}^n \times \mathbb{R} \to \mathbb{C}^n$. The method is often used to solve nonlinear systems $F(u) = 0$ by introducing a homotopy $F(u,v)$ where $F(u,1) = F(u)$ and $F(u,0) = 0$ has known solutions. An introduction to continuation methods is [1], examples of papers that apply this technique to nonlinear eigenvalue problems are [10, 36] and the references cited therein.

In the case of a periodic grating, we already mentioned in Section that a homotopy can be obtained by considering the eigenproblem $G(\beta, v)$ that arises for the refractive index (10). For $v = 0$ the structure is planar and the solutions $\beta(0)$ can be obtained with the methods described in Section 3.5. For $v = 1$ the solutions $\beta(1)$ are the desired eigenvalues the given periodic structure.

Numerical continuation methods are generally of the predictor-corrector type. For an accepted solution (β_0, v_0) a new solution is predicted by following a certain step width Δv in the direction of the tangent vector. The predicted value (β_p, v_p) is in general not on the solution curve $(\beta(v), v)$. Hence a corrected point on the curve is found with Newton's iteration. A popular corrector is the intersection of the curve with the plane through (β_p, v_p) that is normal to the tangent.

In view of the matrix-Newton method one can also consider the following predictor-corrector step. If (β_0, v_0) is an accepted solution, then the predictor is determined from the linearization of $G(\beta(v), v)$ at (β_0, v_0). That is, we find the smallest value of σ such that

$$\frac{\partial G}{\partial \beta}\sigma + \frac{\partial G}{\partial v}$$

is singular. This value is an approximation for $\frac{d\beta}{dv}$. Then the predicted value is

$$\beta_p = \beta_0 + \sigma \Delta v.$$

The value of β_p is the initial guess for the system $G(\beta, \nu_0 + \Delta\nu) = 0$.

8. THE PERFECTLY MATCHED LAYER

The perfectly matched layer (PML) was originally developed by Berenger for scattering problems [6], but the idea can also be used to compute the modes of open dielectric waveguides. There are numerous papers that address various applications, for instance, planar structures are discussed [26] and periodic structures are discussed in [9].

The idea of the PML is to surround the core of the waveguide by a region with lossy materials in such a way that all plane waves are absorbed without any reflection back into the core. Hence the name perfectly matched. For the computation, the absorber is truncated after a finite distance where homogeneous Dirichlet conditions are introduced. On the artificial boundary waves can reflect, but since they have to travel through a lossy medium they have no practical influence when the absorber is large enough.

Consider, for instance, a structure like the one shown in Figure (**1**). This structure requires two absorbers in the regions $x < a$ and $x > b$, where $a < 0$ and $b > w$. We briefly describe the construction of an absorber in the region $x > b$. To simplify notations, we limit ourselves to the TE case, which is governed by the Helmholtz equation (5). The derivation of the TM case is the same after switching $\mathbf{E} \leftrightarrow \mathbf{H}$ and $\mu \leftrightarrow \varepsilon$.

The key idea is introduce a complex electric permittivity $\tilde{\varepsilon}$ and two complex magnetic permittivities $\tilde{\mu}_x$, $\tilde{\mu}_z$ in the absorber. This corresponds to a material with a finite conductivity and an anisotropic (nonphysical) magnetic conductivity. We will see soon that is important to have two different values of the x and z direction in order to achieve a reflection less layer for all angles of incidence.

If the permeability is a diagonal tensor diag$[\mu_x, \mu_z]$ and the permittivity is a scalar, then the equation for the y-component for the electric field is

$$\frac{\partial}{\partial x}\frac{1}{\mu_z}\frac{\partial u}{\partial x} + \frac{\partial}{\partial z}\frac{1}{\mu_x}\frac{\partial u}{\partial z} + \varepsilon k^2 u = 0. \tag{60}$$

Here, μ_x, μ_z and ε are relative permeabilities and permittivities that depend on the position. In the neighborhood of the absorber we have

$$\mu_x(x,z) = \begin{cases} 1, & x < b, \\ \tilde{\mu}_x, & x \geq b, \end{cases} \quad \mu_z(x,z) = \begin{cases} 1, & x < b, \\ \tilde{\mu}_z, & x \geq b, \end{cases}$$

and

$$\varepsilon(x) = \begin{cases} \varepsilon_J & x < b, \\ \tilde{\varepsilon} & x \geq b. \end{cases}$$

Now consider a plane wave that approaches the PML from the (ε, μ)-region in the direction (k_x, k_z). In general, we expect to obtain a reflected and a transmitted wave, hence the total fields are given by

$$\begin{aligned} u(x,z) &= \exp\Big(ik_x(x-b) + ik_z z\Big) \\ &\quad + \Gamma \exp\Big(-ik_x(x-b) + ik_z z\Big), \quad x < b, \\ u(x,z) &= \tau \exp\Big(i\tilde{k}_x(x-b) + i\tilde{k}_z z\Big), \quad x \geq b. \end{aligned}$$

Here, Γ is the reflection- and τ is the transmission coefficient and $(\tilde{k}_x, \tilde{k}_z)$ is the wave vector of the transmitted wave in the PML. These coefficients can be determined from the condition that u and $\frac{1}{\mu_z}\frac{\partial u}{\partial x}$ are continuous across the interface $x = b$. Simple algebra leads to the following conclusions

$$k_z = \tilde{k}_z \tag{61}$$

$$1 + \Gamma = \tau \tag{62}$$

$$k_x(1 - \Gamma) = \frac{\tilde{k}_x}{\tilde{\mu}_z}\tau \tag{63}$$

Suppose that the material constants in the PML are given, then the reflection coefficient can be found by solving (62) and (63), which gives

$$\Gamma = \frac{k_x - \frac{\tilde{k}_x}{\tilde{\mu}_z}}{k_x - \frac{\tilde{k}_x}{\tilde{\mu}_z}}. \tag{64}$$

We now design the material constants in the absorber such that the reflection coefficient vanishes for plane waves of all directions and frequencies. From (5), (60) and (61) it follows that

$$\begin{aligned} k_x^2 &= \varepsilon_J k^2 - k_z^2 \\ \frac{\tilde{k}_x^2}{\tilde{\mu}_z^2} &= \frac{\tilde{\varepsilon}}{\tilde{\mu}_z}k^2 - \frac{k_z^2}{\tilde{\mu}_x\tilde{\mu}_z} \end{aligned}$$

In view of (64), the left hand sides of the last two equations must agree to avoid reflection back into the waveguide region. This can be achieved when

$$\begin{aligned} \frac{\tilde{\varepsilon}}{\tilde{\mu}_z} &= \varepsilon_J, \\ \tilde{\mu}_x\tilde{\mu}_z &= 1. \end{aligned}$$

For a solution we pick some absorption factor $g > 0$ and set

$$\begin{aligned} \tilde{\varepsilon} &= (1 + ig)\varepsilon \\ \tilde{\mu}_z &= (1 + ig)\mu \\ \tilde{\mu}_x &= (1 + ig)^{-1}\mu \end{aligned}$$

Then

$$\tilde{k}_x = \mu_z \mu k_x = (1+ig)k_x$$

thus $u(x,z)$ decays exponentially as $x \to \infty$.

In a numerical scheme the absorber has finite width c, thus the eigenvalue problem with the PML in place is posed in the domain $[a-c, b+c] \times [0, \Lambda]$ as follows

$$\begin{aligned} \frac{\partial}{\partial x}\frac{1}{\mu_z}\frac{\partial u}{\partial x} + \frac{\partial}{\partial z}\frac{1}{\mu_x}\frac{\partial u}{\partial z} &= -\varepsilon k^2 u \\ u_\beta\big|_{\Gamma_\Lambda} &= e^{i\beta\Lambda} u_\beta\big|_{\Gamma_0} \\ \frac{\partial u_\beta}{\partial n}\bigg|_{\Gamma_\Lambda} &= e^{i\beta\Lambda}\frac{\partial u_\beta}{\partial n}\bigg|_{\Gamma_0} \\ u\big|_{x=a-c} = u\big|_{x=b+c} &= 0. \end{aligned}$$

Suppose that β is given and that we seek the singular frequencies, then the discretization with finite elements will result in a non-hermitian, but linear eigenvalue problem in k. Since there are standard packages available to solve such problems iteratively, the problem is considerably simpler than the formulations involving DtN operators. However, it will depend on a case to case basis which approach is more efficient, since the PML formulation involves the discretization of the whole waveguide, whereas in the DtN formulations only the grating region is discretized.

In conclusion, the PML can also be combined with several other techniques that have been developed to solve the eigenvalue problem associated with open waveguides. In this case it is also possible to compute leaky modes to high accuracy with a carefully selected value of the absorption factor g. This is the approach taken in [9]. However, in this setting one ends up solving nonlinear eigenvalue problems and thus it is not clear what computational advantages the PML offers over methods that are based on DtN maps.

REFERENCES

[1] Allgower E, Georg K. An introduction to numerical continuation methods. Classics in Applied Mathematics. SIAM, 2003.

[2] Anemogiannis E, Glytsis E. Multilayer waveguides: efficient numerical analysis of general structures. J Lightwave Tech 1992; 10: 1344-1351.

[3] Anemogiannis E, Glytsis E, Gaylord T. Determination of guided and leaky modes in lossless and lossy planar multilayer optical waveguides: refliction pole method and wavevector density method. J Lightwave Tech 1999; 17: 929-941.

[4] Axmann W, Kuchment P. An efficient finite element method for computing spectra of photonic and acoustic band-gap materials. J Comput Phys 1999; 150: 468-481.

[5] Bao G. Finite element approximation of time harmonic waves in periodic structures. SIAM J Numer Anal 1995; 32: 1155-1169.

[6] Berenger J. A perfectly matched layer for the absorption of electromagnetic waves. J Comput Phys 1994; 114: 185-200.

[7] Bonnet-Bendhia A-S, Starling F. Guided waves by electromagnetic gratings and non-uniqueness examples for the diffration problem. Math Methods Appl Sci 1994; 17: 305-338.

[8] Butler JK, Ferguson WE, Evans GA, Stabile P, Rosen A. A boundary element technique applied to the analysis of waveguides with periodic surface corrugations. IEEE J Quantum Elect 1992; 28: 1701-1707.

[9] Cao Q, Lalanne P, Hugonin J-P. Stable and efficient Bloch-mode computational method for one-dimensional grating waveguides. J Opt Soc Am A 2002; 19: 335-338.

[10] Chan T, Keller H. Arc-length continuation and multi-grid techniques for nonlinear eigenvalue problems. SIAM J Sci Statist Comput 1982; 3: 173-194.

[11] Chang K, Shah V, Tamir T. Scattering and guiding of waves by dielectric gratings with arbitrary profiles. J Opt Soc Amer 1980; 70: 804-813.

[12] Chen C, Berini P, Feng D, Tanev S, Tzolov V. Efficient and accurate numerical analysis of multilayer planar optical waveguides in lossy anisotropic media. Optics Express 2000; 7: 260-272.

[13] Chilwell J, Hodgkinson I. Thin-films field-transfer matrix theory of planar multilayer waveguides and reflection from prism-loaded waveguides. J Opt Soc Amer A 1984; 1: 742-753.

[14] Collin RE, Zucker FJ. Antenna theory 2. Inter-University Electronics Series 7. McGraw-Hill, New York, 1969.

[15] Delves L, Lyness J. A numerical method for locating the zeros of an analytic function. Math Comp 1967, 21: 543-560.

[16] Dobson D. An efficient method for band structure calculations in 2D photonic crystals. J Comput Phys 1999; 149: 363-376.

[17] Dobson DC, Gopalakrishnan J, Pasciak J. An efficient method for band structure calculations in 3D photonic crystals. J Comput Phys 2000; 161: 668-679.

[18] Ehrhardt M, Han H, Zheng C. Numerical simulation of waves in periodic structures. Commun Comput Phys 2009; 5: 849-872.

[19] Elschner J, Schmidt G. Diffraction in periodic structures and optimal design of binary gratings. part I: direct problems and gradient formulas. Math Meth Appl Sci 1998; 21: 1297-1342.

[20] Elschner J, Schmidt G. Numerical solution of optimal design problems for binary gratings. J Comput Phys 1998; 146: 603-626.

[21] Hadjicostas G, Butler J, Evans G, Carlson N, Amantea R. A numerical investigation of wave interactions in dielectric waveguides with periodic surface corrugations. IEEE J Quant Elect 1990; 26: 893-902.

[22] Hanson G, Yakovlev A. An analysis of leaky-wave dispersion phenomena in the vicinty of cutoff using complex frequency plane singularities. Radio Science 1998; 33: 803-819.

[23] Hessel A. General characteristics of traveling-wave antennas. In: Collin R and Zucker F (eds.). Antenna theory 2, Inter-University Electronics Series 7, chapter 19. McGraw-Hill, New York, 1969.

[24] Ho K, Chan C, Soukoulis C, Biswas R, Sigalas M. Photonic band gaps in three dimensions: new layer-by-layer periodic structures. Solid State Comm 1994; 89: 413-416.

[25] Huang W. Coupled-mode theory for optical waveguides: an overview. J Opt Soc Amer A 1994; 11: 963-983.

[26] Huang W, Xu C, Lui W, Yokoyama K. The perfectly matched layer boundary condition for modal analysis of optical waveguides: leaky mode calculations. IEEE Photon Tech Lett 1996; 8: 652-654.

[27] Jacobsen J. Analytical, numerical, and experimental investigation of guided waves on a periodically strip-loaded dielectric slab. IEEE Trans. Antenn Prop 1970; 18: 379-388.

[28] Joannopoulos J, Johnson S, Winn J, Meade R. Photonic crystals, molding the flow of light. Princeton, 2nd ed. 2008.

[29] Johnson S, Mekis A, Fan S, Joannopoulos J. Molding the flow of light. Comput Sci Eng 2001; 3: 38-47.

[30] Petracek J, Singh, K. Determination of leaky modes in planar multilayer waveguides. IEEE Photon Tech Lett 2002; 14: 810-812.

[31] Kuchment P. Floquet theory for partial differential equations. Birkhäuser Verlag, Basel, 1993.

[32] Kuchment P. The mathematics of photonic crystals. In Bao G, Cowsar L, Masters W (eds). Mathematical modelling in optical science. Frontiers in Applied Mathematics 22, pages 207-272. SIAM, 2001.

[33] Lee S, Chung Y, Coldren L, Dagli N. On leaky mode approximations for modal expansion in multilayer open waveguides. IEEE J Quant Electron 1995; 31: 1790-1802.

[34] Marcuse D. Theory of dielectric optical waveguides. Academic Press, New York and London, 1974.

[35] Marcuvitz N. On field representations in terms of leaky modes or eigenmodes. IRE Trans. Antenn Prop 1956; 4: 192-194.

[36] Mittlemann H. A pseudo-arclength continuation method for nonlinear eigenvalue problems. SIAM J Numer Anal 1986; 23: 1007-1016.
[37] Peng S, Tamir T, Bertoni H. Theory of periodic dielectric waveguides. IEEE Trans Microwave Theory Tech 1975; 23: 123-133.
[38] Petit R (ed.), Electromagnetic theory of gratings. Topics in Current Physics. Springer-Verlag, Heidelberg, 1980.
[39] Rayleigh L. On the maintenance of vibrations by forces of double frequency. Phil Mag S 1887; 25: 145-159.
[40] Shipman S. Resonant scattering by open periodic waveguides. Chapter 2 in: Ehrhardt M (ed.), Wave propagation in periodic media - analysis, numerical techniques and practical applications. Progress in Computational Physics 1. Bentham Science Publishers Ltd., 2010.
[41] Smith R, Forbes G, Houde-Walter S. Unfolding the multivalued planar waveguide dispersion relation. IEEE J Quant Elect 1993; 29: 1031-1034.
[42] Smith R, Houde-Walter S, Forbes G. Mode determination for planar waveguides using the four-sheeted dispersion relation. IEEE J Quant Elect 1992; 28: 1520-1526.
[43] Sözüer H, Dowling J. Photonic bandgap calculation for woodpile structures. J Mod Opt 1994; 41: 231-239.
[44] Stowell D, Tausch J. Variational formulation for guided and leaky modes in multilayer dielectric waveguides. Commun Comput Phys 2010; 7: 564-579.
[45] Tamir T, Zhang S. Modal transmission-line theory of multilayered grating structures. J Lightwave Tech 1996; 14: 914-927.
[46] Tausch J, Butler J. Floquet multipliers of periodic waveguides via Dirichlet-to-Neumann maps. J Comput Phys 2000; 159: 90-102.
[47] Tausch J, Butler J. Efficient analysis of periodic dielectric waveguides using Dirichlet-to-Neumann maps. J Opt Soc Amer A 2002; 19: 1120-1128.
[48] Tisseur F, Meerbergen K. The quadratic eigenvalue problem. SIAM Rev 2001; 43: 289-300.
[49] Ulrich R, Prettl W. Planar leaky lightguides and couplers. Appl Phys 1973; 1: 55-68.
[50] Uranus H, Hoekstra H, Groesen EV. Simple high-order Galerkin finite element scheme for the investigation of both guided and leaky modes in ansiotropic planar waveguides. Opt Quant Elect 2004; 36: 239-257.
[51] Wilcox C. Scattering theory for diffraction gratings. Springer, New York, 1984.
[52] Yeh P. Optical waves in layered media. Wiley Series in Pure and Applied Optics. Wiley, 1998.
[53] Zhang S, Tamir T. Rigorous theory of grating-assisted couplers. J Opt Soc Amer A 1996; 13: 2403-2413.

CHAPTER 4

Computational Methods for Multiple Scattering at High Frequency with Applications to Periodic Structure Calculations

X. Antoine[1], C. Geuzaine[2], K. Ramdani[3]

[1] *Institut Elie Cartan Nancy (IECN), Nancy-Université & INRIA (Project-Team CORIDA), Vandoeuvre-lès-Nancy, France.*

[2] *University of Liège, Departement of Electrical Engineering and Computer Science, Montefiore Institute, Liège, Belgium.*

[3] *INRIA (Project-Team CORIDA) & Institut Elie Cartan Nancy (IECN), Nancy-Université, Vandoeuvre-lès-Nancy, France.*

Abstract: The aim of this paper is to explain some recent numerical methods for solving high-frequency scattering problems. Most particularly, we focus on the multiple scattering problem where rays are multiply bounced by a collection of separate objects. We review recent developments for three main families of approaches: Fourier series based methods, Partial Differential Equations approaches and Integral Equations based techniques. Furthermore, for each of these three families of methods, we present original procedures for solving the high-frequency multiple scattering problem. Computational examples are given, in particular for finite periodic structures calculations. Difficulties for solving such problems are explained, showing that many serious simulation problems are still open.

1. INTRODUCTION

Let us define $\Omega^- \subset \mathbb{R}^d$ as a d-dimensional impenetrable bounded domain with boundary $\Gamma := \partial\Omega^-$. Throughout this paper, we suppose that this domain is composed of a collection of M separable domains Ω_p^- with respective $\mathscr{C}^\infty$ boundaries Γ_p, $p = 1,...,M$. Therefore, we have: $\Omega^- = \cup_{p=1}^M \Omega_p^-$ and $\Gamma = \cup_{p=1}^M \Gamma_p$. The associated homogeneous exterior domain of propagation, which is the complementary set of the scatterer $\overline{\Omega^-}$ in $\mathbb{R}^d$, is denoted by Ω^+. Then, the multiple scattering problem of an incident time-harmonic acoustic wavefield u^{inc} by Ω^- can be formulated as the following exterior Boundary Value Problem (BVP): Find the scattered field u solution to

$$\begin{cases} \Delta u + k^2 u = 0, \text{ in } \Omega^+, \\ u = -u^{\text{inc}} \text{ or } \partial_{\mathbf{n}_\Gamma} u = -\partial_{\mathbf{n}_\Gamma} u^{\text{inc}}, \text{ on } \Gamma, \\ \lim\limits_{\|\mathbf{x}\| \to \infty} \|\mathbf{x}\|^{(d-1)/2} (\nabla u \cdot \dfrac{\mathbf{x}}{\|\mathbf{x}\|} - iku) = 0. \end{cases} \quad (1)$$

If $\mathbf{a}$ and $\mathbf{b}$ are two complex-valued vector fields (and $\bar{z}$ denotes the complex conjugate of a complex number $z \in \mathbb{C}$), their inner product is defined by $\mathbf{a} \cdot \mathbf{b} = \sum_{j=1}^d a_j \overline{b_j}$, and the associated norm $\|\cdot\|$ is: $\|\mathbf{a}\|^2 = \mathbf{a} \cdot \mathbf{a}$. Let $\mathbf{x} = (x_1,...,x_d) \in \mathbb{R}^d$, then the gradient ∇ of a complex-valued scalar field f is defined by: $\nabla f = (\partial_{x_1} f,...,\partial_{x_d} f)^{\mathbf{T}}$, designating by $\mathbf{a}^{\mathbf{T}}$ the transposed of $\mathbf{a}$. The Laplace operator Δ is classically defined by: $\Delta = \sum_{j=1}^d \partial_{x_j}^2$. We consider that the incident wave u^{inc} is plane: $u^{\text{inc}}(\mathbf{x}) = e^{ik\mathbf{d}\cdot\mathbf{x}}$. The wavenumber k is related to the wavelength λ by the relation $k = 2\pi/\lambda$. In the two-dimensional case ($d = 2$), the direction of incidence $\mathbf{d}$ is given through the relation: $\mathbf{d} = (\cos(\theta^{\text{inc}}), \sin(\theta^{\text{inc}}))^{\mathbf{T}}$, where θ^{inc} is the scattering angle in the polar coordinates system. In the three-dimensional case ($d = 3$), we have: $\mathbf{d} = (\cos(\theta^{\text{inc}})\sin(\phi^{\text{inc}}), \sin(\theta^{\text{inc}})\sin(\phi^{\text{inc}}), \cos(\phi^{\text{inc}}))^{\mathbf{T}}$. The scattering angles $(\theta^{\text{inc}}, \phi^{\text{inc}})$ are given in the spherical coordinates system. If we define by $\mathbf{n}_\Gamma$ the outwardly directed unit normal to Ω^- at the boundary Γ, then, the sound-soft or Dirichlet (respectively sound-hard or Neumann) boundary condition on Γ corresponds to the second (respectively third) equation of (1). For a given single domain Ω_p^-, the associated outwardly directed unit normal to Γ_p is denoted by $\mathbf{n}_p$. Finally, the last equation which is the well-known Sommerfeld radiation condition allows only outgoing waves at infinity. This thereby guarantees the uniqueness of the solution to the BVP (1).

Other situations of interest that we do not consider here are related to homogeneous or inhomogeneous penetrable scatterers or scatterers with impedance boundary conditions (Fourier-Robin boundary con-

dition). However for the sake of clarity, we will focus on the situation described above.
In the sequel of the paper, we will use material related to functional spaces and operator theory that we introduce now. The Sobolev space of real order s for a domain D is denoted by $H^s(D)$. Its usual inner product and norm will be respectively denoted by $(\cdot,\cdot)_{s,D}$ and $\|\cdot\|_{s,D}$. In the case of two complex-valued vector fields $\mathbf{u}$ and $\mathbf{v}$ defined on a domain D, the inner product on $L^2(D)$ is $(\mathbf{u},\mathbf{v})_{0,D} = \int_D \mathbf{u}\cdot\bar{\mathbf{v}}\,dD$. The space $\mathscr{D}'(D)$ is the space of distributions on D, which is the dual space of $\mathscr{D}(D)$. Furthermore, if D is multiply connected, with $D = \cup_{p=1}^{M} D_p$, then we set: $H^s(D) = H^s(D_1)\times ... \times H^s(D_M)$. Therefore, $u \in H^s(D)$ means $u = (u_1,...,u_M)$, with $u_p \in H^s(D_p)$, $p = 1,...,M$. Let us introduce the functional spaces

$$
\begin{aligned}
&H^1_{\mathrm{loc}}(\overline{\Omega^+}) := \\
&\quad \left\{u \in \mathscr{D}'(\Omega^+) \mid \psi u \in H^1(\Omega^+), \forall \psi \in \mathscr{D}(\mathbb{R}^d)\right\}, \\
&H^1_+(\Delta) := \left\{u \in H^1_{\mathrm{loc}}(\overline{\Omega^+}); \Delta u \in L^2_{\mathrm{loc}}(\overline{\Omega^+})\right\}, \\
&H^1_{-,p}(\Delta) := \left\{u \in H^1(\overline{\Omega^-_p}); \Delta u \in L^2(\overline{\Omega^-_p})\right\}, \\
&H^1_-(\Delta) := H^1_{-,1}(\Delta)\times ... \times H^1_{-,M}(\Delta),
\end{aligned}
\tag{2}
$$

with $p = 1,...,M$. Then, for $u \in H^1_\pm(\Delta)$, the exterior $(+)$ and interior $(-)$ trace operators of order j ($j = 0$ or 1) can be defined by

$$
\begin{aligned}
\gamma_j^\pm : H^1_\pm(\Delta) &\to H^{1/2-j}(\Gamma) \\
u &\mapsto \gamma_j^\pm u = \partial^j_{\mathbf{n}} u_{|\Gamma}.
\end{aligned}
\tag{3}
$$

We will sometimes denote independently by $\gamma_j^{\pm,p} u$ or $(\gamma_j^\pm u)_p$, the p-th component of the trace operator of order j on Γ_p.
Under these notations, a possible functional setting for proving the existence and uniqueness of the solution to our scattering problem (1) is the following [54, 109]

$$
\begin{cases}
\text{Find } u \in H^1_{\mathrm{loc}}(\overline{\Omega^+}) \text{ such that} \\
\Delta u + k^2 u = 0, \quad \text{in } \mathscr{D}'(\Omega^+), \\
\gamma_j^+ u = g := -\gamma_j^+ u^{\mathrm{inc}}, \quad \text{in } H^{1/2-j}(\Gamma), j = 0,1, \\
\lim\limits_{\|x\|\to+\infty} \|\mathbf{x}\|^{(d-1)/2}(\nabla u \cdot \dfrac{\mathbf{x}}{\|\mathbf{x}\|} - iku) = 0.
\end{cases}
\tag{4}
$$

Finally and throughout the paper, $G(\mathbf{x},\mathbf{y})$ denotes the outgoing Green's function of the Helmholtz operator $-\Delta - k^2$ in $\mathbb{R}^d$:

$$
G(\mathbf{x},\mathbf{y}) = \begin{cases}
\dfrac{i}{4} H_0^{(1)}(k\|\mathbf{x}-\mathbf{y}\|) & \text{for } d = 2, \\
\dfrac{1}{4\pi}\dfrac{e^{ik\|\mathbf{x}-\mathbf{y}\|}}{\|\mathbf{x}-\mathbf{y}\|} & \text{for } d = 3,
\end{cases}
\tag{5}
$$

where $H_0^{(1)}$ is the first-kind Hankel's function of order zero.
The paper is organized as follows. In Section , we present one of the main ingredients used in the different algorithms presented in this paper to solve numerically the multiple scattering problem (1) (see Theorem 1). This result essentially shows that the scattered field of the multiple scattering problem can be obtained by superposition of the fictitious scattered fields corresponding to single-obstacle scattering problems. Section is devoted to the investigation of the particular case where the scatterers are disks. Using Fourier series technique, we present an efficient algorithm to solve such multiple scattering problems at high frequency and/or when the number of scatterers is large. The rest of the paper deals with the case of scatterers of arbitrary shapes. Sections 3.6 and 4.3 describe numerical methods based on a PDE approach. A short introduction to these methods is given in Section 3.6, while Section 4.3 is devoted to a detailed presentation of a particular method: the Phase Reduction Finite Element Method (PR-FEM). Sections 5.2 and 6.4 deal with integral equation based strategies. Once again, we start with a short review of these methods (Section 5.2) and then we show on a particular example the efficiency of a high-order high-frequency algorithm (Section 6.4). Finally, Section 7.2 gives a general conclusion.

2. MULTIPLE SCATTERING VIEWED AS COUPLED SINGLE-OBSTACLE SCATTERING PROBLEMS

The multiple scattering problem (4) models the global scattering problem. We will propose in the next sections some possible numerical methods for solving this problem directly. However, at both the theoretical and numerical levels, an interesting alternative is to reformulate the initial multiple scattering problem as M coupled single-obstacle scattering problems. Let us emphasize that this reduction is possible due to the linearity of the problem and holds for arbitrary shapes of the scatterers. We develop this point of view below.
This new formulation of the problem leads to a decomposition of the scattered field of the form $u = \sum_{p=1}^{M} u_p$, where each fictitious scattered wave u_p corresponds to the wave reflected by the scatterer p—and only by it—when it is illuminated simultaneously by the incident wave and the waves u_q, for $q = 1,\ldots,M$, with $q \neq p$. The next theorem provides a precise formulation of the above assertion.

Theorem 1 ([19]). *Let u be the solution of the multiple scattering problem (1). Then, the family of M coupled single-obstacle scattering problems for* $p = 1,...,M$,

$$\begin{cases} \Delta u_p + k^2 u_p = 0, \text{ in } \mathbb{R}^d \setminus \overline{\Omega_p^-}, \\ u_p = -u^{inc} - \sum_{q=1,q\neq p}^{M} u_q, \text{ on } \Gamma_p, \\ \quad \text{or} \\ \partial_{\mathbf{n}_{\Gamma_p}} u_p = -\partial_{\mathbf{n}_{\Gamma_p}} u^{inc} - \sum_{q=1,q\neq p}^{M} \partial_{\mathbf{n}_{\Gamma_p}} u_q, \text{ on } \Gamma_p, \\ \lim_{\|\mathbf{x}\|\to\infty} \|\mathbf{x}\|^{(d-1)/2} (\nabla u_p \cdot \frac{\mathbf{x}}{\|\mathbf{x}\|} - iku_p) = 0, \end{cases} \tag{6}$$

admits a unique solution $(u_1,\ldots,u_M)$. *Furthermore, the following decomposition holds true:*

$$u = \sum_{p=1}^{M} u_p. \tag{7}$$

Proof. For the sake of completeness, we give here a proof for the case of Dirichlet boundary conditions. The proof uses classical results from integral equations theory, that are recalled in Section 6.1.
We start by proving the uniqueness of the solution of the coupled problems (6). Let $u_1,\ldots,u_M$, with $u_p \in H^1_{\text{loc}}(\mathbb{R}^d \setminus \overline{\Omega_p^-})$, solve (6) for Dirichlet boundary condition and $u^{\text{inc}} = 0$, and assume by contradiction that there exists $p \in \{1,\ldots,M\}$ such that u_p does not vanish identically. Then, the function $v = \sum_{q=1}^{M} u_q$ is an outgoing solution of the Helmholtz equation in $\mathbb{R}^d \setminus \overline{\Omega_-}$, which satisfies in addition the boundary condition $v = 0$ on Γ. By classical uniqueness results for Helmholtz equation [55], this implies that $v \equiv 0$ in Ω^+. In particular, we have $v = \partial_{\mathbf{n}} v = 0$ on Γ_p. Let us define then the function

$$w(\mathbf{x}) = \begin{cases} u_p(\mathbf{x}) & \text{for } \mathbf{x} \in \mathbb{R}^d \setminus \overline{\Omega_p^-} \\ -\sum_{q=1,q\neq p}^{M} u_q(\mathbf{x}) & \text{for } \mathbf{x} \in \Omega_p^-. \end{cases}$$

Clearly, w is outgoing and solves the Helmholtz equation in $\mathbb{R}^d \setminus \overline{\Omega_p^-}$ and Ω_p^-. Moreover, w has continuous trace and normal derivative through Γ_p since its jumps are given by $[w] = v$ and $[\partial_{\mathbf{n}_{\Gamma_p}} w] = \partial_{\mathbf{n}_{\Gamma_p}} v$. Consequently, $w \equiv 0$ in $\mathbb{R}^d$, and thus $u_p \equiv 0$ in $\mathbb{R}^d \setminus \overline{\Omega_p^-}$, which provides the desired contradiction.
In order to prove the existence of the solution $(u_1,\ldots,u_M)$ of problems (6), we show that these coupled scattering problems are equivalent to a system of M integral equations of Fredholm type. The announced existence follows then from the uniqueness result proved above. Let us seek the solution u_p in the form of a single-layer potential

$$u_p(\mathbf{x}) = \int_{\Gamma_p} G(\mathbf{x},\mathbf{y})\mu_p(\mathbf{y})d\Gamma_p(\mathbf{y}), \tag{8}$$

for all $\mathbf{x} \in \mathbb{R}^d \setminus \overline{\Omega_p^-}$, where the surface density $\mu_p \in H^{-1/2}(\Gamma_p)$. We define for all $p,q = 1,\ldots,M$, the integral operator

$$\mathscr{L}^{p,q}\mu_q(\mathbf{x}) = \int_{\Gamma_q} G(\mathbf{x},\mathbf{y})\mu_q(\mathbf{y})d\Gamma_q(\mathbf{y}), \quad \forall \mathbf{x} \in \Gamma_p.$$

Then, using the jump relations for the single layer potential (see Section 6.1), the M coupled boundary value problems (6), for $p = 1,...,M$, can be written in the abstract form

$$(\mathscr{L} + \mathscr{K})\mu = F, \tag{9}$$

provided we set $\mu := (\mu_1,\ldots,\mu_M)$ and

$$\mathscr{L} = \begin{bmatrix} \mathscr{L}^{1,1} & \ldots & 0 \\ \vdots & & \vdots \\ 0 & \ldots & \mathscr{L}^{M,M} \end{bmatrix},$$

$$\mathscr{K} = \begin{bmatrix} 0 & \mathscr{L}^{1,2} & \ldots & \mathscr{L}^{1,M} \\ \vdots & & & \vdots \\ \mathscr{L}^{M,1} & \mathscr{L}^{M,2} & \ldots & 0 \end{bmatrix},$$

$$F = -\begin{bmatrix} u^{\text{inc}}_{|\Gamma_1} \\ \vdots \\ u^{\text{inc}}_{|\Gamma_M} \end{bmatrix}.$$

Then, it is well-known (see Theorem 2 of Section 6.1) that the operators $\mathscr{L}^{p,p} : H^{-1/2}(\Gamma_p) \to H^{1/2}(\Gamma_p)$ are isomorphisms provided k is not an irregular frequency of the scatterer p, while, on the other hand, $\mathscr{L}^{p,q} : H^{-1/2}(\Gamma_q) \to H^{1/2}(\Gamma_q)$ are compact for $p \neq q$ (since their kernels are analytic). Therefore, recalling that

$$H^{\pm 1/2}(\Gamma) := H^{\pm 1/2}(\Gamma_1) \times \cdots \times H^{\pm 1/2}(\Gamma_M),$$

the operator $\mathscr{L} : H^{-1/2}(\Gamma) \to H^{1/2}(\Gamma)$ defines an isomorphism when k is not an irregular frequency of any of the scatterers, while $\mathscr{K} : H^{-1/2}(\Gamma) \to H^{1/2}(\Gamma)$ is a compact operator. Therefore, equation (9) is of Fredholm type and the proof is then complete. In the case where k is an irregular frequency of the scatterer p, one can use the trick of Brakhage-Werner to obtain the same result. □

3. A STRATEGY BASED ON FOURIER SERIES DECOMPOSITION FOR CIRCULAR CYLINDERS

After this introduction to multiple scattering, let us come to the possible strategies for solving numerically such problems. In this section, we will consider the particular case where the scatterers are circular cylinders ($d = 2$). Then, it is quite natural to solve the scattering problem using Fourier series technique. This modal approach, which is straightforward in the case of a single scatterer, becomes technically more involved in the case of many scatterers at both the theoretical level (the Fourier series expansions hold in the local system of coordinates associated to each disk) and the numerical level (because of multiple scattering effects). Furthermore, this kind of configuration leads to the introduction of nontrivial fast iterative algorithms which are the keystone for prospecting numerically high frequency multiple scattering problems. The results presented here have been developed in [11]. The extension to the three-dimensional problem is under progress.

3.1 Notation and Problem Formulation

Let us define u as the scattered field resulting from the illumination of M disks $\Omega_1^-,\ldots,\Omega_M^- \subset \mathbb{R}^2$, with boundaries $\Gamma_1,\ldots,\Gamma_M$, by the incident plane wave $u^{\text{inc}}(\mathbf{x}) = e^{ik\mathbf{d}\cdot\mathbf{x}}$, with $\mathbf{d} = (\cos(\theta^{\text{inc}}),\sin(\theta^{\text{inc}}))^{\mathbf{T}}$.

3.1.1 Local Fourier Series Expansions

According to Theorem 1, solving the scattering problem amounts to solving the M coupled single-obstacle scattering problems (6). To achieve this in the case where the scatterers are disks of radius a_p centered at the points $\mathbf{O}_p = (O_{1,p},O_{2,p})$ of a given orthonormal system of coordinates $(\mathbf{O}x_1,\mathbf{O}x_2)$, one can use Fourier series expansions in each local system of coordinates. More precisely, let us set for all $p = 1,...,M$:

$$\mathbf{b}_p = \mathbf{OO}_p \qquad b_p = |\mathbf{b}_p| \qquad \alpha_p = Angle(\mathbf{O}x_1,\mathbf{b}_p)$$

and for all $q = 1,\ldots,M$, with $q \neq p$:

$$\mathbf{b}_{pq} = \mathbf{O}_q\mathbf{O}_p \qquad b_{pq} = |\mathbf{b}_{pq}| \qquad \alpha_{pq} = Angle(\mathbf{O}x_1,\mathbf{b}_{pq}).$$

Any point $\mathbf{M}$ of the plane will be described either by its global polar coordinates

$$\mathbf{r} = \mathbf{OM} \qquad r = |\mathbf{r}| \qquad \theta = Angle(\mathbf{O}x_1,\mathbf{r}),$$

or by its local polar coordinates in the orthonormal system of coordinates associated to the scatterer p:

$$\mathbf{r}_p = \mathbf{O}_p\mathbf{M} \qquad r_p = |\mathbf{r}_p| \qquad \theta_p = Angle(\mathbf{O}x_1,\mathbf{r}_p).$$

Let us introduce for all $m \in \mathbb{Z}$ the global cylindrical wavefunctions

$$\begin{cases} \psi_m(\mathbf{r}) = H_m^{(1)}(kr)e^{im\theta}, \\ \widehat{\psi}_m(\mathbf{r}) = J_m(kr)e^{im\theta}, \end{cases}$$

and the corresponding local cylindrical wavefunctions in the system of coordinates associated with the scatterer p:

$$\begin{cases} \psi_m^p(\mathbf{r}) = \psi_m(\mathbf{r}_p) = H_m^{(1)}(kr_p)e^{im\theta_p}, \\ \widehat{\psi}_m^p(\mathbf{r}) = \widehat{\psi}_m(\mathbf{r}_p) = J_m(kr_p)e^{im\theta_p}, \end{cases} \quad \forall m \in \mathbb{Z}.$$

Since each field u_p is an outgoing solution of a single scattering problem outside a disk, it admits the following modal decomposition in the local cylindrical outgoing wavefunctions:

$$u_p(\mathbf{r}) = \sum_{m\in\mathbb{Z}} c_m^p \psi_m^p(\mathbf{r}), \ \forall r_p > a_p, \tag{10}$$

for $p = 1,...,M$.

3.1.2 Equations for the Fourier Coefficients

The unknown Fourier coefficients $(c_m^p)_{m\in\mathbb{Z}}$ can be determined by imposing that the modal decomposition (10) satisfies the boundary condition given in (6) on each boundary Γ_p. This requires to express the incident wave and the wavefunctions ψ_m^q for $q \neq p$ in the local system of coordinates of the scatterer p, as detailed in the next result (for the proof, we refer to [106, p. 125] and [106, Theorem 2.12]).

Proposition 1. *We have the two following results.*
1. The incident plane wave of direction $\mathbf{d}$ *admits the local Fourier series decomposition :*

$$u^{inc}(\mathbf{r}) = \sum_{m\in\mathbb{Z}} d_m^p \widehat{\psi}_m^p(\mathbf{r}) \tag{11}$$

where $d_m^p = e^{ik\theta^{inc}\cdot\mathbf{b}_p} e^{im(\pi/2-\theta^{inc})}$.
2. Separation Theorem: For all $1 \le p,q \le M$, *with* $p \neq q$, *and for all* $m,n \in \mathbb{Z}$ *set:*

$$\begin{cases} S_{mn}(\mathbf{b}_{pq}) = \psi_{m-n}(\mathbf{b}_{pq}), \\ \widehat{S}_{mn}(\mathbf{b}_{pq}) = \widehat{\psi}_{m-n}(\mathbf{b}_{pq}). \end{cases} \tag{12}$$

Then, for all $m \in \mathbb{Z}$, *we have:*

$$\psi_m^q(\mathbf{r}) = \begin{cases} \sum_{n\in\mathbb{Z}} S_{mn}(\mathbf{b}_{pq})\widehat{\psi}_n^p(\mathbf{r}), \ \text{for } r_p < b_{pq}, \\ \sum_{n\in\mathbb{Z}} \widehat{S}_{mn}(\mathbf{b}_{pq})\psi_n^p(\mathbf{r}), \ \text{for } r_p > b_{pq}. \end{cases} \tag{13}$$

The infinite matrices

$$\mathbb{S}^{p,q} = (S_{mn}(\mathbf{b}_{pq}))_{m,n\in\mathbb{Z}}$$

and

$$\widehat{\mathbb{S}}^{p,q} = (\widehat{S}_{mn}(\mathbf{b}_{pq}))_{m,n\in\mathbb{Z}}$$

are called separation (or transfer) matrices. Using relations (7), (10) and the first equation in (13), straightforward computations show that the unknown Fourier coefficients solve the following M coupled equations:

$$\mathbf{C}^p + \mathbb{D}^p \sum_{q=1,q\neq p}^{M} (\mathbb{S}^{p,q})^{\mathbf{T}} \mathbf{C}^q = \mathbf{B}^p \tag{14}$$

for $p = 1, ..., M$, where

- $\mathbf{C}^p = (c_n^p)_{n\in\mathbb{Z}}$ is the infinite vector containing the coefficients of the cylindrical decomposition (10) of u^p,
- $(\mathbb{S}^{p,q})^{\mathbf{T}}$ denotes the transpose of the separation matrix $\mathbb{S}^{p,q} = (S_{mn}(\mathbf{b}_{pq}))_{m,n\in\mathbb{Z}}$ between the obstacles Ω_p^- and Ω_q^-,
- $\mathbb{D}^p = (\mathbb{D}_{mn}^p)_{mn\in\mathbb{Z}}$ is the diagonal infinite matrix, with diagonal terms

$$\mathbb{D}_{m,m}^p = \begin{cases} \dfrac{J_n(ka_p)}{H_n^{(1)}(ka_p)} & \text{for sound-soft obstacles,} \\ \dfrac{J_n'(ka_p)}{H_n^{(1)'}(ka_p)} & \text{for sound-hard obstacles,} \end{cases}$$

- $\mathbf{B}^p = -\mathbb{D}^p \mathbf{d}^p$, where $\mathbf{d}^p = (d_m^p)_{m\in\mathbb{Z}}$ is the infinite vector containing the coefficients of the cylindrical decomposition (11) of the incident wave.

The M infinite linear systems (14) can equivalently be written in the abstract form

$$\mathbb{A}\mathbf{C} = \mathbf{B} \tag{15}$$

where

$$\mathbb{A} = \begin{bmatrix} \mathbb{I} & \mathbb{D}^1(\mathbb{S}^{1,2})^{\mathbf{T}} & \dots & \mathbb{D}^1(\mathbb{S}^{1,M})^{\mathbf{T}} \\ \mathbb{D}^2(\mathbb{S}^{2,1})^{\mathbf{T}} & \mathbb{I} & \dots & \mathbb{D}^2(\mathbb{S}^{2,M})^{\mathbf{T}} \\ \vdots & & \ddots & \\ \mathbb{D}^M(\mathbb{S}^{M,1})^{\mathbf{T}} & \mathbb{D}^M(\mathbb{S}^{M,2})^{\mathbf{T}} & \dots & \mathbb{I} \end{bmatrix}$$

$$\mathbf{C} = \begin{bmatrix} \mathbf{C}^1 \\ \vdots \\ \mathbf{C}^M \end{bmatrix} \qquad \mathbf{B} = \begin{bmatrix} \mathbf{B}^1 \\ \vdots \\ \mathbf{B}^M \end{bmatrix}$$

and where $\mathbb{I}$ denotes the identity operator on $\ell^2(\mathbf{C})$.

3.2 Approximation

In order to be solved numerically, the infinite linear system (15) must be truncated. In particular, we keep for the scatterer p the modes ψ_m^p corresponding to $-N_p \leq m \leq N_p$. Note that the number $2N_p + 1$ of significant modes might be different for each scatterer, in order to take into account geometrical configurations where the obstacles have different radii. The truncation of (15) leads then to the linear system:

$$\mathscr{A}\mathscr{C} = \mathscr{B} \tag{16}$$

where

- $\mathscr{A} \in \mathbb{C}^{N,N}$ is the full complex square matrix of size $N = \sum_{p=1}^{M} (2N_p + 1)$ given by

$$\mathscr{A} = \begin{bmatrix} \mathscr{I}^1 & \mathscr{D}^1(\mathscr{S}^{1,2})^{\mathbf{T}} & \dots & \mathscr{D}^1(\mathscr{S}^{1,M})^{\mathbf{T}} \\ \mathscr{D}^2(\mathscr{S}^{2,1})^{\mathbf{T}} & \mathscr{I}^2 & \dots & \mathscr{D}^2(\mathscr{S}^{2,M})^{\mathbf{T}} \\ \vdots & & \ddots & \\ \mathscr{D}^M(\mathscr{S}^{M,1})^{\mathbf{T}} & \mathscr{D}^M(\mathscr{S}^{M,2})^{\mathbf{T}} & \dots & \mathscr{I}^M \end{bmatrix}$$

where $\mathscr{I}^p$ denotes the identity matrix of size $2N_p + 1$, $\mathscr{D}^p = (\mathscr{D}_{mn}^p)_{-N_p\leq m\leq N_p, -N_q\leq n\leq N_q}$ is the diagonal finite matrix, with diagonal terms

$$\mathscr{D}_{m,m}^p = \begin{cases} \dfrac{J_m(ka_p)}{H_m^{(1)}(ka_p)} & \text{for sound-soft obstacles,} \\ \dfrac{J_m'(ka_p)}{H_m^{(1)'}(ka_p)} & \text{for sound-hard obstacles,} \end{cases}$$

and $\mathscr{S}^{p,q}$ is the $(2N_p + 1) \times (2N_q + 1)$ finite dimensional separation matrix taking into account only the interactions between the retained modes for the scatterers p and q:

$$\mathscr{S}^{p,q} = (\mathscr{S}_{mn}^{p,q})_{-N_p\leq m\leq N_p, -N_q\leq n\leq N_q}$$

with

$$\mathscr{S}_{mn}^{p,q} = \psi_{m-n}(\mathbf{b}_{pq}).$$

- $\mathscr{C} \in \mathbb{C}^N$ is

$$\mathscr{C} = \begin{bmatrix} \mathscr{C}^1 \\ \vdots \\ \mathscr{C}^M \end{bmatrix}$$

where $\mathscr{C}^p = (c_n^p)_{n=-N_p,...,N_p}$ is the finite vector containing approximations of the first $2N_p + 1$ modal coefficients of the cylindrical decomposition (11) of u_p (that are still denoted c_n^p for the sake of clarity).

- $\mathscr{B} \in \mathbb{C}^N$ is given by

$$\mathscr{B} = \begin{bmatrix} \mathscr{B}^1 \\ \vdots \\ \mathscr{B}^M \end{bmatrix}$$

where

$$\mathscr{B}^p = -\mathscr{D}^p d^p$$

with $d^p = (d_m^p)_{-N_p \leq m \leq N_p}$ is the finite vector containing the $2N_p + 1$ first coefficients of the cylindrical decomposition (10) of the incident wave.

The choice of the number of significant modes N_p to keep in the approximation to get an accurate solution is an important issue. On one hand, N_p must be large enough to catch both the propagating and grazing part of the solution (typically, $N_p \geq ka_p$). On the other hand, taking too many modes for approximating the solution leads to a stagnation of the iterative solver (see Section 3.4.1). Indeed, high order spatial modes $|m|$ correspond to the evanescent part of the field and therefore computing their Fourier coefficients is definitively out of reach using an iterative solver (with a fixed tolerance). For our simulations, we will use the following empirical formula

$$N_p = \left[ka_p + \left(\frac{\ln(2\sqrt{2}\pi ka_p\varepsilon^{-1})}{2\sqrt{2}}\right)^{2/3}(ka_p)^{1/3} + 1\right], \quad (17)$$

where $[x]$ denotes the integer part of a real number x and ε is the desired error bound on the Fourier coefficients. The above formula has been proposed in the literature in the contexts of single scattering [48] and multipole methods [42]. Nevertheless, according to our numerical results, it turns out that it can also be successfully used in the multiple scattering framework investigated here.

3.3 Iterative Solver

We want to be able to solve the linear system (16) not only for simple configurations, but also for complex ones corresponding to a large number M of scatterers and high frequencies ka_p. In this context, the use of direct methods is prohibitive for at least two reasons. First of all, according to (17), the number of modes $2N_p + 1$ needed to approximate the solution u_p with a reasonable precision becomes very large at high frequencies. Therefore, a direct method requires *a priori* a huge memory storage and large computational times to construct the dense matrix $\mathscr{A}$. Secondly, we are led to solve a large scale complex-valued linear system with size $N \times N$ (recall that $N = \sum_{p=1}^{M}(2N_p + 1)$). Using a direct linear solver would yield a prohibitive computational time for high frequencies and/or for large values of M. For these reasons, the use of an iterative solver for such complex configurations is unavoidable. We will use a Generalized Minimal RESidual algorithm, possibly with a restart parameter ρ (denoted by GMRES(ρ)) [111, 112], and the BICGStab algorithm. These choices lead to problems related to fast evaluations of dense matrix-vector products as well as convergence issues. The tolerance error of the iterative solver is set to tol and the number of iterations to get this tolerance is denoted by n^{iter}.

3.3.1 Storage

Although being full, the matrix $\mathscr{A}$ has a particular structure that we will exploit in our method. Indeed, its off-diagonal block (p,q) is obtained by multiplying the diagonal matrix $\mathscr{D}^p \in \mathbb{C}^{2N_p+1,2N_p+1}$ by the matrix $(\mathscr{S}^{p,q})^{\mathbf{T}} \in \mathbb{C}^{2N_p+1,2N_q+1}$ which has a Toeplitz structure [47] since

$$\mathscr{S}_{mn}^{p,q} = \psi_{m-n}(\mathbf{b}_{pq}).$$

Consequently, using the notations from [47], the storage of $(\mathscr{S}^{p,q})^{\mathbf{T}}$ can be optimized using a compressed version based on the root vector

$$\sigma^{p,q} = (\mathscr{S}_{N_q,-N_p}^{p,q}, \ldots, \mathscr{S}_{-N_q+1,-N_p}^{p,q}, \\ \mathscr{S}_{-N_q,-N_p}^{p,q}, \ldots, \mathscr{S}_{-N_q,N_p}^{p,q})^{\mathbf{T}}.$$

This leads us in our algorithms to store both the Toeplitz matrix $(\mathscr{S}^{p,q})^{\mathbf{T}}$ (through the root-vector $\sigma^{p,q}$) and the diagonal matrix $\mathscr{D}^p$. The compressed storage uses then $2(2N_p + N_q + 1)$ entries instead of the $(2N_p + 1)(2N_q + 2)$ complex coefficients required for the full version. For $\mathscr{A}$, this must be repeated for the $M(M-1)$ off-diagonal blocks by summing over p and q. This yields a global storage of $3N(M-1)$ entries, showing a clear saving compared to the $N^2 - NM$ entries needed in the full storage. Furthermore, the computational time involved in the construction of the global matrix is also reduced according to the memory storage. In the case where we have $a_p = a$ for all p, the vector root version of $\mathscr{A}$ leads to a memory storage and a CPU time of the order of $\mathscr{O}(6kaM^2)$ while it is $\mathscr{O}(4k^2a^2M^2)$ for the full version. This is a crucial point for solving a multiple scattering problem for a large wavenumber. To show the improvement induced by the Toeplitz based compressed storage version, we represent on Figure (**1**) the logarithm of the

CPU time[1] scaled by the computational memory requirement versus the wave number ka_p needed for building the global matrix $\mathscr{A}$. These computations have been done for the scattering by two sound-soft disks of radius 1 separated by a distance of 1. As expected, the CPU time for the compressed version is linear according to ka_p while it is quadratic for the full version.

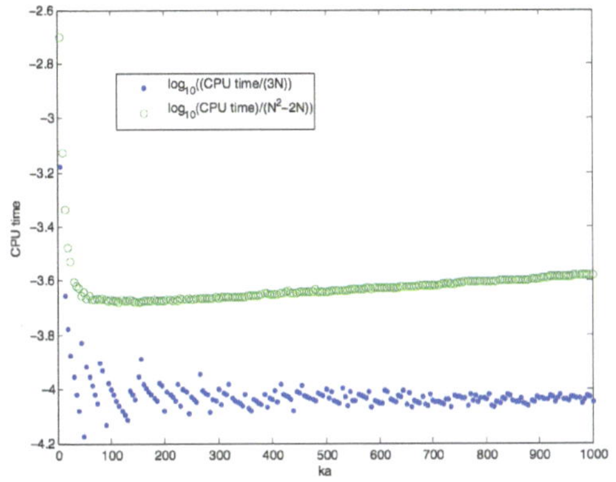

Fig. (**1**): Behaviour of the CPU time according to the wavenumber ka_p for building the matrix $\mathscr{A}$ in the case of the scattering by two circular cylinders fixing N_p by formula (17).

3.3.2 Fast Matrix-Vector Products

As it is well-known, the main CPU cost of each iteration of the GMRES or BICGStab algorithms is due to Matrix-Vector Products (MVPs). Taking advantage of the Toeplitz structure of the off-diagonal blocks of $\mathscr{A}$, the computation of a MVP: $\mathbf{y} = \mathscr{A}\mathbf{x}$ can be easily done by blocks. Let us set: $\mathbf{x} = (\mathbf{x}_1, ..., \mathbf{x}_M)$, with $\mathbf{x}_p \in \mathbb{C}^{2N_p+1}$, and $\mathbf{y} = (\mathbf{y}_1, ..., \mathbf{y}_M)$, with $\mathbf{y}_p \in \mathbb{C}^{2N_p+1}$. Then, we immediately get that

$$\begin{cases} \mathbf{y}_p = \mathbf{x}_p + \mathscr{D}^p \sum_{1 \leq q \neq p \leq M} \mathbf{z}_q, \\ \mathbf{z}_q = (\mathscr{S}^{p,q})^{\mathbf{T}} \mathbf{x}_q, \end{cases} \tag{18}$$

for $1 \leq p \leq M$. In the above evaluation of $\mathbf{y}_p$, the cost is mainly due to the computation of $\mathbf{z}_q$ which is quadratic according to $(2N_q + 1)$. Moreover, this must be repeated for each sub-block and each component of $\mathbf{y}_p$. This is very expensive when the frequency is large since the size of $(\mathscr{S}^{p,q})^{\mathbf{T}}$ is $(2N_p+1) \times (2N_q+1)$. Another way of computing a Toeplitz MVP for a matrix of size $n \times n$ is to use the fast algorithm detailed in [47, pp. 95–96] for MVPs involving Toeplitz matrices. The idea consists in building an associated circulant matrix using the Toeplitz matrix and next applying an FFT-based MVP algorithm for circulant matrices. This algorithm is coded using Matlab FFT function. The resulting total cost in terms of real operations for computing $\mathbf{y}_p$ is composed from: one complex-valued Toeplitz MVP in $15n\log_2(n) + n$ operations (see [35, p. 193]), with $n = 2N_q + 2N_p + 2$, summing up next on $1 \leq q \neq p \leq M$, the computation of the diagonal matrix $\mathscr{D}^p$ which is $6N_p + 3$, and finally adding $\mathbf{x}_p$. Again, summing up on $p = 1, ..., M$ gives the total cost. If $a_p = a$, then, this requires asymptotically $\mathscr{O}(60(M-1)^2 ka \log_2(4ka))$ operations compared to $\mathscr{O}(4(M-1)^2 ka^2)$ for a direct MVP. An example is given on Figure (**2**) showing the CPU time reduction with respect to ka (with $a = a_p$, $1 \leq p \leq M$) using the fast MVP algorithm compared to the direct algorithm for the scattering problem by $M = 30$ aligned disks of radius 1 and separated by a distance of 1 for $k = 100$.

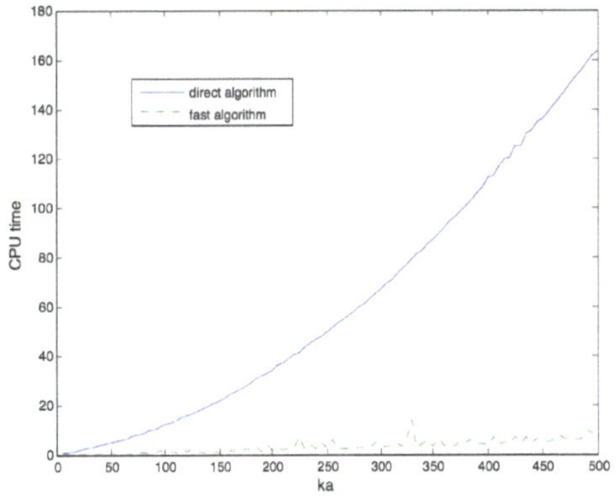

Fig. (**2**): Comparison of the CPU time for computing one MVP using the direct and fast algorithms for the scattering by $M = 30$ aligned disks of radius 1 and separated by a distance of 1 for $k = 100$.

3.4 Computational Tests

For our numerical tests, we consider three kinds of geometrical periodic configurations (see figures below):

Single-row configuration: it is composed of $M_1 = M$ equally spaced disks aligned along the x_1-axis, the distance between the centers of two successive scatterers being denoted $b_1 = b_{12}$.

[1] the computations in this Section have been performed on a Power Mac G4 1.67GHz with 1 Go DDR SDRM. The algorithms are developed under Matlab.

Regular line

Rectangular lattice: we consider a uniform rectangular lattice composed of $M = M_1 \times M_2$ circular disks (the structure is called square lattice if $M_1 = M_2$). For brevity, we restrict our experiments to a regular rectangular lattice which is composed of M_2 uniformly spaced single-rows with respect to $b_2 = b_{1(M_1+1)}$, each row being composed from M_1 equally spaced disks according to $b_1 = b_{12}$.

Rectangular lattice

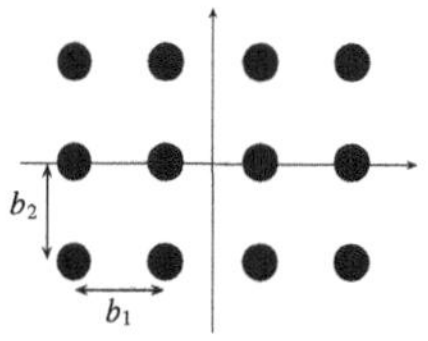

Triangular lattice: it is composed of two parallel horizontal single-rows configurations periodically repeated vertically. The first single-row configuration contains $M_1 \geq 2$ equally spaced disks and the second one $M_1 - 1$. The horizontal distance between two disks is b_1. The first row is then repeated vertically M_2 times with a uniform distance $b_2 = b_{1(2M_1)}$, while the second one is reproduced $M_2 \pm 1$ times with again a separation distance b_2.

Triangular lattice

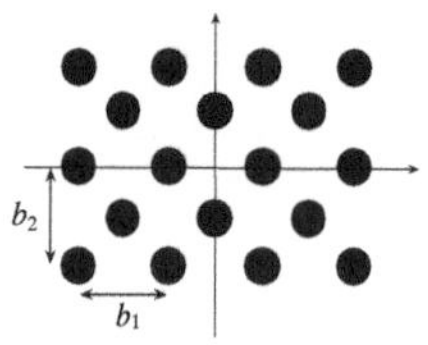

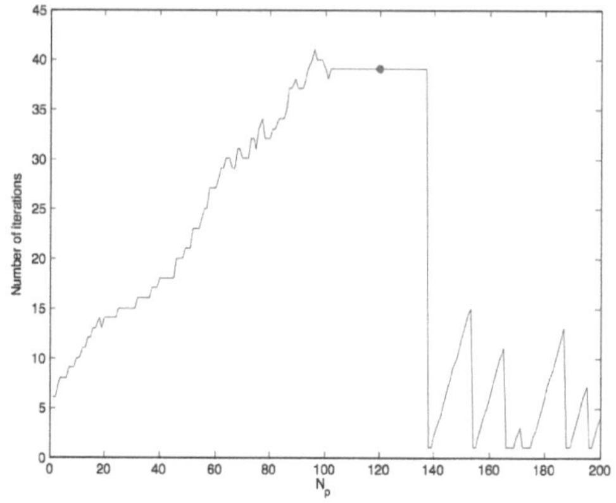

Fig. (**3**): Number of iterations with respect to N_p.

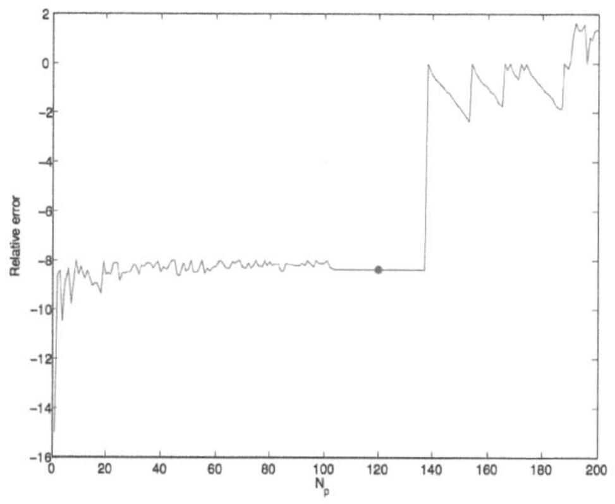

Fig. (**4**): Relative error with respect to N_p.

3.4.1 Influence of the Order of Truncation

As already said, by selecting the order of truncation N_p through formula (17), we expect N_p to be large enough to compute accurately the solution, but not too large to avoid the stagnation of the iterative solver. This statement is confirmed by the numerical experiments. Indeed, let us consider a uniform square lattice with $M_1 = M_2 = 2$ (so that $M = 4$) and $b_1 = b_2 = 3$ for a radius $a_p = 1$, $1 \leq p \leq M$, and $k = 100$. We solve (16) by the GMRES algorithm with $tol = 10^{-8}$. For a given value of the tolerance tol, we will always fix $\varepsilon = tol$ in formula (17). This gives here $N_p = 120$ (represented by a red dot on Figures (**3**) and (**4**)). We report the number of iterations n^{iter} and the relative errors versus N_p respectively on Figures (**3**) and (**4**).

One can distinguish of Figure (**3**) three different zones. First, from $N_p = 1$ to $N_p < ka_p + 2$, the number of iterations n^{iter} increases, showing that the computation of a correct solution requires more harmonics. This is achieved in the second stable zone (for $ka_p + 2 \leq N_p \leq 138$). However, if we include too many harmonics (third zone, $N_p \geq 139$ in our example), then we obtain a break down of the GMRES as it can be seen on Figures (**3**) and (**4**). Indeed, stagnation occurs e.g. for $N_p = 150$ while it does not for $N_p = 120$. In particular, the relative error corresponding to the stagnation at $N_p = 150$ is equal to 10^{-2} while it is tol for $N_p = 120$ (see Figure (**5**)).

3.4.2 Convergence Rate

From a large set of numerical simulations, the GMRES provides the fastest convergence rate (see Figure (**6**)). However, a more important memory storage is required which can significantly limit the possibility of prospecting high-frequencies for example. We present on Figure (**6**) the behaviour in terms of MVPs of the GMRES, BICGStab and GMRES(50) according to ka in the single-row configuration.

Figure (**6**) shows the dependence of the number of

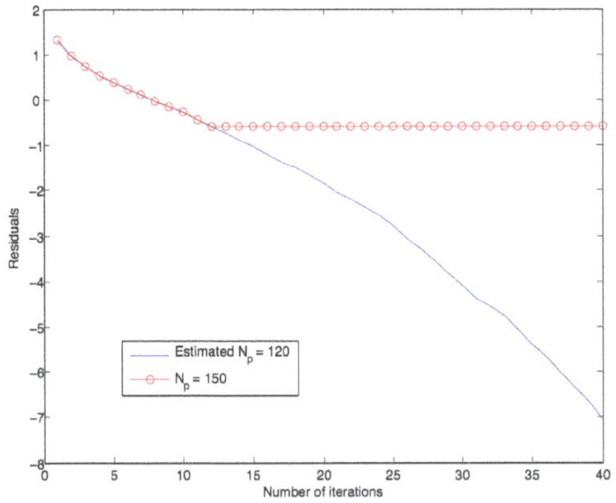

Fig. (**5**): Evolution of the residuals with respect to n^{iter} for $N_p = 120$ and $N_p = 150$.

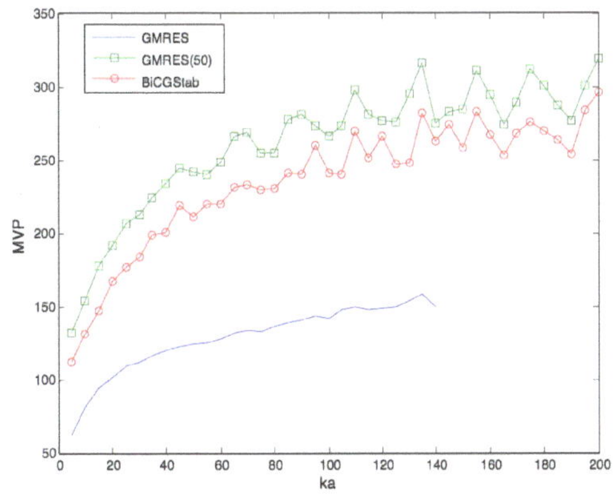

Fig. (**6**): Number of MVPs with respect to the wavenumber ka: single-row configuration with 75 obstacles and $b_1 = 3$ for different iterative solvers. The tolerance is $tol = 10^{-10}$ (Dirichlet problem).

MVPs versus the wavenumber ka. As expected, the number of MVPs increases with ka. Moreover, unlike BICGStab and GMRES(50), we can see that the GMRES algorithm breaks down at $ka = 140$ due to memory limitations. It also appears that the restarted GMRES generally leads to similar or better convergence results than the BICGStab. For these reasons, we choose the GMRES(50) in the sequel.

Single-row configuration. For a fixed wavenumber ($ka = 100$), we show here the dependence with respect to the number of obstacles M and the distance δ between two obstacles: $\delta = b_1 - a$.

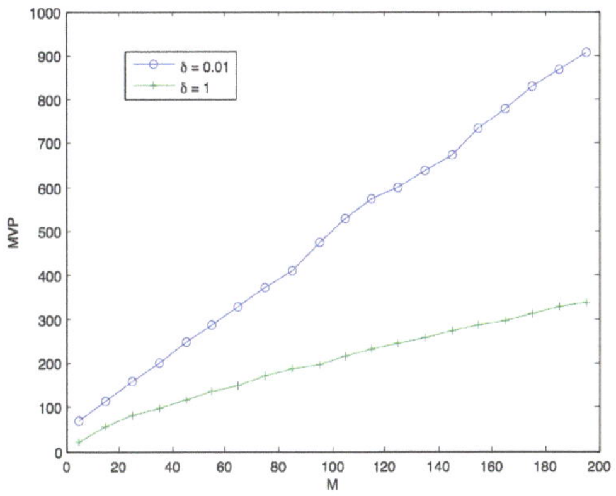

Fig. (**7**): Number of MVPs with respect to the number of obstacles M: single-row configuration with $ka = 100$ and two values of the separation distance δ. The GMRES(50) solver is used fixing $tol = 10^{-8}$ (Dirichlet problem).

We remark on Figure (**7**) that the number of MVPs increases linearly with M. Moreover, the slope of the line is more important as δ tends to zero (closer scatterers) due to stronger interactions. This effect is also observed on Figure (**8**) where the number of MVPs is represented as a function of δ in logarithmic scale for $ka = 100$ and $M = 10$. We see that the number of MVPs strongly decreases as the separation distance δ tends to infinity, i.e. $\delta \gg \lambda$ (weak coupling between the obstacles), while for small values of δ, $\delta \ll \lambda$, the number of MVPs strongly increases, because the linear system becomes ill-conditioned. Finally, we observe an intermediate resonance region for $\delta \approx \lambda$ where we have a few peaks in the number of MVPs.

A physical interpretation of this phenomenon is the following. In this regime, an approximate model is obtained by considering that for $ka \gg 1$, two adjacent scatterers behave like two parallel

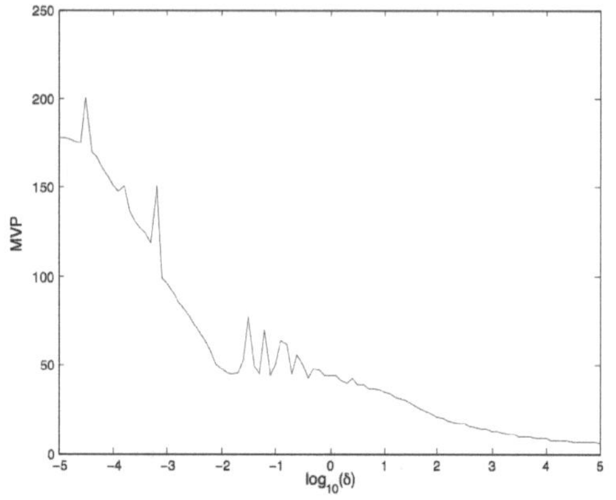

Fig. (**8**): Number of MVPs with respect to the distance δ between two obstacles. We fix: $ka = 100$, $M = 10$ obstacles. The solution is obtained with GMRES(50) for $tol = 10^{-8}$ (Dirichlet problem).

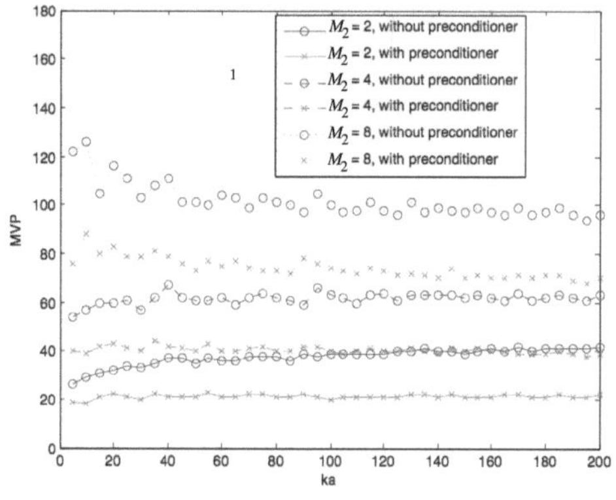

Fig. (**9**): Number of MVPs with respect to the wavenumber ka for the rectangular lattice configuration with $b_1 = 8$ and $b_2 = 13$. We fix $M_1 = 8$ and $M_2 = 2, 4, 8$. The solution is obtained without and with the preconditioned GMRES(50) for $tol = 10^{-8}$ (Dirichlet problem).

planes separated by a distance δ. Then resonances occur for such a configuration when $k\delta = n\pi \in \mathbb{N}^*$. This is confirmed for $n = 1, 2, 3, 4$, on Figure (**8**), corresponding respectively to $\delta = 3.\ 10^{-2}$, $6.\ 10^{-2}$, $1.2\ 10^{-1}$, $2.4\ 10^{-1}$. This generally creates a deterioration of the condition number at these frequencies, due to small eigenvalues in the matrix of the system. A similar problem also arises in the context of integral equations [49, 48, 8, 9].

Rectangular and triangular lattice configurations. A first test-case is given on Figure (**9**). We consider a rectangular lattice with $M_1 = 8$ and increase the size of layers according to $M_2 = 2$, 4, 8. We fix $b_1 = 8$ and $b_2 = 13$. The number of MVPs required by the GMRES(50) is represented as a function of ka.

We observe a stabilization with the frequency but the number of MVPs increases with M_2, and thus, with M. This is consistent with the previous observations in the single-row case. But this situation is, in some sense, not extreme because the distance between the obstacles is large enough.

Figure (**10**) shows the results corresponding to a more delicate situation. We represent the number of iterations for GMRES(50) of an $5 \times M_2$ lattice, for different values of M_2, according to ka for $b_1 = b_2 = 3$ (the separation distance is 1 in this case). We observe that the number of MVPs is again slightly dependent of ka but strongly varies with the number of layers, characterized by M_2. Moreover, some peaks appear at some frequencies, and in particular at $ka \approx 20$. This is more clearly visible on Figure (**11**) where we increase the lattice size according to $M = M_1^2$ at $ka = 50$. We notice that the number of MVPs increases strongly with M_1. In particular, on this example, it can be shown that, on the first values, the number of MVPs behaves like $3M^{3.3}$ and not quadratically with M. Another way of considering particular frequencies where peaks occur consists in modifying δ. Some numerical computations, not reported here, confirm this property.

Concerning the triangular lattice, similar conclusions can be drawn. However, it appears that this situation is less dramatic in terms of MVPs compared to the rectangular lattice case.

3.4.3 Preconditioning

Since many iterations may be necessary in some situations, one way to improve the rate of convergence is to precondition the linear system. In our context, we cannot directly apply an algebraic strategy like the incomplete LU or SPAI preconditioners [111, 47] which would require to reconstruct the full version of the matrix $\mathscr{A}$. An alternative direction is to build a geometrically-based preconditioner. We propose here a simple procedure in two steps.

First, let us set : $\mathscr{A} = \mathscr{I} + \mathscr{F}$, where $\mathscr{I} = \text{diag}((\mathscr{I}_p)_{1 \le p \le M})$ is the identity diagonal block of $\mathscr{A}$ and $\mathscr{F} = \mathscr{A} - \mathscr{I}$ is its complement off-diagonal part. The first-order Neumann series approximation for $\mathscr{A}^{-1}$ yields

$$\mathscr{A}^{-1} \approx \mathscr{I} - \mathscr{F} := \mathscr{P}. \tag{19}$$

In fact, there is no reason to assume that $\mathscr{F}$ satisfies

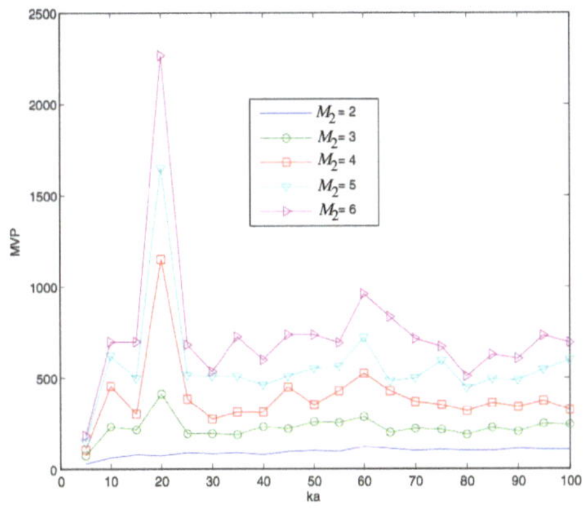

Fig. (**10**): Number of MVPs with respect to the wavenumber *ka*: rectangular lattice configurations with $M_1 = 5, M_2 = 2,...,6$. We consider $b_1 = b_2 = 3$ and solve the linear system using the GMRES(50) solver for $tol = 10^{-8}$ (Dirichlet problem).

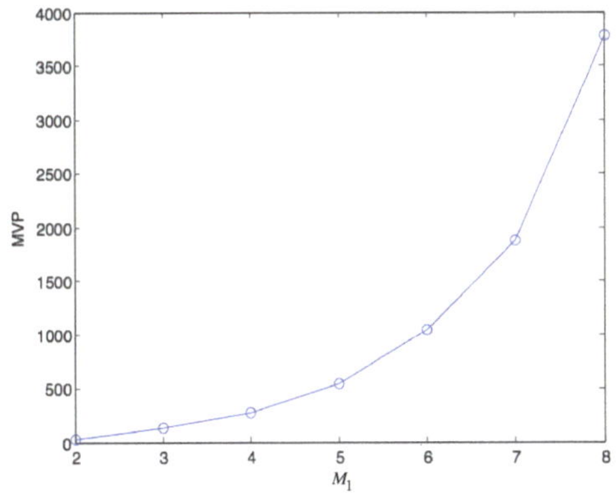

Fig. (**11**): Number of MVPs with respect to $M_1 = \sqrt{M}$ (setting $M_2 = M_1$): rectangular lattice configuration with $ka = 50$ and $b_1 = b_2 = 3$. The GMRES(50) solver is used by fixing $tol = 10^{-8}$ (Dirichlet problem).

$\rho(\mathscr{F}) < 1$, where $\rho(\mathscr{F})$ stands for the spectral radius of $\mathscr{F}$. However, relation (19) must be viewed as a formal way of building an approximation of the inverse of $\mathscr{A}$ and so has also a subjacent limitation range. It could be possible to choose more terms in the approximation. However, extensive numerical tests show that this is not a good strategy, that may even lead to the divergence of the method.

Since $\mathscr{P}$ is still a matrix involving all the interactions between the obstacles, it is interesting to reduce its application cost by considering only the closest interactions. This can be done through a second approximation by introducing a parameter $\mathrm{d} > 0$ representing a maximal coupling interaction distance. Then, the preconditioner, denoted by $\mathscr{P}_{\mathrm{d}}$, only considers the interactions between obstacles with indices $1 \leq p,q \leq M$ satisfying: $b_{pq} < \mathrm{d}$. We must notice that the construction of $\mathscr{P}_{\mathrm{d}}$ is implicit from $\mathscr{A}$ and does not require any extra cost. From intensive numerical experiments, it appears that $\mathrm{d} = b_1$ is an optimal choice for the single-row configuration while $\mathrm{d} = \max(b_1, b_2)$ is the best choice for both the regular rectangular and triangular lattices. Taking a smaller or larger value yields a slower convergence or sometimes divergence. With this choice, the application of the preconditioner requires a negligeable additional cost compared to the version of the solver without preconditioner.

To show the improvement induced by the proposed preconditioner, we present on Figure (**12**) the number of MVPs for a single-row configuration with $M = 75$ obstacles. We observe the improvement in terms of convergence rate if we compare these results to the ones obtained on Figure (**6**). The preconditioner always improves the convergence even for very close scatterers.

The situation is less clear for rectangular or triangular lattices. We report on Figure (**9**) the results obtained for different layers $M_2 = 2,4,8$, setting $M_1 = 8$. The distance between the obstacles is $b_1 = 8$ and $b_2 = 13$ (and so $\mathrm{d} = 13$). We see an interesting gain in terms of reduction of iterations. However, it appears that for situations where the scatterers are close ($b_1 - 2a \approx b_2 - 2a \leq a = \min\{a_p\}_{1 \leq p \leq M}$), the preconditioner is not robust and can even lead to a deterioration of the convergence. This means that, for this kind of configuration, more efforts must be done for building a suitable preconditioner.

We conclude this Section by analyzing the performance of our numerical method for an unstructured geometrical configuration. We consider 60 unit circular cylinders (see Figure (**13**)) which are supposed to be distant enough. More precisely, we assume

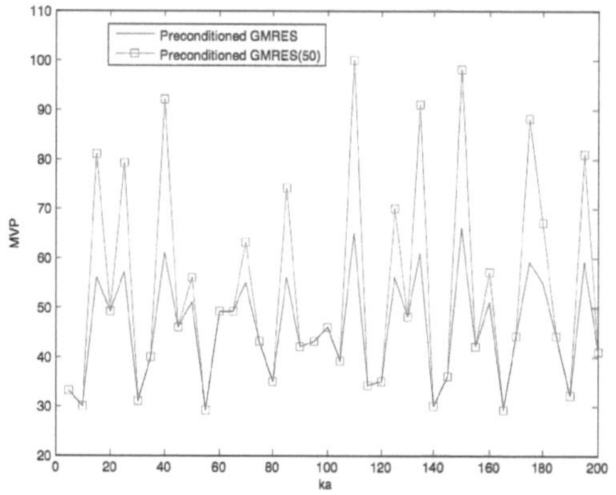

Fig. (**12**): Number of MVPs with respect to the wavenumber ka for the single-row configuration with $M = 75$ and $b_1 = 3$. The solution is obtained with the preconditioned GMRES and GMRES(50) for $tol = 10^{-10}$ (Dirichlet problem).

that $b_{min} := \inf_{1 \leq p < q \leq M} b_{pq} \geq 3$.

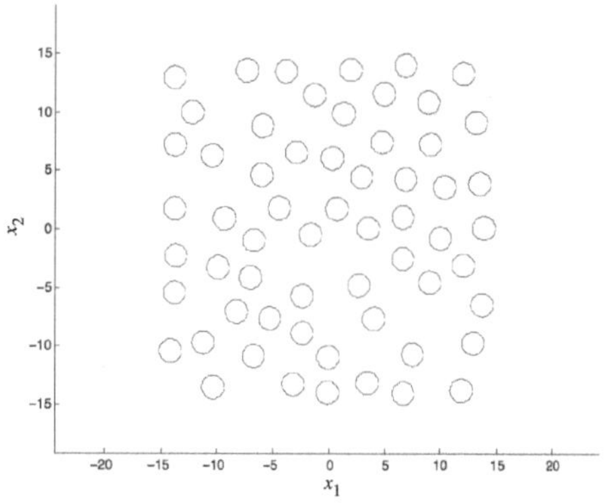

Fig. (**13**): The unstructured configuration.

Figure (**14**) shows the convergence rate of our algorithm without preconditioner and with the preconditioner $\mathscr{P}_{\mathrm{d}}$ for $\mathrm{d} = 1.5 b_{min}$. Once again, we note a faster convergence for the preconditioned algorithm.

3.5 Computation of Physical Fields

Let us emphasize that many physical quantities of interest can be deduced from the solution of (15) or by truncation to its finite dimensional approximation (16). In particular, we can compute the field u in a neighborhood of the obstacles, the normal derivative trace (respectively trace) on each obstacle for the Dirichlet (respectively Neumann) problem or the far-field pattern (see the expressions in [11]). Con-

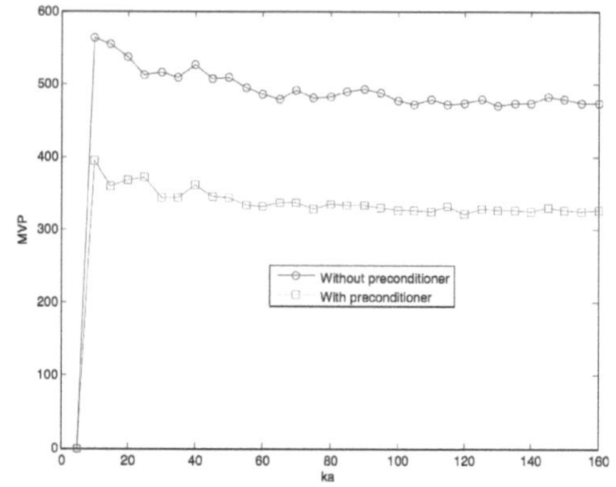

Fig. (**14**): Number of MVPs with respect to the wavenumber ka for an unstructured configuration (with and without preconditioner).

cerning this last quantity, we can prove that the scattering amplitude $a(\theta)$ defined by

$$u(\mathbf{r}) = \frac{e^{ikr}}{\sqrt{r}} a(\theta) + \mathscr{O}\left(\frac{1}{r}\right), \qquad \text{as } r \to +\infty \quad (20)$$

is obtained by

$$\left\{ \begin{array}{l} a(\theta) = e^{-i\pi/4} \sqrt{\dfrac{2}{\pi k}} \displaystyle\sum_{p=1}^{p=M} a_p(\theta), \\ a_p(\theta) = e^{-ib_p k \cos(\theta - \alpha_p)} \left(\displaystyle\sum_{n=-N_p}^{N_p} e^{in(\theta - \frac{\pi}{2})} c_n^p \right). \end{array} \right. \quad (21)$$

In many scattering industrial applications, one is interested in the computation of the far-field, also called Radar (or Sonar) Cross Section (RCS). The RCS is given (in deciBels) by the expression

$$\mathrm{RCS}(\theta) := 10 \log_{10}(2\pi |a(\theta)]^2) \quad (\mathrm{dB}). \quad (22)$$

3.6 Conclusion

We have seen in this Section a first method to solve multiple scattering problems at high frequency. The method is efficient, robust and can be applied to solve problems by complex structures composed of randomly or periodically distributed circular scatterers. We propose to develop in the next Sections some methods and algorithms to solve multiple scattering problems at high frequency for scatterers of arbitrary shapes. More precisely, we review the recent developments and provide one numerical example for the two most commonly used strategies: PDE based methods (Sections 3.6 and 4.3) and integral approaches (Sections 5.2 and 6.4).

4. PDE BASED STRATEGIES

A first strategy, called Partial Differential Equation (PDE) based approach, consists in solving the Helmholtz equation in a finite computational domain through a particular numerical method : finite elements, finite differences, spectral elements or infinite elements. We cannot describe here all these methods in detail. We first propose to decompose these procedures into three basic steps and review the related current developments. We will essentially discuss the following issues:

- Truncation of the infinite domain of computation (through DtN maps, Absorbing Boundary Conditions (ABCs) or Perfectly Matched Layers (PMLs)) (Section 3.4.1)
- Formulations and finite element approximation (Section 4.2)
- Solution of the resulting linear system through an iterative solver and question of preconditioning (Section 4.3).

In a second part, we propose to detail a recent approach called the Phase Reduction Finite Element Method (PR-FEM), originally introduced to solve high frequency single scattering problems. We present a possible extension of this algorithm for scattering by multiple convex obstacles and provide computational examples.

4.1 Domain Truncation

Let us consider problem (1). Since it is set in an unbounded domain, it is clear that a suitable way of truncating Ω^+ must be considered. More precisely, a fictitious boundary Σ enclosing Ω^- is introduced to define a finite domain of computation Ω. Of course, a suitable boundary condition must be written on Σ. Different strategies can be developed to achieve this, but all these approaches require a compromise between

- Accuracy
- Facility of implementation into an existing basic code
- Computational efficiency.

We usually distinguish two classes of methods, depending on the way that the solution computed in the bounded domain approximates the real one. In the first class of methods known as **exact (or transparent) methods** (Dirichlet-to-Neumann boundary condition, artificial or non-reflecting boundary conditions), the solution of the problem set in the bounded domain Ω is exactly the restriction on Ω of the exact solution. On the contrary, for **approximate methods**, the domain truncation generates an approximation error.

4.1.1 Exact (Transparent) Methods

A transparent boundary condition can be expressed as an integral operator set on Σ (for instance through the integral representation formula, see e.g. [93]). Even if this seems a suitable solution, it is not applicable in practice and most particulary if one wishes to solve a high frequency problem. Indeed, while we are trying to solve a local (PDE) equation, the nonlocal form of the integral boundary condition destroys the sparse matrix structure of the resulting discrete version of the PDE. This finally strongly affects the efficiency of the overall algorithm and in fact limits its applicability for realistic practical problems. In the special case of canonical shapes, the integral equation can be rewritten as a Dirichlet-to-Neumann (DtN) map

$$\partial_{\mathbf{n}_\Sigma} u = \Lambda_\Sigma u, \text{on } \Sigma, \tag{23}$$

where Λ_Σ is an operator which depends on the shape of Σ and $\mathbf{n}_\Sigma$ is the outwardly directed unit normal vector to Σ. For example, in the case of the sphere, Givoli and Keller [97] introduced its expression in terms of spherical harmonics for an implementation into a Galerkin Finite Element Method (FEM). The DtN map, represented through an infinite series, is truncated in practice retaining a significant number of modes which strongly increases according to the wavenumber k. Furthermore, even if the expansion is finite, the truncated (nonlocal) DtN map gives a dense contribution in the FEM sparse matrix. More theoretical and implementation details can be obtained e.g. in References [78, 81, 84, 89, 110, 120] but its application to a scattering problem is restricted to moderate wavenumbers k.

4.1.2 Approximate Methods

Instead of an exact and nonlocal boundary condition, an *approximate* but *local* boundary condition is generally preferred. It is referred to as an artificial or **Absorbing Boundary Condition (ABC)**. Among the most widely used ABCs are the Engquist-Majda [64, 65] and Bayliss-Gunzburger-Turkel (BGT) [21, 22] boundary conditions. All these boundary conditions have been derived in the case of a circular or a spherical boundary. An example is given by the second-order BGT ABC derived for a circular boundary $\Sigma = C_R$ of radius R

$$\partial_{\mathbf{n}_\Sigma} u = \mathscr{B}_2 u, \text{on } C_R, \tag{24}$$

where the local operator $\mathscr{B}_2$ is given by

$$\mathscr{B}_2 u = -\frac{\alpha_2}{R^2}\partial_\theta^2 u + \beta_2 u, \tag{25}$$

with

$$\alpha_2 = -\frac{1}{2(R^{-1} - ik)} \tag{26}$$

and

$$\beta_2 = ik - \frac{1}{2R} + \frac{R^{-2}}{8(R^{-1} - ik)}. \tag{27}$$

Function u must be understood here as an approximation of the exact wavefield solution to the BVP (1) since the boundary condition (24)-(27) is not exact. This also means that spurious unphysical reflection arises at the boundary Σ. The boundary condition can be easily implemented into an existing basic code for solving a PDE and yields a sparse matrix representation. Since the pioneering works of Engquist-Majda and Bayliss-Gunzburger-Turkel, many developments have been made. For example, an extension of the BGT2 boundary condition has been developed in [7] for arbitrarily shaped convex boundaries Σ. These boundary conditions have been implemented in the context of the OSRC techniques [5, 6] and as ABCs in [61, 119] using a FEM. An example of such an ABC is given by the second-order BGT (denoted by BGT2 in the sequel)

$$\partial_{\mathbf{n}_\Sigma} u = \mathscr{B}_2 u, \text{on } \Sigma. \tag{28}$$

The surface operator $\mathscr{B}_2$ is expressed as

$$\mathscr{B}_2 u = -\partial_s(\alpha_2 \partial_s u) + \beta_2 u, \tag{29}$$

setting

$$\alpha_2 = -\frac{1}{2(\kappa - ik)} \tag{30}$$

and

$$\beta_2 = ik - \frac{\kappa}{2} + \frac{\kappa^2}{8(\kappa - ik)}. \tag{31}$$

The variable s is the anticlockwise directed curvilinear abscissa along the boundary Σ and κ is the curvature of Σ at a point s. These ABCs, which work well for all frequencies, however require a large domain of computation which can be too prohibitive for high frequencies. Improved high-order local ABCs based on rational square-root approximations [79, 94] can be derived to reduce the computational domain, and most particularly for large wavenumbers. Moreover, these boundary conditions preserve the sparsity pattern of the finite element matrix and can be easily coupled to advanced (plane wave) FEM [95] for two- and three-dimensional problems. Another well-known and standard procedure to bound the computational domain is the so-called **Perfectly Matched Layers (PML)** method introduced by Bérenger [25, 26, 27] in the context of time-domain Maxwell's equations. Unlike the ABCs approach where a fictitious boundary is used, the PMLs technique requires the introduction of a surrounding dissipative volumetric layer to bound the computational domain. Essentially, the two-dimensional Helmholtz equation has the following form in the layer

$$\partial_{x_1}\left(\frac{S_{x_2}}{S_{x_1}}\partial_{x_1} u\right) + \partial_{x_2}\left(\frac{S_{x_1}}{S_{x_2}}\partial_{x_2} u\right) + S_{x_1} S_{x_2} k^2 u = 0,$$

where S_{x_1} and S_{x_2} are two complex-valued functions

$$S_{x_j} = f_{x_j} + \frac{g_{x_j}}{ik}, j = 1,2.$$

Functions f_{x_j} are equal to 1 at the inner layer's interface. Dissipation functions g_{x_j} vary between zero at the inner interface and a maximal value at the outer interface of the layer. Their role is to damp evanescent waves into the layer. At the continuous level, there is no reflection for all wavenumbers and angles of incidence of the scattered field. When a discretization is used, this no longer the case and special care must be paid to the discretization, width of the layer for limiting the memory storage and accuracy and choice of the involved functions. Discussions about PMLs for the Helmhotz equation and the tuning of the parameters can be found e.g. in References [28, 53, 115, 123].

All these approaches can be adapted to multiple scattering problems if the scatterers Ω_p are close enough to be embedded in a single computational domain. In the case where some of the scatterers are far, the size of the bounded domain may be too large for a realistic calculation. Then, it is necessary to bound each group of close scatterers within a separate bounded domain. This can be done as it is proposed e.g. in Reference [82].

4.2 Finite Element Approximations

Once the domain has been truncated, one must solve the resulting bounded BVP problem in Ω using a suitable numerical discretization technique.

4.2.1 Classical Approach

Let us begin by introducing the classical weak formulation and the finite element method to solve (1). We consider a smooth convex fictitious boundary Σ enclosing the scatterer Ω^- and we set Ω as the bounded computational domain delimited by Γ and Σ. On Σ, we consider the BGT2 boundary condition (28)-(31) (where its 3D extension is given in [7]) and restrict the developments to the two-dimensional case (see [14] for the three-dimensional case which is similar). It results that the truncated BVP is given by

$$\begin{cases} \Delta u + k^2 u = 0, \text{ in } \Omega, \\ \partial_{\mathbf{n}_\Gamma} u = -\partial_{\mathbf{n}_\Gamma} u^{\text{inc}} \text{ or } u = -u^{\text{inc}}, \text{on } \Gamma, \\ \partial_{\mathbf{n}_\Sigma} u = \mathscr{B}_2 u, \text{on } \Sigma. \end{cases} \quad (32)$$

For the Neumann problem, the variational formulation leads to computing $u \in H^1(\Omega)$ such that

$$a(u,v) = \ell(v), \quad (33)$$

for any test-function $v \in H^1(\Omega)$, with $a(\cdot,\cdot)$ defined by

$$\begin{aligned} a(u,v) = (\nabla u, \nabla v)_{0,\Omega} - k^2(u,v)_{0,\Omega} \\ + (\alpha_2 \partial_s u, \partial_s v)_{0,\Sigma} + (\beta_2 u, v)_{0,\Sigma} \end{aligned} \quad (34)$$

and ℓ is

$$\ell(v) = (\partial_{\mathbf{n}_\Gamma} u^{\text{inc}}, v)_{0,\Gamma}. \quad (35)$$

The finite element solution consists in introducing a covering Ω_h of Ω using triangles K: $\Omega_h = \cup_{K \in \mathscr{K}_h} K$, where $\mathscr{K}_h$ designates a triangulation of the domain. The corresponding interpolated boundaries associated with Γ and Σ are respectively denoted by Γ_h and Σ_h. The p-finite element version of (33) yields the discrete formulation: find $u_h \in V_h$ such that

$$a_h(u_h, v_h) = \ell_h(v_h), \quad (36)$$

for any test-function v_h of V_h, setting

$$\begin{aligned} a_h(u_h, v_h) = (\nabla u_h, \nabla v_h)_{0,\Omega_h} - k^2(u_h, v_h)_{0,\Omega_h} \\ + (\alpha_{2,h} \partial_{s_h} u_h, \partial_{s_h} v_h)_{0,\Sigma_h} + (\beta_{2,h} u_h, v_h)_{0,\Sigma_h} \end{aligned} \quad (37)$$

and

$$\ell_h(v_h) = (\partial_{\mathbf{n}_{\Gamma_h}} u^{\text{inc}}, v_h)_{0,\Gamma_h}. \quad (38)$$

The classical finite element space of order p is given by

$$V_h := \{v_h \in \mathscr{C}^0(\overline{\Omega_h}) / v_{h\,|K} \in \mathbb{P}_m(K), \forall K \in \mathscr{K}_h\}, \quad (39)$$

where $\mathbb{P}_m$ denotes the space of polynomials of degree less than or equal to m. The approximate fields $\alpha_{2,h}$ and $\beta_{2,h}$ are computed by some suitable schemes based on the surface mesh. We refer to [5] for implementation details. Finally, the solution of (36) leads to the solution of a sparse linear system like (41), with size $n_h \times n_h$, denoting by n_h the number of degrees of freedom associated with the finite element approximation. The whole procedure is referred to as FEM in what follows. Concerning the Dirichlet problem, a similar weak formulation can be obtained in a classical way. We do not detail here this point which is immediate.

4.2.2 Pollution and Discretization for Small Wavelengths

In the situation of high-frequency, a difficult bottleneck occurs concerning the accuracy with which a solution is computed. More specifically, when Galerkin's approximation methods are used, then phase (dispersion) errors appear in the numerical solution u_h. This implies that the size of the mesh step h must be adapted according to the wavenumber k (or wavelength λ). For example, the typical rule of the thumb of "ten points per wavelength" is not valid. This problem has been most particularly studied in detail by Ihlenburg and Babuska [90, 91] who proved that the following error estimate

$$||u - u_h||_{0,\Omega} \leq (C_1 + C_2 kL)(kh)^2 \quad (40)$$

holds between the continuous solution u and its approximation u_h using linear finite elements in the one-dimensional case (L is the characteristic size of the computational domain).

The first error term, related to C_1, is the classical interpolation error of the FEM while the second one, related to C_2, corresponds to the loss of stability of the Helmholtz operator for large wavenumbers k. This last error is usually called "pollution error". In the case of two- and three-dimensional problems, this error leads to considering huge meshes limiting therefore the direct application of usual low-order polynomial basis functions. For this reason, new formulations of the problem or/and the use of alternative basis functions have been prospected over the recent years. Our aim here is not to explain all the methods but rather to refer to some of the most useful and promising approaches, essentially focusing on Finite Element Methods (FEMs).

A direct way to reduce the pollution error is to increase the order p of the polynomial basis functions [2, 60, 90]. It is then proved that the pollution error in the $H^1(\Omega)$-norm for the one-dimensional case behaves like $kL(kh/2p)^{2p}$, showing that the pollution error decreases as the order p increases. This however finds its limitation for higher dimensions and high frequencies since using high-order functions introduces many new degrees of freedom modifying therefore the sparsity structure of the global stiffness and mass matrices. Concerning the use of low-order elements, a simple strategy consists in considering a Galerkin Least Squares (GLS) approach by modifying the initial variational formulation through the addition of a stabilization term [85]. This improves the accuracy of the solution depending on a penalization parameter τ which must be locally adapted according to the FEM [83, 85, 86, 94]. Finally, the GLS methods can be obtained by the variational multiscale approach [87].

Another popular viewpoint considers modified basis functions. The idea is to enrich the approximation space by adding information coming from special analytical solutions like plane waves. Many approaches exist [17, 103, 104]. For instance, the partition-of-unity method [17, 95, 108] considers standard polynomial basis functions multiplied by plane waves. Increased accuracy can be obtained but at the price of being able to efficiently integrate highly oscillatory functions [31]. Let us note also that discontinuous versions exist like the ultra-weak variational formulation [45, 44, 88] or the discontinuous enrichment method [67, 68]. Finally, an important remark is that the resulting linear systems associated with these methods are highly ill-conditioned [104] leading to the breakdown of an iterative solver for large scale problems.

Since we are solving a high-frequency problem, then asymptotic methods are available for computing approximate solutions to the scattering problem (see e.g. [98]). An alternative to solving the diffraction problem in terms of scattered field u is to inject asymptotic and partial informations about the solution in the problem formulation and then to compute the remaining part of the field by a numerical method. Hybrid asymptotic methods in this direction have been introduced by Giladi and Keller [77] or also in [14, 72] by developing a Phase Reduction FEM (PR-FEM) (explained in detail in Section 4.3).

Finally, a last approach is the so-called infinite element method introduced by Bettess [29]. Their goal is to replace the ABC. We do not detail here the method and refer to [15, 16, 30, 120] for a survey of these techniques.

All these groups of methods are developed mainly for single scattering problems. To the best of our knowledge, these techniques have not been adapted yet to problems where multiple scattering occurs. We will see in the sequel an original algorithm extending the PR-FEM to scattering from several separate objects.

4.3 Iterative Solutions – Preconditioning

The use of one of the previous FEM leads to the solution of a linear system of the form

$$[A_h]\mathbf{u}_h = \mathbf{b}_h. \qquad (41)$$

Let us denote by n_h the number of degrees of freedom resulting from one of the above finite element procedures. Then, the matrix $[A_h]$ is sparse, complex-valued and of size $n_h \times n_h$. It can be symmetric or not, according to the ABC or PML. The value of n_h is supposed to be large since we are solving a high-frequency problem, even if it is reduced by a suitable choice of both the ABC/PML and FEM. Finally, the matrix is highly indefinite since the Helmholtz operator is not a positive operator, most particularly for large wavenumbers k.

For our problem of interest, n_h has a value of a few millions or more, even in the two-dimensional case. This implies that a direct solution based on a gaussian elimination solver if out of reach. For this reason, the alternative is to consider a Krylov iterative solution as for example by using the GMRES [111, 112]. Because the system is highly non definite, then the iterative procedure does not generally converge. To avoid the convergence breakdown, the use of a preconditioner $[P_h]$ is required. This preconditioner is built such that we get an explicit approximation of the inverse of $[A_h]$, $[P_h] \approx [A_h]^{-1}$, or an implicit approximation of $[A_h]$ which can be easily "inverted" through the solution to a linear system, $[P_h]^{-1} \approx [A_h]$. Then, instead of (41), we solve

$$[P_h][A_h]\mathbf{u}_h = [P_h]\mathbf{b}_h. \qquad (42)$$

The construction of a robust and efficient preconditioner is a hard task for a highly indefinite linear system. Different solutions have been proposed as ILUT methods like in [96] which however can breakdown for high wavenumbers. Another solution is based on the complex shifted Laplace preconditioner [20, 66, 122], with the aim to reinforce the diagonal dominance of $[A_h]$. Alternative solutions include domain decomposition methods like e.g. the FETI-H methods [23, 52, 59, 69, 70, 118]. Finally, more details and references can also be found in the review papers [120, 121].

Let us remark that all these techniques can be directly applied to multiple scattering problems and most particularly in the case of periodic structures.

5. AN EXAMPLE: THE PHASE REDUCTION FINITE ELEMENT METHOD (PR-FEM)

We present in this Section an advanced algorithm based on a hybrid finite element procedure for computing the field scattered from a collection of separate obstacles. The idea is to use a Gauss-Seidel algorithm to come back to the successive solution of single-scattering problems and to adapt the recent Phase Reduction Finite Element Method (PR-FEM) introduced in [14, 72]. The results presented here have been developed in [74]. The extension to the three-dimensional problem is under progress. Let us remark that other FEM solutions could be applied at this level as well.

5.1 PR-FEM for Single Scattering Configurations

For single scattering configurations (i.e., if Ω^- is a single convex obstacle), the idea of the PR-FEM is to approximate the phase ϕ of the solution $u = \mathbb{A}e^{ik\phi}$ and use this approximate phase $\tilde{\phi}$ to reformulate the problem in terms of a slowly oscillatory envelope $\tilde{\mathbb{A}} = ue^{-ik\tilde{\phi}}$ in order to reduce pollution effects in the future FEM discretization. This approach thus involves two steps:

1) find an approximation $\tilde{\phi}$ of the phase ϕ of u in the whole computational domain;
2) use $\tilde{\phi}$ to solve the scattering problem in terms of a new, slowly varying unknown $\tilde{\mathbb{A}}$.

This has the advantage that the resulting formulation can be easily coded into a classical finite element solver and does not require any integration of new basis functions. Moreover, this technique is not restricted to the finite element method, and can be used in other numerical schemes like finite difference, integral or spectral methods.
Point 1) is solved through the solution of an evolution equation in the exterior domain. It can be decomposed into two steps: the proposition of an initial condition through the OSRC techniques [5, 6, 102] and the construction of a propagator using pseudodifferential operator techniques. In [14] and in the present paper, we only consider the lowest order propagator but higher order propagators could be proposed. These high-order methods would lead to more accurate approximate phases $\tilde{\phi}$ and less oscillating new unknowns $\tilde{\mathbb{A}}$, most particularly for large frequencies. Another solution based on the eikonal equation solution ($||\nabla u||^2 = 1$ in Ω) was also proposed in [14] yielding interesting results and possibly other directions. Point 2) is direct since it is only a change of unknown into the variational formulation (34). This leads to a new variational equation of the following kind: find $\tilde{\mathbb{A}} \in H^1(\Omega)$ such that

$$\tilde{a}(\tilde{\mathbb{A}}, \mathbb{B}) = \tilde{\ell}(\mathbb{B}), \tag{43}$$

for any test-function $\mathbb{B} \in H^1(\Omega)$, where the bilinear form is given by:

$$\begin{aligned}\tilde{a}(\tilde{\mathbb{A}}, \mathbb{B}) &= (\nabla\tilde{\mathbb{A}}, \nabla\mathbb{B})_{0,\Omega} \\ &+ ik\left((\tilde{\mathbb{A}}\nabla\tilde{\phi}, \nabla\mathbb{B})_{0,\Omega} - (\nabla\tilde{\mathbb{A}}, \mathbb{B}\nabla\tilde{\phi})_{0,\Omega}\right) \\ &- k^2((1 - ||\nabla\tilde{\phi}||^2)\tilde{\mathbb{A}}, \mathbb{B}))_{0,\Omega} + (\alpha_2\partial_s\tilde{\mathbb{A}}, \partial_s\mathbb{B})_{0,\Sigma} \\ &+ ik\left(((\alpha_2\partial_s\tilde{\phi})\tilde{\mathbb{A}}, \partial_s\mathbb{B})_{0,\Sigma} - (\alpha_2\partial_s\tilde{\mathbb{A}}, (\partial_s\tilde{\phi})\mathbb{B})_{0,\Sigma}\right) \\ &+ k^2((\alpha_2\partial_s\tilde{\phi})\tilde{\mathbb{A}}, (\partial_s\tilde{\phi})\mathbb{B})_{0,\Sigma} + (\beta_2\tilde{\mathbb{A}}, \mathbb{B})_{0,\Sigma}\end{aligned} \tag{44}$$

and the linear form $\tilde{\ell}$ by

$$\tilde{\ell}(\mathbb{B}) = (\partial_{\mathbf{n}_\Gamma}u^{\text{inc}}e^{ik\tilde{\phi}}, \mathbb{B})_{0,\Gamma}. \tag{45}$$

This bilinear form is quite standard and can be implemented directly in any finite element code. Therefore, we seek an approximate field $\tilde{\mathbb{A}}_h$ in V_h approximating $\tilde{\mathbb{A}}$. Finally, the linear system is solved by a direct Gauss method.
To illustrate the accuracy improvement of the PR-FEM with $\mathbb{P}_1$ elements over the standard $\mathbb{P}_1$-FEM, we consider the model Neuman single-scattering problem for the unit circular cylinder. The fictitious boundary Σ is C_2. We report on Figures (**15**) and (**16**) the evolution of the relative $L^2(\Omega_h)$-error between the exact analytical solution to the BVP (32) and the numerical solution computed by either the standard FEM or the PR-FEM

$$\varepsilon_2 := \frac{||u^{ex} - u_h||_{0,\Omega_h}}{||u^{ex}||_{0,\Omega_h}}.$$

We denote by n_λ the number of discretization points per wavelength for our FEMs. Figure (**15**) shows the evolution of the error for two grids ($n_\lambda = 10$ and 20) and for increasing wavenumbers. We remark that the PR-FEM is much more stable, which is related to smaller pollution error into the way of computing the numerical solution. On figure (**16**), we draw ε_2 with respect to n_λ for $k = 10$, 25 and 50. We observe that we generally gain 10 orders of magnitude on the error when considering the PR-FEM solution over

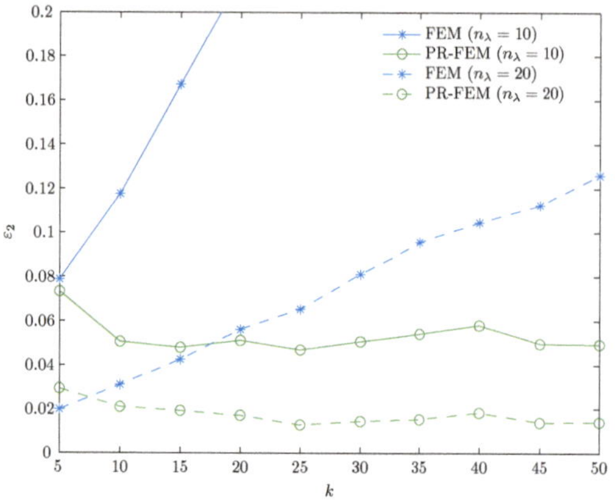

Fig. (**15**): Neumann problem: Evolution of the error ε_2 with respect to the wavenumber k for two discretizations.

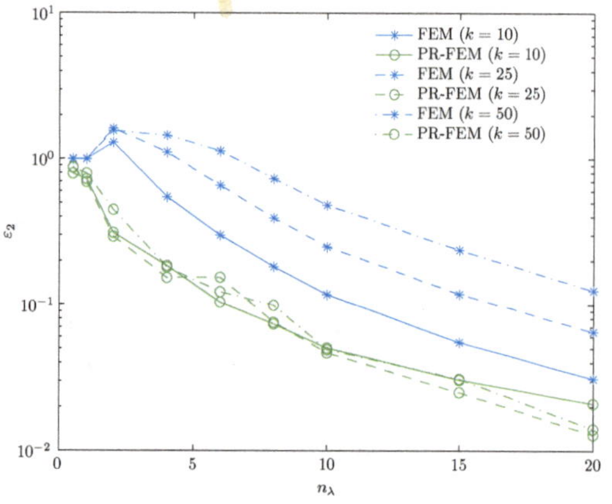

Fig. (**16**): Neumann problem: Evolution of the error ε_2 with respect to the density of discretization points n_λ for three wave numbers k.

the $\mathbb{P}_1$-FEM. The scattering amplitude $a(\theta)$ defined by (20) can be expressed in the polar direction θ as

$$a(\theta) = \frac{e^{i\frac{\pi}{4}}}{\sqrt{8\pi ik}} \int_\Gamma (\partial_{\mathbf{n}_\Gamma} u(\mathbf{x}) + ik\mathbf{n}_\Gamma \cdot \mathbf{d}' u(\mathbf{x})) e^{-ik\mathbf{x}\cdot\mathbf{d}'} d\Gamma_{\mathbf{x}} \tag{46}$$

and the RCS is given by expression (22). Vector $\mathbf{d}' = (\cos(\theta), \sin(\theta))^T$ is the vector of observation in the polar coordinates system (r, θ). We plot on Figure (**17**) the RCS at $k = 25$ for the two methods and a coarse discretization: $n_\lambda = 3$. We see that the RCS calculation is accurate with the PR-FEM while important error fluctuations occur for the standard FEM.

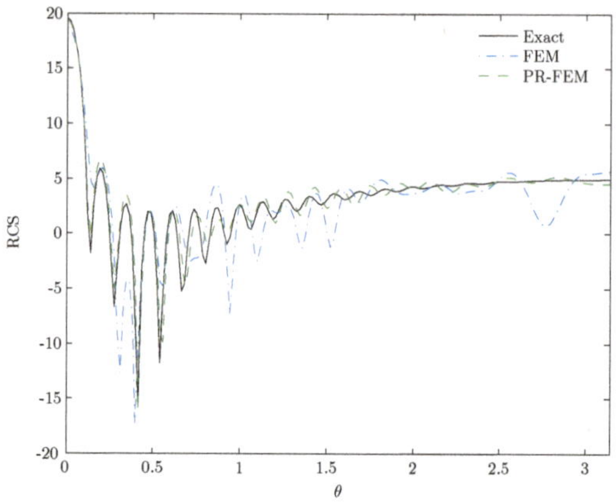

Fig. (**17**): Neumann problem: Bistatic Radar Cross Section for $k = 25$ and $n_\lambda = 3$.

5.2 Extension to Multiple Scattering

Let us recall that the multiple scattering BVP is stated as

$$\begin{cases} \Delta u + k^2 u = 0, \text{ in } \Omega^+, \\ \partial_{\mathbf{n}_\Gamma} u = -\partial_{\mathbf{n}_\Gamma} u^{\text{inc}} \text{ or } u = -u^{\text{inc}}, \text{on } \Gamma, \\ \lim\limits_{\|\mathbf{x}\| \to \infty} \|\mathbf{x}\|^{(d-1)/2} (\nabla u \cdot \frac{\mathbf{x}}{\|\mathbf{x}\|} - iku) = 0. \end{cases} \tag{47}$$

Instead of trying to solve (47) directly, an alternative idea is to solve its coupled single-obstacle version given in Theorem 1. Even if we have now single-obstacle problems, they are naturally still coupled. A suitable way to avoid this problem is to weaken the coupling by an iteration process. More specifically, we look for the solution in terms of the series

$$u = \sum_{m=1}^{\infty} \sum_{p=1}^{M} u_p^{(m)},$$

where $u_p^{(m)}$ is the solution of the single-obstacle scattering problem

$$\begin{cases} \Delta u_p^{(m)} + k^2 u_p^{(m)} = 0, \text{in } \mathbb{R}^d \backslash \overline{\Omega_p^-}, \\ \partial_{\mathbf{n}_p} u_p^{(m)} = \partial_{\mathbf{n}_p} (s^{(m)} - r_p^{(m)}) \text{ or } u_p^{(m)} = s^{(m)} - r_p^{(m)}, \\ \qquad \text{on } \Gamma_p, \\ \lim\limits_{\|\mathbf{x}\| \to \infty} \|\mathbf{x}\|^{(d-1)/2} (\nabla u_p^{(m)} \cdot \frac{\mathbf{x}}{\|\mathbf{x}\|} - iku_p^{(m)}) = 0, \end{cases} \tag{48}$$

with

$$s^{(m)} = \begin{cases} -u^{\text{inc}} & \text{for } m = 1, \\ 0 & \text{for } m > 1, \end{cases} \tag{49}$$

according to the boundary condition, and

$$r_p^{(m)} = \sum_{n=1}^{m-1} \sum_{\substack{q=1 \\ q \neq p}}^{M} u_q^{(n)} + \sum_{q=1}^{p-1} u_q^{(m)}. \tag{50}$$

In other words, we perform a Gauss-Seidel-type iteration where at each step we solve a scattering problem around the single obstacle Ω_p^-, using the fields scattered from the other obstacles as boundary condition [117]. As each correction $u_p^{(m)}$ can be interpreted as the correction introduced by the m-th wave reflection [73, 117], the iteration can be stopped when the norm of all corrections at step m is smaller than a prescribed tolerance.
Note that instead of performing this Gauss-Seidel iteration, other iterative schemes can be used. Indeed, the Dirichlet or Neumann boundary condition in (48) can equivalently written in vector form as:

$$\mathbf{U}^{(m+1)} - A\mathbf{U}^{(m)} := \mathbf{F} = \begin{cases} -\mathbf{U}^{\text{inc}}, & m = 1, \\ \mathbf{0}, & m > 1, \end{cases} \tag{51}$$

where A is the iteration operator acting on the traces of the field, mapping the traces at iteration (m) onto those at iteration $(m+1)$. The desired solution of this problem satisfies

$$(I - A)\mathbf{U}^* = \mathbf{F}, \tag{52}$$

which can be solved iteratively with general Krylov subspace methods. In particular, a preconditioned GMRES could thus be used. Preliminary results suggest that this can lead to a significant acceleration of the convergence. An explicit expression of the iteration operator A in the context of integral equations is given in Section 6.4.
If all the obstacles Ω_p^- are convex, all the problems in (48) become single-scattering problems, and are thus amenable to the PR-FEM. Applying the PR-FEM procedure for multiple scattering amounts to computing the amplitude $\tilde{\mathbb{A}}_p^{(m)}$ and the phase $\tilde{\phi}_p^{(m)}$ at each iteration m, for each scattering obstacle p, $p = 1, \ldots, M$, such that

$$u_p^{(m)} = \tilde{\mathbb{A}}_p^{(m)} e^{ik\tilde{\phi}_p^{(m)}}. \tag{53}$$

For our numerical tests we consider the same geometrical periodic configurations as in Section 3.4: a single-row configuration and a rectangular lattice of circular cylinders, with an incoming plane wave wave e^{ikx}. For each configuration the radius of the circular cylinders is set to $R = 1$ and the cylinders are placed radius apart, and we use a Bayliss-Gunzburger-Turkel-like radiation condition to truncate the infinite domain. The following mesh size field is used [76]:

$$h(\mathbf{x}) = \min\left\{\frac{\lambda}{2}, \frac{\lambda}{15} + \text{dist}_\Gamma(\mathbf{x})\right\}. \tag{54}$$

Indeed, for the solution of point 1) above (the computation of an initial condition for the one-dimensional propagation equation), we need to extract the phase of the field u, at each iteration m, on the boundary of each scatterer. This requires a refined mesh on the scale of the wavelength on Γ (we chose 15 points per wavelength). Away from Γ, a much coarser discretization can be used (here we use 2 points per wavelength). Note that independent volume and surface grids could also be used (a coarse volume grid and a refined surface grid) and the fields projected from one onto the other, e.g. with an L^2 projection [75].
Figure (**18**) shows the solution obtained for $kR = 25$ in a two-cylinder single-row configuration. The real part of the amplitude $a_p^{(m)}$ and the phase $\phi_p^{(m)}$ for $m = 1, 2, 6$ and $p = 1, 2$ are displayed on the left, while the real part of the final solution u is displayed on the right. Of particular notice is how each term in the series is not highly oscillatory and can thus be computed on a coarser grid than the final solution. Using the mesh size field given in (54) the mesh required to compute the approximate phase and the slowly oscillating amplitude contains 4300 nodes. The original problem would have required a mesh density of at least 10 points per wavelength, leading to about 25 times more unknowns.
Figure (**19**) shows the convergence of the Gauss-Seidel iterative process for a two-cylinder single-row configuration as well as for a four-cylinder rectangular lattice, for $kR = 1 \ldots 35$. The Gauss-Seidel iteration was stopped after 100 iterations or at iteration m if

$$\frac{\max_p ||a_p^{(m)}||_{0,\Omega}}{\max_p ||a_p^{(1)}||_{0,\Omega}} < 10^{-3}, \tag{55}$$

i.e., for a relative tolerance of 10^{-3} on the maximum of the norm of the corrections at iteration m.
For the two-cylinder single-row configuration the Gauss-Seidel process required between 6 and 7 iterations to converge. For the four-cylinder rectangular lattice the series diverges (or converges very slowly) for small values of the wavenumber, which is linked to the resonant frequencies of the structure. At higher frequencies the convergence improves and the convergence rates are in agreement with the theoretical rates derived in [62] in the asymptotic (infinite frequency) case.

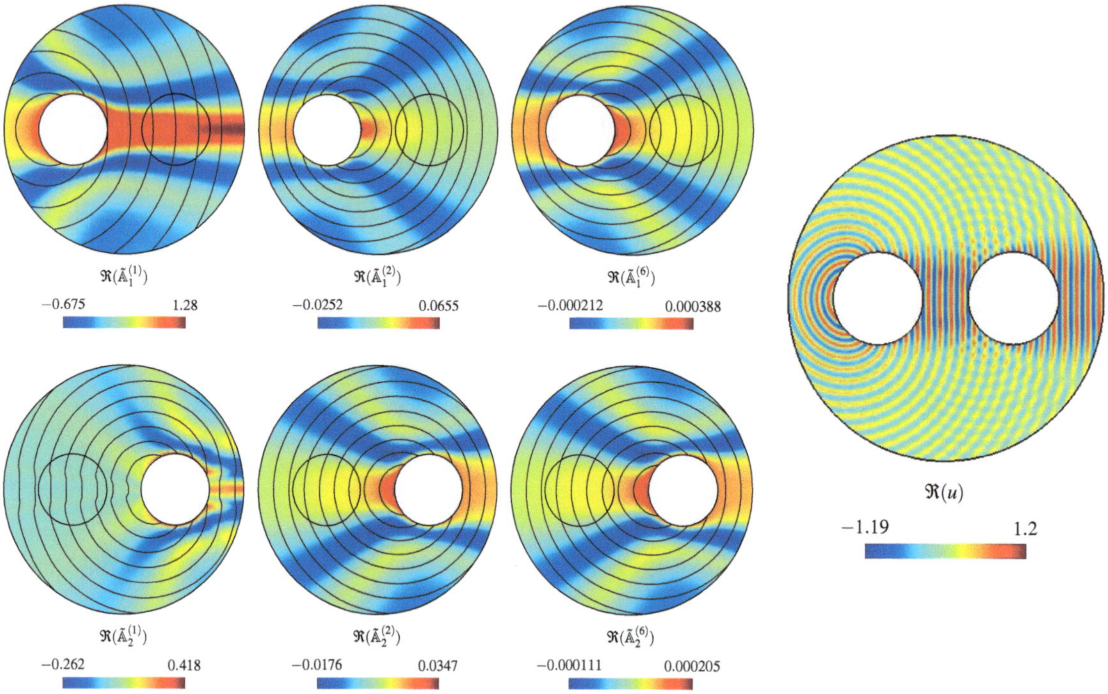

Fig. (**18**): Iterative solution around two circular cylinders of unit radius R for an incident plane wave arriving from the left, with $kR = 25$. Left: Real part of the amplitudes $\tilde{\mathbb{A}}_p^{(m)}$ for $p = 1$ and $p = 2$ (top to bottom) and $m = 1,2,6$ (left to right). On each graph iso-curves of the approximate phase $\tilde{\phi}_p^{(m)}$ are superimposed. Right: Real part of the final solution $u = \sum_{m=1}^{6} \sum_{p=1}^{2} \tilde{\mathbb{A}}_p^{(m)} e^{ik\tilde{\phi}_p^{(m)}}$.

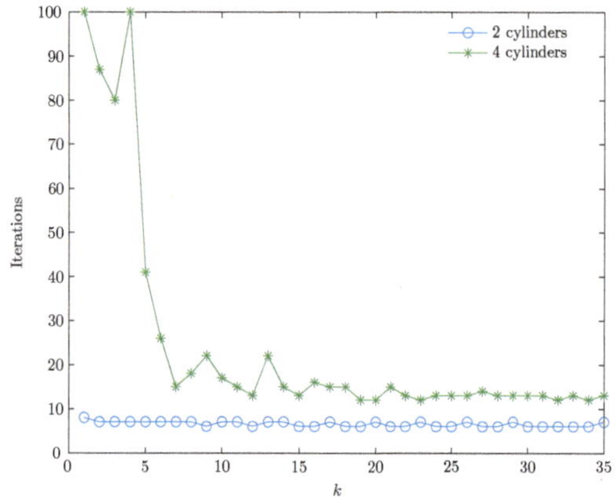

Fig. (**19**): Convergence of the Gauss-Seidel iterative process for the 2- and 4-cylinder case, for $k = 1 \ldots 35$.

6. INTEGRAL EQUATIONS BASED STRATEGIES

One of the main difficulties arising in the numerical solution of an exterior boundary value problem is related to the unboundedness of the domain Ω^+. An alternative solution to the volume truncation described in the previous two sections is to use integral equations methods [48, 54]. This technique is based on the reformulation of the initial exterior problem (4) as an integral equation written on the boundary Γ of the scatterer. In this Section, we propose a short review of this method, going from the theoretical analysis to the effective resolution using iterative solvers.

6.1 Background on Potential Theory

The first ingredient to derive an integral equation is provided by the following classical integral representation formula for solutions of interior and exterior Helmholtz equations.

Proposition 2. *Let us define by G the Green's kernel*

(5). Let $(v^-, v^+) \in H^1(\Omega^-) \times H^1_{loc}(\overline{\Omega^+})$ *satisfying*

$$\Delta v^- + k^2 v^- = 0 \quad (\Omega^-),$$

and

$$\begin{cases} \Delta v^+ + k^2 v^+ = 0 & (\Omega^+), \\ v^+ \text{ outgoing.} \end{cases}$$

Then, we have

$$L([\partial_{\mathbf{n}} v(\mathbf{y})])(\mathbf{x}) - D([v(\mathbf{y})])(\mathbf{x}) = \begin{cases} v^-(\mathbf{x}) & \text{for } \mathbf{x} \in \Omega^- \\ v^+(\mathbf{x}) & \text{for } \mathbf{x} \in \Omega^+, \end{cases} \tag{56}$$

where we have set for all $(\xi, \mu) \in H^{1/2}(\Gamma) \times H^{-1/2}(\Gamma)$:

$$\begin{cases} L\mu(\mathbf{x}) = \int_\Gamma G(\mathbf{x},\mathbf{y})\mu(\mathbf{y})d\Gamma(\mathbf{y}), \\ D\xi(\mathbf{x}) = \int_\Gamma \partial_{\mathbf{n}(\mathbf{y})} G(\mathbf{x},\mathbf{y})\xi(\mathbf{y})d\Gamma(\mathbf{y}), \end{cases} \quad \mathbf{x} \notin \Gamma, \tag{57}$$

and

$$[v] = v^- - v^+ \qquad [\partial_{\mathbf{n}} v] = \partial_{\mathbf{n}} v^- - \partial_{\mathbf{n}} v^+.$$

The operators L and D defined by (57) are respectively known as the single-layer and double-layer potentials. Note that in the definition of these potentials, the integrals are meaningful only if the surface densities ξ and μ are regular enough (typically bounded densities, since the kernels are integrable). Otherwise, they should be interpreted as dualities between the spaces $H^{-1/2}(\Gamma)$ and $H^{1/2}(\Gamma)$. Nevertheless, for the sake of simplicity we will keep the integral notation.

The second important ingredient to obtain integral equations is given by the trace formula for these above potentials (see for instance [109, Theorem 3.1.2.]).

Proposition 3. *We have the following identities:*

$$\begin{cases} \gamma_0^- \circ L = \gamma_0^+ \circ L = \mathscr{L} \\ \gamma_1^- \circ L = (\frac{\mathscr{I}}{2} + \mathscr{N}) \\ \gamma_1^+ \circ L = (-\frac{\mathscr{I}}{2} + \mathscr{N}) \\ \\ \gamma_0^- \circ D = (-\frac{\mathscr{I}}{2} + \mathscr{D}) \\ \gamma_0^+ \circ D = (\frac{\mathscr{I}}{2} + \mathscr{D}) \\ \gamma_1^- \circ D = \gamma_1^+ \circ D = \mathscr{S} \end{cases} \tag{58}$$

where $\mathscr{I}$ *is the identity operator and where* $\mathscr{L}, \mathscr{N}, \mathscr{D}$ *and* $\mathscr{S}$ *are the boundary integral operators defined by for all* $\mathbf{x} \in \Gamma$ *by*

$$\begin{cases} \mathscr{L}\mu(\mathbf{x}) = \int_\Gamma G(\mathbf{x},\mathbf{y})\mu(\mathbf{y})d\Gamma(\mathbf{y}) \\ \mathscr{N}\mu(\mathbf{x}) = \int_\Gamma \partial_{\mathbf{n}(\mathbf{x})} G(\mathbf{x},\mathbf{y})\mu(\mathbf{y})d\Gamma(\mathbf{y}) \\ \mathscr{D}\xi(\mathbf{x}) = \int_\Gamma \partial_{\mathbf{n}(\mathbf{y})} G(\mathbf{x},\mathbf{y})\xi(\mathbf{y})d\Gamma(\mathbf{y}) \\ \mathscr{S}\xi(\mathbf{x}) = \oint_\Gamma \frac{\partial^2 G}{\partial \mathbf{n}(\mathbf{x})\partial \mathbf{n}(\mathbf{y})}(\mathbf{x},\mathbf{y})\xi(\mathbf{y})d\Gamma(\mathbf{y}). \end{cases} \tag{59}$$

Here, the $\circ$ *denotes the composition operator.*

Note that the expression defining $\mathscr{S}$ is not an integral (its singularity is not in integrable) but a finite part expression associated with a hypersingular kernel. We preferred to keep formally the integral expression for the sake of clarity. We also need for the sequel the continuity properties of the above integral operators, which are summarized in the next result (see for instance [109, Theorem 4.4.1] or Theorems 7.1 and 7.2 in [107]).

Proposition 4. *For smooth boundary* Γ*, the boundary integral operators given in Proposition 3 define the following continuous mappings*

$$\begin{array}{lcllll} \mathscr{L} & : & H^s(\Gamma) & \longrightarrow & H^{s+1}(\Gamma), \\ \mathscr{N} & : & H^s(\Gamma) & \longrightarrow & H^s(\Gamma), \\ \mathscr{D} & : & H^s(\Gamma) & \longrightarrow & H^s(\Gamma), \\ \mathscr{S} & : & H^s(\Gamma) & \longrightarrow & H^{s-1}(\Gamma) \end{array} \tag{60}$$

for all $s \in \mathbb{R}$*. Moreover, the operators* $\mathscr{N}$ *and* $\mathscr{D}$ *are compact from* $H^s(\Gamma)$ *onto itself for all* $s \in \mathbb{R}$.

In the case of Lipschitz boundary [56, 107], the above continuity properties still hold for $-1 \leq s \leq 0$ (respectively for $0 \leq s \leq 1$) for operators $\mathscr{L}$ and $\mathscr{N}$ (respectively $\mathscr{D}$ and $\mathscr{S}$), while the compactness properties of $\mathscr{N}$ and $\mathscr{D}$ fail. A possible approach to rigorously extend the following developments is to use e.g. some regularizing techniques [40].

In the case of multiply connected domains, we will use the following notations concerning the integral operators acting from Γ_q to Γ_p, $p, q = 1, ..., M$,

$$\begin{array}{lcllll} \mathscr{L}^{p,q} & : & H^s(\Gamma_q) & \longrightarrow & H^{s+1}(\Gamma_p), \\ \mathscr{N}^{p,q} & : & H^s(\Gamma_q) & \longrightarrow & H^s(\Gamma_p), \\ \mathscr{D}^{p,q} & : & H^s(\Gamma_q) & \longrightarrow & H^s(\Gamma_p), \\ \mathscr{S}^{p,q} & : & H^s(\Gamma_q) & \longrightarrow & H^{s-1}(\Gamma_p), \end{array} \tag{61}$$

for all $s \in \mathbb{R}$.

6.2 Integral Equations Formulations

The scattering problem (4) can be reformulated as an integral equation. For the sake of clarity, let us consider first the case of Dirichlet boundary condition. We will come back later to the Neumann case in Section 6.2.3. In fact, an infinite number of integral equations can be written in the case of a closed surface. Each integral equation has its own mathematical properties, having a direct impact on the numerical solution in terms of accuracy, solvability, spectrum... We usually distinguish direct and indirect integral equations.

6.2.1 Direct Integral Formulations

The starting point of direct integral formulations is to express the total field

$$w := u^+ + u^{\text{inc}}$$

using its classical *integral representation formula* based on its Cauchy data $w_{|\Gamma}$ and $\partial_{\mathbf{n}} w_{|\Gamma}$. To achieve this, let us apply Proposition 2 to the particular choice of solutions $(v^-, v^+) := (-u^{\text{inc}}, u^+)$. Since

$$[v] = 0 \qquad [\partial_{\mathbf{n}} v] = -\partial_{\mathbf{n}} w_{|\Gamma} := \mu$$

formula (56) reads

$$L\mu(\mathbf{x}) = \begin{cases} -u^{\text{inc}}(\mathbf{x}) & \text{for } \mathbf{x} \in \Omega^- \\ u^+(\mathbf{x}) & \text{for } \mathbf{x} \in \Omega^+ \end{cases}$$

Adding u^{inc} to the above expressions yields the following single-layer potential representation of the total field:

$$L\mu(\mathbf{x}) + u^{\text{inc}}(\mathbf{x}) = \begin{cases} 0 & \text{for } \mathbf{x} \in \Omega^- \\ w(\mathbf{x}) & \text{for } \mathbf{x} \in \Omega^+ \end{cases}$$

Clearly, this formulation is completely equivalent to extending artificially the total field by zero inside Ω^-, which explains that this approach is also referred sometimes to as the *null field method.*
The next step is to obtain an integral equation for this unknown density $\mu = -\partial_{\mathbf{n}} w_{|\Gamma} \in H^{-1/2}(\Gamma)$. To achieve this, the idea is to apply a boundary operator to the relation

$$L\mu(\mathbf{x}) + u^{\text{inc}}(\mathbf{x}) = 0, \qquad \forall \mathbf{x} \in \Omega^-. \qquad (62)$$

Many choices are here possible, leading to different integral equations:

- **EFIE :** This equation is obtained by applying the trace operator γ_0^- to (62). Thanks to the trace relations of Proposition 3, this leads to the so-called Electric Field Integral Equation:

$$\mathscr{L}\mu = -u^{\text{inc}}{}_{|\Gamma}. \qquad (63)$$

- **MFIE :** This equation is obtained by applying the normal trace operator γ_1^- to (62). Thanks to the trace relations of Proposition 3, this leads to the so-called Magnetic Field Integral Equation:

$$(\frac{\mathscr{I}}{2} + \mathscr{N})\mu = -\partial_{\mathbf{n}} u^{\text{inc}}{}_{|\Gamma}. \qquad (64)$$

- **CFIE :** This equation is obtained by applying to (62) the Fourier type operator $\gamma_1^- + \eta\gamma_0^-$, with $\eta \neq 0$. Once again, the trace relations of Proposition 3 give the so-called Combined Field Integral Equation:

$$\begin{aligned} &\{(\tfrac{\mathscr{I}}{2} + \mathscr{N}) + \eta\mathscr{L}\}\,\mu \\ &\quad = -(\partial_{\mathbf{n}} u^{\text{inc}}{}_{|\Gamma} + \eta u^{\text{inc}}{}_{|\Gamma}). \end{aligned} \qquad (65)$$

Let us emphasize here that if μ solves one of the above integral equations ((63), (64) or (65)), then the scattered field can be recovered through its single layer integral representation $u^+(\mathbf{x}) = L\mu(\mathbf{x})$. Moreover, physical quantities of interest like the scattering amplitude can also be easily recovered from μ. For instance, in the two-dimensional case, we have

$$a(\theta) = \frac{e^{i\frac{\pi}{4}}}{\sqrt{8\pi i k}} \int_\Gamma \mu(\mathbf{x}) e^{-ik\mathbf{x}\cdot\mathbf{d}'} d\Gamma(\mathbf{x}).$$

A natural question to be investigated is that of the well-posedness of the above integral equations. Let us introduce some notation:

- $F_D(\Omega^-) = \{k_m^D, m \in \mathbb{N}\}$, the set of Dirichlet irregular frequencies, is the set of values of k such that the boundary value problem

$$\begin{cases} -\Delta v = k^2 v, & \text{in } \Omega^-, \\ \gamma_0^- v = 0, & \text{on } \Gamma, \end{cases}$$

admits a non vanishing solution.

- $F_N(\Omega^-) = \{k_m^N, m \in \mathbb{N}\}$, the set of Neumann irregular frequencies, is the set of values of k such that the boundary value problem

$$\begin{cases} -\Delta v = k^2 v, & \text{in } \Omega^-, \\ \gamma_1^- v = 0, & \text{on } \Gamma, \end{cases}$$

admits a non vanishing solution.

Clearly, such irregular frequencies are nothing but the square-roots of the eigenvalues of the self-adjoint positive operator with compact resolvent $-\Delta$ respectively with domains $H^2(\Omega^-) \cap H_0^1(\Omega^-)$ and $\{v \in H^2(\Omega^-) \mid \gamma_1^- v = 0\}$. Moreover, note that in the case of multiple scattering, these irregular frequencies are simply given by those corresponding to each obstacle Ω_p^-:

$$F_D(\Omega^-) = \cup_{p=1}^M F_D(\Omega_p^-),$$
$$F_N(\Omega^-) = \cup_{p=1}^M F_N(\Omega_p^-).$$

In other words, there is no creation of new irregular frequencies by the scattering coupling process between separate obstacles.
The next result provides existence and uniqueness results for the integral equations (63), (64) and (65).

Theorem 2.

1. *The operator $\mathscr{L}$ defines an isomorphism from $H^{-1/2}(\Gamma)$ onto $H^{1/2}(\Gamma)$ if and only if $k \notin F_D(\Omega^-)$. Under this condition, the EFIE (63) is uniquely solvable in $H^{-1/2}(\Gamma)$.*

2. *The operator*

$$(\frac{\mathscr{I}}{2} + \mathscr{N})$$

defines an isomorphism from $H^{-1/2}(\Gamma)$ onto $H^{-1/2}(\Gamma)$ if and only if $k \notin F_N(\Omega^-)$. Under this condition, the MFIE (64) is uniquely solvable in $H^{-1/2}(\Gamma)$.

3. *The operator*

$$(\frac{\mathscr{I}}{2} + \mathscr{N}) + \eta \mathscr{L}$$

defines an isomorphism from $H^{-1/2}(\Gamma)$ onto $H^{-1/2}(\Gamma)$ for all $k > 0$ provided $\Im(\eta) \neq 0$. Under this condition, the CFIE (65) is uniquely solvable in $H^{-1/2}(\Gamma)$ for all frequency $k > 0$.

In the case where k is an irregular frequency, the integral equations EFIE and MFIE have non zero kernels. Nevertheless, it can be shown that the spurious modes of the EFIE do not modify the solution of the scattering problem (we say that the spurious modes of the EFIE do not radiate):

$$\mathscr{L}\mu = 0 \text{ on } \Gamma \implies L\mu = 0 \text{ in } \Omega^+.$$

Hence, the integral equation (63) provides accurate computations e.g. in the far-field region and is useful for practical calculations. On the contrary, the spurious solutions of the MFIE integral equation (64) do radiate now, leading to a wrong solution.
According to Theorem 2, the CFIE (65) does not suffer from the existence of spurious solutions if we take for instance $\eta = -ik\alpha/(1-\alpha)$, $\alpha \in]0,1[$. A commonly used choice of α is $\alpha = 0.2$ which gives an almost minimal condition number for the CFIE. Finally, recent versions, called Generalized CFIE consider coupling complex-valued pseudodifferential surface operators η instead of complex numbers for conditioning improvement and eigenvalue clustering (see e.g. [12, 13]). But this last solution is much more to consider for Neumann problems rather than Dirichlet problems. This point is discussed in Section 6.2.3.

6.2.2 The Indirect Integral Formulation of Burton-Miller (or Brakhage-Werner)

In indirect formulations, the unknowns are not obtained *via* the integral representation formula of the scattered field. Consequently, the unknowns are generally not related to quantities of physical interest. Here, we restrict our presentation to the most commonly used indirect integral formulation, which has been independently derived by Burton-Miller [41] and Brakhage-Werner [36]. The idea is to seek the exterior field as a superposition of the single- and double-layer potentials acting on a fictitious surface density ψ:

$$u^+(\mathbf{x}) = (D + \eta L)\psi(\mathbf{x}), \ \forall \mathbf{x} \in \Omega^+, \tag{66}$$

where η is a complex-valued coupling parameter to determine. In the case of the Dirichlet boundary condition, the above expression leads thanks to the trace relations (58) to the following integral equation :

$$\left\{(\frac{\mathscr{I}}{2} + \mathscr{D}) + \eta \mathscr{L}\right\} \psi = -u^{\text{inc}}{}_{|\Gamma}. \tag{67}$$

We consider the above integral equation in the space $H^{1/2}(\Gamma)$. According to Proposition 4, the operators

$$\mathscr{D} : H^{1/2}(\Gamma) \to H^{1/2}(\Gamma)$$

and

$$\mathscr{L} : H^{1/2}(\Gamma) \to H^{1/2}(\Gamma)$$

are both compact for smooth Γ (for $\mathscr{L}$, this follows from the continuity of $\mathscr{L} : H^{1/2}(\Gamma) \to H^{3/2}(\Gamma)$ and the compactness of the embedding of $H^{3/2}(\Gamma)$ into $H^{1/2}(\Gamma)$). Therefore, the operator

$$(\frac{\mathscr{I}}{2} + \mathscr{D}) + \eta \mathscr{L} : H^{1/2}(\Gamma) \to H^{1/2}(\Gamma)$$

is of Fredholm type of index 0. Moreover, it can be easily checked that

$$(\frac{\mathscr{I}}{2}+\mathscr{D})+\eta\mathscr{L}$$

is injective if $\Im(\eta)\neq 0$, yielding the following result.

Theorem 3. *The operator*

$$(\frac{\mathscr{I}}{2}+\mathscr{D})+\eta\mathscr{L}$$

defines an isomorphism from $H^{1/2}(\Gamma)$ onto $H^{1/2}(\Gamma)$ for all $k>0$ provided $\Im(\eta)\neq 0$. Under this condition, (67) is uniquely solvable in $H^{1/2}(\Gamma)$ for all frequency $k>0$.

This integral equation is uniquely solvable if and only if $\Im(\eta)>0$. An almost optimal value of η has been obtained in [101, 99, 4] as: $\eta = ik$. In [12, 13], generalized Burton-Miller integral equations have been derived by the introduction of a coupling pseudodifferential operator η with the aim of minimizing the condition number of the integral operator. Again, this improvement is essentially crucial in the case of a Neumann boundary condition (see Section 6.2.3).

6.2.3 The Neumann Problem

For the sake of completeness, let us briefly list the integral equations obtained in the case of a Neumann boundary condition.

- **EFIE :**

$$\mathscr{S}\xi = -\partial_{\mathbf{n}}u^{\text{inc}}{}_{|\Gamma}. \tag{68}$$

- **MFIE :**

$$(-\frac{\mathscr{I}}{2}+\mathscr{D})\xi = -u^{\text{inc}}{}_{|\Gamma}. \tag{69}$$

- **CFIE :**

$$\begin{aligned}\{(-\tfrac{\mathscr{I}}{2}+\mathscr{D})+\eta\mathscr{S}\}\,\xi \\ = -(\eta\partial_{\mathbf{n}}u^{\text{inc}}{}_{|\Gamma}+u^{\text{inc}}{}_{|\Gamma}).\end{aligned} \tag{70}$$

- **Burton-Miller :**

$$\left\{(-\frac{\mathscr{I}}{2}+\mathscr{N})+\eta\mathscr{S}\right\}\psi = -\partial_{\mathbf{n}}u^{\text{inc}}{}_{|\Gamma}. \tag{71}$$

The existence and uniqueness results for the above integral equations are summarized in the next result.

Theorem 4.

1. *The operator $\mathscr{S}$ defines an isomorphism from $H^{1/2}(\Gamma)$ onto $H^{-1/2}(\Gamma)$ if and only if $k\notin F_N(\Omega^-)$. Under this condition, the EFIE (68) is uniquely solvable in $H^{1/2}(\Gamma)$.*
2. *The operator*

$$(-\frac{\mathscr{I}}{2}+\mathscr{D})$$

defines an isomorphism from $H^{1/2}(\Gamma)$ onto $H^{1/2}(\Gamma)$ if and only if $k\notin F_D(\Omega^-)$. Under this condition, the MFIE (69) is uniquely solvable in $H^{1/2}(\Gamma)$.

3. *The operator*

$$(-\frac{\mathscr{I}}{2}+\mathscr{D})+\eta\mathscr{S}$$

defines an isomorphism from $H^{1/2}(\Gamma)$ onto $H^{-1/2}(\Gamma)$ for all $k>0$ provided $\Im(\eta)\neq 0$. Under this condition, the CFIE (70) is uniquely solvable $H^{1/2}(\Gamma)$ for all frequency $k>0$.

4. *The operator*

$$(-\frac{\mathscr{I}}{2}+\mathscr{N})+\eta\mathscr{S}$$

defines an isomorphism from $H^{1/2}(\Gamma)$ onto $H^{-1/2}(\Gamma)$ for all $k>0$ provided $\Im(\eta)\neq 0$. Under this condition, (71) is uniquely solvable in $H^{1/2}(\Gamma)$ for all frequency $k>0$.

6.2.4 First- vs. Second-Kind Fredholm Integral Equations

A first difference between the integral equations described in the previous section is their well-posedness for all frequencies or not. Another important difference, which is of particular interest for their numerical resolution, is related to their general structure. Let us recall that given an integral operator $\mathscr{A}\in\mathscr{L}(X)$ on a Hilbert space X, an integral equation is called of first-kind if it is of the form

$$\mathscr{A}\rho = f$$

of second-kind if it is of the form

$$(\mathscr{I}+\mathscr{A})\rho = f. \tag{72}$$

Moreover, if $\mathscr{A}: X\to X$ is compact, the above equations are respectively called Fredholm integral equations of the first-kind and second-kind. As classically known [100], the spectrum of compact operators is composed in the infinite dimensional case of 0 and a sequence of discrete eigenvalues possibly accumulating at the origin. Therefore, second-kind Fredholm integral equation have large clusters of eigenvalues accumulating at the real value point 1. This is an important feature in view of a numerical resolution of the integral equation using a Krylov solver like the GMRES. Indeed, in these iterative methods, clustering in the complex plane generally implies fast convergence.

Because $\mathscr{L}$ is bounded from $H^s(\Gamma)$ onto $H^{s+1}(\Gamma)$, it is compact from $L^2(\Gamma)$ onto itself. Therefore, the EFIE (63) is a Fredholm first-kind integral equation on $L^2(\Gamma)$.

On the contrary, the MFIE (64), the CFIE (65) and Burton-Miller (67) equations are all second-kind Fredholm integral equations. Due to their additional well-posedness properties, the CFIE integral equation and Burton-Miller formulation are particularly well-adapted for the numerical resolution of scattering problems. In the case of a Neumann boundary condition, the situation is more complex. Indeed, only the MFIE which is ill-posed is a second-kind Fredholm integral equation. The EFIE, CFIE and Burton-Miller integral equations involve the operator $\mathscr{S}$ which is a first-order, strongly singular and non-compact operator. This implies that these equations are first-kind integral equations. To obtain eigenvalue clustering, then two solutions are possible. The first one is based on the introduction of a suitable regularizing operator η of order -1 which multiplies $\mathscr{S}$ to get a second-kind equation [12, 13]. A second possibly is to precondition the integral operator [8, 9]. We will come back to this question in Section 6.4.

6.3 Approximation

Different techniques can be considered for discretizing an integral equation but the most standard one is the Boundary Element Method (BEM). We will also present another approach in Section 6.4 which is based on the use of a Partition Of Unity of the surface in conjunction with the Fast Fourier Transform (FFT).

6.3.1 Boundary Element Methods

In the case of the BEM, we first introduce some discrete surfaces $\Gamma_{p,h}$ approximating each separate surface Γ_p, $p = 1,...,M$. These surfaces are constructed using polygonal (2D) or polyhedral (3D) surfaces respectively with the help of segments (2D) or triangles (3D) $K_{p,\ell}$, $\ell = 1,\dots,NK_p$. Higher-order surface representations could also be used for improving the accuracy of the approximation. The discretization step h_p for each boundary $\Gamma_{p,h}$ is defined by $h_p = \max_{1\leq\ell\leq NK_p}|K_{p,\ell}|$. Therefore, we have:

$$\Gamma_{p,h} = \cup_{\ell=1}^{NK_p} K_{p,\ell}.$$

With these notation, the global discretized scatterer is now $\Gamma_h = \cup_{p=1}^{M}\Gamma_{p,h}$, with: $h := \max_{1\leq p\leq M} h_p$. The total number of nodes is then $N_h = \sum_{p=1}^{M} N_p$, where N_p is the number of nodes of the triangularization of the p-th discrete surface. Let us introduce now the approximation spaces of interpolating polynomials of order m

$$W_{p,h}^m := \Big\{ v \in \mathscr{C}^0(\Gamma_{p,h})/v|_{K_{p,\ell}} \in \mathbb{P}_m, \quad \forall K_{p,\ell}, 1 \leq \ell \leq NK_p \Big\}. \tag{73}$$

for $p = 1,...,M$, $m \in \mathbb{N}$ and the product space

$$W_h^m := W_{1,h_1}^m \times ... \times W_{M,h_M}^m. \tag{74}$$

We denote by $\mathscr{A}$ the integral operator and by f the corresponding right hand side of one of the integral equations (63), (64), (65) or (67) written in $L^2(\Gamma)$. To proceed to the discretization of the integral equation

$$\mathscr{A}\rho = f \quad \text{in}\, L^2(\Gamma), \tag{75}$$

we first consider a weak formulation as

$$\int_\Gamma \mathscr{A}\rho\bar{q}d\Gamma = \int_\Gamma f\bar{q}d\Gamma, \tag{76}$$

for any test-function $q \in L^2(\Gamma)$. This leads to the linear system

$$[\mathscr{A}_h]\rho_h = [M_h]\mathbf{f}_h, \tag{77}$$

where $[\mathscr{A}_h]$ is a dense complex-valued matrix of size $N_h \times N_h$, and ρ_h and $\mathbf{f}_h$ are two complex-valued vectors of $\mathbb{C}^{N_h}$ which are composed of the nodal values of p and f through the linear interpolation in W_h^1. The matrix $[M_h]$ is the $N_h \times N_h$ mass matrix associated with the linear interpolation. The matrix $[\mathscr{A}_h]$ is composed of $M \times M$ block matrices $[\mathscr{A}_h^{p,q}]$ of discrete interactions from surface $\Gamma_{q,h}$ to $\Gamma_{p,h}$, for $1 \leq p,q \leq M$. In particular, another way of writting the integral equation (77) is

$$\forall 1 \leq p \leq M, \sum_{q=1}^{M}[\mathscr{A}_h^{p,q}]\rho_{q,h} = [M_h^{p,p}]\mathbf{f}_{p,h}, \tag{78}$$

where $\rho_{q,h}$ (respectively $\mathbf{f}_{p,h}$) is the vector of nodal values of ρ_h (respectively $\mathbf{f}_p$) on $\Gamma_{q,h}$ (respectively $\Gamma_{p,h}$): $\rho_h = (\rho_{q,h})_{1\leq q\leq M}$. In fact, equation (78) is equivalent to writing system (6) of Theorem 1 in an integral form.

Let us come now to the computation of the matrix coefficients. They are defined through double integrals over two triangles of Γ_h (not necessarily on the same separate discrete surface $\Gamma_{p,h}$). These integrals involve singular and hyper-singular kernels to integrate. The strategy is the following. A numerical quadrature rule is used for computing the outer integral according to an integration variable, for example $\mathbf{x}$ to fix the ideas. Then, if for the second inner integral in $\mathbf{y}$, the kernel is regular then we employ a numerical quadrature rule. If it is however singular, then the kernel is split into a singular and a regular parts. The regular part is evaluated numerically while exact integration formulas exist for the singular part. Let us remark that considering singularity here means that $\|\mathbf{x}-\mathbf{y}\|$ is less than a few wavelengths λ for high frequency. In particular, $\|\mathbf{x}-\mathbf{y}\|$ can be small even if $\mathbf{x}$ and $\mathbf{y}$ are not on the same discrete surface. Indeed, they can be very close for sticky or almost sticky elementary scatterers.

Concerning the choice of the discretization step, it is particularly important to choose it carefully when solving high frequency problems. Unlike the FEM method, the BEM does not suffer from pollution. Therefore, generally a constant density of discretization points per wavelength $n_{p,\lambda}$ can be chosen for the j-th scatterer: $n_{p,\lambda} = \lambda/h_p$. Let us consider that we are working in the d-dimensional space. Then, an estimate of the number of points N_p for the p-th surface is: $N_p \sim C_p k^d n_{p,\lambda}^d |\Gamma_p|$, $1 \le p \le M$, where C_p are some positive constants. For example, for $d=2$, and for M disks of radius a, we get: $N_p \sim C_p k a n_{p,\lambda}$. The linear system (77) has finally a size: $N_h = CkaMn_\lambda$, if we assume that $h \sim h_p$, $1 \le p \le M$ and that C is a positive constant. In the case of M spheres of radius a, we obtain: $N_h \sim Ck^2a^2Mn_\lambda^2$. If we have in mind that the matrix $[\mathscr{A}_h]$ is dense, then the construction cost of the matrix is a first expensive step since we have to evaluate accurately N_h^2 coefficients. This cost quickly grows for high frequencies $k \gg 1$, large numbers $M \gg 1$ of large-size elementary objects $|\Gamma_p| \gg 1$ as well as small densities of discretization points $n_{p,\lambda} \gg 1$. Furthermore, in addition to the computational cost, the matrix requires a storage of M_h^2 complex values which for high frequency and/or multiple scattering leads to strong limitations. For example, considering $M=75$ unit circular cylinders at $k=200$ for $n_\lambda = 20$ gives $N_h \approx 3 \times 10^5$ (see examples in Section 3.4.3). Finally, the only solution at this step for solving the linear system is to use a direct Gauss solver which requires $\mathscr{O}(N_h^3)$ operations.

Instead of polynomial basis functions and like in the FEM, oscillating polynomials could be considered for approximating the physical unknown of the problem

$$\mu|_{\Gamma_p} \approx \mu_{h,p} = \sum_{\ell=1}^{N_p}\sum_{m=1}^{nb} \mu_{\ell,m} A_{\ell,m}(\mathbf{x}) e^{ik\varphi_m(\mathbf{x})}, \quad (79)$$

where the nb phases φ_m are known: $\varphi_m(\mathbf{x}) = \mathbf{x}\cdot\mathbf{d}_m$ for example, with fixed directions $\mathbf{d}_m$ (see also related methods in Section 4.2.2). For such methods, we refer in particular to the microlocal discretization techniques developed in [1, 57] (possibly coupled to the FMM). Phase prediction based on asymptotic techniques can be very useful at this point for large wavenumbers. Let us remark that this approach requires the efficient computation of integrals with highly oscillating kernels. About all these points and inherent difficulties, we refer to [92, 38, 37, 71, 58]. Finally, the algorithm presented in Section 6.3.2 goes in this way but by using high-order representations, and can be adapted to multiple scattering computations as proposed in Section 6.4 (see also [71]).

6.3.2 Partition Of Unity Methods

For smooth obstacles collocation methods based on overlapping patches and smooth partitioning can present a very efficient alternative to the variational boundary element method presented above.

The aim of this approach is to strongly satisfy the integral equation on a set of points on the boundary of the scattering obstacle Γ. To that effect, the unknowns of the problem are chosen as the punctual value of the unknown field at these points, and the integral equation is solved iteratively with the help of a iterative solver like GMRES. Each iteration of the linear solver requires the application of the integral operator on the discretized scattered field, and most of the algorithm thus consists in the numerical computation of the integrals appearing in the integral equation.

Below we present the patching, discretization and smooth-partitioning strategies proposed in [39] in a 3D context. We assume that the scattering surface Γ is covered by a number K of overlapping patches $\mathscr{P}_j$, $j=1,\dots,K$, each one of which is mapped to a two-dimensional coordinate set $\mathscr{H}_j$ on the unit square $[0,1]\times[0,1]$, via a smooth invertible parameterization

$$\mathbf{x} = \mathbf{x}(u,v), \quad \text{for } (u,v) \in \mathscr{H}_j, \quad j = 1,\dots,K. \quad (80)$$

Further, we use a partition of unity subordinated to this covering, i.e., a set of non-negative smooth functions $\{w_j, j = 1,\dots,K\}$, such that

1. w_j is defined, smooth and non-negative in Γ and vanishes outside $\mathscr{P}_j$, and
2. $\sum_{j=1}^{K} w_j = 1$ throughout Γ.

Such a partition of unity can be constructed numerically for a given surface Γ without difficulty [37]. The unknown field is discretized on Cartesian sets of $L_j \times M_j$ nodes (u_ℓ, v_m) (with $\ell = 1,\dots,L_j$ and $m = 1,\dots,M_j$) on each coordinate set $\mathscr{H}_j$. These nodal values are the unknowns we seek to obtain when we solve the discretized scattering problem, i.e., they constitute the data we are given to compute the integrals in the boundary integral formulation. Thanks to the smooth partition of unity, the integrals in (85) can be efficiently evaluated, away from the singularities introduced by the Green function, using the Trapezoidal rule on the aforementioned Cartesian grid. In a neighborhood of the singularity, we use a smooth cutoff and a change in polar coordinates to regularize the integrand. This last point requires the off-grid interpolation of the scattered wave, which can be performed using a hybrid FFT-polynomial scheme [39, 37]. In 2D, a single patch can be used for each scattering surface, and the singularity of the Green function can be treated with a specialized quadrature rule. The resulting Nystrom scheme can be found in [55].

6.4 Iterative Solution – Acceleration – Preconditioning

As seen above, solving multiple scattering problems at high frequency by integral equations quickly leads to the numerical solution of large scale linear systems. The recent years have seen the introduction of alternative procedures based on Krylov subspace iterative solvers. The GMRES is often used to this aim. Essentially, the computational cost of such a procedure is related to

i) The total number of iterations N^{iter} required to reach an *a priori* fixed tolerance ε

ii) The cost of one iteration which is mainly matrix-vector products $\mathbf{y} = [\mathscr{A}_h]\mathbf{x}$, where $\mathbf{x}$ is a given complex-valued vector of $\mathbb{C}^{N_h}$ and $\mathbf{y}$ is the result of the matrix-vector product.

If this algorithm is directly used, then the total cost is $\mathscr{O}(N^{\text{iter}}N_h^2)$. In terms of memory storage, the algorithm still needs $\mathscr{O}(N_h^2)$ entries. Furthermore, since the problem is highly undefinite, most particularly when high frequencies are involved, then convergence breakdown can arise.

Let us begin by discussing point ii). Algorithms for computing efficiently matrix-vector products and for reducing memory storage have received much attention among the last two decades. The idea is to develop fast algorithms for evaluating the application of the map

$$\begin{aligned} \mathbb{C}^{N_h} &\to \mathbb{C}^{N_h} \\ \mathbf{x} &\mapsto \mathbf{y} = [\mathscr{A}_h]\mathbf{x} \end{aligned} \tag{81}$$

The most well-known method is the so-called multilevel Fast Multipole Method (FMM) introduced by Greengard and Rokhlin [80]. At some point, this technique can be considered as the extension to non uniform discretizations of the FFT algorithm used in Section 3.3.2 for the approach based on Toeplitz matrices. Indeed, in the case of a general boundary, the surface mesh is not built on uniform grids. Roughly speaking, the idea of the FMM is to compute in a fast but approximate way the far interactions while directly evaluating the near interactions. This can be further accelerated by using mutilevel ideas, leading to the multilevel FMM. The method then yields a matrix-vector evaluation in $\mathscr{O}(N_h \log N_h)$ operations and requires $\mathscr{O}(N_h)$ entries. This is a drastic reduction (both in computational cost and memory) for one iteration of the iterative scheme which gives expectations for simulating high frequency problems. Other possible fast algorithms exist like e.g. the Panel Clustering [113] or the fast high-order methods explained in Section 6.4 and based on Partition of Unity Methods in conjunction with the FFT (see also [39]).

Let us now come to point i). As said above, the GMRES is usually used as an iterative scheme. For memory reasons, the restarted version GMRES(ρ) must be considered, with a restart parameter ρ (e.g. $\rho = 50,\dots,100$). For moderate or large wavenumbers k, the scheme requires many iterations and often does not converge, similarly to the FEM methods studied in Section 4.3 (label). This means that preconditioning is needed. In the case where the matrix $[\mathscr{A}_h]$ is at hand, then algebraic methods [63] can be used but suffer from the same difficulties as the FEM (see also [3, 43] for preconditioning of integral equations in electromagnetism). However the BEM does not generally provide the matrix $[\mathscr{A}_h]$ but rather a fast evaluation of map (81) if we think in terms of FMM. For this reason, matrix-free preconditioning techniques have recently been proposed. Before giving insight into these methods, let us remark that as for the FEM, we expect large clustering of eigenvalues in the complex plane for our preconditioned algorithm. This remark focuses on the fact that it is generally preferable to work with a second-kind Fredholm integral equation like (72) since the eigenvalues accumulate at point 1. However, it can be shown that this eigenvalue clustering improves the convergence rate of the iterative solver according to the density of discretization points per wavelength. However, the convergence remains penalized by both the frequency parameter k and the geometry of the scatterer and therefore requires special care. As a tentative explanation of this phenomenon, we refer to [12, 46] for condition number estimates and spectral analysis. A first class of preconditioners are built algebraically by using for instance SPAI or ILUT techniques for the near-field matrix of integral equations (see e.g. [114]). Even if these techniques gives improved convergence, it does not lead to a convergence rate independent of the frequency. Alternative solutions to algebraic approaches have been proposed since a few years. Unlike the purely algebraic approaches which require at some points the knowledge of the matrix $[\mathscr{A}]_h$, analytical preconditioning methods have been introduced following the two-steps principle

- First, if an integral operator $\mathscr{A}$ is given at the continuous level, we propose to build an approximate operator $\mathscr{P}$ such that $\mathscr{P}\mathscr{A}$ is second-kind Fredholm,
- then, in a second step, we proceed to the discretization of $\mathscr{P}\mathscr{A}$.

These techniques that lead to analytical preconditioners have the advantage to consider the mathematical structure of the integral operator $\mathscr{A}$. As an example, let us consider the EFIE solution of the Dirichlet scattering problem

$$\mathscr{L}\mu = -u^{\mathrm{inc}}{}_{|\Gamma}. \tag{82}$$

Because we have the so-called integral Calderòn relation

$$\mathscr{S}\mathscr{L} = \frac{\mathscr{I}}{4} - \mathscr{D}^2, \tag{83}$$

then, we could also solve (82) as

$$\mathscr{S}\mathscr{L}\mu = -\mathscr{S}u^{\mathrm{inc}}{}_{|\Gamma}, \tag{84}$$

which is now a second-kind integral equation of Fredholm kind for smooth surfaces. This idea has been introduced by Steinbach and Wendland in [116] and applied to the acoustics (and electromagnetics) EFIE integral equation by Christiansen and Nédélec [50, 51]. An extension to transmission scattering problems can be found in [10]. Another idea for obtaining analytical preconditioners is to build approximate surface/volumetric DtN maps in a small domain surrounding the scatterer. This technique has been introduced in [8, 9]. All these methods provide improved convergence, in particular according to the discretization parameters but some problems remain concerning for example the convergence dependence with respect to the frequency parameter k or resonance phenomena linked to multiple scattering.

Another recent direction has been directed towards building generalized Combined Field Integral Equations and Burton-Miller formulations. Basically, the idea is to consider a coupling parameter η which is an integral or pseudodifferential operator to get new second-kind Fredholm formulations by regularization. This is particularly useful for Neumann scattering problems since the standard combined formulations are first-kind equations. Different strategies have been developed for example in [12, 13, 105, 33]. In particular, in [12, 13], it is proved that the convergence rate of the Generalized CFIE and Burton-Miller formulations is independent of the wavenumber k when the obstacles are convex. A small dependence can occur when some rays are multiply bounced. The extension to multiple scattering problems by convex obstacles has not been treated yet but ideas using Theorem 1 should lead to a possible extension, using e.g. Gauss-Seidel-type algorithms.

Finally, let us remark that efforts have also been made to propose Domain Decomposition Methods for integral equations. We refer to [34, 24] for such methods. However, it does not directly concern multiple scattering problems but rather decomposition in smaller surface subdomains of a single surface.

7. AN EXAMPLE: A HIGH-ORDER, HIGH-FREQUENCY INTEGRAL SOLVER

In this Section we present a high-order, high-frequency integral equation method for computing the field scattered from a collection of separate obstacles. The method is based on the CFIE (65) where a high-frequency ansatz is introduced to avoid discretizing the scattering surfaces at the scale of the wavelength, and on an iterative approach analogous to the Gauss-Seidel algorithm presented in section 5.2 to handle multiple scattering. The results presented here have been developed in [38, 73].

The relevant frequency-domain problem is modeled by the scalar combined-field integral equation formulation (65), that can be rewritten as:

$$\frac{1}{2}\mu(\mathbf{x}) + \int_\Gamma H(\mathbf{x},\mathbf{y})\mu(\mathbf{y})\,d\Gamma(\mathbf{y}) = \partial_{\mathbf{n}(\mathbf{x})}u^{\text{inc}}(\mathbf{x}) + i\gamma u^{\text{inc}}(\mathbf{x}), \quad \mathbf{x}\in\Gamma, \quad (85)$$

with $H(\mathbf{x},\mathbf{y}) := \partial_{\mathbf{n}(\mathbf{x})}G(\mathbf{x},\mathbf{y}) + i\gamma G(\mathbf{x},\mathbf{y})$.

7.1 High-Frequency Integral Solver for Single Scattering Configurations

For very large frequencies the integral equation techniques described above become computationally intractable. Indeed, even with the most efficient acceleration techniques, such formulations require to discretize the fields at the level of the wavelength and thus exhibit a computational complexity of at least $\mathcal{O}(k^d)$ for an d-dimensional discretization.

Current state-of-the-art simulation technology for high-frequency scattering thus usually relies on methods that are based on asymptotic ("infinite frequency") approximations of the scattering problem, such as the geometrical theory of diffraction [98]. A most attractive feature of these procedures is that they can bypass the resolution of the wavelength and work with frequency-independent discretizations. However these methods are not error-controllable, since the most accurate solution they can produce exhibits an error on the order of the wavelength.

An "ideal" solver for the high-frequency regime then would be one that retains this feature without compromising error-controllability. Further, they would exhibit high-order convergence, which would minimize the computational effort to achieve a given error.

For single-scattering configurations (which arise e.g. when Γ is convex) such an algorithm was presented in [38, 37], based on the observation that, away from shadow regions, the unknown current $\mu(\mathbf{x})$ oscillates like the incoming radiation [98], that is,

$$\mu(\mathbf{x}) = \mu_{\text{slow}}(\mathbf{x})e^{ik\alpha\cdot\mathbf{x}}. \quad (86)$$

It follows that (85) can be rewritten as

$$\frac{1}{2}\mu_{\text{slow}}(\mathbf{x}) + \int_\Gamma H(\mathbf{x},\mathbf{y})e^{ik[\varphi^{(0)}(\mathbf{y})-\varphi^{(0)}(\mathbf{x})]}\mu_{\text{slow}}(\mathbf{y})\,d\Gamma(\mathbf{y}) = g^{(0)}_{\text{slow}}(\mathbf{x}), \quad \mathbf{x}\in\Gamma, \quad (87)$$

with

$$\varphi^{(0)}(\mathbf{x}) = \alpha\cdot\mathbf{x} \quad \text{and} \quad g^{(0)}_{\text{slow}}(\mathbf{x}) = i(k\alpha\cdot n(\mathbf{x})+\gamma).$$

Note that, for $\mu_{\text{slow}}(r) = 2$, this ansatz actually corresponds to the physical optics (PO) current. For non-convex scatterers (or, more generally, in presence of multiple reflections), a more elaborate ansatz will be constructed using ray-tracing (GO) techniques; see Section 7.2. As it happens, only the solution of certain types of integral equations can be represented through an ansatz of this type. As a rule, an integral equation whose unknown is a *physical quantity* can be represented by an ansatz of this form—the unknown in (85) is the normal derivative of the solution, and it therefore admits such a representation. In contrast, the density in the indirect integral equations presented in Section 6.2.2 do not admit such a representation: see [38] for a discussion.

Throughout the illuminated region of Γ, the variations in the envelope μ_{slow} in (87) *do not* accentuate with increasing frequency and thus, for arbitrarily short wavelengths, μ_{slow} can be represented, to any prescribed accuracy, with a *fixed* number of discretization points [38].

In an effort to produce an algorithm that can solve (85) with fixed accuracy and with a frequency-independent computational cost two main challenges arise, namely:

1. The ansatz of (86) is not valid in the vicinity of shadow boundaries (where $\alpha\cdot n(\mathbf{x}) = 0$); and
2. Even if μ_{slow} could be represented with a fixed number of degrees of freedom, the numerical evaluation of the integrals in (87) would require a number of quadrature points large enough to resolve the wavelength, leading to a computational complexity that increases with frequency.

As was shown in [38] the first problem can be overcome by using frequency-dependent changes of variables within the boundary layers around the shadow boundaries where the ansatz (86) breaks down. Beyond these boundary layers and toward the illuminated region the ansatz holds true, while toward the deep shadow the current vanishes exponentially and its contribution to the integral can be controllably neglected. Moreover, the oscillations in μ_{slow} within the boundary layer occur precisely on the lengthscale of the transition region and, therefore, the current there can again be resolved with a *fixed* number of discretization points.

To solve the second problem, as is explained next, we use a localized integration scheme around the critical points of the oscillatory integral in (87).

For a sufficiently large wavenumber k and for $\mathbf{x} \neq \mathbf{y}$ the kernel $H(\mathbf{x},\mathbf{y})e^{ik\left[\varphi^{(0)}(\mathbf{y})-\varphi^{(0)}(\mathbf{x})\right]}$ behaves like the kernel

$$e^{ik\left[|\mathbf{x}-\mathbf{y}|+\varphi^{(0)}(\mathbf{y})-\varphi^{(0)}(\mathbf{x})\right]} \equiv e^{ik\phi}$$

of a generalized Fourier integral. It follows that, *asymptotically*, the only significant contributions to the oscillatory integral in (87) arise from values of μ_{slow} and its derivatives at the critical points [32]; in the present context, these critical points are the target point (where the kernel is singular) and the stationary phase points (where the gradient of the phase ϕ vanishes).

In order to obtain a *convergent* (not merely asymptotic) method for *arbitrary frequencies* which runs in *frequency independent* computing times, we introduced the following localized integration procedure *around* the critical points in [38]: For each target point $\mathbf{x}$ on the surface Γ the corresponding set of critical points is covered by a number of small regions:

1. The target point is covered by a region of radius proportional to the wavelength $\lambda = 2\pi/k$;
2. The ℓ-th stationary phase point is covered by a region of radius proportional to $\sqrt[3]{\lambda}$ (at the shadow boundaries) or $\sqrt{\lambda}$ (away from the shadow boundaries).

A partition of unity [39] is then used to reduce the integral over Γ into a number of integrals over these small regions, each of which can be evaluated numerically to high order using appropriate quadrature rules [38]. In practice, this localized integration method can be thought of as an error-controllable version of the "asymptotic method of stationary phase" [32].

Since the density μ_{slow} can be represented with a fixed number of degrees of freedom for arbitrary k and since the size of the regions associated with the critical points shrinks as the frequency increases (so that the number of oscillations of the integrand in each interval remains constant), the overall procedure results, as desired, in a convergent integration method whose computational complexity is independent of frequency. The numerical method is then completed through the use of the iterative linear algebra solver GMRES.

7.2 Multiple Scattering Extension Using an Iteratively Computable Neumann Series

The extension of our method to multiple-scattering configurations is based on three main elements [73]:

1. An iteratively computable Neumann series for the currents induced on the scattering surfaces, which accounts rigorously for multiple scattering;
2. A generalized ansatz that allows for *a priori* determination of the highly oscillatory phase of the currents in each term of the series; and
3. Use of the single-scattering algorithm mentioned above for the $\mathscr{O}(1)$ evaluation of each one of the terms in this series.

Let us consider a surface $\Gamma = \Gamma_1 \cup \Gamma_2$ composed of two convex, closed sub-surfaces Γ_1 and Γ_2. In this case (85) takes on the form

$$\frac{1}{2}\mu_p(\mathbf{x}) + \sum_{q=1}^{2}\int_{\Gamma_q} H(\mathbf{x},\mathbf{y})\mu_q(\mathbf{y})\,d\Gamma_q(\mathbf{y}) = g^{(0)}_{p,\text{slow}}(\mathbf{x})e^{ik\varphi^{(0)}_p(\mathbf{x})}, \quad \mathbf{x}\in\Gamma_p,\ p=1,2,$$

where $\mu(\mathbf{x}) \equiv \mu_p(\mathbf{x})$, $\varphi^{(0)}(\mathbf{x}) \equiv \varphi^{(0)}_p(\mathbf{x})$ and $g^{(0)}_{\text{slow}}(\mathbf{x}) \equiv g^{(0)}_{p,\text{slow}}(\mathbf{x})$ in Γ_p. Alternatively, this equation can be written as

$$\begin{bmatrix} I-T_{11} & -R_{12} \\ -R_{21} & I-T_{22} \end{bmatrix}\begin{bmatrix} \mu_1 \\ \mu_2 \end{bmatrix} = \begin{bmatrix} 2g^{(0)}_{1,\text{slow}}(\mathbf{x})e^{ik\varphi^{(0)}_1(\mathbf{x})} \\ 2g^{(0)}_{2,\text{slow}}(\mathbf{x})e^{ik\varphi^{(0)}_2(\mathbf{x})} \end{bmatrix} \tag{88}$$

where the operators T_{pp} and R_{pq} are defined as

$$T_{pp}(\mu_p)(\mathbf{x}) = -2\int_{\Gamma_p} H(\mathbf{x},\mathbf{y})\mu_p(\mathbf{y})\,d\Gamma_p(\mathbf{y}), \quad \mathbf{x}\in\Gamma_p,$$

$$R_{pq}(\mu_q)(\mathbf{x}) = -2\int_{\Gamma_q} H(\mathbf{x},\mathbf{y})\mu_q(\mathbf{y})\,d\Gamma_q(\mathbf{y}), \quad \mathbf{x}\in\Gamma_p.$$

The operators in the diagonal of the matrix B correspond precisely to the scattering problems for each isolated sub-surface and are therefore invertible. Thus, equation (88) is equivalent to

$$(I-A)\begin{bmatrix}\mu_1\\ \mu_2\end{bmatrix}=\begin{bmatrix}[I-T_{11}]^{-1}2g^{(0)}_{1,\text{slow}}(\mathbf{x})e^{ik\varphi_1^{(0)}(\mathbf{x})}\\ [I-T_{22}]^{-1}2g^{(0)}_{2,\text{slow}}(\mathbf{x})e^{ik\varphi_2^{(0)}(\mathbf{x})}\end{bmatrix} \tag{89}$$

where A is the iteration operator defined as:

$$A=\begin{bmatrix}0 & [I-T_{11}]^{-1}R_{12}\\ [I-T_{22}]^{-1}R_{21} & 0\end{bmatrix}.$$

The series solution of (89) is given by

$$\begin{bmatrix}\mu_1\\ \mu_2\end{bmatrix}=\sum_{m=0}^{\infty}\begin{bmatrix}\mu_1^{(m)}\\ \mu_2^{(m)}\end{bmatrix} \tag{90}$$

where the terms are inductively defined as

$$\begin{bmatrix}\mu_1^{(0)}\\ \mu_2^{(0)}\end{bmatrix}=\begin{bmatrix}[I-T_{11}]^{-1}2g^{(0)}_{1,\text{slow}}(\mathbf{x})e^{ik\varphi_1^{(0)}(\mathbf{x})}\\ [I-T_{22}]^{-1}2g^{(0)}_{2,\text{slow}}(\mathbf{x})e^{ik\varphi_2^{(0)}(\mathbf{x})}\end{bmatrix} \tag{91}$$

and

$$\begin{bmatrix}\mu_1^{(m)}\\ \mu_2^{(m)}\end{bmatrix}=A\begin{bmatrix}\mu_1^{(m-1)}\\ \mu_2^{(m-1)}\end{bmatrix},\quad m\geq 1. \tag{92}$$

More explicitly, relations (91) and (92) can be expressed as

$$\frac{1}{2}\mu_p^{(0)}(\mathbf{x})+\int_{\Gamma_p}H(\mathbf{x},\mathbf{y})\mu_p^{(0)}(\mathbf{y})\,d\Gamma_p(\mathbf{y}) = g^{(0)}_{p,\text{slow}}(\mathbf{x})e^{ik\varphi_p^{(0)}(\mathbf{x})},\quad \mathbf{x}\in\Gamma_p,\ p=1,2, \tag{93}$$

and

$$\frac{1}{2}\mu_p^{(m)}(\mathbf{x})+\int_{\Gamma_p}H(\mathbf{x},\mathbf{y})\mu_p^{(m)}(\mathbf{y})\,d\Gamma_p(\mathbf{y}) = \sum_{\substack{q=1\\ q\neq p}}^{2}\int_{\Gamma_q}H(\mathbf{x},\mathbf{y})\mu_q^{(m-1)}(\mathbf{y})\,d\Gamma_q(\mathbf{y}),\quad m\geq 1,\ \mathbf{x}\in\Gamma_p,\ p=1,2, \tag{94}$$

respectively.

>From (93) and (94) we see that the m-th order correction $(\mu_1^{(m)},\mu_2^{(m)})$ in (90) corresponds precisely to the current generated on each sub-surface by the m-th order reflection—that is, by the field produced on the sub-surface after m bounces of the original incident plane wave. Indeed, on the one hand, (93) corresponds to the solution of the scattering problems for each isolated sub-surface in response to the incoming radiation, ignoring interactions. On the other hand, the right-hand side of (94) is precisely the field scattered by each sub-surface at the $(m-1)$-st stage, evaluated on the complementary part of the structure. Each one of the problems (93) and (94) can be tackled by means of the methods described in Section 7.1; to do this, however, and in order to derive a representation analogous to (86), we must first identify, as shown in the following section, the phase of each one of the subsequent corrections $(\mu_1^{(m)},\mu_2^{(m)})$.

To produce the phase of each correction to the current we appeal to the interpretation of the corrections as corresponding to successive wave reflections—which suggests that the phases must coincide with those arising in a geometrical optics solution.

More precisely, the geometrical optics solution provides a sequence of functions $\delta_p^{(m)}(\mathbf{x})$ measuring the optical distance traveled by a ray arriving at $\mathbf{x}\in\Gamma_p$ after m reflections. The overall phase is then defined as

$$\varphi_p^{(m)}(\mathbf{x})=\begin{cases}\alpha\cdot\mathbf{x} & m=0,\ \mathbf{x}\in\Gamma_p,\\ \varphi_p^{(0)}(\mathbf{x})+\delta_p^{(m)}(\mathbf{x}) & m\geq 1,\ \mathbf{x}\in\Gamma_p.\end{cases} \tag{95}$$

Using (95), equations (93)–(94) read:

$$\frac{1}{2}\mu^{(m)}_{p,\text{slow}}(\mathbf{x}) + \int_{\Gamma_p}H(\mathbf{x},\mathbf{y})e^{ik(\varphi_p^{(m)}(\mathbf{y})-\varphi_p^{(m)}(\mathbf{x}))}\mu^{(m)}_{p,\text{slow}}(\mathbf{y})\,d\Gamma_p(\mathbf{y}) = g^{(m)}_{p,\text{slow}}(\mathbf{x}),\quad m\geq 0,\ \mathbf{x}\in\Gamma_p,\ p=1,2, \tag{96}$$

where

$$g^{(m)}_{p,\text{slow}}(\mathbf{x})=\begin{cases}i(k\alpha\cdot\mathbf{n}(\mathbf{x})+\gamma) & m=0,\\ e^{-ik\varphi_p^{(m)}(\mathbf{x})}\displaystyle\sum_{\substack{q=1\\ q\neq p}}^{2}\int_{\Gamma_q}H(\mathbf{x},\mathbf{y})e^{ik\varphi_q^{(m-1)}(\mathbf{y})}\mu^{(m-1)}_{q,\text{slow}}(\mathbf{y})\,d\Gamma_q(\mathbf{y}) & m\geq 1.\end{cases}$$

As we said, the slowly oscillatory character of the latter quantities follows from the interpretation of the right-hand side of (94) as the field scattered by Γ_p *after* $(m-1)$ reflections, so that its phase is precisely given by $\varphi_p^{(m)}(\mathbf{x})$. Equation (96) is then amenable to the treatment described in Section 7.1—the only difference being that the evaluation of the right hand side of (96), for $m\geq 1$, entails an integral of a highly oscillatory function.

This, however, can again be treated with the aforementioned strategies of localized integration. In fact, in this case the integrand is regular and only integrations around stationary points of the overall phase must be performed. Moreover, as is to be expected from the asymptotic limit, for any given target point $\mathbf{x} \in \Gamma_p$ there will be exactly one stationary point $\mathbf{y} \in \Gamma \setminus \Gamma_p$ of the corresponding integral. Indeed, this point will coincide with the point in $\Gamma \setminus \Gamma_p$ from which a geometrical ray that has experienced $(m-1)$ reflections goes through $\mathbf{x}$ upon an additional reflection at $\mathbf{y}$.

To test this approach we consider a scatterer composed of two circular sub-surfaces Γ_1 (with center $\mathbf{O}_1 = (0,0)$ and radius $R_1 = 1$) and Γ_2 (with center $\mathbf{O}_2 = (1,-2.5)$ and radius $R_2 = 1.5$). The wavenumber is $k = 200$, so that there are 200 and 300 wave oscillations in the perimeter of Γ_1 and Γ_2 respectively.

Figure (**20**) illustrates the convergence properties of the series (90). More precisely, there we display the values of truncated approximations

$$\widetilde{\mu}_{1,M}(\mathbf{x}) = \frac{1}{k} \sum_{m=0}^{M} \mu_1^{(m)}(\mathbf{x}) \qquad (97)$$

and of the error

$$\text{Max. Error} = \max |\widetilde{\mu}_{1,M}(\mathbf{x}) - \mu_1(\mathbf{x})| / k, \qquad (98)$$

where $\mu_1(\mathbf{x})$ is the solution (converged to machine precision) obtained by means of the algorithm proposed in [55, p. 66], and where the error is normalized by the wavenumber since the solution $\mu_1(\mathbf{x})$ grows linearly with k. The hollow circles demonstrate the spectral rate of convergence of the series, which translates for instance in an error of 1% in less than 15 iterations. The dots show how the convergence of the series can be improved even further by analytic continuation using Padé approximants [18]—in this case leading to machine accuracy in 23 iterations. As mentioned in Section 5.2, using Krylov subspace methods can lead to further improvements, as can also be seen on Figure (**20**) (stars): using ORTHODIR leads to machine accuracy in 17 iterations. The efficient application of Krylov subspace methods, and in particular the design of preconditioners, is the object of current research.

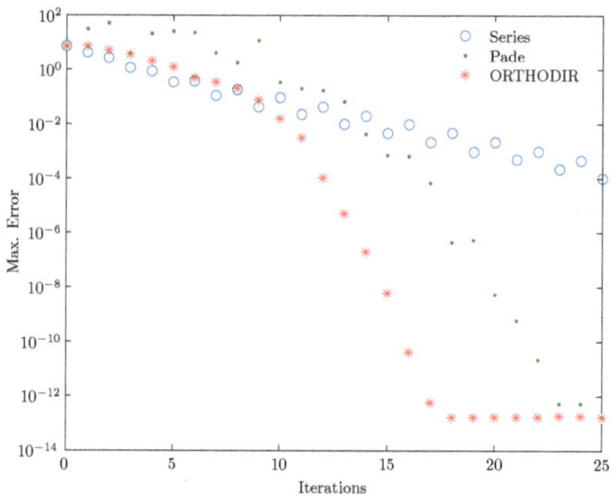

Fig. (**20**): Error as a function of the number of reflections for $k = 200$, by direct summation of the series (hollow circles), by using Padé approximants (dots) and ORTHIDIR (stars).

GENERAL CONCLUSION

This contribution gives an introduction to some mathematical and numerical aspects related to high-frequency scattering by multiple single scattering configurations. Discussions about possible recent strategies for solving single scattering problems are developed. Mainly, Fourier-based, PDE-based and integral equation-based solutions are considered. For each of these techniques, specific efficient and original algorithms are detailed to illustrate the particular difficulties arising in the numerical solution of multiple scattering configurations.

ACKNOWLEDGEMENTS

The first and third authors are partially supported by the French ANR fundings under the project MicroWave and the INRIA project CORIDA.

The second author is partially supported by the Belgian Science Policy under grant IAP P6/21 and by NSF under grant DMS-0609824.

REFERENCES

[1] Abboud T, Nédélec J-C, Zhou B. Integral equation method for high-frequencies. CR Acad Sci Paris Sér I 1994; 318: 165-170.

[2] Ainsworth M. Discrete dispersion relation for *hp*-version finite element approximation at high wave number. SIAM J Numer Anal 2004; 42: 553-575.

[3] Alleon G, Benzi M, Giraud L. Sparse approximate inverse preconditioning for dense linear systems arising in computational electromagnetics. Numer Alg 1997; 16: 1-15.

[4] Amini S. On the choice of the coupling parameter in boundary integral formulations of the exterior acoustic problem. Appl Anal 1990; 35: 75-92.

[5] Antoine X. Fast approximate computation of a time-harmonic scattered field using the on-surface radiation condition method IMA J Appl Math 2001; 66: 83-110.

[6] Antoine X. Advances in the on-surface radiation condition method: theory, numerics and Applications, Chapter in: Computational methods for acoustics problems, Magoulès F. (ed.), Saxe-Coburg Publications, 2008, pp.169-194.

[7] Antoine X, Barucq H, Bendali A. Bayliss-Turkel-like radiation condition on surfaces of arbitrary shape. J Math Anal Appl 1999; 229: 184-211.

[8] Antoine X, Bendali A, Darbas M. Analytic preconditioners for the electric field integral equation. Int J Numer Meth Eng 2004; 61: 1310-1331.
[9] Antoine X, Bendali A, Darbas M. Analytic preconditioners for the boundary integral solution of the scattering of acoustic waves by open surfaces. J Comput Acoust 2005; 13:477-498.
[10] Antoine X, Boubendir Y. An integral preconditioner for solving the two-dimensional scattering transmission problem using integral equations. Int J Comput Math 2008; 85: 1473-1490.
[11] Antoine X, Chniti C, Ramdani K. On the numerical approximation of high-frequency acoustic multiple scattering problems by circular cylinders. J Comput Phys 2008; 227: 1754-1771.
[12] Antoine X, Darbas M. Alternative integral equations for the iterative solution of acoustic scattering problems. Q J Mech Appl Math 2005; 58: 107-128.
[13] Antoine X, Darbas M. Generalized combined field integral equations for the iterative solution of the three-dimensional Helmholtz equation. Math Model Numer Anal 2007; 41: 147-167.
[14] Antoine X, Geuzaine C. Phase reduction models for improving the accuracy of the finite element solution of time-harmonic scattering problems I: general approach and low-order models. J Comput Phys 2009; 228: 3114-3136.
[15] Astley RJ. Infinite elements for wave problems: a review of current formulations and an assessment of accuracy. Int J Num Meth Eng 2000; 49: 951-976.
[16] Astley J. Infinite elements. Chapter in computational methods for acoustics problems, Editor F. Magoulès, Saxe-Coburg Publications, 2008, pp. 37-68.
[17] Babuska I, Melenk JM. The partition of unity method. Int J Num Meth Eng 1997; 40(29): 727-758.
[18] Baker GA Jr, Graves-Morris P. Padé approximants. 2nd ed. Cambridge University Press, 1996.
[19] Balabane M. Boundary decomposition for Helmholtz and Maxwell equations I. Disjoint sub-scatterers. Asymptot Anal 2004; 38: 1-10.
[20] Bayliss A, Goldstein CI, Turkel E. An iterative method for the Helmholtz equation. J Comput Phys 1983; 49: 443-457.
[21] Bayliss A, Gunzburger M, Turkel E. Boundary conditions for the numerical solution of elliptic equations in exterior regions. SIAM J Appl Math 1982; 42: 430-451.
[22] Bayliss A, Turkel E. Radiation boundary conditions for wave-like equations. Comm Pure Appl Math 1980; 33: 707-725.
[23] Benamou JD, Desprès B. A domain decomposition method for the Helmholtz equation and related optimal control problems. J Comput Phys 1997; 136: 68-82.
[24] Bendali A, Boubendir B, Fares M. A FETI-like domain decomposition method for coupling finite elements and boundary elements in large-size problems of acoustic scattering. Comput & Struct 2007; 85: 526-535.
[25] Bérenger JP. A perfectly matched layer for the absorption of electromagnetic waves. J Comput Phys 1994; 114: 185-200.
[26] Bérenger JP. Three-dimensional perfectly matched layer for the absorption of electromagnetic waves. J Comput Phys 1996; 127: 363-379.
[27] Bérenger JP. Perfectly matched layer for the FDTD solution of wave-structure interaction problems. IEEE Trans Antennas Prop 1996; 44: 110-117.
[28] Bermudez A, Hervella-Nieto L, Prieto A, Rodriguez R. An exact bounded perfectly matched layer for time-harmonic scattering problems. SIAM J Sci Comput 2007; 30: 312-338.
[29] Bettess P. Infinite elements. Int J Num Meth Eng 1977; 11: 53-64.
[30] Bettess P. Infinite elements. Penshaw Press, Sunderland, UK, 1992.
[31] Bettess P, Shirron J, Laghrouche O, Peseux B, Sugimoto R, Trevelyan J. A numerical integration scheme for special finite elements for the Helmholtz equation. Int J Num Meth Eng 2003; 56: 531-552.
[32] Bleistein N, Handelsman RA. Asymptotic expansions of integrals. Dover 1986.
[33] Borel S, Levadoux D, Alouges F. A new well-conditioned integral formulation for Maxwell equations in three dimensions. IEEE Trans Antennas Prop 2005; 53: 2995-3004.
[34] Boubendir Y, Bendali A Collino F. Domain decomposition methods and integral equation for solving Helmholtz diffraction problem. in: 5th International Conference on Mathematical and Numerical Aspects of Wave Propagation, Bermudez A, Gomez D, Hazard C, Roberts JE (Eds), pp. 760-764, 2000.
[35] Boyd JP. Chebychev and Fourier spectral methods. 2nd Ed. Dover Publications, 2000.
[36] Brakhage H, Werner P. Über das Dirichletsche Aussenraumproblem für die Helmholtzsche Schwingungsgleichung. Numer Math 1965; 16: 325-329.
[37] Bruno O, Geuzaine C. An $\mathcal{O}(1)$ integration scheme for three-dimensional surface scattering problems. J Comput Appl Math 2007; 204: 463-476.
[38] Bruno O, Geuzaine C, Monro J Jr, Reitich F. Prescribed error tolerances within fixed computational times for scattering problems of arbitrarily high frequency: the convex case. Phil Trans Royal Soc Ser A 2004; 362: 629-645.
[39] Bruno O, Kunyansky L. High-order algorithm for the solution of surface scattering problems: basic implementation, tests, and applications. J Comput Phys 2001; 169: 80-110.
[40] Buffa A, Hiptmair R. Regularized combined field integral equations. Numer Math 2005; 100: 1-19.
[41] Burton AJ and Miller GF. The application of integral equation methods to the numerical solution of some exterior boundary-value problems. Proc Roy Soc Lond Ser A 1971; 323: 201-210.
[42] Carayol Q, Collino F. Error estimates in the fast multipole method for scattering problems. Part 1: truncation of the Jacobi-Anger series. M2AN 2004; 38: 371-394.
[43] Carpinteri B, Duff IS, Giraud L. Sparse pattern selection strategies for robust Frobenius norm minimization preconditioners in electromagnetism. In: Preconditioning techniques for large sparse matrix problems in industrial applications (Minneapolis, MN, 1999), Numer Lin Algebra Appl 2000; 7: 667-685.
[44] Cessenat O, Desprès B. Application of an ultra-weak variational formulation of elliptic PDEs to the two-dimensional Helmholtz problem. SIAM J Numer Anal 1998; 35: 255-299.
[45] Cessenat O, Desprès B. Using plane waves as base functions for solving time harmonic equations with the ultra-weak variational formulation. J Comput Acoust 2003; 11: 227-238.
[46] Chandler-Wilde SN, Monk P. Wavenumber explicit bounds in time-harmonic scattering. SIAM J Math Anal 2007; 39: 1428-1455.
[47] Chen K. Matrix preconditioning techniques and applications. Cambridge Monographs on Applied and Computational Mathematics, 2005.
[48] Chew WC, Jin JM, Michielssen E, Song J. Fast and efficient algorithms in computational electromagnetics. Artech House Antennas and Propagation Library, Norwood, 2001.
[49] Chew WC, Warnick KF. On the spectrum of the electric field integral equation and the convergence of the moment method. Int J Numer Meth Eng 2001; 51: 475-489.
[50] Christiansen SH, Nédélec J-C. Preconditioners for the numerical solution of boundary integral equations from acoustics. CR Acad Sci Paris Sér I 2000; 330: 617-622.
[51] Christiansen SH, Nédélec J-C. A preconditioner for the electric field integral equation based on Calderon formulas. SIAM J Numer Anal 2002; 40: 1100-1135.
[52] Collino F, Ghanemi S, Joly P. Domain decomposition method for harmonic wave propagation: a general presentation. Comput Methods Appl Mech Eng 2000; 184: 171-211.
[53] Collino F, Monk P. The perfectly matched layer in curvilinear coordinates. SIAM J Sci Comput 1998; 19: 2061-2090.
[54] Colton DL, Kress R. Integral equation methods in scattering theory. Pure and Applied Mathematics, John Wiley & Sons Inc., 1983.
[55] Colton DL, Kress R. Inverse acoustic and electromagnetic scattering theory. 2nd Ed., Applied Mathematical Sciences 93, Springer-Verlag, 1998.
[56] Costabel M. Integral equation methods in scattering theory. SIAM J Math Anal 1988; 19: 613-626.
[57] Darrigrand E. Coupling of fast multipole method and microlocal discretization for the 3-D Helmholtz equation. J Comput Phys 2002; 181: 126-154.
[58] De La Bourdonnaye A. High-frequency approximation of integral equations modeling scattering phenomena. RAIRO Math Model Numer Anal 1994; 28: 223-241.
[59] Desprès B. Domain decomposition method and the Helmholtz problem (part II). In: 2nd International Conference on Mathematical and Numerical Aspects of Wave Propagation, Newark, DE, 1993. SIAM, Philadelphia.
[60] Dey S. Evaluation of p-FEM approximations for mid-frequency elasto-acoustics. J Comput Acoust 2003; 11: 195-225.
[61] Djellouli R, Farhat C, Macedo A, Tezaur R. Finite element solution of two-dimensional acoustic scattering problems using arbitrarily shaped convex artificial boundaries. J Comput Acoust 2000; 8: 81-99.

[62] Ecevit F, Reitich F. Analysis of multiple scattering iterations for high-frequency scattering problems. I: the two-dimensional case. Preprint, School of Mathematics, University of Minnesota 2006.
[63] Egidi N, Gobbi R, Maponi P. The efficient solution of electromagnetic scattering for inhomogeneous media. J Comput Appl Math 2007; 210: 175-182.
[64] Engquist B, Majda A. Absorbing boundary conditions for the numerical simulation of waves. Math Comp 1977; 31: 629-651.
[65] Engquist B, Majda A. Radiation boundary conditions for acoustics and elastic wave calculations. Comm Pure Appl Math 1979; 32: 313-357.
[66] Erlangga YA, Vuik C, Oosterlee CW. On a class of preconditioners for the Helmholtz equation. Appl Numer Math 2004; 50: 409-425.
[67] Farhat C, Harari I, Franca LP. The discontinuous enrichment method. Comput Methods Appl Mech Eng 2001; 190: 6455-6479.
[68] Farhat C, Harari I, Hetmaniuk U. A discontinuous Galerkin method with Lagrange multipliers for the solution of Helmholtz problems in the mid-frequency regime. Comput Methods Appl Mech Eng 2003; 192: 1389-1419.
[69] Farhat C, Macedo A, Lesoinne M, Roux F, Magoulès F, De La Bourdonnaye A. Two-level domain decomposition methods with lagrange multipliers for the fast iterative solution of acoustic scattering problems. Comput Meth Appl Mech Eng 2000; 184: 213-239.
[70] Gander MJ, Magoulès F, Nataf F. Optimized Schwarz methods without overlap for the Helmholtz equation. SIAM J Sci Comput 2002; 24: 38-60.
[71] Ganesh M, Hawkins SC. Simulation of acoustic scattering by multiple obstacles in three dimensions. Aust NZ Indus Appl Math J 2008; 50: 31-45.
[72] Geuzaine C, Bedrossian J, Antoine X. An amplitude formulation to reduce the pollution error in the finite element solution of time-harmonic scattering problems. IEEE Trans Magn 2008; 44: 782-785.
[73] Geuzaine C, Bruno O, Reitich F. On the $\mathcal{O}(1)$ solution of multiple-scattering problems. IEEE Trans Magn 2005; 41: 1488-1491.
[74] Geuzaine C, Dular P, Gaignaire R, Sabariego R. An amplitude finite element formulation for multiple-scattering by a collection of convex obstacles. In: 17th Conference on the Computation of Electromagnetic Fields, Florianopolis, Brazil, 2009.
[75] Geuzaine C, Meys B, Dular P, Henrotte F, Legros W. A Galerkin projection method for mixed finite elements. IEEE Trans Magn 1999; 35: 1438-1441.
[76] Geuzaine C, Remacle JF. Gmsh: a three-dimensional finite element mesh generator with built-in pre- and post-processing facilities. Int J Numer Meth Eng 2009; 79: 1309-1331.
[77] Giladi E, Keller JB. A hybrid numerical asymptotic method for scattering problems. J Comput Phys 2001; 174: 226-247.
[78] Givoli D. Recent advances in the DtN FE method. Arch Comput Meth Eng 1999; 6: 71-116.
[79] Givoli D. High-order non-reflecting boundary conditions: a review. Wave Motion 2004; 39: 319-326.
[80] Greengard L, Rokhlin V. A fast algorithm for particle simulations. J Comput Phys 1987; 73: 325-348.
[81] Grote MJ, Keller JB. On nonreflecting boundary conditions. J Comput Phys 1995; 122: 231-243.
[82] Grote MJ, Kirsch C. Dirichlet-to-Neumann boundary conditions for multiple scattering problems. J Comput Phys 2004; 201: 630-650.
[83] Harari I, Grosh K, Hughes TJR, Malhotra M, Pinsky PM, Stewart JR, Thompson LL. Recent developments in finite element methods for structural acoustics. Arch Comput Meth Eng 1996; 3: 132-311.
[84] Harari I, Hughes TJR. A cost comparison of boundary element and finite element methods for problems of time-harmonic acoustics. Comput Meth Appl Mech Eng 1992; 97: 77-102.
[85] Harari I, Hughes TJR. Galerkin/least-squares finite element methods for the reduced wave equation with non-reflecting boundary conditions in unbounded domains. Comput Meth Appl Mech Eng 1992; 98: 411-454.
[86] Harari I, Nogueira CL. Reducing dispersion of linear triangular elements for the Helmholtz equation. J Eng Mech 2002; 128: 351-358.
[87] Hughes TJR, Feijoo GR, Mazzei L, Quincy JB. The variational multiscale method: a paradigm for computational mechanics. Comput Meth Appl Mech Eng 1998; 166: 3-24.
[88] Huttunen T, Monk P, Kaipio J. Computational aspects of the ultra-weak variational formulation. J Comput Phys 2002; 182: 27-46.
[89] Ianculescu C, Thompson LL. Parallel iterative solution for the Helmholtz equation with exact non-reflecting boundary conditions. Comp Methods Appl Mech Eng 2006; 195: 3709-3741.
[90] Ihlenburg F. Finite Element Analysis of Acoustic Scattering. Springer-Verlag, 1998.
[91] Ihlenburg F, Babuska I. Dispersion analysis and error estimation of Galerkin finite element metods for the Helmholtz equation. Int J Num Meth Eng 1995; 38: 3745-3774.
[92] Iserles A, Norsett SP. On quadrature methods for highly oscillatory integrals and their implementation. BIT 2004; 44: 755-772.
[93] Jami A, Lenoir M. A variational formulation for exterior problems in linear hydrodynamics. Comput Methods Appl Mech Eng 1978; 16: 341-359.
[94] Kechroud R, Antoine X, Soulaimani A. Numerical accuracy of a Padé-type non-reflecting boundary condition for the finite element solution of acoustic scattering problems at high-frequency. Int J Numer Meth Eng 2005; 64: 1275-1302.
[95] Kerchroud R, Soulaimani A, Antoine X. Performance study of plane wave finite element methods with a Padé-type artificial boundary condition in acoustic scattering. Adv Engrg Softw 2009; 40: 738-750.
[96] Kerchroud R, Soulaimani A, Saad Y. Preconditioning techniques for the solution of the Helmholtz equation by the finite element method. Math Comput Simul 2004; 65: 303-321.
[97] Keller JB, Givoli D. Exact non-reflecting boundary conditions. J Comput Phys 1989; 81: 172-192.
[98] Keller JB, Lewis RM. Asymptotic methods for partial differential equations: the reduced wave equation and Maxwell's equations. Surveys in Applied Mathematics 1, Plenum Press, New York, 1995, pp. 1-82.
[99] Kress R. Minimizing the condition number of boundary integral operators in acoustic and electromagnetic scattering. Q J Mech Appl Math 1985; 38: 323-341.
[100] Kress R. Linear integral equations. 2nd ed. Applied Mathematical Sciences 82, New York, Springer-Verlag, 1999.
[101] Kress R, Spassov WT. On the condition number of boundary integral operators for the exterior Dirichlet problem for the Helmholtz equation. Numer Math 1983; 42: 77-95.
[102] Kriegsmann GA, Taflove A, Umashankar KR. A new formulation of electromagnetic wave scattering using the on-surface radiation condition method. IEEE Trans Antennas Prop 1987; 35: 153-161.
[103] Laghrouche O, Bettess P. Short wave modelling using special finite elements. J Comput Acoust 2000; 8: 189-210.
[104] Laghrouche O, Bettess P, Astley RJ. Modeling of short wave diffraction problems using approximating systems of plane waves. Int J Num Meth Eng 2002; 54: 1501-1533.
[105] Levadoux D, Michielsen B. Nouvelles formulations intégrales pour les problèmes de diffraction d'ondes. M2AN Math Model Numer Anal 2004; 38: 157-175.
[106] Martin PA. Multiple scattering, interaction of time-harmonic waves with N obstacles. Encyclopedia of Mathematics and its Applications 107, Cambridge, 2006.
[107] McLean W. Strongly elliptic systems and boundary integral equations. Cambridge University Press, Cambridge, 2000.
[108] Melenk JM, Babuska I. The partition of unity method: basic theory and applications. Comput Meth Appl Mech Eng 1996; 139: 289-314.
[109] Nédélec J-C. Acoustic and electromagnetic equations. Applied Mathematical Sciences 144, New York, Springer-Verlag, 2001.
[110] Oberai AA, Malhotra M, Pinsky PM. On the implementation of the Dirichlet-to-Neumann radiation condition for iterative solution of the Helmholtz equation. Appl Numer Math 1998; 27: 443-464.
[111] Saad Y. Iterative methods for sparse linear systems. Second Edition, SIAM, 2003.
[112] Saad Y, Schultz MH. A generalized minimal residual algorithm for solving nonsymmetric linear systems. SIAM J Sci Comput 1986; 7: 856-869.
[113] Sauter SA. Variable order panel clustering. Computing 2000; 64: 223-261.
[114] Sertel K, Volakis JL. Incomplete LU preconditioner for FMM implementation Microwave Opt Technol Letters 2000; 26: 265-267.
[115] Singer I, Turkel E. A perfectly matched layer for the Helmholtz equation in a semi-infinite strip. J Comput Phys 2004; 201: 439-465.
[116] Steinbach O, Wendland WL. The construction of some efficient preconditioners in the boundary element method. Adv Comput Math 1998; 9: 191-216.
[117] Tai CT. An iterative method of solving a system of linear equations and its physical interpretation from the point of view of scattering theory. IEEE Trans Antennas Prop 1970; 18: 713-714.
[118] Tezaur R, Macedo A, Farhat C. Iterative solution of large-scale acoustic scattering problems with multiple right hand-sides by a domain decomposition method with lagrange multipliers. Int J Num Meth Eng 2001; 51: 1175-1193.

[119] Tezaur R, Macedo A, Farhat C, Djellouli R. Three-dimensional finite element calculations in acoustic scattering using arbitrarily shaped artificial boundaries. Intern J Numer Meth Eng 2002; 53: 1461-1476.
[120] Thompson LL. A review of finite-element methods for time-harmonic acoustics. J Acoust Soc Amer 2006; 119: 2272-2293.
[121] Turkel E. Boundary conditions and iterative schemes for the Helmholtz equation in unbounded regions. Chapter in: Computational methods for acoustics problems, Magoulès F (ed), Saxe-Coburg Publications, 2008, pp. 127-159.
[122] Turkel E, Erlangga Y. Preconditioning for the Helmholtz equation. In: Proc. ECCOMAS CFD Conf. 2006, Wesseling P, Onate E, Periaux J (Eds). The Netherlands, 2006.
[123] Turkel E, Yefet A. Absorbing PML boundary layers for wave-like equations. Appl Numer Math 1998; 27: 533-557.

CHAPTER 5

Exact Boundary Conditions for Wave Propagation in Periodic Media Containing a Local Perturbation

S. Fliss, P. Joly and J-R. Li

INRIA POEMS Project, Rocquencourt, Le Chesnay Cedex, France.

Abstract : We present in this chapter a review of some recent research work about a new approach to the numerical simulation of time harmonic wave propagation in infinite periodic media including a local perturbation. The main difficulty lies in the reduction of the effective numerical computations to a bounded region enclosing the perturbation. Our objective is to extend the approach by Dirichlet-to-Neumann (DtN) operators, well known in the case of homogeneous media (as non local transparent boundary conditions). The new difficulty is that this DtN operator can no longer be determined explicitly and has to be computed numerically. We consider successively the case of a periodic waveguide and the more complicated case of the whole space. We show that the DtN operator can be characterized through the solution of local PDE cell problems, the use of the Floquet-Bloch transform and the solution of operator-valued quadratic or linear equations. In our text, we shall outline the main ideas without going into the rigorous mathematical details. The non standard aspects of this procedure will be emphasized and numerical results demonstrating the efficiency of the method will be presented.

1. INTRODUCTION

Periodic media play a major role in applications, in particular in optics for micro and nano-technology [18, 20, 23, 28]. From the point of view of applications, one of the main interesting features is the possibility offered by such media of selecting ranges of frequencies for which waves can or cannot propagate. Mathematically, this property is linked to the gap structure of the spectrum of the underlying differential operator appearing in the model. For a complete, mathematically oriented presentation, we refer the reader to [23, 24]. There is a need for efficient numerical methods for computing the propagation of waves inside such structures. In real applications, the media are not perfectly periodic but differ from periodic media only in bounded regions (which are small with respect to the total size of the propagation domain). In this case, a natural idea is to reduce the pure numerical computations to these regions and to try to take advantage of the periodic structure of the problem outside : this is particularly of interest when the periodic regions contain a large number of periodicity cells.

In the case where the unperturbed medium is homogeneous (in some sense, a periodic medium with an arbitrarily small period), this is a very old problematic. Various methods can be used to restrict the computation around the perturbation. A first class of methods consists in applying an artificial boundary condition which is transparent or approximately transparent. Let us cite :

(i) the coupling techniques between volumic methods and integral representations or integral equation techniques [19, 25, 14, 17]

(ii) the DtN approaches which consists in computing exactly the Dirichlet-to-Neumann operator associated to the exterior medium, provided that the geometry of the boundary is properly chosen (typically a circle in 2D)

(iii) the local radiation conditions at finite distance [8, 2], constructed as a local approximation of the exact non local condition at various orders with respect to a small parameter, typically the inverse of the frequency.

Methods (i) and (ii) are exact (up to numerical approximation). The method (iiii) is approximate and its accuracy improves where the order of the condition increases or the artificial boundary goes to infinity. However, none of these methods can be applied or extended directly to general exterior periodic media because they use the homogeneous nature of the exterior medium (explicit formulas are used for the solution of the exterior problem in (i), (ii) and (iii), the knowledge of the Green function is used in case (ii) and separation of variables is used in case (iii)).

The second approach consists in surrounding

the computational domain by an absorbing layer in which the PML technique [3] is applied. Physically the method can be interpreted as letting an incident wave from the computational domain enters the layer without reflexion and absorbs the wave inside the layer preventing it to come back in the computational domain. This principle is not adapted a priori to periodic media for which a wave leaving the computational domain will interact with heterogeneities of the medium up to infinity. That is why the standard PML technique cannot work in this case (see however the pole condition techniques that can be seen as a generalization of the PML method in the case of non-homogeneous media [15, 16]).

It seems that there are very few works in the same spirit in the mathematical literature for the case of periodic perturbed media. A problem similar that have some similarities with the one we consider in this paper is the numerical computation of localized modes (non trivial solutions of the propagation model in the absence of any source term) that may appear for specific frequencies due to the presence of a local perturbation of the periodic media (see [9, 10, 11] for existence results). The supercell method analyzed [29] has similarities with the radiation condition at finite distance (i) : it consists in making computations in a bounded domain of large size, the resulting solution converging to the true solution when the size goes to infinity. Note however that in this case as the localized modes are exponentially decreasing, this convergence is exponentially fast with respect to the size of the truncated domain.

The notion of DtN maps already appears for instance, in the works of T. Abboud [1] for the diffraction problem by periodic gratings or of J. Tausch [30] for periodic open waveguides. However in these two cases the DtN map is used to deal with the unboundedness of the propagation medium in the direction(s) transverse to the periodicity direction(s).

In a first paper [21], we treated the case of locally perturbed periodic waveguide : typically the unperturbed propagation medium is bounded in one direction and periodic in the other. We proposed a numerical method for determining DtN operators by solving local cell problems an operator valued stationary Ricatti equation. In a second paper [13], we proposed an extension of the above work to the case where the unperturbed media is periodic in the two directions. We presented the conceptual aspects of the method for the construction of the DtN operator and we exposed the main theoretical issues and results.

This chapter is devoted to a general presentation of the method of [13, 12] adapted to the case of a RtR (for Robin-to-Robin) boundary condition : instead of relating the Dirichlet and the Neumann traces of the solution, we want to relate two different Robin traces of the solution, typically

$$\frac{\partial u}{\partial \mathbf{n}} \pm \imath Z u,$$

where Z is a non-zero impedance, that are related through a transparent RtR operator. Such Robin traces naturally appear in non overlapping domain decomposition methods for solving the Helmholtz equation [4, 5, 6]. They are also used in the context of periodic media in [7]. From the numerical point of view, one of the interest of RtR operators is that, contrary to DtN operators for instance, they are bounded operators with bounded inverse and their discretization leads to well-conditioned matrices.

For the sake of conciseness of the presentation, we shall restrict ourselves to the exposition of the main ideas and results and to illustrate the method through numerical results. On the other hand, for the sake of rigor, we have chosen to make precise the functional framework inside which the arguments we shall develop can be completely justified. However, most theorems will be stated without proofs. The reader can find these proofs together additional details in [21, 13, 12]. Moreover, the functional analysis makes the presentaton of the method involved but it can be omitted at the first reading.

Let us also mention that, in order to avoid mathematical difficulties, we consider the case where the propagation medium is slightly absorbing, the absorption being quantified by a small positive parameter $\varepsilon > 0$ (see Section 2). The challenging question of studying the limit case when ε tends to 0 (i.e. the limiting absorption principle) is still an open question to our knowledge. However the method that we present here can be formally extended to non absorbing media by using the heuristics proposed in [21, 12] for the case of periodic waveguides. Another advantage of RtR conditions is that they seem better adapted to the development of a numerical limiting absorption principle, which is the object of a future work.

2. MODEL PROBLEM

The model problem that we consider is the propagation of a time harmonic scalar wave in a 2D periodic medium, $\Omega = \mathbb{R}^2$, with a local perturbation. More precisely, we shall assume that the geometry as well as the material properties of the plane are x and $y-$ periodic except in a bounded region (see Figure (**1**) for an example). Specifically, we want to solve the

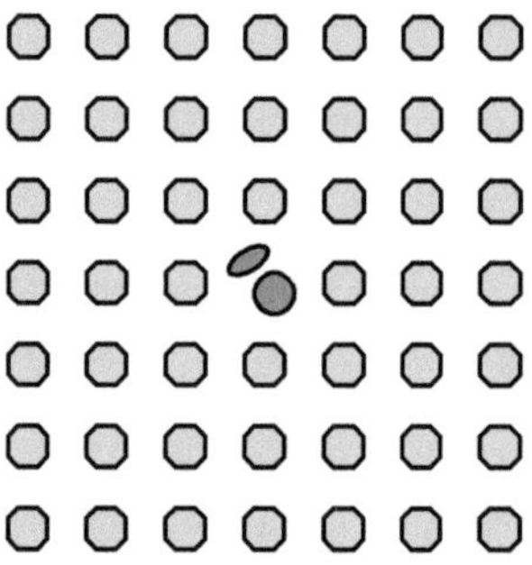

FIGURE (**1**): Example of domain of propagation Ω : the function ρ_p takes different values in the white, the light gray and the gray parts.

2D Helmholtz equation

$$-\Delta u - \rho\,(\omega^2 + \imath\varepsilon\omega)\,u = f, \quad \text{in } \Omega = \mathbb{R}^2 \qquad (\mathscr{P})$$

under the following hypotheses :

1. The local perturbation to the periodicity in the domain of propagation $\Omega = \mathbb{R}^2$ is contained inside the bounded region Ω^i. In $\Omega^e = \Omega \setminus \Omega^i$ the periodicity is along the x and y directions with the same period L. We denote by $\mathscr{C}$ the reference unit periodicity cell
$$\mathscr{C} = \left] -\frac{L}{2}, \frac{L}{2} \right[^2.$$
Without loss of generality, Ω^i is chosen to be a square of the same size as $\mathscr{C}$:
$$\Omega^i = \left] -\frac{L}{2}, \frac{L}{2} \right[^2.$$
The boundary of Ω^i is denoted by $\Sigma^i = \partial\Omega^i$. See Figure (**2**) for notation.
2. The index of refraction satisfies the following conditions :
(H1) $0 < \rho_- \leq \rho(x) \leq \rho_+$,
(H2) $\left| \begin{array}{l} \rho_p(x \pm L, y \pm L) = \rho_p(x,y),\ \forall (x,y), \\ \text{Supp}(\rho - \rho_p) \subset \Omega^i. \end{array} \right.$

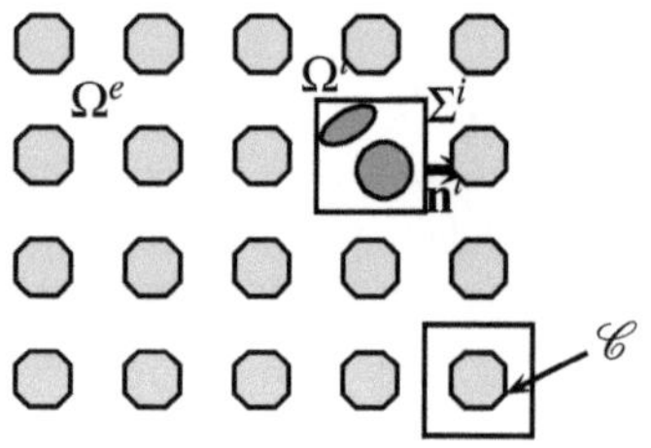

FIGURE (**2**): Geometry and notation.

3. The support of the source f is contained in Ω^i.
4. ε is a physical parameter, typically small, that represents a slight absorption in the medium. We shall assume that $\varepsilon > 0$.

It is well known that the problem $(\mathscr{P})$ admits a unique solution in $H^1(\triangle, \Omega)$, the closed subspace of functions in $H^1(\Omega)$ whose laplacian is in $L^2(\Omega)$.

In order to compute numerically the solution, our goal is to characterize the restriction of the solution u to Ω^i, that we denote u^i via a transparent boundary condition on Σ^i. Instead of looking for an exact DtN condition as it is done in [13], we want to write an impedance-like boundary condition. To be more precise, let us define the impedance

$$Z \equiv Z(x) := \omega\sqrt{\rho(x)}\sqrt{1 + \imath\frac{\varepsilon}{\omega}} \qquad (1)$$

with the convention $\text{Im}\sqrt{z} > 0$ and $\mathbf{n}^i$ the unit normal vector which is outgoing with respect to Ω^i. We wish to relate two « Robin traces » of u^i along σ^i, namely

$$\left| \begin{array}{lll} \dfrac{\partial u^i}{\partial \mathbf{n}^i} + \imath Z u^i, & : & \text{the outgoing trace of } u^i \\ -\dfrac{\partial u^i}{\partial \mathbf{n}^i} + \imath Z u^i, & : & \text{the incoming trace of } u^i \end{array} \right.$$

Troughout the rest of this paper, this kind of boundary traces will be symbolized by arrows whose directions indicates what type of trace (outgoing or incoming) is considered.

We look for an operator Λ, Robin-to-Robin or RtR, acting on functions defined on Σ^i such as the transparent boundary condition for u^i is written

$$-\frac{\partial u^i}{\partial \mathbf{n}^i} + \imath Z u^i = \Lambda\Big[\frac{\partial u^i}{\partial \mathbf{n}^i} + \imath Z u^i\Big] \text{ on } \Sigma^i. \qquad (2)$$

To characterize, at least in an abstract way, the operator Λ, it suffices to write $(\mathscr{P})$ as the transmission between u^i to u^e, the restriction of u to Ω^e and rewrite the standard transmission conditions using Robin traces of u^e and u^i, namely

$$\left| \begin{array}{rcl} \dfrac{\partial u^e}{\partial \mathbf{n}^e} + \imath Z u^e & = & -\dfrac{\partial u^i}{\partial \mathbf{n}^i} + \imath Z u^i \\ -\dfrac{\partial u^e}{\partial \mathbf{n}^e} + \imath Z u^e & = & \dfrac{\partial u^i}{\partial \mathbf{n}^i} + \imath Z u^i \end{array} \right. \tag{3}$$

(u^e and u^i exchange their outgoing and incoming Robin traces) where $\mathbf{n}^e = -\mathbf{n}^i$ is the unit normal vector of Σ^i outgoing with respect to Ω^e.

From (2) and (3), Λ be defined as

$$\begin{array}{rcl} \Lambda : H^{-\frac{1}{2}}(\Sigma^i) & \rightarrow & H^{-\frac{1}{2}}(\Sigma^i) \\ \varphi & \mapsto & \Lambda\varphi = \dfrac{\partial}{\partial \mathbf{n}^e} u^e(\varphi) + \imath Z u^e(\varphi)\Big|_{\Sigma^i}, \end{array}$$

where $u^e(\varphi) \in H^1(\triangle, \Omega^e)$ is the unique solution of the exterior problem posed on Ω^e with non homogeneous incoming Robin conditions on Σ^i, namely

$$\left| \begin{array}{rcl} -\triangle u^e - \rho_p(\omega^2 + \imath\varepsilon\omega)\, u^e = & 0 & \text{in } \Omega^e \\ -\dfrac{\partial u^e}{\partial \mathbf{n}^e} + \imath Z u^e = & \varphi & \text{on } \Sigma^i. \end{array} \right. \tag{$\mathscr{P}^e$}$$

The notation $u^e(\varphi)$ indicates the dependence of the solution to ($\mathscr{P}^e$) on the incoming Robin boundary data. The RtR operator Λ maps the incoming Robin boundary data of $u^e(\varphi)$ onto its outgoing Robin boundary data.

We derive a method for the characterization of the RtR operator Λ and we show that it can be characterized through the solution of local PDE cell problems, the use of analytical tools such as the Floquet-Bloch transform and the solution of operator-valued quadratic or linear equations.

We shall restrict ourselves to a particular situation that makes the presentation of our method simpler. More precisely, we shall consider the case where the periodicity cell $\mathscr{C}$ and the boundary Σ^i are squares (as indicates previously), that implies that they are doubly symmetric and

(H3) the restriction of the function ρ_p to $\mathscr{C}$ is a function doubly symmetric.

The notion of double symmetry will be explained in Section 3. This situation is often met in the applications. The method can be extended to the cases where (H3) is relaxed (see the appendix B of [13]).

3. DOUBLE SYMMETRY AND RELATED RESULTS

We consider now a specific situation where the geometry and material properties, in addition to satisfying periodicity along the lines x and y, must satisfy some symmetry properties.

3.1 General Definitions

We introduce the two diagonal lines $D_1 = \{(x,y) : x = y\}$ and $D_{-1} = \{(x,y) : x = -y\}$. The symmetry across D_1 will be denoted by $\mathbf{S}_1$ and across D_{-1} by $\mathbf{S}_{-1}$. See Figure (**3**).

A subset $\mathscr{O}$ is doubly symmetric if, the origin being chosen as the barycenter of $\mathscr{O}$, it is invariant through the transformations S_1 and S_{-1} .

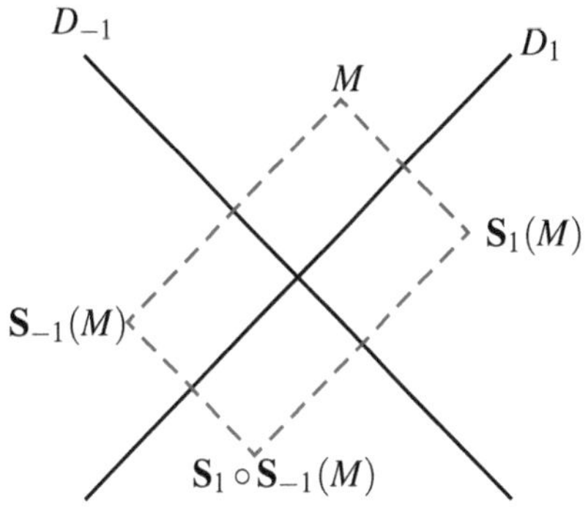

FIGURE (**3**): The symmmetries S_1 and S_{-1}

We will denote $\mathscr{O}_i, i \in \{0,1,2,3\}$ the four isomorphic sets :

$$\mathscr{O}_i = \mathscr{O} \bigcap \mathscr{Q}_i$$

where $\mathscr{Q}_i, i \in \{0,1,2,3\}$ are the four quadrants of $\mathbb{R}^2$ represented in Figure (**4**).

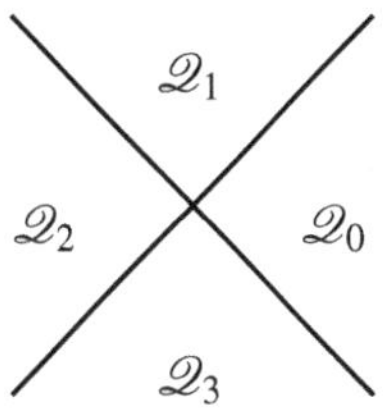

FIGURE (**4**): The four quadrants of $\mathbb{R}^2$, $\mathscr{Q}_i, i \in \{0,1,2,3\}$

A function $n : \mathscr{O} \rightarrow \mathbb{C}$ is called doubly symmetric when it keeps unchanged after reflection across D_1 and D_{-1} :

$$n = n \circ \mathbf{S}_1 = n \circ \mathbf{S}_{-1}.$$

3.2 Functional Spaces on Doubly Symmetric Sets

Let $\mathscr{O}$ be a doubly symmetric open set in $\mathbb{R}^2$ and $H = H^p(\mathscr{O})$, $p \in \mathbb{N}$ or $H = H^1(\triangle, \mathscr{O})$. The space H can be decomposed as the sum of four subspaces of functions that are symmetric or antisymmetric across D_1 and D_{-1} (See Figure (**5**) for examples)

$$H(\mathscr{O}) = H_{(s,s)} \oplus H_{(s,a)} \oplus H_{(a,s)} \oplus H_{(a,a)}, \quad (4)$$

where

$$\left|\begin{array}{lllll} v \in H_{(s,s)} & \Leftrightarrow & v = & +v \circ \mathbf{S}_1 = & +v \circ \mathbf{S}_{-1}, \\ v \in H_{(a,a)} & \Leftrightarrow & v = & -v \circ \mathbf{S}_1 = & -v \circ \mathbf{S}_{-1}, \\ v \in H_{(s,a)} & \Leftrightarrow & v = & +v \circ \mathbf{S}_1 = & -v \circ \mathbf{S}_{-1}, \\ v \in H_{(a,s)} & \Leftrightarrow & v = & -v \circ \mathbf{S}_1 = & +v \circ \mathbf{S}_{-1}. \end{array}\right.$$

Note that this spaces are orthogonal in L^2.

Given $(p,q) \in \{s,a\}^2$, we define the closed subspace $H^{1/2}_{(p,q)}(\partial\mathscr{O})$ of $H^{1/2}(\partial\mathscr{O})$ by

$$H^{1/2}_{(p,q)}(\partial\mathscr{O}) = \left\{ u\Big|_{\partial\mathscr{O}}, \quad u \in H^1_{(p,q)}(\mathscr{O}) \right\}$$

and the closed subspace $H^{-1/2}_{(p,q)}(\partial\mathscr{O})$ of $H^{-1/2}(\partial\mathscr{O})$ by

$$H^{-1/2}_{(p,q)}(\partial\mathscr{O}) = \left\{ \frac{\partial u}{\partial \mathbf{n}}\Big|_{\partial\mathscr{O}}, \quad u \in H^1_{(p,q)}(\triangle, \mathscr{O}) \right\}.$$

REMARK 3.1
Despite the notation,

$$H^{-1/2}_{(p,q)}(\partial\mathscr{O}) \subsetneq H^{1/2}_{(p,q)}(\partial\mathscr{O})'.$$

By definition, we can prove that (4) holds for $H = H^{1/2}(\partial\mathscr{O})$ and $H = H^{-1/2}(\partial\mathscr{O})$ with

$$H_{(p,q)} = H^{1/2}_{(p,q)}(\partial\mathscr{O}) \quad \text{and} \quad H_{(p,q)} = H^{-1/2}_{(p,q)}(\partial\mathscr{O}).$$

In the sequel it will be useful to define restriction to a subset $\partial\mathscr{O}_0 \subset \partial\mathscr{O}$ of functions defined in $H^{\pm 1/2}(\partial\mathscr{O})$ or $H^{\pm 1/2}_{(p,q)}(\partial\mathscr{O})$. This is a slightly delicate issue for an analytical point of view.

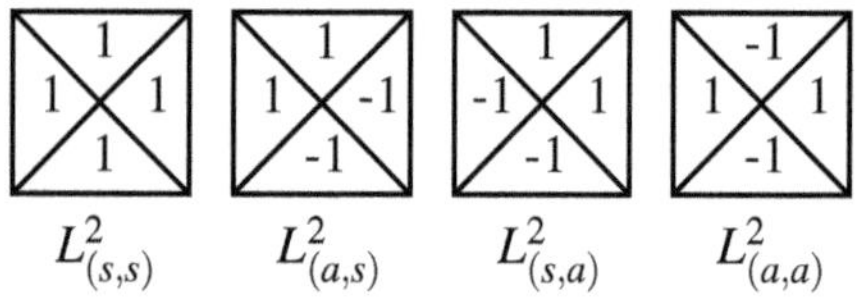

FIGURE (**5**): Functions in $L^2_{(p,q)}(\mathscr{O})$, $\mathscr{O} = [-1,1]^2$.

Let R be the restriction operator defined by

$$\begin{array}{rccc} R & : L^2(\partial\mathscr{O}) & \to & L^2(\partial\mathscr{O}_0) \\ & \phi & \mapsto & \phi|_{\partial\mathscr{O}_0}, \end{array}$$

For all $(p,q) \in \{s,a\}^2$, one checks that R is an isomorphism from $L^2_{(p,q)}(\partial\mathscr{O})$ onto $L^2(\partial\mathscr{O}_0)$.

Let denote $R_{(p,q)}$ the restriction of R to $L^2_{(p,q)}(\partial\mathscr{O})$.

Its inverse is an extension operator, which can be given explicitly thanks to the symmetries. Given $\phi_0 \in L^2(\partial\mathscr{O}_0)$:

$$\left|\begin{array}{ll} E_{(p,q)}\phi_0|_{\partial\mathscr{O}_0} = & \phi_0, \\ E_{(p,q)}\phi_0|_{\partial\mathscr{O}_1} = & \varepsilon_p \phi_0 \circ S_1, \\ E_{(p,q)}\phi_0|_{\partial\mathscr{O}_3} = & \varepsilon_q \phi_0 \circ S_{-1}, \\ E_{(p,q)}\phi_0|_{\partial\mathscr{O}_2} = & \varepsilon_p \varepsilon_q \phi_0 \circ S_1 \circ S_{-1}, \end{array}\right.$$

with $\varepsilon_s = 1$ and $\varepsilon_a = -1$. For each $(p,q) \in \{s,a\}^2$, we define the following spaces

$$H^{1/2}_{(p,q)}(\partial\mathscr{O}_0) := \left\{ R_{(p,q)}\phi, \quad \phi \in H^{1/2}_{(p,q)}(\partial\mathscr{O}) \right\}$$

equipped with the graph norm. One can show that $R_{(p,q)}$ is an isomorphism from

$$H^{1/2}_{(p,q)}(\partial\mathscr{O}) \quad \text{onto} \quad H^{1/2}_{(p,q)}(\partial\mathscr{O}_0)$$

whose inverse is $E_{(p,q)}$ and that

$$H^{1/2}_{(p,q)}(\partial\mathscr{O}_0) \equiv$$
$$\left\{ \phi_0 \in H^{1/2}(\partial\mathscr{O}_0),\ E_{(p,q)}\phi_0 \in H^{1/2}_{(p,q)}(\partial\mathscr{O}) \right\}.$$

Next, we extend the operator $R_{(p,q)}$ to $H^{-1/2}_{(p,q)}(\partial\mathscr{O})$.

This can be done by duality, noticing that $((\cdot,\cdot)_\Gamma$ is the inner product in $L^2(\Gamma))$

$$\Big(R_{(p,q)}\phi, \psi_0\Big)_{\partial\mathscr{O}_0} = \frac{1}{4}\Big(\phi, E_{(p,q)}\psi_0\Big)_{\partial\mathscr{O}}, \quad (5)$$

where $\phi \in L^2_{(p,q)}(\partial\mathscr{O}), \psi_0 \in L^2(\partial\mathscr{O}_0)$.

This suggests an extension of $R_{(p,q)}$ as an operator

$$R_{(p,q)} \in \mathscr{L}\Big(H^{-1/2}_{(p,q)}(\partial\mathscr{O}), [H^{1/2}_{(p,q)}(\partial\mathscr{O}_0)]'\Big)$$

as follows

$$\forall\, \phi \in H^{-1/2}_{(p,q)}(\partial\mathscr{O}),\ \forall \psi_0 \in H^{1/2}_{(p,q)}(\partial\mathscr{O}_0),$$

$$\Big\langle R_{(p,q)}\phi, \psi_0\Big\rangle_{\partial\mathscr{O}_0} = \frac{1}{4}\Big\langle \phi, E_{(p,q)}\psi_0\Big\rangle_{\partial\mathscr{O}}. \qquad (6)$$

Thus, introducing the subspace of $\big[H^{1/2}_{(p,q)}(\partial\mathscr{O}_0)\big]'$:

$$H^{-1/2}_{(p,q)}(\partial\mathscr{O}_0) = R_{(p,q)}\Big(H^{-1/2}_{(p,q)}(\partial\mathscr{O})\Big), \qquad (7)$$

equipped with the graph norm, $R_{(p,q)}$ is an isomorphism from

$$H^{-1/2}_{(p,q)}(\partial\mathscr{O}) \quad \text{onto} \quad H^{-1/2}_{(p,q)}(\partial\mathscr{O}_0).$$

Conversely, $E_{(p,q)}$ is extended as an operator

$$E_{(p,q)} \in \mathscr{L}\Big(H^{-1/2}_{(p,q)}(\partial\mathscr{O}_0), H^{-1/2}_{(p,q)}(\partial\mathscr{O})\Big),$$

using :

$$\forall\, \psi \in H^{-1/2}_{(p,q)}(\partial\mathscr{O}_0),\ \forall\, \phi \in H^{1/2}_{(p,q)}(\partial\mathscr{O}),$$

$$< E_{(p,q)}\psi, \phi >_{\partial\mathscr{O}} = 4 < \psi, R_{(p,q)}\phi >_{\partial\mathscr{O}_0}. \qquad (8)$$

REMARK 3.2
We can easily see that

$$\begin{aligned} H^{1/2}_{(s,s)}(\partial\mathscr{O}_0) &= H^{1/2}(\partial\mathscr{O}_0), \\ \text{and}\ \ H^{1/2}_{(a,a)}(\partial\mathscr{O}_0) &= H^{1/2}_{00}(\partial\mathscr{O}_0) \end{aligned}$$

using the notation of [26] and

$$\begin{aligned} & H^{1/2}_{(a,a)}(\partial\mathscr{O}_0) \subset H^{1/2}_{(a,s)}(\partial\mathscr{O}_0) \subset H^{1/2}_{(s,s)}(\partial\mathscr{O}_0) \\ \text{and}\ \ & H^{1/2}_{(a,a)}(\partial\mathscr{O}_0) \subset H^{1/2}_{(s,a)}(\partial\mathscr{O}_0) \subset H^{1/2}_{(s,s)}(\partial\mathscr{O}_0) \end{aligned}$$

with continuous injections. Finally, we have $\forall (p,q) \in \{s,a\}^2$

$$H^{-1/2}_{(p,q)}(\partial\mathscr{O}_0) \subset H^{1/2}_{(p,q)}(\partial\mathscr{O}_0)' \subset H^{1/2}_{(a,a)}(\partial\mathscr{O}_0)'$$

3.3 Physical Assumptions and Related Decomposition of Λ

The sets $\mathscr{C}$, Ω^i and Ω^e are doubly symmetric in the sense of Section 3.2. In the following, we will suppose that the restriction of ρ_p to $\mathscr{C}$ is a function satisfying double symmetry. Figure (**6**) presents some examples of media possessing double symmetry.

FIGURE (**6**): Examples of media possessing double symmetry : the function ρ_p is a constant respectively in the dashed and the white regions.

Using properties of the laplace operator, we can show that if the incoming Robin data φ has given symmetry or antisymmetry properties, the solution $u^e(\varphi)$ of the exterior problem $(\mathscr{P}^e)$ satisfies the same symmetries. Then, it is easy to conclude that Λ preserves symmetry and antisymmetry :

THEOREM 3.3
For $(p,q) \in \{s,a\}^2$, Λ is a continuous map from $H^{-1/2}_{(p,q)}(\Sigma^i)$ to $H^{-1/2}_{(p,q)}(\Sigma^i)$.

This means that if $\phi \in H^{-1/2}(\Sigma^i)$ has the decomposition :

$$\phi = \sum_{(p,q)\in\{s,a\}^2} \phi_{(p,q)}, \quad \phi_{(p,q)} \in H^{-1/2}_{(p,q)}(\Sigma^i),$$

then

$$\Lambda\,\phi = \Sigma_{i,j}\Lambda_{(p,q)}\phi_{(p,q)},$$

where

$$\Lambda_{(p,q)} = \Lambda|_{H^{-1/2}_{(p,q)}(\Sigma^i)} \in \mathscr{L}(H^{-1/2}_{(p,q)}(\Sigma^i), H^{-1/2}_{(p,q)}(\Sigma^i)).$$

Hence we can write Λ in block diagonal form :

$$\Lambda = \begin{bmatrix} \Lambda_{(s,s)} & 0 & 0 & 0 \\ 0 & \Lambda_{(s,a)} & 0 & 0 \\ 0 & 0 & \Lambda_{(a,s)} & 0 \\ 0 & 0 & 0 & \Lambda_{(a,a)} \end{bmatrix}. \qquad (9)$$

Therefore characterizing and computing Λ amounts to characterize and to compute each of the $\Lambda_{(p,q)}$'s. For this, we will use a factorization of $\Lambda_{(p,q)}$ that we construct in the next section.

4. A FACTORIZATION OF $\Lambda_{(p,q)}$

The characterization of $\Lambda_{(p,q)}$ is based on the factorization as the product of two operators. To do so, we need to introduce some additional notation, summarized by Figure (7).

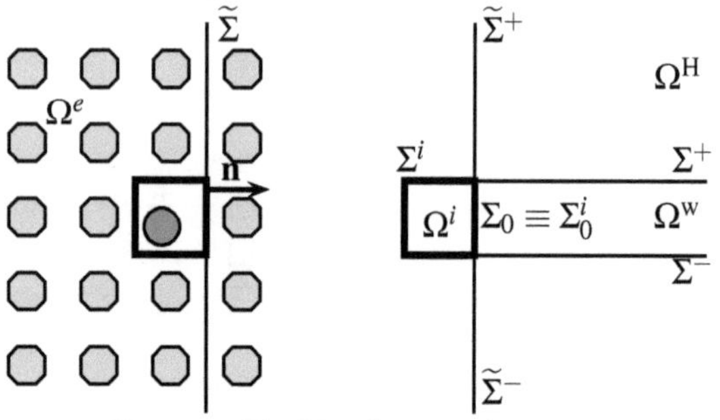

FIGURE (7): 2D-plane medium.

We shall divide the boundary $\widetilde{\Sigma} = \partial\Omega^{\mathrm{H}}$ of the half-space Ω^{H} into three parts :

$$\widetilde{\Sigma} = \widetilde{\Sigma}^- \cup \Sigma_0 \cup \widetilde{\Sigma}^+ \tag{10}$$

$$\text{where} \quad \left| \begin{array}{l} \widetilde{\Sigma}^- = \{x = \frac{L}{2}\} \times]-\infty, -\frac{L}{2}[, \\ \Sigma_0 = \{x = \frac{L}{2}\} \times]-\frac{L}{2}, \frac{L}{2}[\equiv \Sigma_0^i, \\ \widetilde{\Sigma}^+ = \{x = \frac{L}{2}\} \times]\frac{L}{2}, +\infty[. \end{array} \right.$$

4.1 The RtR Operators $\widetilde{I}_{(p,q)}$.

We introduce the operator which maps the incoming trace φ on the square Σ^i of $u^e(\varphi)$ (i. e. the boundary data of $(\mathscr{P}^e)$ into another Robin data along another boundary, namely the incoming trace of $u^e(\varphi)$ on $\widetilde{\Sigma}$, as illustrated by Figure (8). Since the incoming normal vector of Ω^{H} is nothing but e_x, this gives :

$$\begin{array}{rcl} \widetilde{I} : H^{-1/2}(\Sigma^i) & \to & H^{-1/2}(\widetilde{\Sigma}) \\ \varphi & \mapsto & \left[\frac{\partial}{\partial x} u^e(\varphi) + \imath Z u^e(\varphi)\right]\Big|_{\widetilde{\Sigma}}, \end{array}$$

and the associated operators :

$$\widetilde{I}_{(p,q)} = \widetilde{I}\Big|_{H^{-1/2}_{(p,q)}(\Sigma^i)} \in \mathscr{L}(H^{-1/2}_{(p,q)}(\Sigma^i), H^{-1/2}(\widetilde{\Sigma})).$$

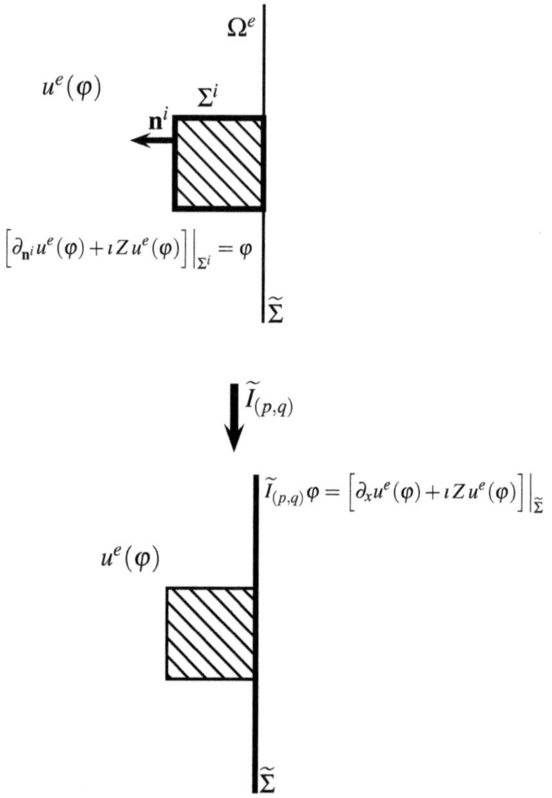

FIGURE (8): Definition of the RtR operator $\widetilde{I}_{(p,q)}$.

The computation of these operators will be discussed in Section 6. Since $\Sigma_0 = \Sigma^i \cap \widetilde{\Sigma}$, we have formally

$$\forall\, \varphi \in H^{-1/2}_{(p,q)}(\Sigma^i), \quad \widetilde{I}_{(p,q)}\varphi = \varphi \quad \text{on} \quad \Sigma_0.$$

which can be stated more rigorously

$$\widetilde{R}\widetilde{I}_{(p,q)}\varphi = R_{(p,q)}\,\varphi \in H^{-1/2}_{(p,q)}(\Sigma_0), \tag{11}$$

where we have defined another restriction operator $\widetilde{R} : L^2(\widetilde{\Sigma}) \to L^2(\Sigma_0)$:

$$\widetilde{R}\,\psi := \psi\big|_{\Sigma_0}.$$

extended to an operator

$$\widetilde{R} \in \mathscr{L}\left(H^{-1/2}(\widetilde{\Sigma}), H^{1/2}_{(a,a)}(\Sigma_0)'\right)$$

by duality as we did for $R_{(p,q)}$ (See Section 3.2).

4.2 The Halfspace RtR Operator $\Lambda^{\rm H}$.

We define the halfspace RtR operator $\Lambda^{\rm H}$ as follows

$$\begin{aligned}\Lambda^{\rm H} : H^{-1/2}(\widetilde{\Sigma}) &\to H^{-1/2}(\widetilde{\Sigma})\\ \psi &\mapsto \Big[-\frac{\partial}{\partial x}u^{\rm H}(\psi)+\imath Z u^{\rm H}(\psi)\Big]\Big|_{\widetilde{\Sigma}},\end{aligned}$$

associated with the half space problem with incoming Robin boundary conditions

$$(\mathscr{P}^{\rm H})\quad\left|\begin{aligned}&-\Delta u^{\rm H}-\rho_p\,(\omega^2+\imath\varepsilon\omega)\,u^{\rm H}=0, \quad \text{in } \Omega^{\rm H},\\ &\frac{\partial u^{\rm H}}{\partial x}+\imath Z u^{\rm H}=\psi, \quad \text{on } \widetilde{\Sigma},\end{aligned}\right.$$

where $u^{\rm H}(\psi)$ is the unique solution in $H^1(\triangle,\Omega^{\rm H})$. See Figure (**9**).

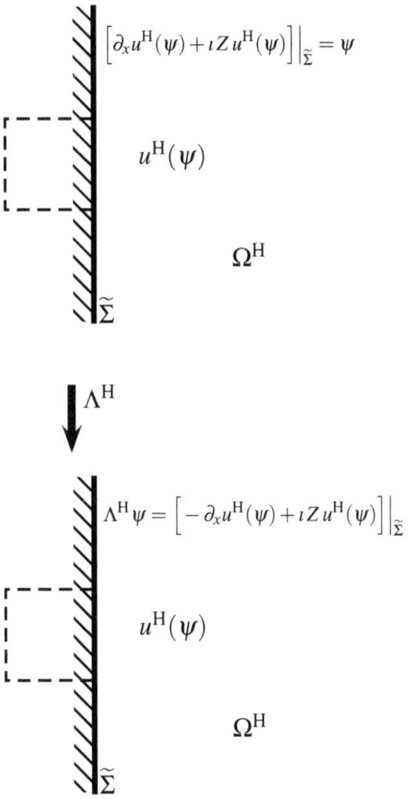

FIGURE (**9**): Definition of NtD operator Λ^H.

4.3 The Factorization of $\Lambda_{(p,q)}$

The important result concerning the factorization of the desired RtR operator $\Lambda_{(p,q)}$ is the following theorem whose proof is formally a consequence of the construction of the operators

$$\Lambda^{\rm H} \quad \text{and} \quad \widetilde{I}_{(p,q)}$$

(see Figures (**8**) and (**9**)). Technically, the proof relies on the uniqueness of the solution of $(\mathscr{P}^{\rm H})$, see [13, 12] for more details.

THEOREM 4.1
The operator $\widetilde{R}\circ\Lambda^H\circ\widetilde{I}_{(p,q)}$ maps

$$H^{-1/2}_{(p,q)}(\Sigma^i) \quad \textit{into} \quad H^{-1/2}_{(p,q)}(\Sigma^i_0)$$

and one has the factorisation

$$\Lambda_{(p,q)}=E_{(p,q)}\circ\widetilde{R}\circ\Lambda^H\circ\widetilde{I}_{(p,q)}. \tag{12}$$

In the factorisation (12), the two extension and restriction operators $E_{(p,q)}$ and $\widetilde{R}$ are trivial operators. However, we need to have a numerical method for computing, at least approximately, the operators $\Lambda^{\rm H}$and $\widetilde{I}_{(p,q)}$.
The next two sections are devoted to a characterization of these operators that lead to a numerical method.

5. COMPUTATION OF THE HALF-SPACE RTR OPERATOR $\Lambda^{\rm H}$

5.1 Technical Tools

We recall the definition and more useful properties of the Floquet Bloch Transform (see [22] for a more complete exposition) of function of one variable. This transform maps a function of $y\in\mathbb{R}$ into a function $(y,k_y)\in\mathbb{K}$ where

$$\mathbb{K}=\Big]-\frac{L}{2},\frac{L}{2}\Big[\times\Big]-\frac{\pi}{L},\frac{\pi}{L}\Big[.$$

By defintion, k_y is a dual variable or a wave number.

DEFINITION 5.1
The Floquet Bloch Transform (FBT) with period L is defined by (see [22])

$$\begin{aligned}\mathscr{F}:\mathscr{C}_0^\infty(\mathbb{R}) &\to L^2(\mathbb{K})\\ f(y) &\to \hat{f}(y;k),\end{aligned}$$

with

$$\hat{f}(y;k)=\sqrt{\frac{L}{2\pi}}\sum_{n\in\mathbb{Z}} f(y+nL)e^{-inkL} \tag{13}$$

where $\mathscr{C}_0^\infty(\mathbb{R})$ is the set of $\mathscr{C}^\infty$-functions with compact support.

Note that the sum in the definition of the FB Transformation is finite because of the compact support of f.

PROPOSITION 5.2
The FB Transformation extends to an isometry between $L^2(\mathbb{R})$ and $L^2(\mathbb{K})$:

$$\forall (f,g) \in L^2(\mathbb{R})^2, \quad (\mathscr{F} f, \mathscr{F} g)_{L^2(\mathbb{K})} = (f,g)_{L^2(\mathbb{R})}$$

This identity allows us to extend uniquely the operator $\mathscr{F}$ by density to an isometry from $L^2(\mathbb{R})$ to $L^2(\mathbb{K})$.

The most important formula for us is the inversion formula :

PROPOSITION 5.3
The operator $\mathscr{F}$ is an isometric isomorphism from $L^2(\mathbb{R})$ into $L^2(\mathbb{K})$ whose inverse is given by

$$\forall \hat{f} \in L^2(\mathbb{K}), \quad p.p.\, y \in \Big[-L/2, L/2\Big], \quad \forall n \in \mathbb{Z},$$

$$\left(\mathscr{F}^{-1}\hat{f}\right)(y+nL) = \sqrt{\frac{L}{2\pi}} \int_{-\frac{\pi}{L}}^{\frac{\pi}{L}} \hat{f}(y;k) e^{inkL}\, dk.$$

We shall use in the sequel the following properties of the FB transformation (the proofs are straightforward and left to the reader. See also [22]). These properties make the FB transformation a priviligied tool for the analysis of linear PDEs with periodic coefficients.

PROPOSITION 5.4
The FB tranformation has the following properties

1. *it commutes with the differential operators, in the sense that*

$$\mathscr{F}\left(\frac{d\psi}{dy}\right) = \frac{\partial}{\partial y}\Big(\mathscr{F}\psi\Big).$$

2. *It diagonalizes the translation operators*

$$(\tau_q \psi)(y) = \psi(y+qL)$$
$$\Downarrow$$
$$\mathscr{F}(\tau_q \psi)(y;k) = e^{-iqkL} \mathscr{F}\psi(y;k), \quad (y,k) \in \mathbb{K}.$$

3. *It commutes with the multiplication by a periodic function, in the sense that if μ is a L-periodic function*

$$\mathscr{F}(\mu\psi)(y;k) = \mu(y) \mathscr{F}\psi(y;k), \quad (y,k) \in \mathbb{K}.$$

Next we define the partial Floquet Bloch Transform in the $y-$direction in the halfspace Ω^{H}

$$\begin{array}{rccc} \mathscr{F}_y : & L^2(\Omega^{\mathrm{H}}) & \rightarrow & L^2(\Omega^{\mathrm{w}} \times \Big] -\dfrac{\pi}{L}, \dfrac{\pi}{L}\Big[) \\ & u(x,y) & \mapsto & \mathscr{F}_y(x,y;k_y) \end{array}$$

such that

$$\text{p.p.}\, x \in [\frac{L}{2}, +\infty[, \quad (\mathscr{F}_y u)(x, \cdot\, ; \cdot) = \mathscr{F}\big[u(x,\cdot)\big].$$

It is easy to see that $\mathscr{F}_y$ is an isomorphism from $L^2(\Omega^{\mathrm{H}})$ into $L^2(\Omega^{\mathrm{w}} \times]-\pi/L, \pi/L])$.

We want to know, now, how the Floquet Bloch Transform can extend to every spaces appearing naturally in the study. We need then to introduce spaces of functions on the domain Ω^{w} of so-called k_y quasi-periodic functions (k_y being a parameter in $]-\pi/L, \pi/L[$. We start from smooth quasi-periodic functions in Ω^{w} :

$$C^{\infty}_{k_y}(\Omega^{\mathrm{w}}) = \Big\{ u = \tilde{u}|_{\Omega_w}, \tilde{u} \in C^{\infty}(\Omega^{\mathrm{H}}),$$
$$\tilde{u}(x, y+L) = e^{ik_y L}\, \tilde{u}(x,y) \Big\}.$$

Let $H^1_{k_y}(\Omega^{\mathrm{w}})$ be the closure of $C^{\infty}_{k_y}(\Omega^{\mathrm{w}})$ in $H^1(\Omega^{\mathrm{w}})$.

$$H^1_{k_y}(\Omega^{\mathrm{w}}) = \Big\{ u \in H^1(\Omega^{\mathrm{w}}), \quad u|_{\Sigma^+} = e^{ik_y L}\, u|_{\Sigma^-} \Big\}.$$

where in the last equality we have identified the spaces $H^{1/2}(\Sigma^+)$ and $H^{1/2}(\Sigma^-)$.

Let $H^1_{k_y}(\triangle, \Omega^{\mathrm{w}})$ be the closure of $C^{\infty}_{k_y}(\Omega^{\mathrm{w}})$ of

$$H^1(\triangle, \Omega^{\mathrm{w}}) = \Big\{ u \in H^1(\Omega^{\mathrm{w}}), \Delta u \in L^2(\Omega^{\mathrm{w}}) \Big\}.$$

or equivalently

$$H^1_{k_y}(\triangle, \Omega^{\mathrm{w}}) = \Big\{ u \in H^1(\triangle, \Omega^{\mathrm{w}}),$$
$$\frac{\partial u}{\partial y}(x, y+L) = e^{ik_y L}\, \frac{\partial u}{\partial y}(x,y) \Big\}.$$

where in the last equality we have identified the spaces $H^{1/2}_{00}(\Sigma^+)'$ and $H^{1/2}_{00}(\Sigma^-)'$.

The space $H^{1/2}_{k_y}(\Sigma_0)$ is defined by

$$H^{1/2}_{k_y}(\Sigma_0) = \gamma_0\Big(H^1_{k_y}(\Omega^{\mathrm{w}})\Big)$$

where $\gamma_0 \in \mathscr{L}(H^1(\Omega^{\mathrm{w}}), H^{1/2}(\Sigma_0))$ is the trace map on Σ_0 :

$$\forall u \in H^1(\Omega^{\mathrm{w}}), \quad \gamma_0 u = u\Big|_{\Sigma_0}.$$

$H^{1/2}_{k_y}(\Sigma_0)$ is then a dense subspace of $H^{1/2}(\Sigma_0)$.

Moreover, the injection from $H^{1/2}_{k_y}(\Sigma_0)$ onto $H^{1/2}(\Sigma_0)$ is continuous.

We define $H^{-1/2}_{k_y}(\Sigma_0)$ as the dual of $H^{1/2}_{k_y}(\Sigma_0)$.

Finally, the trace application

$$\gamma_1 \in \mathscr{L}(H^1(\triangle,\Omega^{\mathrm{w}}), H^{1/2}_{(a,a)}(\Sigma_0)')$$

defined by :

$$\forall u \in H^1(\triangle,\Omega^{\mathrm{w}}), \quad \gamma_1 u = \frac{\partial u}{\partial x}\Big|_{\Sigma_0}.$$

is a continous application from

$$H^1_{k_y}(\triangle,\Omega^{\mathrm{w}}) \text{ onto } H^{-1/2}_{k_y}(\Sigma_0).$$

Moreover, we can show that

$$H^{-1/2}_{k_y}(\Sigma_0) = \gamma_1\left(H^1_{k_y}(\triangle,\Omega^{\mathrm{w}})\right)$$

We can now state the following results.

THEOREM 5.5

$\mathscr{F}_y$ is an isomorphism from $H^1(\Omega^H)$ into

$$\left| \begin{array}{l} H^1_{QP}\left(\left]-\frac{\pi}{L},\frac{\pi}{L}\right[\times\Omega^{\mathrm{w}}\right) = \\ \left\{\hat{u} \in L^2\left(-\frac{\pi}{L},\frac{\pi}{L};H^1(\Omega^{\mathrm{w}})\right) / \right. \\ \quad \left. \text{for a. e. } k_y \in \left]-\frac{\pi}{L},\frac{\pi}{L}\right[,\ \hat{u}(\cdot;k_y) \in H^1_{k_y}(\Omega^{\mathrm{w}}) \right\}, \end{array}\right.$$

equipped with the norm of $L^2\left(-\frac{\pi}{L},\frac{\pi}{L};H^1(\Omega^{\mathrm{w}})\right)$.

$\mathscr{F}_y$ is an isomorphism from $H^1(\triangle,\Omega^H)$ into

$$\left| \begin{array}{l} H^1_{QP}\left(\triangle;\left]-\frac{\pi}{L},\frac{\pi}{L}\right[\times\Omega^{\mathrm{w}}\right) = \\ \left\{\hat{u} \in L^2\left(-\frac{\pi}{L},\frac{\pi}{L};H^1(\triangle;\Omega^{\mathrm{w}})\right) / \right. \\ \quad \left. \text{for a. e. } k_y \in \left]-\frac{\pi}{L},\frac{\pi}{L}\right[,\ \hat{u}(\cdot;k_y) \in H^1_{k_y}(\triangle,\Omega^{\mathrm{w}}) \right\}. \end{array}\right.$$

equipped with the norm of $L^2\left(-\frac{\pi}{L},\frac{\pi}{L};H^1(\triangle;\Omega^{\mathrm{w}})\right)$.

$\mathscr{F}_y$ is an isomorphism from $H^{1/2}(\widetilde{\Sigma})$ into

$$\left| \begin{array}{l} H^{1/2}_{QP}\left(\left]-\frac{\pi}{L},\frac{\pi}{L}\right[\times\Sigma_0\right) = \\ \left\{\hat{\varphi} \in L^2\left(-\frac{\pi}{L},\frac{\pi}{L};H^{1/2}(\Sigma_0)\right) / \right. \\ \quad \left. \text{for a. e. } k_y \in \left]-\frac{\pi}{L},\frac{\pi}{L}\right[,\ \hat{\varphi}(\cdot;k_y) \in H^{1/2}_{k_y}(\Sigma_0) \right\}. \end{array}\right.$$

equipped with the norm

$$\|\hat{\varphi}\|^2_{H^{1/2}_{QP}} = \int_{-\pi/L}^{\pi/L} \|\hat{\varphi}(\cdot;k_y)\|^2_{H^{1/2}_{k_y}(\Sigma_0)}\, dk_y.$$

Finally, we can extend by duality the definition of $\mathscr{F}_y$ to the space $H^{-1/2}(\widetilde{\Sigma})$ introducing the dual of $H^{1/2}_{QP}\left(\left]-\frac{\pi}{L},\frac{\pi}{L}\right[\times\Sigma_0\right)$

$$\left| \begin{array}{l} H^{-1/2}_{QP}\left(\left]-\frac{\pi}{L},\frac{\pi}{L}\right[\times\Sigma_0\right) = \\ \left\{\hat{u} \in L^2\left(-\frac{\pi}{L},\frac{\pi}{L};H^{1/2}_{aa}(\Sigma_0)\right) / \right. \\ \quad \left. \text{for a. e. } k_y \in \left]-\frac{\pi}{L},\frac{\pi}{L}\right[,\ \hat{u}(\cdot;k_y) \in H^{\pm 1/2}_{k_y}(\Sigma_0) \right\}. \end{array}\right. ,$$

DEFINITION 5.6

Let ψ be in $H^{-1/2}(\widetilde{\Sigma})$, the following application ($<\cdot,\cdot>$ is the duality product between $H^{-1/2}(\widetilde{\Sigma})$ and $H^{1/2}(\widetilde{\Sigma})$)

$$\hat{\varphi} \mapsto \langle \psi, \mathscr{F}_y^{-1}\hat{\varphi}\rangle$$

is a continuous linear application of

$$H^{1/2}_{QP}\left(\left]-\frac{\pi}{L},\frac{\pi}{L}\right[\times\Sigma_0\right)$$

because of Theorem 5.5. The theorem of Riesz representation implies then

$$\exists \hat{\psi} \in H^{-1/2}_{QP}\left(\left]-\frac{\pi}{L},\frac{\pi}{L}\right[\times\Sigma_0\right), \quad \langle\hat{\psi},\hat{\varphi}\rangle = \langle\psi,\varphi\rangle$$

where the first duality product is between

$$H^{-1/2}_{QP}\left(\left]-\frac{\pi}{L},\frac{\pi}{L}\right[\times\Sigma_0\right) \text{ and } H^{1/2}_{QP}\left(\left]-\frac{\pi}{L},\frac{\pi}{L}\right[\times\Sigma_0\right)$$

and the second one is between

$$H^{-1/2}(\widetilde{\Sigma}) \text{ and } H^{1/2}(\widetilde{\Sigma}).$$

Finally, we define the FBT $\mathscr{F}_y$ in $H^{-1/2}(\widetilde{\Sigma})$ by $\forall\psi \in H^{-1/2}(\widetilde{\Sigma})$,

$$\mathscr{F}_y\psi = \hat{\psi} \in H^{-1/2}_{QP}\left(\left]-\frac{\pi}{L},\frac{\pi}{L}\right[\times\Sigma_0\right)$$

which coincides with the classical definition in $L^2(\widetilde{\Sigma})$ (see Proposition 5.2). Similarly, for any

$$\hat{\psi} \in H^{-1/2}_{QP}\left(\left]-\frac{\pi}{L},\frac{\pi}{L}\right[\times\Sigma_0\right),$$

we define by duality $\mathscr{F}_y^{-1}\hat{\psi}$ in $H^{-1/2}(\widetilde{\Sigma})$.

5.2 Reduction of the Halfspace Problem to Half-Waveguide Problems

In this section, we reduce the solution of the halfspace problem $(\mathscr{P}^{\mathrm{H}})$ and thus the characterization of Λ^{H} to the solution of a family of a half-waveguide problem in Ω^{w} (See Figure (**7**)) parametrized by the wavenumber k_y.

Let $u^{\mathrm{H}}(\psi)$ be the solution of $(\mathscr{P}^{\mathrm{H}})$ and $\mathscr{F}_y(u^{\mathrm{H}}(\psi))$ its FBT in the variable y. Applying $\mathscr{F}_y$ to $(\mathscr{P}^{\mathrm{H}})$ and using Proposition 5.4 one easily sees that each $\mathscr{F}_y u^{\mathrm{H}}(\psi)\,(\cdot,k_y)$ is the solution of a waveguide problem. More precisely,

THEOREM 5.7
For each $k_y \in]-\pi/L,\pi/L[$,

$$\hat{u}^H_{k_y} := \mathscr{F}_y u^H(\psi)\,(\cdot,k_y)$$

is the unique solution in $H^1_{k_y}(\triangle,\Omega^{\mathrm{w}})$ *of the half-waveguide problem*

$$\left|\begin{array}{rcl} -\Delta\hat{u}^H_{k_y} - \rho_p(\omega^2+\imath\varepsilon\omega)\,\hat{u}^H_{k_y} &=& 0,\ in\,\Omega^{\mathrm{w}} \\ \left[\dfrac{\partial}{\partial x}\hat{u}^H_{k_y} + \imath Z\,\hat{u}^H_{k_y}\right]\Big|_{\Sigma_0} &=& \hat{\psi}_{k_y}, \end{array}\right. \qquad (\hat{\mathscr{P}}^{\mathrm{H}}_{k_y})$$

where $\forall \psi \in H^{-1/2}(\widetilde{\Sigma})$

$$\left|\begin{array}{l} \hat{\psi} = \mathscr{F}_y\,\psi \in H^{-1/2}_{QP}\Big(\big]-\dfrac{\pi}{L},\dfrac{\pi}{L}\big[\times\Sigma_0\Big) \\ \hat{\psi}_{k_y} = \hat{\psi}(\cdot;k_y) \in H^{-1/2}_{k_y}(\Sigma_0). \end{array}\right.$$

Hence to determine the solution $u^{\mathrm{H}}(\psi)$ of $(\mathscr{P}^{\mathrm{H}})$, we compute for all $k_y \in]-\pi/L,\pi/L[$ the solution $\hat{u}^{\mathrm{H}}_{k_y}(\hat{\psi}_{k_y})$ of $(\hat{\mathscr{P}}^{\mathrm{H}}_{k_y})$ and use the inversion formula in Proposition 5.3 : $\forall\,(x,y)\in\Omega^{\mathrm{w}}$, $\forall n\in\mathbb{Z}$,

$$u^{\mathrm{H}}(\psi)(x,y+nL) = \sqrt{\frac{L}{2\pi}}\int_{-\frac{\pi}{L}}^{\frac{\pi}{L}} \hat{u}^{\mathrm{H}}_{k_y}(\hat{\psi}_{k_y})\,(x,y)\,e^{\imath nk_yL}\,dk_y. \qquad (14)$$

Let us remind that the halfspace RtR operator Λ^{H} is defined by : $\forall\psi\in H^{-1/2}(\widetilde{\Sigma})$,

$$\Lambda^{\mathrm{H}}\,\psi = \left[-\frac{\partial}{\partial x}u^{\mathrm{H}}(\psi) + \imath Z\,u^{\mathrm{H}}(\psi)\right]\Big|_{\widetilde{\Sigma}}.$$

Let us introduce the RtR operator on Σ_0 for the k_y quasi periodic half-waveguide problem, namely

$$\Lambda^{\mathrm{w}}(k_y) \in \mathscr{L}\big(H^{-1/2}_{k_y}(\Sigma_0)\big)$$

such that for any $\hat{\psi}_{k_y} \in H^{-1/2}_{k_y}(\Sigma_0)$

$$\Lambda^{\mathrm{w}}(k_y)\,\hat{\psi}_{k_y} := \left[-\frac{\partial}{\partial x}\hat{u}^{\mathrm{H}}_{k_y} + \imath Z\,\hat{u}^{\mathrm{H}}_{k_y}\right]\Big|_{\Sigma_0}.$$

where $\hat{u}_{k_y}$ is the solution of $\hat{\mathscr{P}}^{\mathrm{H}}_{k_y}$. Let

$$\widehat{\Lambda}^{\mathrm{H}} \in \mathscr{L}\Big(H^{-1/2}_{\mathrm{QP}}\big(]-\pi/L,\pi/L[\times\Sigma_0\big)\Big)$$

such that, for $\hat{\psi}$ in $H^{-1/2}_{\mathrm{QP}}\big(]-\pi/L,\pi/L[\times\Sigma_0\big)$,

$$(\widehat{\Lambda}^{\mathrm{H}}\,\hat{\psi})\,(\cdot;k_y) = \Lambda^{\mathrm{w}}(k_y)\,\hat{\psi}\,(\cdot;k_y). \qquad (15)$$

The link between Λ^{H} and $\widehat{\Lambda}^{\mathrm{H}}$ is given by the

THEOREM 5.8
The halfspace RtR operator Λ^H *is given by :*

$$\Lambda^H = \mathscr{F}_y^{-1}\circ\widehat{\Lambda}^H\circ\mathscr{F}_y \qquad (16)$$

where $\widehat{\Lambda}^H$ *is given by* (15).

5.3 Solving the Half-Waveguide Problems $(\hat{\mathscr{P}}^{\mathrm{H}}_{k_y})$

Here we discuss the determination of $\Lambda^{\mathrm{w}}(k_y)$. We shall use the division of the half-waveguide into periodicity cells separated by vertical segments (See Figure (**10**)) :

$$\Omega^{\mathrm{w}} = \bigcup_{n=0}^{+\infty}\mathscr{C}_n, \quad \mathscr{C}_n := \mathscr{C} + (nL,0), \qquad (17)$$

The segments $\Sigma_n = \Sigma_0 + (nL,0)$ can all be identified to the leftmost one $\Sigma_0 \sim [-L/2,L/2]$ and the cells $\mathscr{C}_n$ can all be identified to the first cell $\mathscr{C}_1 = \mathscr{C}$.

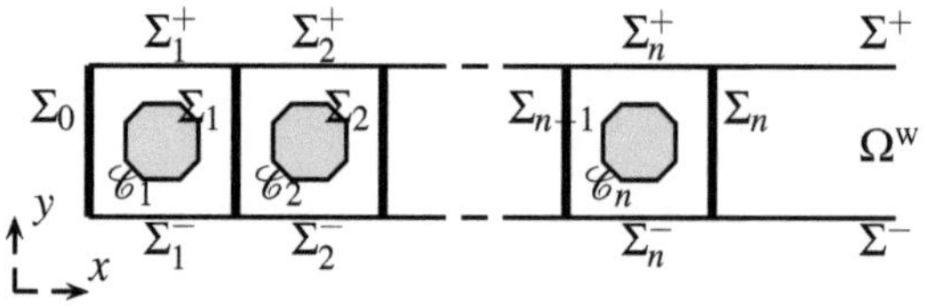

FIGURE (**10**): Notation for a half guide

By periodicity in x, the construction of $\hat{u}^{\mathrm{H}}_{k_y}$ in Ω^{w} will reduced to the knowledge of two linear operators (see Figure (**11**) for a schematic definition). The first one, called the propagation operator, is denoted $P(k_y)$ and defined by

$$\begin{array}{rccl} P(k_y): & H^{-1/2}_{k_y}(\Sigma_0) & \rightarrow & H^{-1/2}_{k_y}(\Sigma_0) \\ & \hat{\psi}_{k_y} & \mapsto & \left[\dfrac{\partial}{\partial x}\hat{u}^{\mathrm{H}}_{k_y} + \imath Z\hat{u}^{\mathrm{H}}_{k_y}\right]\Big|_{\Sigma_1}. \end{array} \qquad (18)$$

One can show that $P(k_y)$ is compact (using interior elliptic regularity for $\hat{u}^{\mathrm{H}}_{k_y}$), injective (using an argument of unique continuation) and has a spectral radius less than 1 (due to the L^2 nature of $\hat{u}^{\mathrm{H}}_{k_y}$). See [21, 12] for a detailed proof of these results.

The second operator, $D(k_y)$, is defined by

$$\begin{array}{rcl} D(k_y) : H^{-1/2}_{k_y}(\Sigma_0) & \rightarrow & H^{-1/2}_{k_y}(\Sigma_0) \\ \hat{\psi}_{k_y} & \mapsto & \left[-\dfrac{\partial}{\partial x}\hat{u}^{\mathrm{H}}_{k_y} + \imath Z \hat{u}^{\mathrm{H}}_{k_y}\right]\Big|_{\Sigma_1}. \end{array} \qquad (19)$$

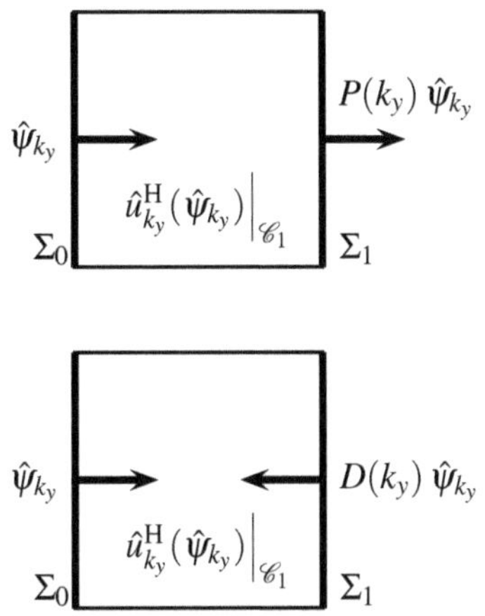

FIGURE (**11**): The operators $D(k_y)$ and $P(k_y)$.

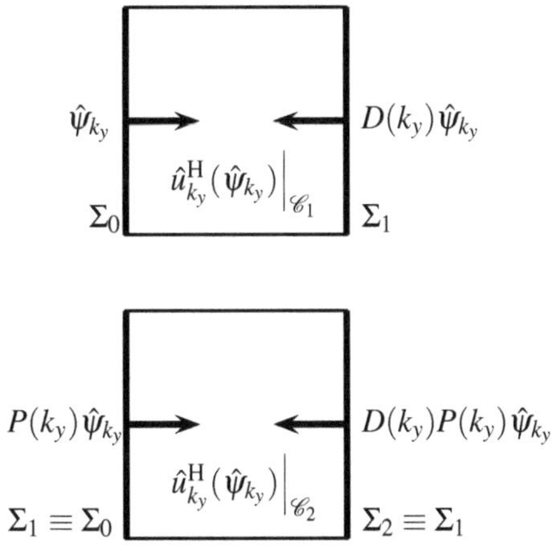

FIGURE (**12**): $\hat{u}^{\mathrm{H}}_{k_y}(\hat{\psi}_{k_y})$ in the first two cells.

Using the periodicity of the problem, one easily that

$$\left[\frac{\partial}{\partial x}\hat{u}^{\mathrm{H}}_{k_y} + \imath Z \hat{u}^{\mathrm{H}}_{k_y}\right]\Big|_{\Sigma_{j-1}} = P(k_y)^{j-1}\,\hat{\psi}_{k_y}$$

$$\left[-\frac{\partial}{\partial x}\hat{u}^{\mathrm{H}}_{k_y} + \imath Z \hat{u}^{\mathrm{H}}_{k_y}\right]\Big|_{\Sigma_j} = D(k_y)\,P(k_y)^{j-1}\,\hat{\psi}_{k_y}$$

Then, by linearity, we have for any $j \geq 1$,

$$\begin{aligned} \hat{u}^{\mathrm{H}}_{k_y}(\hat{\psi}_{k_y})\Big|_{\mathscr{C}_j} &= e^0\big(k_y; P(k_y)^{j-1}\,\hat{\psi}_{k_y}\big) \\ &\quad + e^1\big(k_y; D(k_y)P(k_y)^{j-1}\,\hat{\psi}_{k_y}\big)\big), \end{aligned} \qquad (20)$$

where for any $\hat{\psi}_{k_y} \in H^{-1/2}_{k_y}(\Sigma_0)$, the two functions

$$e^0(k_y;\hat{\psi}_{k_y}) \text{ and } e^1(k_y;\hat{\psi}_{k_y}),$$

namely the unique solutions in $H^1(\mathscr{C})$ of the following elementary cell problems posed on $\mathscr{C}$:

$$-\Delta e^\ell - \rho_p\,(\omega^2 + \imath\varepsilon\omega)\,e^\ell = 0, \text{ in } \mathscr{C}, \qquad (21)$$

satisfying k_y quasi-periodic boundary conditions on Σ_1^+ et Σ_1^- :

$$\left|\begin{array}{rcl} e^\ell\Big|_{\Sigma_1^+} & = & e^{\imath k_y L}\,e^\ell\Big|_{\Sigma_1^-}, \\ \dfrac{\partial e^\ell}{\partial y}\Big|_{\Sigma_1^+} & = & e^{\imath k_y L}\,\dfrac{\partial e^\ell}{\partial y}\Big|_{\Sigma_1^-}, \end{array}\right. \qquad (22)$$

and nonhomogeneous incoming Robin conditions on Σ_0 et Σ_1 (see Figure (**13**) for an illustration) :

$$\left|\begin{array}{l} \left[\dfrac{\partial e^0}{\partial x} + \imath Z e^0\right]\Big|_{\Sigma_0} = \hat{\psi}_{k_y},\ \left[-\dfrac{\partial e^0}{\partial x} + \imath Z\,e^0\right]\Big|_{\Sigma_1} = 0, \\ \left[\dfrac{\partial e^1}{\partial x} + \imath Z e^1\right]\Big|_{\Sigma_0} = 0,\ \left[-\dfrac{\partial e^1}{\partial x} + \imath Z\,e^1\right]\Big|_{\Sigma_1} = \hat{\psi}_{k_y}, \end{array}\right. \qquad (23)$$

Formula (20) shows that the computation of the so-

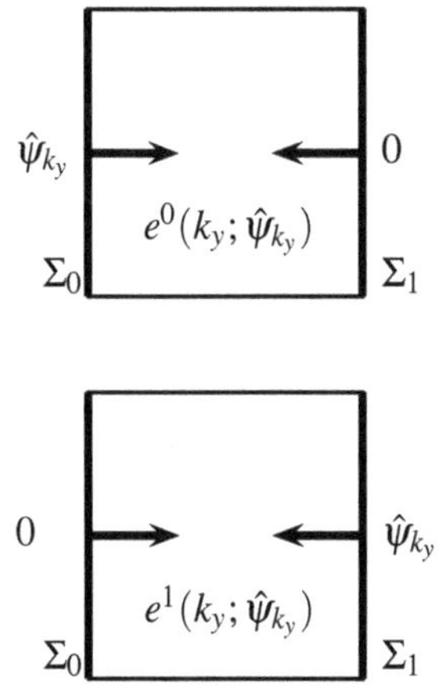

FIGURE (**13**): The functions $e^j(k_y;\hat{\psi}_{k_y}), j = 0, 1$

lution $\hat{u}^{\mathrm{H}}_{k_y}$ is achieved through the characterization of the two operators $D(k_y)$ and $P(k_y)$. At this stage of the exposition, the definitions of these operators rely on $\hat{u}^{\mathrm{H}}_{k_y}(\hat{\psi}_{k_y})$ which is a solution of a problem posed in an unbounded domain.

We shall see in the following how to determine these operators by solely solving local problems of the type (21, 22, 23), which is one key point of the method.
Note that the relation (20) ensures that $\hat{u}^{\mathrm{H}}_{k_y}(\hat{\psi}_{k_y})$ is the solution of the Helmholtz equation inside each cell $\mathscr{C}_j$.
To make the characterization complete, we have to write that the correct transmission condition across Σ_j, that we can write as

$$\left|\begin{array}{l}\left[\dfrac{\partial \hat{u}^{\mathrm{H}}_{k_y}}{\partial x} + \imath Z \hat{u}^{\mathrm{H}}_{k_y}\right]\Big|_{\mathscr{C}_j} = \left[\dfrac{\partial \hat{u}^{\mathrm{H}}_{k_y}}{\partial x} + \imath Z \hat{u}^{\mathrm{H}}_{k_y}\right]\Big|_{\mathscr{C}_{j+1}}, \\[2ex] \left[-\dfrac{\partial \hat{u}^{\mathrm{H}}_{k_y}}{\partial x} + \imath Z \hat{u}^{\mathrm{H}}_{k_y}\right]\Big|_{\mathscr{C}_j} = \left[-\dfrac{\partial \hat{u}^{\mathrm{H}}_{k_y}}{\partial x} + \imath Z \hat{u}^{\mathrm{H}}_{k_y}\right]\Big|_{\mathscr{C}_{j+1}}.\end{array}\right. \qquad (24)$$

If we define the local RtR operators such that for any incoming Robin data (see Figure (**14**) for a schematic illustration)

$$\hat{\psi}_{k_y} \in H^{-1/2}_{k_y}(\Sigma_0),$$

$$\left|\begin{array}{l} T^{00}_{k_y}\hat{\psi}_{k_y} = \left[-\dfrac{\partial}{\partial x} e^0(k_y;\hat{\psi}_{k_y}) + \imath Z e^0(k_y;\hat{\psi}_{k_y}\right]\Big|_{\Sigma_0}, \\[2ex] T^{01}_{k_y}\hat{\psi}_{k_y} = \left[\dfrac{\partial}{\partial x} e^0(k_y;\hat{\psi}_{k_y}) + \imath Z e^0(k_y;\hat{\psi}_{k_y})\right]\Big|_{\Sigma_1}, \\[2ex] T^{10}_{k_y}\hat{\psi}_{k_y} = \left[-\dfrac{\partial}{\partial x} e^1(k_y;\hat{\psi}_{k_y}) + \imath Z e^1(k_y;\hat{\psi}_{k_y}\right]\Big|_{\Sigma_0}, \\[2ex] T^{11}_{k_y}\hat{\psi}_{k_y} = \left[\dfrac{\partial}{\partial x} e^1(k_y;\hat{\psi}_{k_y}) + \imath Z e^1(k_y;\hat{\psi}_{k_y})\right]\Big|_{\Sigma_1}, \end{array}\right. \qquad (25)$$

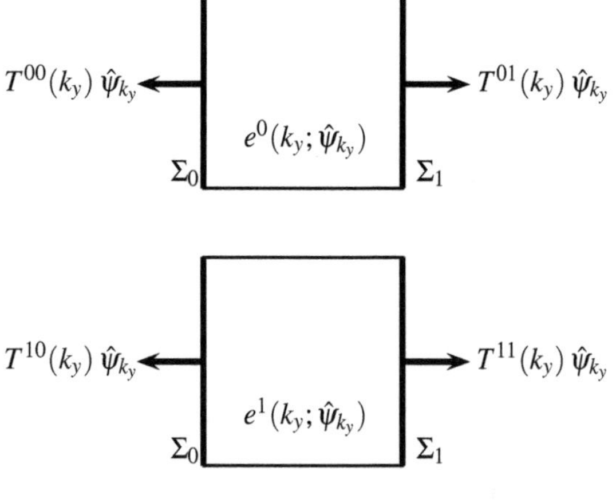

FIGURE (**14**): The local RtR operators $T^{ij}_{k_y}$.

the reader can easily that the relations of continuity (24) for $j = 1$ are equivalent to

$$\left|\begin{array}{rcl} T^{01}_{k_y} + T^{11}_{k_y} D(k_y) &=& P(k_y) \\ D(k_y) &=& T^{00}_{k_y} P(k_y) + T^{10}_{k_y} D(k_y) P(k_y). \end{array}\right. \qquad (26)$$

Eliminating $D(k_y)$, the operator $P(k_y)$ is then a solution of the stationary Riccati equation (the operator $T^{11}_{k_y}$ is invertible, see [12])

$$T^{10}_{k_y}\left(T^{11}_{k_y}\right)^{-1} P(k_y)^2 + \left(T^{00}_{k_y} - T^{01}_{k_y}\left(T^{11}_{k_y}\right)^{-1} T^{10}_{k_y} - \left(T^{11}_{k_y}\right)^{-1}\right)P(k_y) + \left(T^{11}_{k_y}\right)^{-1} T^{01}_{k_y} = 0.$$

Actually, this equation characterizes uniquely the operator $P(k_y)$:

THEOREM 5.9 (CARACTERISTIC EQUATION)
The operator $P(k_y)$ is the unique compact operator

$$X \in \mathscr{K}\left(H^{-1/2}_{k_y}(\Sigma_0)\right)$$

satisfying the condition

$$\rho(P(k_y)) < 1 \qquad (27)$$

which solves the stationary Riccati equation :

$$\mathscr{T}(k_y, X) = 0, \qquad (\mathscr{E}^R_{k_y})$$

where

$$\mathscr{T}(k_y,\cdot) : \mathscr{L}\left(H^{-1/2}_{k_y}(\Sigma_0)\right) \to \mathscr{L}\left(H^{-1/2}_{k_y}(\Sigma_0)\right)$$

and is the quadratic map given by

$$\mathscr{T}(k_y,X) = T^{10}_{k_y}\left(T^{11}_{k_y}\right)^{-1} X^2 + \left(T^{11}_{k_y}\right)^{-1} T^{01}_{k_y} + \left(T^{00}_{k_y} - T^{01}_{k_y}\left(T^{11}_{k_y}\right)^{-1} T^{10}_{k_y} - \left(T^{11}_{k_y}\right)^{-1}\right)X.$$

Once $P(k_y)$ is determined solving the stationary Riccati equation, $D(k_y)$ is obtained using the first relation of (26) :

$$D(k_y) = \left(T^{11}_{k_y}\right)^{-1}\left(P(k_y) - T^{01}_{k_y}\right),$$

we build cell by cell the solution $\hat{u}^{\mathrm{H}}_{k_y}$ using (20) and finally using again (20) for $j = 0$, we see that

$$\Lambda^{\mathrm{w}}(k_y) = T^{00}_{k_y} + T^{10}_{k_y} D(k_y). \qquad (28)$$

6. CHARACTERIZATION OF THE RTR OPERATORS $\widetilde{I}_{(p,q)}$

In this section, we establish a linear equation that characterized the operator $\widetilde{I}_{(p,q)}$ and is adapted for numerical computation. This requires to introduce again new operators.

6.1 New RtR Operators associated to the Half Space Problem

Here we define some incoming RtR operators associated with the halfspace problem $(\mathscr{P}^{\mathrm{H}})$. Let Σ be the boundary of Ω^{w} :

$$\Sigma = \Sigma^- \cup \Sigma_0 \cup \Sigma^+ \tag{29}$$

where $\Sigma^{\pm} = \left] +\infty, \frac{L}{2} \right[\times \{y = \pm\frac{L}{2}\}$.

We can identify $\widetilde{\Sigma}$ with Σ $(\equiv \mathbb{R})$, as well as $\widetilde{\Sigma}^{\pm}$ with $\Sigma^{\pm}$ via the bijection Φ (see Figure (**15**)) :

$$\Phi : \left| \begin{array}{lll} (L/2,s) \in \widetilde{\Sigma}^- & \mapsto & (-s, L/2) \in \Sigma^-, \\ (L/2,s) \in \Sigma_0 & \mapsto & (L/2,s) \in \Sigma_0, \\ (L/2,s) \in \widetilde{\Sigma}^+ & \mapsto & (s, L/2) \in \Sigma^+. \end{array} \right. \tag{30}$$

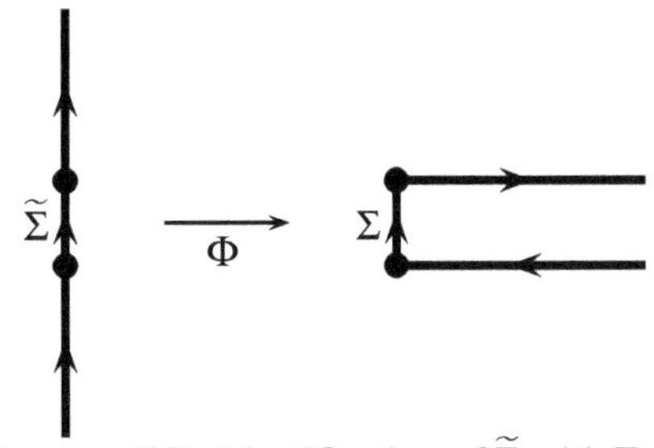

FIGURE (**15**): Identification of $\widetilde{\Sigma}$ with Σ.

Accordingly, functions defined on Σ can be associated to functions on Σ. More precisely, for any $\psi : \Sigma \to \mathbb{C}$, we will note in the following :

$$\psi\Big|_{\overrightarrow{\Sigma}^{\pm}} = \psi \circ \Phi^{-1}\Big|_{\Sigma^{\pm}}$$

The arrow in the notation $\overrightarrow{\Sigma}^{\pm}$ simply emphasizes the fact that he identification between $\Sigma^{\pm}$ and $\widetilde{\Sigma}^{\pm}$ is coherent with the orientations indicated of Figure (**15**). Next we introduce additional operators the interest of which will appear later.

To define rigorously these operators, we first recall that taking the restriction of a function defined on $\widetilde{\Sigma}$ respectively to Σ_0, $\widetilde{\Sigma}^+$ and $\widetilde{\Sigma}^-$, can be extended, when properly defined by duality, as a continuous linear operator from

$$H^{-1/2}(\widetilde{\Sigma})$$

repectively in

$$H^{1/2}_{aa}(\widetilde{\Sigma}_0)', H^{1/2}_{00}(\widetilde{\Sigma}^+)' \text{ and } H^{1/2}_{00}(\widetilde{\Sigma}^-)'.$$

We need an analogous framework for the concatenation operation : construct a function on $\widetilde{\Sigma}$ by concatenating functions defined on Σ_0, $\widetilde{\Sigma}^+$ and $\widetilde{\Sigma}^-$. For this we need new functional spaces :

$$H^{1/2}_{aa}(\widetilde{\Sigma}) = \left\{ u \in H^{1/2}(\widetilde{\Sigma}), \quad u\Big|_{\Sigma_0} \in H^{1/2}_{aa}(\Sigma_0) \right\}$$

equipped with its natural Hilbert structure and its dual (this is a definition)

$$H^{-1/2}_{aa}(\widetilde{\Sigma}) := \left(H^{1/2}_{aa}(\widetilde{\Sigma}) \right)'.$$

The interest of this space is that the concatenation operator, naturally defined for L^2 functions, can be extended continuously into a continuous operator from the product

$$H^{1/2}_{aa}(\widetilde{\Sigma}_0)' \times H^{1/2}_{00}(\widetilde{\Sigma}^+)' \times H^{1/2}_{00}(\widetilde{\Sigma}^-)'$$

into $H^{-1/2}_{aa}(\widetilde{\Sigma})$.

DEFINITION 6.1
Given $\psi \in H^{-1/2}(\widetilde{\Sigma})$ and $u^H(\psi)$ be the solution of $(\mathscr{P}^{\mathrm{H}})$, we define the RtR operators :

$$I^H_{(s,s)} : H^{-1/2}(\widetilde{\Sigma}) \longrightarrow H^{-1/2}_{aa}(\widetilde{\Sigma})$$

such that for all $\psi \in H^{-1/2}(\widetilde{\Sigma})$

$$\left| \begin{array}{l} I^H_{(s,s)}\psi|_{\widetilde{\Sigma}^-} \equiv + \left[-\dfrac{\partial}{\partial y} u^H(\psi) + \imath Z u^H(\psi) \right] \Big|_{\overrightarrow{\Sigma}^-}, \\ I^H_{(s,s)}\psi|_{\Sigma_0} = \psi|_{\Sigma_0}, \\ I^H_{(s,s)}\psi|_{\widetilde{\Sigma}^+} \equiv + \left[+\dfrac{\partial}{\partial y} u^H(\psi) + \imath Z u^H(\psi) \right] \Big|_{\overrightarrow{\Sigma}^+}, \end{array} \right.$$

$$I^H_{(s,a)} : H^{-1/2}(\widetilde{\Sigma}) \longrightarrow H^{-1/2}_{aa}(\widetilde{\Sigma})$$

such that for all $\psi \in H^{-1/2}(\widetilde{\Sigma})$

$$\left| \begin{array}{l} I^H_{(s,a)}\psi|_{\widetilde{\Sigma}^-} \equiv - \left[-\dfrac{\partial}{\partial y} u^H(\psi) + \imath Z u^H(\psi) \right] \Big|_{\overrightarrow{\Sigma}^-}, \\ I^H_{(s,a)}\psi|_{\Sigma_0} = \psi|_{\Sigma_0}, \\ I^H_{(s,a)}\psi|_{\widetilde{\Sigma}^+} \equiv + \left[+\dfrac{\partial}{\partial y} u^H(\psi) + \imath Z u^H(\psi) \right] \Big|_{\overrightarrow{\Sigma}^+}, \end{array} \right.$$

$$I^H_{(a,s)} \; : H^{-1/2}(\widetilde{\Sigma}) \longrightarrow H_{aa}^{-1/2}(\widetilde{\Sigma})$$

such that for all $\psi \in H^{-1/2}(\widetilde{\Sigma})$

$$\left| \begin{array}{l} I^H_{(a,s)} \psi|_{\widetilde{\Sigma}^-} \equiv + \left[-\dfrac{\partial}{\partial y}\; u^H(\psi) + \imath Z u^H(\psi) \right] \Big|_{\overrightarrow{\Sigma}^-}, \\ I^H_{(a,s)} \psi|_{\Sigma_0} = \psi|_{\Sigma_0}, \\ I^H_{(a,s)} \psi|_{\widetilde{\Sigma}^+} \equiv - \left[+\dfrac{\partial}{\partial y}\; u^H(\psi) + \imath Z u^H(\psi) \right] \Big|_{\overrightarrow{\Sigma}^+}, \end{array} \right.$$

$$I^H_{(a,a)} \; : H^{-1/2}(\widetilde{\Sigma}) \longrightarrow H_{aa}^{-1/2}(\widetilde{\Sigma})$$

such that for all $\psi \in H^{-1/2}(\widetilde{\Sigma})$

$$\left| \begin{array}{l} I^H_{(a,a)} \psi|_{\widetilde{\Sigma}^-} \equiv - \left[-\dfrac{\partial}{\partial y}\; u^H(\psi) + \imath Z u^H(\psi) \right] \Big|_{\overrightarrow{\Sigma}^-}, \\ I^H_{(a,a)} \psi|_{\Sigma_0} = \psi|_{\Sigma_0}, \\ I^H_{(a,a)} \psi|_{\widetilde{\Sigma}^+} \equiv - \left[+\dfrac{\partial}{\partial y}\; u^H(\psi) + \imath Z u^H(\psi) \right] \Big|_{\overrightarrow{\Sigma}^+}. \end{array} \right.$$

6.2 Characterization of the Incoming RtR Operators $\widetilde{I}_{(p,q)}$

By definition of $\widetilde{I}_{(p,q)}$ and thanks to (11), it is easy to see that each incoming RtR operator belongs to the following affine space :

$$\mathscr{L}^0_{(p,q)} = \Big\{ L \in \mathscr{L}\Big(H^{-1/2}_{(p,q)}(\Sigma^i), H^{-1/2}(\widetilde{\Sigma}) \Big), \\ \forall\, \widetilde{d} \in H^{-1/2}_{(p,q)}(\Sigma^i),\; \widetilde{R}(L\widetilde{d}) = R_{(p,q)}\, \widetilde{d} \Big\}.$$

where $\widetilde{R}$ is defined in 4.1. Now we present the fundamental relation satisfied by $\widetilde{I}_{(p,q)}$.

THEOREM 6.2
For each $(p,q) \in \{s,a\}^2$, *the operator* $\widetilde{I}_{(p,q)}$ *is the unique solution to the following problem :*

$$\text{Find}\, \widetilde{I} \in \mathscr{L}^0_{(p,q)}, \quad \widetilde{I} = I^H_{(p,q)} \circ \widetilde{I}. \qquad (\mathscr{E}_{(p,q)})$$

■ **PROOF:** We give the proof for $\widetilde{I}_{(s,s)}$. The other cases are treated similarly.

We first prove that $\widetilde{I}_{(s,s)}$ is a solution of $\mathscr{E}_{(p,q)}$.

Since $\varphi \in H^{-1/2}_{(s,s)}(\Sigma^i)$, $u^e(\varphi) \in H^1_{(s,s)}(\Omega^e)$. Thus, from

$$u^e(\varphi)\Big|_{\Omega^{\mathrm{H}}} = u^{\mathrm{H}}(\widetilde{I}_{(s,s)}\, \varphi),$$

and $\widetilde{\Sigma}^+ = S_1 \Sigma^+$ we deduce that

$$\left| \begin{array}{l} \left[\dfrac{\partial}{\partial x} u^e(\varphi) + \imath Z u^e(\varphi) \right] \Big|_{\widetilde{\Sigma}^+} \\ \equiv \left[\dfrac{\partial}{\partial y} u^e(\varphi) + \imath Z u^e(\varphi) \right] \Big|_{\overrightarrow{\Sigma}^+} \\ = \left[\dfrac{\partial}{\partial y} u^{\mathrm{H}}(\widetilde{I}^{\varepsilon}_{(s,s)}\, \varphi) + \imath Z u^{\mathrm{H}}(\widetilde{I}^{\varepsilon}_{(s,s)}\, \varphi) \right] \Big|_{\Sigma^+}, \end{array} \right.$$

In the same way, from $\widetilde{\Sigma}^- = S_{-1}\Sigma^-$, we deduce that

$$\left| \begin{array}{l} \left[\dfrac{\partial}{\partial x} u^e(\varphi) + \imath Z u^e(\varphi) \right] \Big|_{\widetilde{\Sigma}^-} \\ \equiv \left[-\dfrac{\partial}{\partial y} u^e(\varphi) + \imath Z u^e(\varphi) \right] \Big|_{\overrightarrow{\Sigma}^-} \\ = \left[-\dfrac{\partial}{\partial y} u^{\mathrm{H}}(\widetilde{I}^{\varepsilon}_{(s,s)}\, \varphi) + \imath Z u^{\mathrm{H}}(\widetilde{I}^{\varepsilon}_{(s,s)}\, \varphi) \right] \Big|_{\Sigma^-}, \end{array} \right.$$

Completing the above equalities with

$$\left[\frac{\partial}{\partial x} u^e(\varphi) + \imath Z u^e(\varphi) \right] \Big|_{\Sigma_0} = \varphi \Big|_{\Sigma_0}$$

we easily conclude, using the definition of $\widetilde{I}_{(s,s)}$.

For the uniqueness of the solution, one combines symmetry arguments with uniqueness results for quarter plane problems with incoming RtR conditions. The idea are are similar to the ones developed in [13, 12], respectively for Dirichlet or Neumann conditions. ■

The equation $\mathscr{E}_{(p,q)}$ is quite abstract for he moment but we are going to see how it can be solved numerically.

6.3 About the Resolution of the Affine Equation ($\mathscr{E}_{(p,q)}$)

To be able to solve $\mathscr{E}_{(p,q)}$ we need to be able

to compute the operators $I^{\mathrm{H}}_{(p,q)}$.

The computation of these operators relies on the solution $u^{\mathrm{H}}(\psi)$ of the halfspace problem ($\mathscr{P}^{\mathrm{H}}$) which can be computed via their Floquet Bloch transforms and the method described in section 5.

To express our result, we need to introduce some local RtR operators depending on the elementary solutions

e^0 and e^1

of the cell problems (21)-(22)-(23) (see Figure (**10**)), namely :

$$I^{0,\pm}(k_y) \in \mathscr{L}\big(H^{-1/2}_{k_y}(\Sigma_0), H^{1/2}_{aa}(\Sigma_1^{\pm})'\big)$$
$$I^{1,\pm}(k_y) \in \mathscr{L}\big(H^{-1/2}_{k_y}(\Sigma_0), H^{1/2}_{aa}(\Sigma_1^{\pm})'\big),$$

defined for all $\hat{\psi}_{k_y} \in H_{k_y}^{-1/2}(\Sigma_0)$ by

$$
\begin{aligned}
I^{0,-}(k_y)\,\hat{\psi}_{k_y} &= \Big[-\frac{\partial}{\partial y}e^0(k_y;\hat{\psi}_{k_y}) + \imath Z\, e^0(k_y;\hat{\psi}_{k_y})\Big]\Big|_{\overrightarrow{\Sigma}_1^-},\\
I^{0,+}(k_y)\,\hat{\psi}_{k_y} &= \Big[+\frac{\partial}{\partial y}e^0(k_y;\hat{\psi}_{k_y}) + \imath Z\, e^0(k_y;\hat{\psi}_{k_y})\Big]\Big|_{\overrightarrow{\Sigma}_1^+},\\
I^{1,-}(k_y)\,\hat{\psi}_{k_y} &= \Big[-\frac{\partial}{\partial y}e^1(k_y;\hat{\psi}_{k_y}) + \imath Z\, e^1(k_y;\hat{\psi}_{k_y})\Big]\Big|_{\overrightarrow{\Sigma}_1^-},\\
I^{1,+}(k_y)\,\hat{\psi}_{k_y} &= \Big[+\frac{\partial}{\partial y}e^1(k_y;\hat{\psi}_{k_y}) + \imath Z\, e^1(k_y;\hat{\psi}_{k_y})\Big]\Big|_{\overrightarrow{\Sigma}_1^+},
\end{aligned}
\tag{31}
$$

where

$$
\overrightarrow{\Sigma}_1^+ = \left[\tfrac{L}{2},\tfrac{3L}{2}\right]\times\left\{\pm\tfrac{L}{2}\right\}
\quad\text{and}\quad
\overrightarrow{\Sigma}_1^- = \left[\tfrac{3L}{2},\tfrac{L}{2}\right]\times\left\{-\tfrac{L}{2}\right\}
$$

are oriented segments (the trace on $\overrightarrow{\Sigma}_1^+$ is taken in the direction of increasing x whereas the trace on $\overrightarrow{\Sigma}_1^-$ is taken in the direction of decreasing x). See Figure (**16**).

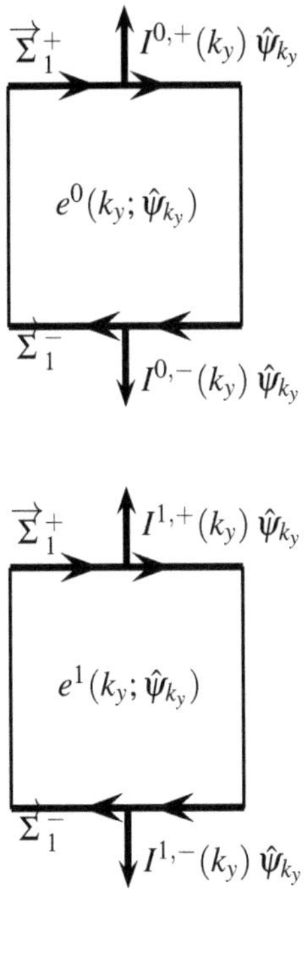

FIGURE (**16**): Other local RtR operators $I^{i,\pm}(k_y)$.

Next, we form new operators in

$$
\mathscr{L}\big(H_{k_y}^{-1/2}(\Sigma_0), H_{aa}^{1/2}(\Sigma_0)'\big)
$$

namely

$$
\mathbf{K}^{\pm}(k_x,k_y) = e^{\mp\imath Lk_x}\, I^{\pm}(k_y)\Big(\mathbf{I} - P(k_y)e^{\mp\imath Lk_x}\Big)^{-1}
$$

where

$$
I^{\pm}(k_y) = I^{0,\pm}(k_y) + I^{1,\pm}(k_y)D(k_y)
$$

and, setting $\varepsilon_s = 1$ ans $\varepsilon_a = -1$,

$$
\begin{aligned}
\mathbf{K}_{(p,q)}(k_x,k_y) \;=\; \mathbf{I} \;&+\; \varepsilon_p\,\mathbf{K}^{+}(k_x,k_y)\\
&+\; \varepsilon_q\,\mathbf{K}^{-}(k_x,k_y).
\end{aligned}
\tag{32}
$$

It can be shown that, for each $(p,q)\in\{a,s\}^2$,

$$
\mathbf{K}_{(p,q)}(k_x,k_y) \in \mathscr{L}\big(H_{k_y}^{-1/2}(\Sigma_0), H_{k_x}^{-1/2}(\Sigma_0)\big), \tag{33}
$$

and leads to a useful characterization of $I^{\mathrm{H}}_{(p,q)}$:

PROPOSITION 6.3
For all $(p,q)\in\{s,a\}^2$, *let*

$$
\hat{I}^{H}_{(p,q)} \in \mathscr{L}\Big(H_{QP}^{-1/2}\big(\big]-\frac{\pi}{L},\frac{\pi}{L}\big[\times\Sigma_0\big)\big)\Big)
$$

defined by :

$$
\hat{I}^{H}_{(p,q)} = \mathscr{F}_y \circ I^{H}_{(p,q)} \circ \mathscr{F}_y^{-1}.
$$

For any $\hat{\psi}\in H_{QP}^{-1/2}\Big(\big]-\frac{\pi}{L},\frac{\pi}{L}\big[\times\Sigma_0\big)\Big)$, *one has*

$$
\hat{I}^{H}_{(p,q)}\,\hat{\psi}(\cdot,k_x) = \frac{L}{2\pi}\int_{-\pi/L}^{\pi/L} \mathbf{K}_{(p,q)}(k_x,k_y)\,\hat{\psi}(\cdot;k_y)\,dk_y.
$$

■ **PROOF:** We write the proof for $(p,q) = (s,s)$, the other cases follow similarly.

For simplicity, we restrict ourselves to the case of sufficiently smooth ψ for instance. In this case,

$$
I^{\mathrm{H}}_{(s,s)}\,\psi \in L^2(\widetilde{\Sigma})
$$

and its restriction to any subset of $\widetilde{\Sigma}$ is defined directly. The extension to the more general case can be done by duality.

In the case $n=0$, Expression (14) gives :

$$
u^{\mathrm{H}}(\psi)|_{\overrightarrow{\Sigma}^{\pm}} = \sqrt{\frac{L}{2\pi}}\int_{-\pi/L}^{\pi/L} \hat{u}^{\mathrm{H}}_{k_y}(\hat{\psi}_{k_y})|_{\overrightarrow{\Sigma}^{\pm}}\,dk_y
$$

where $\hat{\psi}_{k_y} = \mathscr{F}_y\psi(\cdot;k_y)$. We denote by $\overrightarrow{\Sigma}_n^-$ and $\overrightarrow{\Sigma}_n^+$ the following sequence of intervals of length L :

$$
\forall n\geq 1,\quad
\left|\begin{aligned}
\overrightarrow{\Sigma}_n^- &= \big]\frac{3L}{2}+nL, \frac{L}{2}+nL\big[\times\{\frac{-L}{2}\},\\
\overrightarrow{\Sigma}_n^+ &= \big]\frac{L}{2}+nL, \frac{3L}{2}+nL\big[\times\{\frac{L}{2}\}.
\end{aligned}\right.
$$

Using the relations (20), we see easily that the outgoing Robin data of the solution $\hat{u}^{\mathrm{H}}_{k_y}(\hat{\psi}_{k_y})$ of $(\hat{\mathscr{P}}^{\mathrm{H}}_{k_y})$ on $\overrightarrow{\Sigma}^{+}_{n}$ and $\overrightarrow{\Sigma}^{-}_{n}$ can be expressed via the operators $I^{\{0,1\},\pm}(k_y)$ defined in (31), for all $k_y \in]-\pi/L;\pi/L[$ and $n \geq 1$:

$$\left|\begin{array}{l} [+\dfrac{\partial \hat{u}^{\mathrm{H}}_{k_y}(\hat{\psi}_{k_y})}{\partial y} + \imath Z\, \hat{u}^{\mathrm{H}}_{k_y}(\hat{\psi}_{k_y})]\Big|_{\overrightarrow{\Sigma}^{+}_{n}} = I^{+}(k_y)\, P(k_y)^{n-1}\, \hat{\psi}_{k_y}, \\ [-\dfrac{\partial \hat{u}^{\mathrm{H}}_{k_y}(\hat{\psi}_{k_y})}{\partial y} + \imath Z\, \hat{u}^{\mathrm{H}}_{k_y}(\hat{\psi}_{k_y})]\Big|_{\overrightarrow{\Sigma}^{-}_{n}} = I^{-}(k_y)\, P(k_y)^{n-1}\, \hat{\psi}_{k_y}. \end{array}\right.$$

where $I^{\pm}(k_y) = I^{0,\pm}(k_y) + I^{1,\pm}(k_y)D(k_y)$. Using these relations, we obtain

- On $\overrightarrow{\Sigma}^{-} : \forall n \geq 1$,

$$I^{\mathrm{H}}_{(s,s)}\psi|_{\overrightarrow{\Sigma}^{-}_{n}} = \sqrt{\frac{L}{2\pi}} \int_{-\pi/L}^{\pi/L} I^{-}(k_y)\, P(k_y)^{n-1}\, \hat{\psi}_{k_y}\, dk_y$$

- On Σ_0 :

$$I^{\mathrm{H}}_{(s,s)}\psi|_{\Sigma_0} = \psi|_{\Sigma_0}$$

- On $\overrightarrow{\Sigma}^{+} : \forall n \geq 1$,

$$I^{\mathrm{H}}_{(s,s)}\psi|_{\overrightarrow{\Sigma}^{+}_{n}} = \sqrt{\frac{L}{2\pi}} \int_{-\pi/L}^{\pi/L} I^{+}(k_y)\, P(k_y)^{n-1}\, \hat{\psi}_{k_y}\, dk_y$$

We apply then the FB-Transform to $I^{\mathrm{H}}_{(s,s)}\psi$ using the identification $\Sigma \sim \mathbb{R}$,

$$\mathscr{F}_y(I^{\mathrm{H}}_{(s,s)}\psi)(\cdot,k_x) = \sqrt{\frac{L}{2\pi}}\Big(\sum_{n=\infty}^{1} I^{\mathrm{H}}_{(s,s)}\psi|_{\overrightarrow{\Sigma}^{-}_{n}} e^{\imath n k_x L} + I^{\mathrm{H}}_{(s,s)}\psi|_{\Sigma_0} + \sum_{n=1}^{+\infty} I^{\mathrm{H}}_{(s,s)}\psi|_{\overrightarrow{\Sigma}^{+}_{n}} e^{-\imath n k_x L}\Big).$$

By inverting the integrals over $[-\pi/L, \pi/L]$ and the sum over n, we are led to using the following formula

$$\sum_{n=1}^{+\infty} P(k_y)^{n-1} e^{\pm \imath n L k_x} = e^{\pm \imath L k}\Big(\mathbf{I} - P(k_y) e^{\pm \imath L k_x}\Big)^{-1}. \quad (34)$$

In fact for every k_y, $P(k_y)$ is compact and its spectral radius is strictly less than 1. Actually, we could prove (see [12] for more details) that for $\varepsilon > 0$ small enough, the spectral radius of $P(k_y)$ is uniformly bounded in k_y by a constant that is strictly less than 1 :

$$\exists \tau > 0, \quad \forall k_y \in \Big]-\frac{\pi}{L}, \frac{\pi}{L}\Big], \quad \rho(P(k_y)) \leq e^{-\tau\varepsilon}.$$

The property :

$$\lim_{n\to+\infty} \|P(k_y)^n\|^{1/n} = \rho(P(k_y))$$

for the norm of $\mathscr{L}(L^2(\Sigma_0))$ ([31]), implies that for some $\rho_* \in]e^{-\tau\varepsilon}, 1[$, n large enough we have for all k_y

$$\|P(k_y)^j\| \leq \rho_*^j$$

which yields the absolute convergence of the series (34). Therefore, for each k_y and k_x, $\mathbf{I} - P(k_y)e^{\pm \imath L k_x}$ is inversible and the sum (34) converges uniformely in the norm of $\mathscr{L}(L^2(\Sigma_0))$. The inversion of the integral and the sum is then possible. ∎

Let us come back now to the resolution of the affine equations $(\mathscr{E}_{(p,q)})$. Since the operators $I^{\mathrm{H}}_{(p,q)}$ (see Definition 6.1 and Proposition 6.3) are defined via their Floquet Bloch tranforms, it makes sense that we will try to formulate $(\mathscr{E}_{(p,q)})$ using the FB tranform as well.

COROLARY 6.4
For any $\varphi \in H^{-1/2}_{(p,q)}(\Sigma^i)$, the function

$$\hat{\psi}_{ij} = \mathscr{F}_y\big(\widetilde{I}_{(p,q)}\varphi\big) \quad \in H^{-1/2}_{QP}\Big(\Big]-\frac{\pi}{L}, \frac{\pi}{L}\Big[\times \Sigma_0\Big)\Big)$$

is the unique solution to the following problem :

$$\text{Find } \hat{\psi} \in H^{-1/2}_{QP}\Big(\Big]-\frac{\pi}{L}, \frac{\pi}{L}\Big[\times \Sigma_0\Big)\Big), \text{ such that}$$

$$\left|\begin{array}{ll} (i) & \hat{\psi} - \hat{I}^{H}_{(p,q)}\hat{\psi} = 0, \\ (ii) & \sqrt{\dfrac{L}{2\pi}} \displaystyle\int_{-\pi/L}^{\pi/L} \hat{\psi}(\cdot,k_x)\, dk_x = R_{(p,q)}\varphi, \end{array}\right. \quad (35)$$

where $R_{(p,q)}$ is the restriction operator on Σ_0 defined in section 3 and $\hat{I}^{H}_{(p,q)}$ is given in Proposition 6.3.

∎ **PROOF:** To obtain (35-(i)), we apply the FB tranform to the equation

$$\widetilde{I}_{(p,q)}\varphi = I^{\mathrm{H}}_{(p,q)}\Big(\widetilde{I}_{(p,q)}\varphi\Big),$$

and use proposition 6.3. The relation (35-(ii)) expresses in terms of the FB variable the condition :

$$\sqrt{\frac{L}{2\pi}} \int_{-\pi/L}^{\pi/L} \mathscr{F}_y\big(\widetilde{I}_{(p,q)}\varphi\big)(\cdot,k_x)\, dk_x = \widetilde{I}_{(p,q)}\varphi|_{\Sigma_0} = \varphi|_{\Sigma_0}.$$

∎

Numerically, we wish to apply a Galerkin procedure to approximate the problem (35). This means that we would like to look for the unknown function in a vector space which amounts to make the condition (35-(ii)) homogeneous. To do so, we introduce an extension operator

$$\widetilde{E}\varphi \in \mathscr{L}\Big(H^{-1/2}_{(p,q)}(\Sigma^i), H^{-1/2}(\widetilde{\Sigma})\Big)$$

such that

$$\widetilde{E}\varphi\Big|_{\Sigma_0} = R_{(p,q)}\varphi, \quad (36)$$

so that $\widetilde{I}_{(p,q)}\varphi - \widetilde{E}\varphi = 0$ on Σ_0.

Introducing the new unknown

$$\hat{\psi}^0_{pq} = \mathscr{F}_y\Big(\widetilde{I}_{(p,q)}\varphi - \widetilde{E}\varphi\Big),$$

we easily see that it is the unique solution to the problem :

$$\text{Find } \hat{\psi}^0 \in H^{-1/2}_{\mathrm{QP}}\Big(\Big]-\frac{\pi}{L},\frac{\pi}{L}\Big[\times\Sigma_0\Big)\Big), \text{ such that}$$

$$\left|\begin{array}{ll} (i) & \hat{\psi}^0 - \hat{I}^{\varepsilon,\mathrm{H}}_{(p,q)}\hat{\psi}^0 = \hat{g}_\varphi, \\ (ii) & \displaystyle\int_{-\pi/L}^{\pi/L} \hat{\psi}^0(\cdot,k_x)\,dk_x = 0, \end{array}\right. \qquad (37)$$

with $\hat{g}_\varphi \in H^{-1/2}_{\mathrm{QP}}\Big(\Big]-\frac{\pi}{L},\frac{\pi}{L}\Big[\times\Sigma_0\Big)\Big)$ given by

$$\hat{g}_\varphi = -\mathscr{F}_y\widetilde{E}\,\varphi + \hat{I}^{\varepsilon,\mathrm{H}}_{(p,q)}\mathscr{F}_y\widetilde{E}\,\varphi \qquad (38)$$

In practice, we will solve (37) instead of (35).

6.4 Variational Formulation

Before discussing a variational formulation of (37) we define some function spaces :

$$V := H^{-1/2}_{\mathrm{QP}}\Big(\Big]-\frac{\pi}{L},\frac{\pi}{L}\Big[\times\Sigma_0\Big)\Big), \qquad (39)$$

$$W := \Big\{\hat{\psi}^0 \in V, \quad \int_{-\pi/L}^{\pi/L}\hat{\psi}^0(\cdot,k)\,dk = 0\Big\}, \qquad (40)$$

$$V' := H^{1/2}_{\mathrm{QP}}\Big(\Big]-\frac{\pi}{L},\frac{\pi}{L}\Big[\times\Sigma_0\Big)\Big), \qquad (41)$$

and

$$W' := \Big\{\hat{\psi}^0 \in V', \quad \int_{-\pi/L}^{\pi/L}\hat{\psi}^0(\cdot,k)\,dk = 0\Big\}. \qquad (42)$$

REMARK 6.5
According to proposition 5.3, it is clear that

$$W = \mathscr{F}_y\Big(\big\{\psi \in H^{-1/2}(\widetilde{\Sigma}), \quad \psi \equiv 0 \quad \text{on } \Sigma_0\big\}\Big),$$

where we recall that the restriction of a function in $H^{-1/2}(\widetilde{\Sigma})$ *is in* $[H^{1/2}_{00}(\Sigma_0)]'$.

Similarly,

$$W' = \mathscr{F}_y\Big(\big\{\psi \in H^{+1/2}(\widetilde{\Sigma}), \quad \psi \equiv 0 \quad \text{on } \Sigma_0\big\}\Big).$$

Note that by definition, V' is the dual of V. In what follows, we shall denote the duality product between V and V'

$$\langle\hat{\psi},\hat{\theta}\cdot\rangle_V := \int_{-\pi/L}^{\pi/L}\langle\hat{\psi}(\cdot;k_x);\hat{\theta}(\cdot;k_x)\rangle$$

where the duality product inside the integral is the one between $H^{-1/2}_{k_x}(\Sigma_0)$ and $H^{1/2}_{k_x}(\Sigma_0)$. This duality product simply extends the standard inner product in

$$L^2\Big(\Big]-\frac{\pi}{L},\frac{\pi}{L}\Big[\times\Sigma_0\Big)\Big).$$

We choose to solve by a Galerkin method the problem (37), which according to proposition 6.3 corresponds to the integral equation : $\forall k_x \in \big]-\frac{\pi}{L},\frac{\pi}{L}\big]$,

$$\hat{\psi}^0(\cdot;k_x) - \int_{-\pi/L}^{\pi/L}\mathbf{K}_{(p,q)}(k_x,k_y)\,\hat{\psi}^0(\cdot;k_y)\,dk_y = \hat{g}_\varphi(\cdot;k_x),$$

with the linear constraint (37-(ii)). The kernel $\mathbf{K}_{(p,q)}$ is defined in (32).

Now we describe the variational formulation where we look for a solution in W with test functions also in W'.

PROPOSITION 6.6 (VARIATIONAL FORMULATION)
For $(p,q) \in \{s,a\}^2$ *and* $\varphi \in H^{-1/2}_{(p,q)}(\Sigma^i)$, *we have*

$$\widetilde{I}_{(p,q)}\varphi = \widetilde{E}\varphi + \mathscr{F}_y^{-1}\hat{\psi}^0_{pq},$$

where $\widetilde{E}\varphi$ *is defined by* (36) *and* $\hat{\psi}^0_{pq} \in W$ *is the unique solution to the following problem :*

$$\text{Find } \hat{\psi}^0 \in W, \text{ such that for all } \hat{\theta}^0 \in W'$$

$$\Big\langle\big(\mathbf{I} - \hat{I}^H_{(p,q)}\big)\hat{\psi}^0,\hat{\theta}^0\Big\rangle = \Big\langle\hat{g}_\varphi,\hat{\theta}^0\Big\rangle, \qquad (\mathscr{Q}^\varphi_{(p,q)})$$

with $\hat{g}_\varphi$ *defined by* (38) *and* $<\cdot,\cdot>$ *the dual product between* V *and* V'.

■ **PROOF:** It is easy to see that if $\hat{\psi}^0_{pq}$ is the solution of (37), it is also the solution of $(\mathscr{Q}^\varphi_{(p,q)})$.

Let us show now the uniqueness. Let $\hat{\psi}^0$ be a solution of

$$\Big\langle\big(\mathbf{I} - \hat{I}^{\mathrm{H}}_{(p,q)}\big)\hat{\psi}^0,\hat{\theta}^0\Big\rangle = 0, \quad \hat{\theta}^0 \in W' \qquad (43)$$

and let us show that $\hat{\psi}^0 = 0$. The proof can be done in three steps.

(1) We prove first that $(\mathbf{I}-\hat{I}^{\mathrm{H}}_{(p,q)})\hat{\psi}^0 \in W$.
Let us remind that by definition

$$I^{\mathrm{H}}_{(p,q)})\psi\Big|_{\Sigma^0} = \psi\Big|_{\Sigma^0}$$

then

$$\sqrt{\frac{L}{2\pi}} \int_{-\pi/L}^{\pi/L} \hat{I}^{\mathrm{H}}_{(p,q)} \mathscr{F}_y \psi(\cdot;k_y)\, dk_y = \psi\Big|_{\Sigma^0}.$$

The inversion formula (5.3) gives

$$\sqrt{\frac{L}{2\pi}} \int_{-\pi/L}^{\pi/L} \mathbf{I}\mathscr{F}_y \psi(\cdot;k_y)\, dk_y = \psi\Big|_{\Sigma^0};$$

which implies

$$\mathrm{Im}(\mathbf{I}-\hat{I}^{\mathrm{H}}_{(p,q)}) \subset W$$

and it is true in particular for $\hat{\psi}^0$.

(2) We prove now that $(\mathbf{I}-\hat{I}^{\mathrm{H}}_{(p,q)})\hat{\psi}^0 = 0$.

Given the step (1) and the relation (43), the function

$$\hat{v}^0 = (\mathbf{I}-\hat{I}^{\mathrm{H}}_{(p,q)})\hat{\psi}^0 \quad \in W$$

satisfies

$$\forall \hat{\theta}^0 \in W', \quad \left\langle \hat{v}^0, \hat{\theta}^0 \right\rangle = 0. \qquad (44)$$

By definition of V' and W'

$$\forall \hat{\theta}^0 \in V', \quad \hat{\theta}^0 - \frac{L}{2\pi}\int_{-\pi/L}^{\pi/L} \hat{\theta}^0(\cdot,k)\, dk \in W'$$

and then since (44), we have $\forall \hat{\theta}^0 \in V'$,

$$\begin{aligned} \left\langle \hat{v}^0, \hat{\theta}^0 \right\rangle &= \left\langle \hat{v}^0, \hat{\theta}^0 - \frac{L}{2\pi}\int_{-\pi/L}^{\pi/L} \hat{\theta}^0(\cdot,k)\, dk \right\rangle \\ &\quad + \left\langle \hat{v}^0, \frac{L}{2\pi}\int_{-\pi/L}^{\pi/L} \hat{\theta}^0(\cdot,k)\, dk \right\rangle \\ &= \left\langle \hat{v}^0, \frac{L}{2\pi}\int_{-\pi/L}^{\pi/L} \hat{\theta}^0(\cdot,k)\, dk \right\rangle. \end{aligned}$$

We conclude writing that

$$\begin{aligned} &\left\langle \hat{v}^0, \int_{-\pi/L}^{\pi/L} \hat{\theta}^0(\cdot,k)\, dk \right\rangle \\ &= \int_{-\pi/L}^{\pi/L} \left(\hat{v}^0(\cdot;\xi), \int_{-\pi/L}^{\pi/L} \hat{\theta}^0(\cdot,k)\, dk \right) d\xi \\ &= \left(\int_{-\pi/L}^{\pi/L} \hat{v}^0(\cdot;\xi)\, d\xi, \int_{-\pi/L}^{\pi/L} \hat{\theta}^0(\cdot,k)\, dk \right) = 0 \end{aligned}$$

by definition of W. $\hat{v}^0$ is an element of $W \subset V$, for which the scalar product with any element of V' vanishes, that implies that $\hat{v}^0 = 0$.

(3) We can prove now that $\hat{\psi}^0 = 0$.

Indeed, the uniqueness of the problem (35) gives

$$\left| \begin{array}{ll} (i) & \hat{\psi} - \hat{I}^{\mathrm{H}}_{(p,q)}\hat{\psi} = 0, \\ (ii) & \sqrt{\frac{L}{2\pi}} \int_{-\pi/L}^{\pi/L} \hat{\psi}(\cdot,k_x)\, dk_x = 0 \end{array} \right.$$

$$\Downarrow$$

$$\hat{\psi} = 0.$$

which can be summarized by

$$\mathrm{Ker}(\mathbf{I}-\hat{I}^{\mathrm{H}}_{(i,j)}) \cap W = \{0\}.$$

We know that $\hat{\psi}^0 \in W$ and the result of the step (2) gives

$$\hat{\psi}^0 \in \mathrm{Ker}(\mathbf{I}-\hat{I}^{\mathrm{H}}_{(p,q)}).$$

We conclude that $\hat{\psi}^0 = 0$.

■

According to proposition 6.3, $(\mathscr{Q}^{\varphi}_{(p,q)})$ can be rewritten as

$$\text{Find } \hat{\psi}^0 \in W, \text{ such that for all } \hat{\theta}^0 \in W'$$

$$a_{(p,q)}(\hat{\psi}^0, \hat{\theta}^0) = \ell(\hat{\theta}^0), \qquad (\mathscr{Q}^{\varphi}_{(p,q)})$$

where

$$\left| \begin{array}{l} a_{(p,q)}(\hat{\psi}^0, \hat{\theta}^0) := \displaystyle\int_{-\pi/L}^{\pi/L} dk_x \Big[\left\langle \hat{\psi}^0(\cdot;k_x); \hat{\theta}^0(\cdot;k_x) \right\rangle \\ \quad - \displaystyle\int_{-\pi/L}^{\pi/L} \left\langle \mathbf{K}_{(p,q)}(k_x,k_y)\, \hat{\psi}^0(\cdot;k_y); \hat{\theta}^0(\cdot;k_x) \right\rangle dk_y \Big] \\ \ell(\hat{\theta}^0) := \displaystyle\int_{-\pi/L}^{\pi/L} \left\langle \hat{g}_{\varphi}(\cdot;k_x); \hat{\theta}^0(\cdot;k_x) \right\rangle dk_x \end{array} \right. \qquad (45)$$

where $\mathbf{K}_{(p,q)}(k_x,k_y)$ is defined by (32) and $< \cdot;\cdot >$ is the duality bracket between

$$H^{-1/2}_{k_x}(\Sigma_0) \text{ and } H^{1/2}_{k_x}(\Sigma_0).$$

This is the problem we solve in practice.

7. ALGORITHM FOR THE RESOLUTION OF $(\mathscr{P})$

We summarize the method presented in the previous sections for the computation of the RtR operator Λ in the following algorithm :

1. Construction of the halfspace RtR operator Λ^{H}
   (i) For each $k_y \in [-\pi/L, \pi/L]$, resolution of the cell problems (21)-(22)-(23) and computation of local RTR operators (25), $T^{ij}(k_y)$. Computation of the other RtR operators (31), $I^{i,\pm}(k_y)$ which will be useful for the step 2-(i).
   (ii) For each $k_y \in [-\pi/L, \pi/L]$, determination of the propagative operator $P(k_y)$ solving the stationay Riccati $(\mathscr{E}^R_{k_y})$
   (iii) Construction of $\hat{\Lambda}^{\mathrm{H}}$ using the expression (28)
   (iv) Computation of Λ^{H} using the Floquet Bloch Transformation and its inverse by (16),

2. Construction of $\Lambda_{(p,q)}$ for each $(p,q) \in \{s,a\}^2$
   (i) Build the incoming RtR operator $\widetilde{I}_{(p,q)}$ solving for each $\varphi \in H^{-1/2}_{(p,q)}(\Sigma^i)$ the variationnal problem $(\mathscr{Q}^{\varphi}_{(p,q)})$,
   (ii) Apply the relation

$$\Lambda_{(p,q)} = E_{(p,q)} \circ \widetilde{R} \circ \Lambda^{\mathrm{H}} \circ \widetilde{I}_{(p,q)},$$

   where $\widetilde{R}$ is the restriction operator from $\widetilde{\Sigma}$ onto Σ^i_0 and $E_{(p,q)}$ is the extension operator introduced at the end of Section 3.
3. Determination of the RtR operator Λ from (9) .

Once the RtR operator Λ is computed, the interior problem posed in the bounded domain Ω^i

$$(\mathscr{P}^i)\quad \left| \begin{array}{l} -\triangle u^i - \rho\,(\omega^2 + \imath\varepsilon\omega)\, u^i = f \text{ dans } \Omega^i, \\ -\dfrac{\partial u^i}{\partial \mathbf{n}^i} + \imath Z u^i = \Lambda\Big[\dfrac{\partial u^i}{\partial \mathbf{n}^i} + \imath Z u^i\Big] \text{ sur } \Sigma^i, \end{array} \right.$$

can be solved.

We want now to compute the solution u of $(\mathscr{P})$ outside the bounded region Ω^i, defined, thanks to the solutions of the interior and exterior problems, by

$$\left| \begin{array}{ll} u = u^i, & \text{in } \Omega^i \\ u = u^e(\varphi^i), & \text{in } \Omega^e, \text{ with } \varphi^i = \Big[\dfrac{\partial u^i}{\partial \mathbf{n}^i} + \imath Z u^i\Big]\Big|_{\Sigma^i}. \end{array} \right.$$

where u^i is the solution of $(\mathscr{P}^i)$ and u^e is the solution of $(\mathscr{P}^e)$.
It suffices to use the following algorithm of reconstruction using essentially the solution u^{H} of halfspace problems $(\mathscr{P}^{\mathrm{H}})$ and the incoming RtR operators $\widetilde{I}_{(p,q)}$ involved in the caracterization of Λ.

1. Thanks to the results of Section 3, φ can be decomposed by

$$\varphi^i = \sum_{(p,q)\in\{s,a\}^2} \varphi^i_{(p,q)}$$

   with

$$\forall (p,q) \in \{s,a\}^2, \quad \varphi^i_{(p,q)} \in H^{-1/2}_{(p,q)}(\Sigma^i)$$

   and by linearity :

$$u^e(\varphi^i) = \sum_{(p,q)\in\{s,a\}^2} u^e(\varphi^i_{(p,q)})$$

   with

$$\forall (p,q) \in \{s,a\}^2, \quad u^e(\varphi^i_{(p,q)}) \in H^1_{(p,q)}(\triangle, \Omega^e)$$

2. By definition of the RtR operators $\widetilde{I}_{(p,q)}$, $\forall (p,q) \in \{s,a\}^2$

$$\Big[\frac{\partial}{\partial x} u^e(\varphi^i_{(p,q)}) + \imath Z\, u^e(\varphi^i_{(p,q)})\Big]\Big|_{\widetilde{\Sigma}} = \widetilde{I}_{(p,q)}\varphi^i_{(p,q)}.$$

3. The solution $u^{\mathrm{H}}(\psi)$ of the halfspace problem $(\mathscr{P}^{\mathrm{H}})$ can be compute semi-analytically for all $\psi \in H^{-1/2}(\widetilde{\Sigma})$, thanks to (20), so we have : $\forall (p,q) \in \{s,a\}^2$,

$$u^e(\varphi^i_{(p,q)})\Big|_{\Omega^{\mathrm{H}}} = u^{\mathrm{H}}\big(\widetilde{I}_{(p,q)}\varphi^i_{(p,q)}\big).$$

4. By symmetry arguments, we have finally $\forall (p,q) \in \{s,a\}^2$

$$u^e(\varphi^i_{(p,q)}) = \left| \begin{array}{ll} u^{\mathrm{H}}(\tilde{\psi}_{(p,q)}), & \text{in } \Omega^{\mathrm{H}} \\ \varepsilon_p S_1\left(u^{\mathrm{H}}(\tilde{\psi}_{(p,q)})\right), & \text{in } S_1\Omega^{\mathrm{H}} \\ \varepsilon_q S_{-1}\left(u^{\mathrm{H}}(\tilde{\psi}_{(p,q)})\right), & \text{in } S_{-1}\Omega^{\mathrm{H}} \\ \varepsilon_p\varepsilon_q S_1 \circ S_{-1}\big(u^{\mathrm{H}}(\tilde{\psi}_{(p,q)})\big), & \text{in } S_1 \circ S_{-1}\Omega^{\mathrm{H}}. \end{array} \right.$$

   where we have posed $\tilde{\psi}_{(p,q)} = \widetilde{I}_{(p,q)}\varphi^i_{(p,q)}$

8. NUMERICAL RESULTS

8.1 Discretization

From a numerical point of view, it seems that two steps of the discretization of the problem are non classical :

1. the approximation of the operator $\widetilde{I}_{(p,q)}$ (for the computation of each RtR operator $\Lambda_{(p,q)}$) ;
2. the approximation of the operators $P(k_y)$ and $D(k_y)$ for each k_y (for the computation of the halfspace RtR operator Λ^{H})

8.1.1 Discretization of ($\mathscr{Q}^{\varphi}_{(p,q)}$)

The choice for the discretization of the problem is taken mainly because of the resolution of the non standard integral equation ($\mathscr{Q}^{\varphi}_{(p,q)}$). Indeed, a priori, the discretization of ($\mathscr{Q}^{\varphi}_{(p,q)}$) relies on

- the choice of an appropriate finite dimensional approximation space for W ;
- the construction of an appropriate approximation of the bilinear form $a(\cdot,\cdot)$.

One has to take into account that

- The operator $\mathbf{K}_{(p,q)}(k_x,k_y)$ is not known analytically and must be approximated numerically (this is related to the resolution of the cell problems ans the approximation of the operators $I^{0,\pm}(k_y)$, $I^{1,\pm}(k_y)$ defined in 31, $D(k_y)$) defined in (19) and $P(k_y)$ defined in (18) ;
- the approximation of $\mathbf{K}_{(p,q)}(k_x,k_y)$ (which depends smoothly on (k_x,k_y)) can be done only for discrete values of (k_x,k_y) : quadrature in (k_x,k_y) is required.

For this reason it seems to us that it is easier to work in a space of functions generated by basis functions which are tensor product, as we shall detail later. Moreover, we need in principle to deal with two constraints :

- the constrain of zero-mean value in the k-variable appearing in the definition of W ;
- the k_y-quasi-periodicity in the $y-$variable condition for the operators $P(k_y)$, $D(k_y)$, $K_{(p,q)}(k_x,k_y)$,...

It appears difficult to take into account these two constraints strongly in the approximation space, especially if we want to work in the "(x,k)-tensor product space".

We have chosen to take into account the quasi-periodicity condition weakly by using a mixed variational formulation and a mixed finite element approximation of the cell problems, which allows us to construct approximate operators

$$P_h(k_y) \quad \text{and} \quad D_h(k_y) \in \mathscr{L}(\mathscr{V}_h)$$

where $\mathscr{V}_h$ is a finite dimensional subspace of $L^2(\Sigma_0)$ (typically with piecewise polynomial functions). Using appropriate discretization of the periodicity cell problems, we can construct

$$I_h^{0,\pm}(k_y) \quad \text{and} \quad I_h^{1,\pm}(k_y) \in \mathscr{L}(\mathscr{V}_h).$$

See Section 8.1.2 for more details. Finally, by construction,

$$K^h_{(p,q)}(k_x,k_y) \in \mathscr{L}(\mathscr{V}_h).$$

Then, we shall construct the approximation subspace of V as a subspace of :

$$V_h = L^2(-\frac{\pi}{L},\frac{\pi}{L},\mathscr{V}_h)$$

and then the approximation subspace of W as a subspace of :

$$W_h = \{\hat{\psi}_h \in V_h, \quad \int \hat{\psi}_h(\cdot,k)\,dk = 0\}.$$

It suffices then to choose the same approximation subspaces for V' and W' as respectively the ones for V and W and to replace the duality product in (45) by the scalar product in $L^2(\Sigma_0)$.

The semi-discrete (in y) problem is :

Find $\hat{\psi}^0_h \in W_h$, such that for all $\hat{\theta}^0_h \in W_h$

$$a^h_{(p,q)}(\hat{\psi}^0_h,\hat{\theta}^0_h) = \ell(\hat{\theta}^0_h), \tag{46}$$

where

$$a^h_{(p,q)}(\hat{\psi}^0_h,\hat{\theta}^0_h) := \int_{-\pi/L}^{\pi/L} dk_x\Big[\Big(\hat{\psi}^0_h(\cdot;k_x);\hat{\theta}^0_h(\cdot;k_x)\Big)_{L^2} - \int_{-\pi/L}^{\pi/L}\Big(\mathbf{K}^h_{(p,q)}(k_x,k_y)\,\hat{\psi}^0_h(\cdot;k_y);\hat{\theta}^0_h(\cdot;k_x)\Big)_{L^2} dk_y\Big]$$

$$\ell(\hat{\theta}^0_h) := \int_{-\pi/L}^{\pi/L}\Big(\hat{g}_\varphi(\cdot;k_x);\hat{\theta}^0_h(\cdot;k_x)\Big)_{L^2} dk_x.$$

For the discretization in k, we divide the interval

$$\Big]-\frac{\pi}{L},\frac{\pi}{L}\Big]$$

into N equal intervals of length

$$\Delta k = \Delta k_x = \Delta k_y = \frac{2\pi}{NL}, \quad k_l = l\Delta k$$

Given $q \in \mathbb{N}$, we introduce

$$\mathbf{P}_{N,q} = \{w \in L^2(-\frac{\pi}{L},\frac{\pi}{L}),\ \forall l,\ w|_{[k_l,k_{l+1}]} \in \mathscr{P}_q, \int_{-\pi/L}^{\pi/L} w(k)\,dk = 0\}$$

where $\mathscr{P}_q$ is the set of polynomials of degree q.

Let us now introduce

$$\left| \begin{array}{l} \{\theta_i^h(y),\ 1 \le i \le N_h\} \text{ a basis of } \mathscr{V}_h \\ \{w_j(k),\ 1 \le j \le N_k\} \text{ a basis of } \mathbf{P}_{N,q}, \end{array} \right.$$

with $N_k = N(q+1)-1$ and consider the approximation space of W

$$W_{h,N,q} = \mathbf{P}_{N,q} \otimes \mathscr{V}_h = \mathrm{span}\left[w_j(k)\theta_i^h(y)\right]_{1\le i\le N_h,\,1\le j\le N_k}$$

with dimension $N_t = N_h \times N_k$.

Finally, we consider a quadrature formula in $[0,1]$

$$\int_0^1 f(\tau)\,d\tau \equiv \sum_{m=1}^{M} \omega_m f(\tau_m), \quad 0 \le \tau_1 < \ldots < \tau_M \le 1$$

and introduce the quadrature points

$$k_l^m = k_l + \tau_m \Delta k.$$

The fully-discret problem that we solve is :

Find $\hat{\psi}_h^0 \in W_{h,N,q}$, such that $\forall i[[1,N_h]]$, $\forall j \in [[1,N_k]]$

$$a_{(p,q)}^{h,N,q,M}(\hat{\psi}_h^0, w_j\theta_i^h) = \ell(w_j\theta_i^h), \qquad (47)$$

where

$$\left| \begin{array}{l} a_{(p,q)}^{h,N,q,M}(\hat{\psi}_h^0, w_j\theta_i^h) := \\ (\frac{2\pi}{NL})^2 \sum_{l_x=1}^{N}\sum_{m_x=1}^{M} \omega_{m_x} w_j(k_{l_x}^{m_x})\Big[\frac{NL}{2\pi}(\hat{\psi}_h^0(\cdot;k_{l_x}^{m_x});\theta_i^h)_{L^2} \\ -\sum_{m_y=1}^{M}\omega_{m_y}\sum_{l_y=1}^{N}(\mathbf{K}_{(p,q)}^h(k_{l_x}^{m_x},k_{l_y}^{m_y})\,\hat{\psi}_h^0(\cdot;k_{l_y}^{m_y});\theta_i^h)_{L^2}\Big] \\ \ell(w_j\theta_i^h) := \\ \frac{2\pi}{NL}\sum_{l_x=1}^{N}\sum_{m_x=1}^{M}\omega_{m_x}\left(\hat{g}_\varphi(\cdot;k_{l_x}^{m_x});\theta_i^h\right)_{L^2} w_j(k_{l_x}^{m_x}). \end{array} \right.$$

8.1.2 Discretization of the Half Waveguide Problem $(\hat{\mathscr{P}}_{k_y}^{\mathrm{H}})$

For each quasi-period k_y (typically the quadrature points introduced in the previous section), we need to construct discrete approximations of

$$P(k_y), D(k_y) \text{ and also } I^{0,\pm}(k_y) \text{ and } I^{1,\pm}(k_y)$$

The first two quantities require approximations to the operators $T_{k_y}^{ij}$, $i = 0,1$, $j = 0,1$.
We introduce a regular (for simplicity) 1D mesh of Σ_0 (and then Σ_1 by periodicity) made of N_h equal segments of length $h > 0$. We introduce the same mesh for Σ_1^+ and Σ_1^- to keep the double symmetry property of the periodicity cell. We approximate

$$H_k^{-1/2}(\Sigma_0) \quad (\text{resp. } H_{aa}^{1/2}(\Sigma_1^{\pm})')$$

by the subspace $\mathscr{V}_h$ of piecewise constant functions on this mesh.

The approximate operators

$$T_{k_y,h}^{ij}, P_h(k_y), D_h(k_y) \text{ and also } I_h^{0,\pm}(k_y) \text{ and } I_h^{1,\pm}(k_y)$$

will be constructed as operators in $\mathscr{L}(\mathscr{V}_h)$ and thus are represented by $N_h \times N_h$ matrices.

For solving the cell problems (21-22-23) (we have $2N_h$ problems of this type), we first rewrite them as a (∇, div) first order system, use $H(\mathrm{div}) \times L^2$ mixed formulation and discretize the resulting variational problem with the lowest order Raviart-Thomas mixed finite elements [27] on the doubly-symmetric periodic mesh of $\mathscr{C}$ (i.e. the "traces" of this mesh on Σ_0, Σ_1 and $\Sigma_1^{\pm}$ coïncide with the 1D mesh introduced above).

The advantage of such a choice is that both the traces of the scalar unknown and of its normal derivative are degrees of freedom of the method and both belong to $\mathscr{V}_h$, so that the operators

$$T_{k_y,h}^{ij}, I_h^{0,\pm}(k_y) \text{ and } I_h^{1,\pm}(k_y)$$

are naturally in $\mathscr{L}(\mathscr{V}_h)$.

To determine $P_h(k_y)$, we solve the discrete problem

$$\text{Find } X \in \mathscr{L}(\mathscr{V}_h) \text{ such that } \left| \begin{array}{l} \mathscr{T}_h(k_y, X) = 0 \\ \rho(X) < 1 \end{array} \right. \qquad (48)$$

where

$$\mathscr{T}_h(k_y,X) = T^{10}_{k_y,h}\left(T^{11}_{k_y,h}\right)^{-1}X^2 + \left(T^{11}_{k_y,h}\right)^{-1}T^{01}_{k_y,h} + \left(T^{00}_{k_y,h} - T^{01}_{k_y,h}\left(T^{11}_{k_y,h}\right)^{-1}T^{10}_{k_y,h} - \left(T^{11}_{k_y,h}\right)^{-1}\right)X.$$

is a matrix quadratic equation.

The resolution of (48) can be done using one of the two following methods (see [21] for more details about these methods) :
(i) a spectral approach ;
(ii) a modified Newton method.
The spectral approach (i) leads to solve the matrix quadratic eigenvalue problem :

$$\lambda^2 T^{10}_{k_y,h}\left(T^{11}_{k_y,h}\right)^{-1} + \left(T^{11}_{k_y,h}\right)^{-1}T^{01}_{k_y,h} + \lambda\left(T^{00}_{k_y,h} - T^{01}_{k_y,h}\left(T^{11}_{k_y,h}\right)^{-1}T^{10}_{k_y,h} - \left(T^{11}_{k_y,h}\right)^{-1}\right) = 0$$

with the condition

$$|\lambda| < 1.$$

We can show that the solutions of this matrix quadratic eigenvalue problem are associated by pairs $(\lambda, 1/\lambda)$. We keep exactly those N_h eigenvalue and eigenvector pairs, $\{(\lambda_1,\varphi_1^k),\cdots,(\lambda_{N_h},\varphi_{N_h}^k)\}$, for which $|\lambda_i| < 1$, and discarding the rest (also numbering N_h).

One can also think about solving directly the nonlinear equation (48) using a Newton's algorithm for instance. The difficulty is to take into account the constraint about the spectral radius. That is why we have proposed a heuristic modified Newton's algorithm where a projection step is applied at each step of the algorithm. The algorithm we suggest consists in constructing, from the initial guess $P_h^0(k_y) = 0$, the sequence $P_h^n(k_y)$ defined by :

- Compute δP_h^{n+1} solution of the Lyapunov equation :

$$T^{10}_{k_y,h}\left(T^{11}_{k_y,h}\right)^{-1}\left(P_h^n(k_y)\delta P_h^{n+1} + \delta P_h^{n+1}P_h^n(k_y)\right) + \left(T^{00}_{k_y,h} - T^{01}_{k_y,h}\left(T^{11}_{k_y,h}\right)^{-1}T^{10}_{k_y,h} - \left(T^{11}_{k_y,h}\right)^{-1}\right)\delta P_h^{n+1} = \mathscr{T}_h(k_y, P_h^n(k_y)).$$

- Compute $\tilde{P}_h^{n+1}(k_y) = P_h^n(k_y) - \delta P_h^{n+1}$,
- If $\rho(\tilde{P}_h^{n+1}(k_y)) < 1$, keep

$$P_h^{n+1}(k_y) = \tilde{P}_h^{n+1}(k_y),$$

if $\rho(\tilde{P}_h^{n+1}(k_y)) \geq 1$, take

$$P_h^{n+1}(k_y) = \tilde{P}_h^{n+1}(k_y)/\rho(\tilde{P}_h^{n+1}(k_y))$$

- Stop the algorithm when :

$$\frac{\|\delta P_h^{n+1}\|}{\|P_h^n(k_y)\|} \quad \text{is small enough.}$$

The solution $P_h(k_y)$ is expected to be the limit of the sequence $P_h^n(k_y)$.

Once, $P_h(k_y)$ is determined, the operator $D_h(k_y)$ is obtained using the discrete version of the first relation of (26)

$$D_h(k_y) = \left(T^{11}_{k_y,h}\right)^{-1}\left(P_h(k_y) - T^{01}_{k_y,h}\right)$$

Finally the approximation of the RtR operator $\Lambda_h^{\mathrm{w}}(k_y)$ is obtained by

$$\Lambda_h(k_y) = T^{00}_{k_y,h} + T^{10}_{k_y,h}D_h(k_y)$$

and the approximation of the solution of $(\hat{\mathscr{P}}^{\mathrm{H}}_{k_y})$ can be constructed cell by cell using a discrete version of (20) once we have the $2N_h$ basic solutions of the cell problems (21-22-23), $P_h(k_y)$ and $D_h(k_y)$.

For the approximation of the halfspace RtR operator Λ^{H} (resp. the approximation of the solution of $(\mathscr{P}^{\mathrm{H}})$), it suffices to apply the theorem 5.8 (resp. the relation (14)) taking into account the discretization of the k_y-variables introduced in Section 8.1.1.

Since we do not have actual solutions to the problem of wave propagation in a general locally perturbed periodic media, we cannot say for certain that the solution we obtain by following the procedure described in the previous sections is indeed the solution. However, we make comparisons to check self-consistency in all the numerical examples.

8.2 Particular Case of Homogeneous Media

We first apply the procedure described in the previous sections to the case of a locally perturbed homogenous media, namely the case where ρ_p is a constant. Note that for an homogeneous media, it is quite original to use a square boundary to construct the DtN operator.

We present in Figure 17(a) the source f whose support is included in $\Omega^i = [-0.5, 0.5]^2$, the index $\rho_p = 1$ and we suppose the period of the media equal to 1 (which means that all the computations will be done in a cell of periodicity 1). Using the algorithm described previously, the RtR operator can be computed and the interior problem can be solved. We show the interior solution in Figure 17(b) with $\omega = 5$ and $\varepsilon = 1$. Thus, even with a squared artificial boundary, we recover the resolution symmetry of the solution.

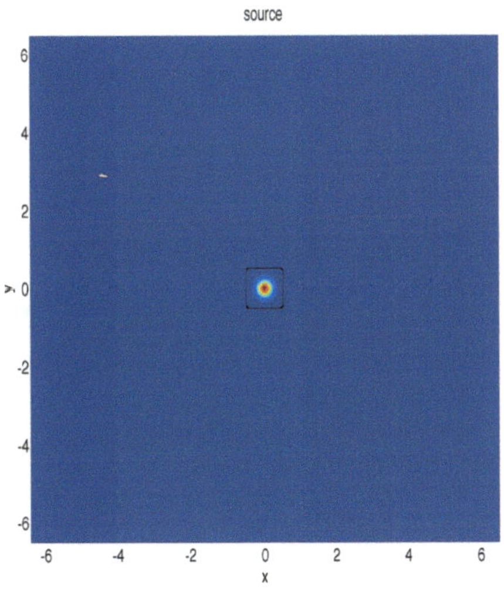

(a) The source and the artificial boundary Σ^i

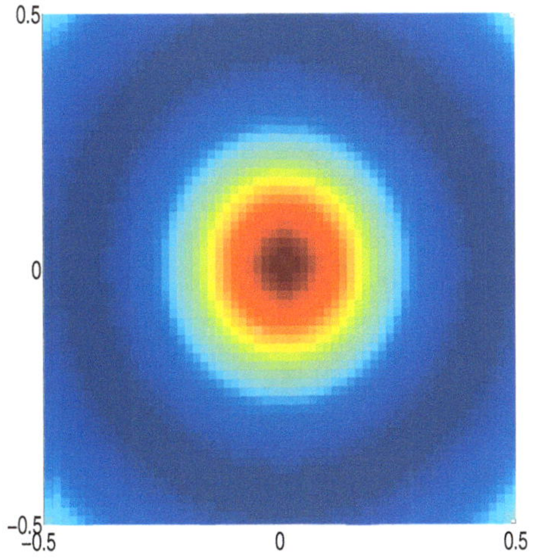

(b) The solution u^i in Ω^i

FIGURE (**17**): The source f on $[-6.5, 6.5]^2$ (its support is included in $\Omega^i = [-0.5, 0.5]^2$) and the solution u^i of in $\Omega^i(\mathscr{P}^i)$ (computed thanks to the construction of RtR operator Λ)

Finally to build the solution everywhere, we use the algorithm presented in Section 7. Note that in this case, since the source is with double symmetry, φ^i is too : $\varphi^i = \varphi^i_{(s,s)}$. Here again, we recover the resolution symmetry of the solution.

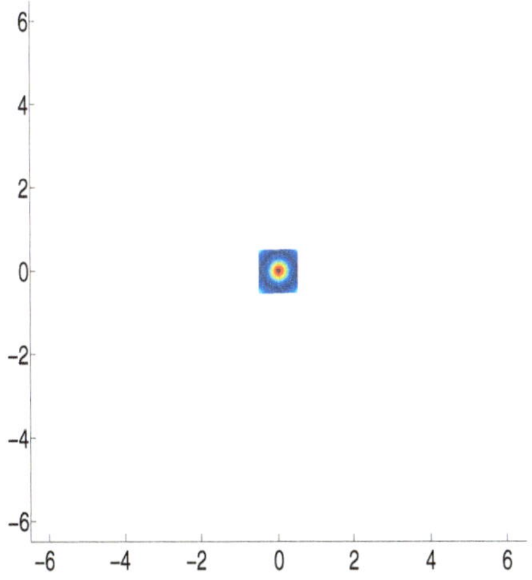

FIGURE (**18**): The interior solution u^i

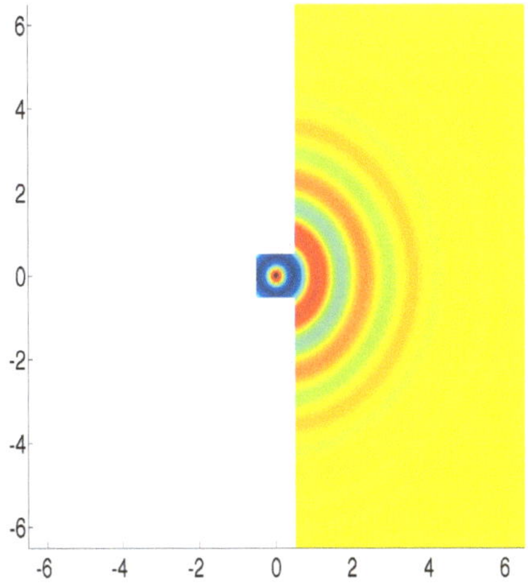

FIGURE (**19**): Its restriction to Ω^H is computed thanks to a halfspace problem.

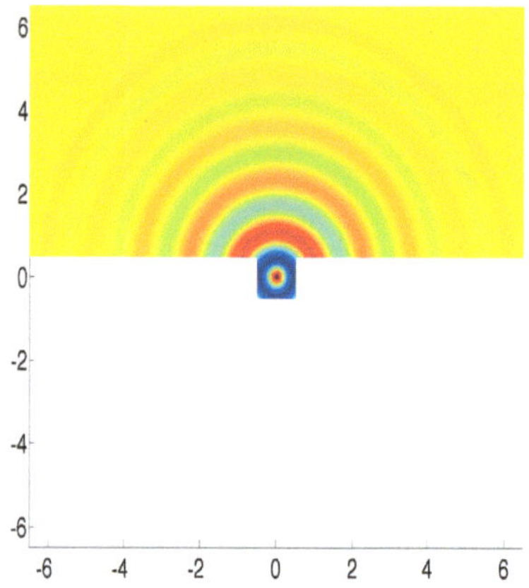

FIGURE (**20**): Its restriction to this halfspace is deduced by symmetry.

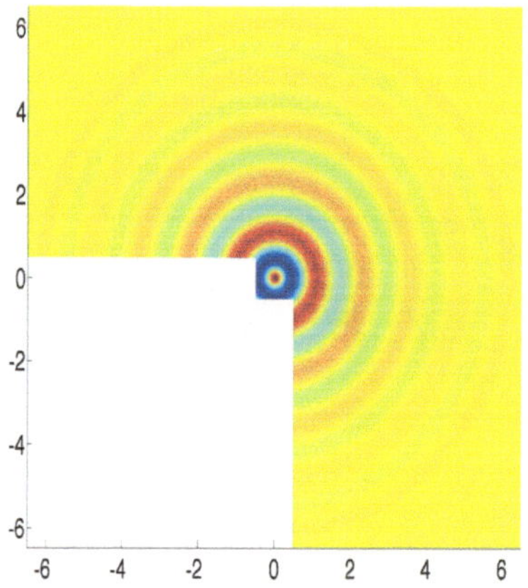

FIGURE (**21**): The restrictions to the quarter plane corresponds thanks to $\mathscr{E}_{(s,s)}$.

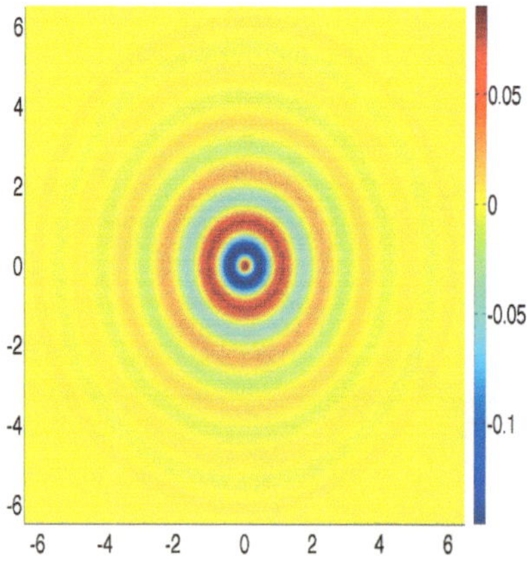

FIGURE (**22**): The solution in the whole space.

8.3 More General Periodic Media

We can apply now our algorithm to a general periodic media, whose refraction index is represented Figure (**23**), the source is given in Figure (**24**). The period here is equal to 1. After computing the DtN operator, the interior problem can be solved and we represent the interior solution Figure (**25**). We use finally the same algorithm as previously for the reconstruction of the solution outside Ω^i (see Figure (**26**)).

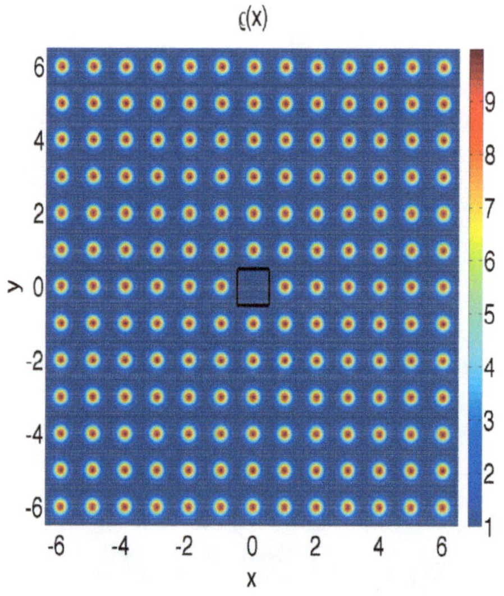

FIGURE (**23**): The locally perturbed periodic media with the artificial boundary Σ^i.

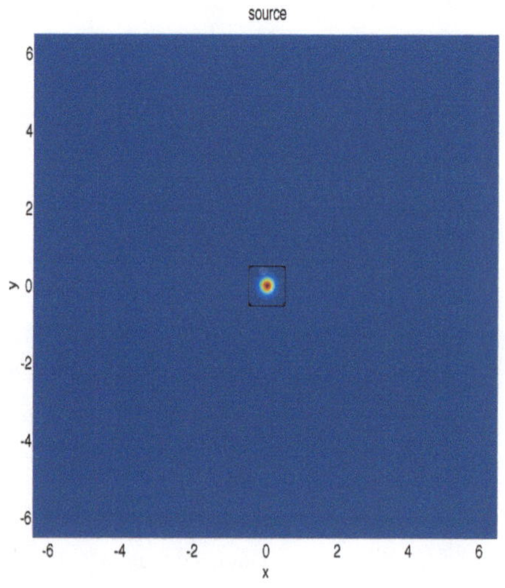

FIGURE (**24**): The source whose support is compact, with the artificial boundary Σ^i.

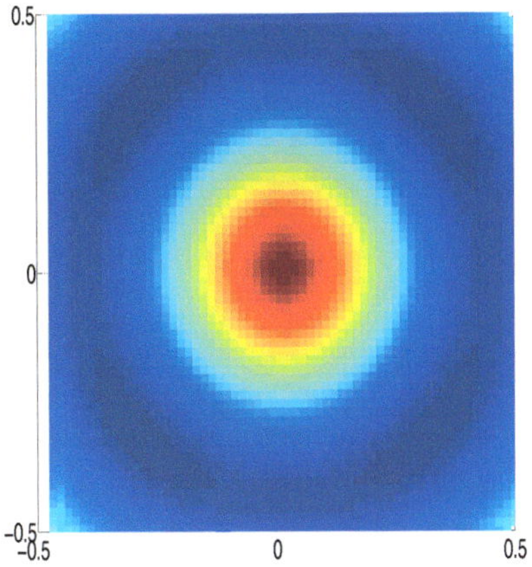

FIGURE (**25**): The interior solution u^i in $\Omega^i = [-0.5, 0.5]^2$ for $\omega = 10$ and $\varepsilon = 0.1$.

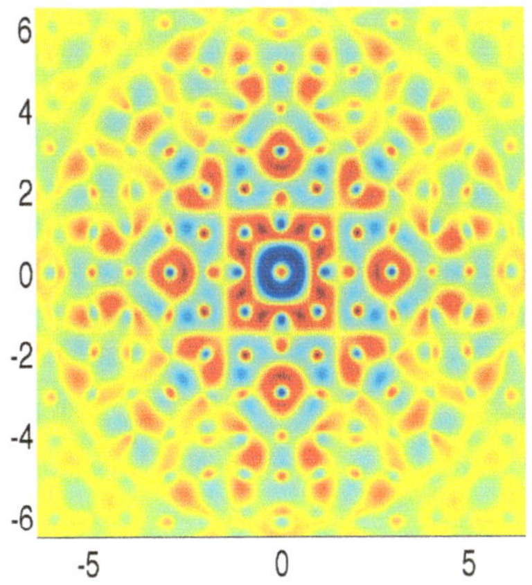

FIGURE (**26**): The solution in the whole space in the case for $\Omega^i = [-0.5, 0.5]^2$, $\omega = 10$, $\varepsilon = 0.1$.

8.4 Invariance with Respect to the Choice of Σ^i and $\mathscr{C}$

The solution of the whole problem has to be independant of the choice of the artificial boundary Σ^i and the periodicity cell $\mathscr{C}$. One easy way to validate the method is to change their size and check that the solution is the same. For the same media as previously, we choose a bigger boundary Σ^i as shown in Figure (**27**). All the computations are done in a periodicity cell whose side is equal to 2. The new interior solution u^i is represented Figure (**28**) and the solution is finally reconstructed in the region $[-6.5, 6.5]^2$ as shown in Figure (**29**). We recover the solution computed in the previous section and shown in Figure (**26**).

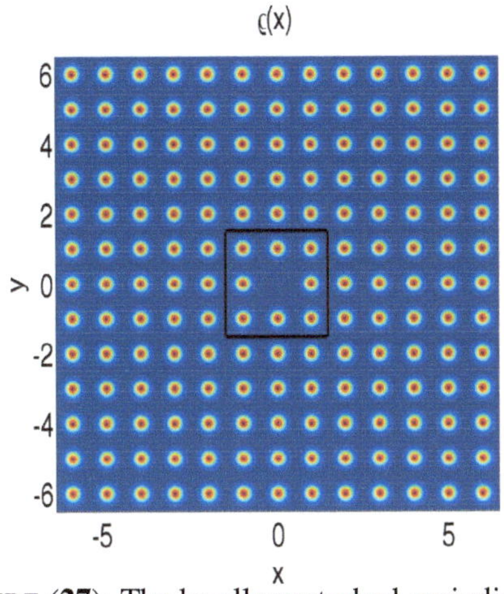

FIGURE (**27**): The locally perturbed periodic media with the artificial boundary Σ^i.

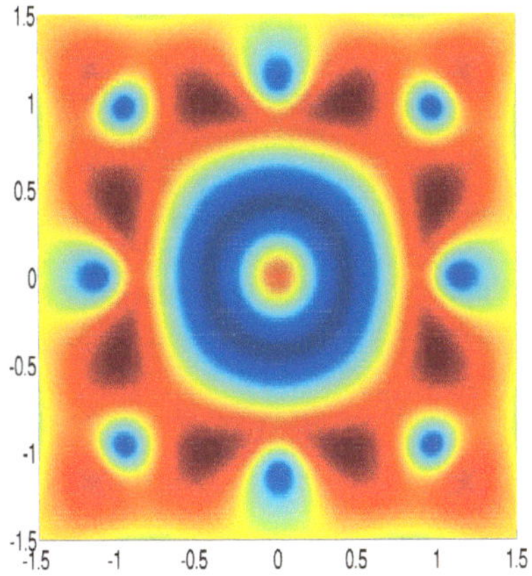

FIGURE (**28**): The interior solution u^i in $\Omega^i = [-0.5, 1.5]^2$ with $\omega = 10$, $\varepsilon = 0.1$.

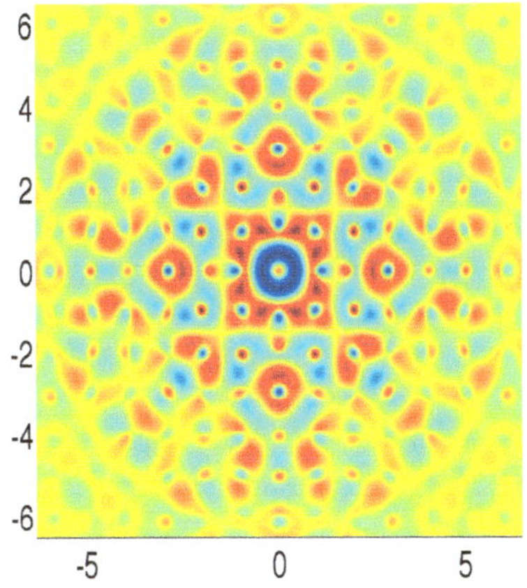

FIGURE (**29**): The solution in the whole space with $\Omega^i = [-0.5, 1.5]^2$, $\omega = 10$, $\varepsilon = 0.1$.

REFERENCES

[1] Abboud T. Electromagnetic waves in periodic media. In : Second International Conference on Mathematical and Numerical Aspects of Wave Propagation (Newark, DE, 1993), pp. 1-9. SIAM, Philadelphia, PA, 1993.

[2] Bayliss A, Turkel E. Radiation boundary conditions for wave-like equations. Comm Pure Appl Math 1980 ; 33 : 707-725.

[3] Bérenger JP. A perfectly matched layer for the absorption of electromagnetic waves. J Comput Phys 1994 ; 114 :185-200.

[4] Collino F, Ghanemi S, Joly P. Domain decomposition method for harmonic wave propagation : a general presentation. Comput Methods Appl Mech Eng 2000 ; 184 : 171-211.

[5] Després B. Décomposition de domaine et problème de Helmholtz. C R Acad Sci Paris Sér I Math 1990 ; 311 : 313-316.

[6] Després B. Domain decomposition method and the Helmholtz problem. In : Mathematical and numerical aspects of wave propagation phenomena (Strasbourg, 1991). SIAM, Philadelphia, PA, 1991, pp. 44-52.

[7] Ehrhardt M, Zheng C. Exact artificial boundary conditions for problems with periodic structures. J Comput Phys 2008 ; 227 : 6877-6894.

[8] Engquist B, Majda A. Radiation boundary conditions for acoustic and elastic wave calculations. Comm Pure Appl Math 1979 ; 32 : 314-358.

[9] Figotin A, Klein A. Localized classical waves created by defects. J Stat Phys 1997 ; 86 : 165-177.

[10] Figotin A, Klein A. Localization of light in lossless inhomogeneous dielectrics. J Opt Soc Am A 1998 ; 15 : 1423-1435.

[11] Figotin A, Klein A. Midgap defect modes in dielectric and acoustic media. SIAM J Appl Math 1998 ; 58 : 1748-1773.

[12] Fliss S. Etude mathématique et numérique de la propagation des ondes dans des milieux périodiques localement perturbés. Palaiseau, 2009.

[13] Fliss S, Joly P. Exact boundary conditions for time-harmonic wave propagation in locally perturbed periodic media. Appl Numer Math 2009 ; 59 : 2155-2178.

[14] Hazard C, Lenoir M. On the solution of time-harmonic scattering problems for Maxwell's equations. SIAM J Math Anal 1996 ; 27 : 1597-1630.

[15] Hohage T, Schmidt F, Zschiedrich L. Solving time-harmonic scattering problems based on the pole condition I. Theory. SIAM J Math Anal 2003 ; 35 : 183-210.

[16] Hohage T, Schmidt F, Zschiedrich L. Solving time-harmonic scattering problems based on the pole condition II. Convergence of the PML method. SIAM J Math Anal 2003 ; 35 : 547-560.

[17] Jami A, Lenoir M. Formulation variationnelle pour le couplage entre une méthode d'éléments finis et une représentation intégrale. C R Acad Sci Paris Sér A-B 1977 ; 285 : A269-A272.

[18] Joannopoulos JD, Meade RD, and Winn JN. Photonic crystal – molding the flow of light. Princeton University Press, 1995.

[19] Johnson C, Nédélec JC. On the coupling of boundary integral and finite element methods. Math Comp 1980, 35 : 1063-1079.

[20] Johnson SG, Joannopoulos JD. Photonic crystal – The road from theory to practice. Kluwer Academic Publications, 2002.

[21] Joly P, Li J-R, Fliss S. Exact boundary conditions for periodic waveguides containing a local perturbation. Commun Comput Phys 2006 ; 1 : 945-973.

[22] Kuchment P. Floquet theory for partial differential equations, Operator Theory : Advances and Applications 60, Birkhäuser Verlag, Basel, 1993.

[23] Kuchment P. The mathematics of photonic crystals. Chapter 7 In : Mathematical modeling in optical science, Frontiers in applied mathematics 22. SIAM, Philadelphia, 2001.

[24] Kuchment P. On some spectral problems of mathematical physics. In : Partial differential equations and inverse problems, Contemp Math 2004 ; 362 : 241-276.

[25] Levillain V. Coupling integral equation methods and finite volume elements for the resolution of time harmonic maxwell's equations in three dimensional heterogeneous medium. Technical Report 222, Centre de Mathematiques appliquees, Ecole Polytechnique, 1990.

[26] Lions JL Magenes E. Problèmes aux limites non homogènes et applications. Vol. 1. Travaux et Recherches Mathématiques 17. Dunod, Paris, 1968.

[27] Raviart PA, Thomas JM. A mixed finite element method for 2nd order elliptic problems. In : Mathematical aspects of finite element methods (Proc. Conf., Consiglio Naz. delle Ricerche (C.N.R.), Rome, 1975), Lecture Notes in Math 606, pp. 292-315. Springer, Berlin, 1977.

[28] Sakoda K. Optical properties of photonic crystals. Springer Verlag Berlin, 2001.

[29] Soussi S. Convergence of the supercell method for defect modes calculations in photonic crystals. SIAM J Numer Anal 2005 ; 43 : 1175-1201.

[30] Tausch J, Butler J. Floquet multipliers of periodic waveguides via Dirichlet-to-Neumann maps. J Comput Phys 2000 ; 159 : 90-102.

[31] Weidmann J. Linear operators in Hilbert spaces, Graduate Texts in Mathematics 68. Springer-Verlag, New York, 1980.

CHAPTER 6

Fast Numerical Methods for Waves in Periodic Media

M. Ehrhardt[1] **and C. Zheng**[2]

[1] *Bergische Universität Wuppertal, Fachbereich Mathematik und Naturwissenschaften, Lehrstuhl für Angewandte Mathematik und Numerische Mathematik, Gaußstrasse 20, 42119 Wuppertal, Germany.*
[2] *Department of Mathematical Sciences, Tsinghua University, Beijing 100084, P.R. China.*

Abstract: Periodic media problems widely exist in many modern application areas like semiconductor nanostructures (e.g. quantum dots and nanocrystals), semi-conductor superlattices, photonic crystals (PC) structures, meta materials or Bragg gratings of surface plasmon polariton (SPP) waveguides, etc. Often these application problems are modeled by partial differential equations with periodic coefficients and/or periodic geometries.
In order to numerically solve these periodic structure problems efficiently one usually confines the spatial domain to a bounded computational domain (i.e. in a neighborhood of the region of physical interest). Hereby, the usual strategy is to introduce so-called *artificial boundaries* and impose suitable boundary conditions. For wave-like equations, the ideal boundary conditions should not only lead to well-posed problems, but also mimic the perfect absorption of waves traveling out of the computational domain through the artificial boundaries.
In the first part of this chapter we present a novel analytical impedance expression for general second order ODE problems with periodic coefficients. This new expression for the kernel of the Dirichlet-to-Neumann mapping of the artificial boundary conditions is then used for computing the bound states of the Schrödinger operator with periodic potentials at infinity. Other potential applications are associated with the exact artificial boundary conditions for some time-dependent problems with periodic structures. As an example, a two-dimensional hyperbolic equation modeling the TM polarization of the electromagnetic field with a periodic dielectric permittivity is considered.
In the second part of this chapter we present a new numerical technique for solving periodic structure problems. This novel approach possesses several advantages. First, it allows for a fast evaluation of the Sommerfeld-to-Sommerfeld operator for periodic array problems. Secondly, this computational method can also be used for bi-periodic structure problems with local defects. In the sequel we consider several problems, such as the exterior elliptic problems with strong coercivity, the time-dependent Schrödinger equation and the Helmholtz equation with damping.
Finally, in the third part we consider periodic arrays that are structures consisting of geometrically identical subdomains, usually called periodic cells. We use the Helmholtz equation as a model equation and consider the definition and evaluation of the exact boundary mappings for general semi-infinite arrays that are periodic in one direction for any real wavenumber. The well-posedness of the Helmholtz equation is established via the *limiting absorption principle* (LABP).
An algorithm based on the doubling procedure of the second part of this chapter and an extrapolation method is proposed to construct the exact Sommerfeld-to-Sommerfeld boundary mapping. This new algorithm benefits from its robustness and the simplicity of implementation. But it also suffers from the high computational cost and the resonance wave numbers. To overcome these shortcomings, we propose another algorithm based on a conjecture about the asymptotic behaviour of limiting absorption principle solutions. The price we have to pay is the resolution of some generalized eigenvalue problem, but still the overall computational cost is significantly reduced. Numerical evidences show that this algorithm presents theoretically the same results as the first algorithm. Moreover, some quantitative comparisons between these two algorithms are given.

Matthias Ehrhardt (ED)

1. INTRODUCTION

Nowadays periodic media problems exist in many modern application areas like semiconductor nanostructures (e.g. quantum dots and nanocrystals), semi-conductor superlattices [11], [68], photonic crystals (PC) structures [10], [43], [53], meta materials [60] or Bragg gratings of surface plasmon polariton (SPP) waveguides [31], [61]. In many cases these problems are modeled by partial differential equations (PDEs) on unbounded domains with periodic coefficients and / or periodic geometries.

The most interesting property of these periodic media, especially in optical applications, is the capability to select the ranges of frequencies of the waves that are allowed to pass or blocked in the waveguide ('frequency filter'). Waves in (infinite) periodic media only exist if their frequency lies inside these allowed continuous bands separated by forbidden gaps. This fact corresponds mathematically to the gap structure of the differential operator having so-called *pass bands* and *stop bands*. Numerical simulations are necessary for the design, analysis and finally optimization of the waveguiding periodic structures. E.g. in a typical application the wanted frequencies of defect modes are the eigenvalues of a PDE eigenvalue problem posed on an *unbounded domain* [28].

In order to numerically solve these equations efficiently it is a standard practice to confine the spatial domain to a bounded computational region (usually in the neighborhood of the domain of physical interest). Hence it is necessary to introduce so-called *artificial boundaries* and impose adequate boundary conditions. Note that even in the case of a bounded but large domain, it is a common practice to reduce the original domain to a smaller one by introducing artificial boundaries, for example, see [44]. This technique is especially beneficial if these generated *exterior domains* consist of a huge number of periodicity cells. For wave-like equations, the ideal boundary conditions should not only lead to well-posed problems, but also mimic the perfect absorption of waves leaving the computational domain through the artificial boundaries. Moreover, these boundary conditions should allow for an easy implementation and a fast, efficient and accurate evaluation of the *Dirichlet-to-Sommerfeld* (DtS) mapping is essential. In the literature these boundary conditions are usually called *artificial* (or transparent, non-reflecting in the same spirit). The interested reader is referred to the review papers [6], [26], [29], [30], [65] on this fundamental research topic.

Artificial boundary conditions (ABCs) for the Schrödinger equation and related problems has been a hot research topic for many years, cf. [6] and the references therein. Since the first exact ABC for the Schrödinger equation was derived by Papadakis [48] 25 years ago in the context of underwater acoustics, many developments have been made on the designing and implementing of various ABCs, also for multi-dimensional and nonlinear problems. However, the question of exact ABCs for periodic structures still remained open, and it is a very up-to-date research topic, cf. the current papers [23], [25], [40], [56], [63], [64], [72], [73].

Let us note that recently Zheng [78] derived exact ABCs for the Schrödinger equation of the form

$$iu_t + u_{xx} = V(x)u, \quad x \in \mathbb{R}, \tag{1a}$$

$$u(x,0) = u_0(x), \quad x \in \mathbb{R}, \tag{1b}$$

$$u(x,t) \to 0, \quad x \to \pm\infty. \tag{1c}$$

Here, the initial function $u_0 \in L^2(\mathbb{R})$ is assumed to be compactly supported in an interval $[x_L, x_R]$, with $x_L < x_R$, and the real potential function $V \in L^\infty(\mathbb{R})$ is assumed to be sinusoidal on the interval $(-\infty, x_L]$ and $[x_R, +\infty)$. It is well-known that the system (1) has a unique solution $u \in C(\mathbb{R}^+, L^2(\mathbb{R}))$ for bounded potentials (cf. [50], e.g.):

Theorem 1. *Let $u_0 \in L^2(\mathbb{R})$ and $V \in L^\infty(\mathbb{R})$. Then the system (1) has a unique solution $u \in C(\mathbb{R}^+, L^2(\mathbb{R}))$. Moreover, it is a unitary evolution i.e. the "energy" is preserved:*

$$\|u(.,t)\|_{L^2(\mathbb{R})} = \|u_0\|_{L^2(\mathbb{R})}, \forall t \geq 0.$$

We remark that a recent paper [22] derives approximate ABCs for (1) with a more general class of possibly unbounded potentials.

In [78] Zheng considered the periodic potentials

$$V(x) = V_L + 2q_L \cos\frac{2\pi(x_L - x)}{S_L}, \ \forall x \in (-\infty, x_L],$$

$$V(x) = V_R + 2q_R \cos\frac{2\pi(x - x_R)}{S_R}, \ \forall x \in [x_R, +\infty),$$

where S_L and S_R are the periods, V_L and V_R are the average potentials, and the nonnegative numbers q_L and q_R relate to the amplitudes of sinusoidal part of the potential function V on $(-\infty, x_L]$ and $[x_R, +\infty)$, respectively.

Though *absorbing boundary conditions* (ABCs) for wave-like equations have been a hot research issue for many years and many developments have been made on their designing and implementing, the question of exact ABCs for periodic structure problems is not fully settled yet. Some progresses can be found in the recent research articles [19], [20], [21], [22], [23], [25], [40], [56], [58], [63], [64], [72], [73], [74] and [78]. For a comprehensive review on the theory of waves in locally periodic media including a survey on physical applications we refer the interested reader to [27].

In the existing literature *frequency domain methods* (FDMs) are usually considered for wave problems with periodic structures [39]. These methods are able to exploit the special geometric structure and are based on an eigenmode expansion in every longitudinally uniform cell. Frequently, the FDMs are used in conjunction with the *perfectly matched layer* (PML) [12] technique for dealing with unbounded domains. Afterwards the *bidirectional beam propagation methods* (BiBPMs) [34] were introduced. Like the FDMs, they can utilize the periodic geometry but additionally they (and also the *eigenmode expansion methods* in [12] and [34]) are able to resolve the multiple reflections at the longitudinal interfaces.

The methods of Jacobsen [36] and Yuan & Lu [72] were developed to be more efficient than the eigenmode expansion methods, because it turns out that solving the eigenmodes in each segment is quite time consuming. Recently, a *DtN mapping method* [71] was developed by Yuan and Lu that is more accurate than the BiBPMs, since this approach works (mostly) without any approximation. In [73] the efficiency of this sequential DtN approach was further improved by a *recursive doubling process* for the DtN map.

In this chapter we study a numerical method for the *Helmholtz equation*

$$-\Delta u(\mathbf{x}) + (V - z)u(\mathbf{x}) = f(\mathbf{x}). \qquad (2)$$

Here z denotes a complex parameter, and $V = V(\mathbf{x})$ is a sufficiently smooth real function bounded from both below and above. The domain of definition and the function V are assumed to be periodic at least on some part of the region.

The Helmholtz equation is one of the fundamental equations of mathematical physics and models time-harmonic wave propagation. In many cases, the Helmholtz equation (2) is posed on the unbounded domain $\mathbb{R}^2$ and solved as a boundary value problem with some radiation boundary conditions, for example, see [41].

In some special cases [36] it is possible to obtain analytic expressions of the solution, but in general, the Helmholtz equation (2) has to be solved numerically. However, If the number of periodic cells is large, then a direct discretization of the whole domain involves a huge number of unknowns which makes it costly and even impractical from an implementational point of view. In this chapter our goal is to find a smart resolution without naively solving the whole domain problem.

The most interesting property of periodic arrays, especially in optical applications in nano- and microtechnology, is the capability of selecting waves in a range of frequencies that are allowed to pass or blocked through the media. Waves in periodic arrays only exist when their frequency lies inside some allowed continuous bands separated by forbidden gaps. This fact corresponds mathematically to the dispersion diagram of suitable differential operator having so-called *pass bands* and *stop bands*. Since the governing wave equation is either of periodic variable coefficient, or defined on a domain consisting of periodic subregions, theoretical analysis is very limited, and numerical simulation is a fundamental tool for the design, analysis and finally optimization of the periodic arrays.

In many cases some defect cells are artificially introduced into a perfect periodic array for some additional interesting property. For example, if the defect cells are properly designed, some defect modes [59] can exist for certain frequencies in the band gaps. This phenomena has many important applications, e.g. in light emitting devices (LEDs) and photonic circuits [52].

The organization of this chapter is as follows. In Section 2, we present an elegant analytical expression of the impedance operator for problems with periodic coefficients. In Section 3 we use this result to compute bound states for the Schrödinger operator. In Section 4 we show how the results can be generalized to the time-dependent Schrödinger equation, a diffusion equation and a second order hyperbolic equation and present a concise numerical example.

In the sequel of the chapter we turn our considerations to more complicated periodic structures. We consider in Section 5 the Helmholtz equation (2) without the source term $f(\mathbf{x})$ on an array that is periodic in one direction and perform a cell analysis. Next, we explain how an improvement to the approach of Yuan and Lu [73] can be achieved by introducing *Sommerfeld-to-Sommerfeld (StS) mappings*. Moreover, we construct an efficient and robust method for numerically evaluating these StS operators. In Section 6 we present an application of the methods of Section 5 to waveguide problems discussing concisely the so-called pass and stop bands. We consider in Section 7 the transient Schrödinger equation on a semi-infinite array periodic in one direction and show how our fast evaluation method of Section 5 computes the exact StS mapping very efficiently. In Section 8 we discuss the numerical simulation of the time-dependent Schrödinger equation in two space dimensions with a bi-periodic potential function containing a defect. In Section 9 we return to the model problem of the Helmholtz equation now posed on a semi–infinite periodic array. Afterwards, we propose two different methods: the extrapolation method (Section 10) that is based on the limiting absorption principle (LABP) and the asymptotic method (Section 11) based on a conjecture about the asymptotic behavior of an LABP solution. Our proposed algorithm combines the doubling technique of Section 5 (now for evaluating the operator related to infinite arrays) and the limiting procedure (letting $\varepsilon \to 0$) with the extrapolation technique. The numerical tests in Section 12 supports the validity of our basic conjecture on how to identify the traveling Bloch waves which are compatible with the LABP, since from the numerical point of view the asymptotic method presents the same results as the extrapolation method does.

2. THE IMPEDANCE EXPRESSION

We consider the *general second order ODE*

$$-\frac{d}{dx}\left(\frac{1}{m(x)}\frac{dy}{dx}\right)+V(x)y=\rho(x)zy, \quad \forall x\geq 0, \tag{3}$$

where z denotes a complex parameter whose value space is to be determined. We assume that the functions $m(x)$, $V(x)$ and $\rho(x)$ are all *S-periodic* in $[0,+\infty)$ and *centrally symmetric* in each period, i.e.,

$$\begin{aligned} m(x)=m(S-x), \quad V(x)=V(S-x), \\ \rho(x)=\rho(S-x), \quad \text{a.e.} x\in[0,S]. \end{aligned} \tag{4}$$

The symmetry condition (4) simply implies that the *even* extensions of these functions to the whole real axis are still S-periodic. Moreover, we assume that the functions $m(x)$, $V(x)$ and $\rho(x)$ are sufficiently smooth and bounded, i.e. there exist several constants M_0, M_1, V_0 and ρ_0, such that

$$\begin{aligned} 0<M_0\leq m(x)\leq M_1<+\infty, \quad V(x)\geq V_0, \\ \rho(x)\geq\rho_0>0, \quad \forall x\in[0,S]. \end{aligned}$$

By introducing the new variable

$$w=\frac{1}{m(x)}\frac{dy}{dx},$$

the second order ODE (3) is transformed into a *first order ODE system*

$$\frac{d}{dx}\begin{pmatrix} w \\ y \end{pmatrix}=\begin{pmatrix} 0 & V(x)-\rho(x)z \\ m(x) & 0 \end{pmatrix}\begin{pmatrix} w \\ y \end{pmatrix}, \tag{5}$$

for $x\geq 0$. The first part of this chapter deals with the L^2-solution of (3) in $[0,+\infty)$. To be more precise, we want to analyze

1. for which parameter z does the general ODE (3) possess a nontrivial L^2-solution $y(x)$?
2. and in this case, is it possible to formulate a closed form of the impedance $I := y'(0)/y(0)$, i.e. the quotient of Neumann data over Dirichlet data evaluated at $x=0$?

For any two points x_1 and x_2, the ODE system (5) uniquely determines a linear transformation from the two-dimensional vector space associated with x_1, to the same space associated with x_2. We identify this transformation with the 2-by-2 matrix $T(x_1,x_2)$, which is an orthogonal change of basis matrix with periodicity properties. This matrix T satisfies the same form of equation as (5), namely:

$$\frac{d}{dx}T(x_1,x)=\begin{pmatrix} 0 & V(x)-\rho(x)z \\ m(x) & 0 \end{pmatrix}T(x_1,x), \tag{6}$$

for all $x_1\geq 0$ and $x\geq 0$.

Lemma 2. *The* transformation matrix T *possess the following properties:*

$$T(x,x)=I_{2\times 2}, \tag{7a}$$

$$\det T(x_1,x_2)=\det T(x_1,x_1)=1, \tag{7b}$$

$$T(x_2,x_3)T(x_1,x_2)=T(x_1,x_3), \tag{7c}$$

$$T(x_1+S,x_2+S)=T(x_1,x_2). \tag{7d}$$

Proof. We prove a more general result. Suppose

$$T'=AT,$$

where T is an n-by-n matrix-valued function. Then

$$(\det T)'=\mathrm{trace}(A)\det T.$$

We write T into a column of row vectors $T=(t_1^\top,\cdots,t_n^\top)^\top$.Then

$$(\det T)'=\sum_{k=1}^{n}\det((t_1^\top,\cdots,t_{k-1}^\top,(t_k^\top)',t_{k+1}^\top,\cdots,t_n^\top)^\top).$$

Since

$$(t_k^\top)'=\sum_{l=1}^{n}a_{kl}t_l^\top,$$

we have

$$
\begin{aligned}
&(\det T)' \\
&= \sum_{k=1}^{n} \det((t_1^\top, \cdots, t_{k-1}^\top, \sum_{l=1}^{n} a_{kl} t_l^\top, t_{k+1}^\top, \cdots, t_n^\top)^\top) \\
&= \sum_{k=1}^{n}\sum_{l=1}^{n} \det((t_1^\top, \cdots, t_{k-1}^\top, a_{kl} t_l^\top, t_{k+1}^\top, \cdots, t_n^\top)^\top) \\
&= \sum_{k=1}^{n}\sum_{l=1}^{n} \delta_{lk} \det((t_1^\top, \cdots, t_{k-1}^\top, a_{kl} t_l^\top, t_{k+1}^\top, \cdots, t_n^\top)^\top) \\
&= \sum_{k=1}^{n} a_{kk} \det((t_1^\top, \cdots, t_{k-1}^\top, t_k^\top, t_{k+1}^\top, \cdots, t_n^\top)^\top) \\
&= \mathrm{trace}(A) \det T.
\end{aligned}
$$

Here, A is a 2-by-2 matrix with zero diagonal. According to the above result we have

$$\det T(x_1, x_2) = \det T(x_1, x_1) = 1. \quad \square$$

We proceed with a small illustrating example showing that there might not exist any nontrivial L^2 solutions of the second order ODE (3) for some z in the complex plane.

Example 1. We assume for simplicity that all coefficients $m(x)$, $\rho(x)$ and $V(x)$ are constant, hence the problem is periodic with any period. E.g. setting

$$V = 0,\ \rho = 1,\ m = 1,\ z = 1$$

leads to a constant system matrix A in (5) or (6)

$$A = \begin{pmatrix} 0 & -1 \\ 1 & 0 \end{pmatrix}$$

Then the matrix $T(x_1, x_2)$ is simply $\mathrm{Exp}((x_2 - x_1)A)$, where Exp denotes the matrix exponential. The two eigenvalues of $T(0, S)$ have modulus 1 and thus prevent any nontrivial L^2 solution.
In fact, as revealed later in this section (see Figs. (**1**)-(**3**) and the related discussion, if z lies in one of the so-called *pass bands*, then there exists no nontrivial L^2 solution. In this constant coefficient example there is only one pass band $(0, +\infty)$ and $z = 1$ lies exactly in this interval.

2.1 The Impedance Expression

The next step of the construction of the ABC is to consider the polar form of the eigenvalue σ with modulus lower than 1 and express the L^2-bounded solution in order to finally extract the impedance condition. According to (7a), the matrix $T(0, S)$ has two eigenvalues $\sigma(\neq 0)$ and $1/\sigma$ with $|\sigma| \leq 1$. Their associated eigenvectors are denoted by $(c_+, d_+)^\top$ and $(c_-, d_-)^\top$. If $|\sigma| < 1$, then $T(0, x)(c_\pm, d_\pm)^\top$ yields two linearly independent solutions of the ODE system (5). By setting $\sigma = e^{\mu S}$ with $\mathrm{Re}\, \mu < 0$ it is straightforward to verify that $e^{\mp \mu x} T(0, x)(c_\pm, d_\pm)^\top$ are periodic functions. Therefore, we conclude that

$$y_+ := T(0, x)(c_+, d_+)^\top = e^{\mu x} e^{-\mu x} T(0, x)(c_+, d_+)^\top$$

is L^2-bounded, while

$$y_- := T(0, x)(c_-, d_-)^\top = e^{-\mu x} e^{\mu x} T(0, x)(c_-, d_-)^\top$$

is not. For the L^2-bounded solution y_+, the impedance I is thus given as

$$I := \frac{y_+'(0)}{y_+(0)} = m(0) \frac{c_+}{d_+}. \tag{8}$$

We remark that σ and $(c_+, d_+)^\top$ depend on z, and hence the impedance I also depends on z. In the sequel we will refer to σ as the *Floquet's factor* [9, 42, 51]. It typically reflects how fast the L^2-bounded solution of the ODE (3) decays to zero when x tends to $+\infty$: the smaller its modulus, the faster. Also note that $\sigma(\bar{z}) = \overline{\sigma(z)}$ and $I(\bar{z}) = \overline{I(z)}$ holds.
For any fixed z, the impedance $I = I(z)$ in (8) can be computed numerically with arbitrary high accuracy. First we solve the ODE system (5) to get $T(0, S)$. Then we compute σ and its associated eigenvector (c_+, d_+). Finally we use (8) to determine the impedance (cf. the impedance plots in Figs. (**5**), (**6**) for some values of z).
In general, the matrix $T(0, S)$ cannot be represented with a simple analytical expression in terms of the functions $m(x)$, $V(x)$ and $\rho(x)$. However, it can be computed sufficiently accurately by integrating the ODE (6) numerically (setting $x_1 = 0$) in the interval $[0, S]$ with the initial data $T(0, 0) = I_{2\times 2}$. Since this is a standard task, the detailed discussion is omitted here.

2.2 Numerical Tests A,B and C

In the sequel we present three numerical examples

$$
\begin{aligned}
&\textbf{Case A:} \quad m(x) = \rho(x) = 1, \quad V(x) = 2\cos(2x); \\
&\textbf{Case B:} \quad m(x) = \rho(x) = 1 + \cos(2x)/5, \\
&\qquad\qquad V(x) = \cos(2x); \\
&\textbf{Case C:} \quad m(x) = \rho(x) = 1 + \cos(2x)/5, \\
&\qquad\qquad V(x) = \sin(2x).
\end{aligned}
$$

that provide an illustration of what can be expected for the computation of the eigenvalues showing some of its expected properties. This is an important issue since it shows where the solution to the periodic equation is bounded in L^2.

Figs. (**1**)-(**3**) show the modulus of σ, which denotes the eigenvalue of $T(0,S)$ with a smaller modulus. We observe that apart from some intervals in the real axis, for any z in the complex plane, σ has a modulus less than 1, thus the second order ODE (3) has a nontrivial L^2-solution. Furthermore, it turns out that the ending points of these intervals are exactly the eigenvalues of the following *characteristic problem*:

Find $\lambda \in \mathbb{R}$ and a nontrivial $y \in C^1_{\text{per}}[0,2S]$, such that

$$-\frac{d}{dx}\left(\frac{1}{m(x)}\frac{dy}{dx}\right) + V(x)y = \rho(x)\lambda y. \quad (9)$$

We note that the symmetry condition (4) is not necessary for the above statements (In fact Case C does not satisfy (4)). We admit that the above statements have not been proven up to this time, but a vast number of other numerical evidences also support their validity.

If the coefficient functions $m(x)$, $V(x)$ and $\rho(x)$ satisfy the symmetry condition (4), then the characteristic problem (9) has a nice property: all the eigenvalues can be classified into two different groups

$$a_1 < a_2 < a_3 < \ldots \quad \text{and} \quad b_1 < b_2 < b_3 < \ldots,$$

where the eigenvalues a_r are associated with even eigenfunctions, and b_r with odd eigenfunctions. Besides, it holds that

$$a_1 < \min(a_2,b_1) \leq \max(a_2,b_1) < \min(a_3,b_2) < \ldots$$

For the Schrödinger equation (SE) with a periodic cosine potential, a special case of (3) with $m(x) = \rho(x) = 1$ and $V(x) = 2q\cos(2x)$, Zheng formulated in [78] a *conjecture* upon the impedance expression

$$I_{SE}(z) = -\sqrt[+]{-z+a_1}\prod_{r=1}^{+\infty}\frac{\sqrt[+]{-z+a_{r+1}}}{\sqrt[+]{-z+b_r}}, \quad (10)$$

$\operatorname{Im} z > 0$, where $\sqrt[+]{\cdot}$ denotes the branch of the square root with positive real part and the branch cut is set along the negative real axis. While the validity (10) was checked numerically in [78] the analytical proof was done recently by Zhang and Zheng in [76] Since formally $I_{SE}(\bar{z}) = \overline{I_{SE}(z)}$ for any z with $\operatorname{Im} z \neq 0$, it is thus tempting to generalize the above conjecture to our general second order ODE (3), i.e.,

$$I(z) = -\sqrt{m(0)\rho(0)}\,\sqrt[+]{-z+a_1}\cdot \cdot\prod_{r=1}^{+\infty}\frac{\sqrt[+]{-z+a_{r+1}}}{\sqrt[+]{-z+b_r}}, \quad \operatorname{Im} z \neq 0. \quad (11)$$

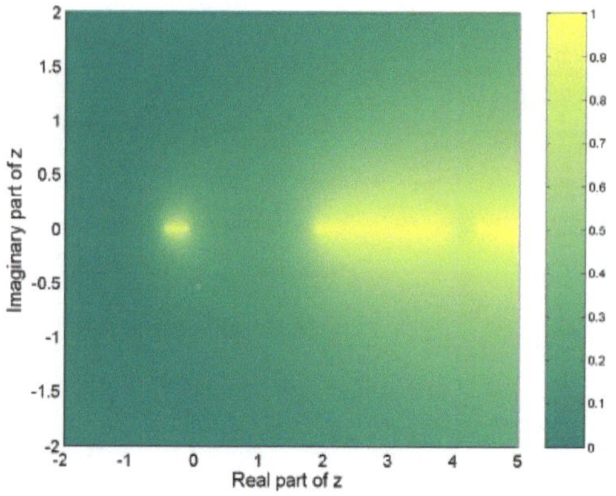

Fig. (**1**): **Case A:** Modulus of σ with respect to z.

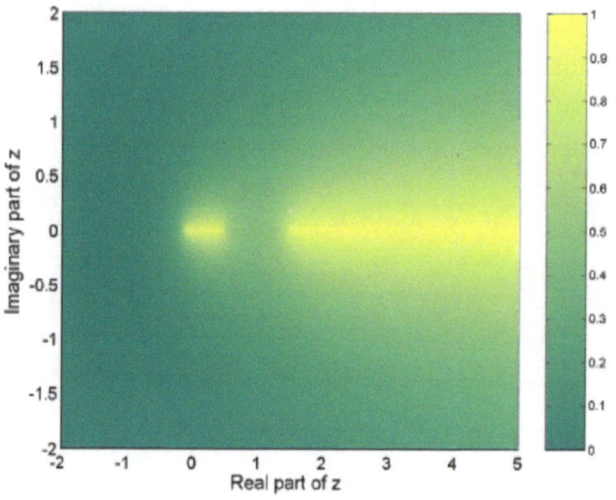

Fig. (**2**): **Case B:** Modulus of σ with respect to z.

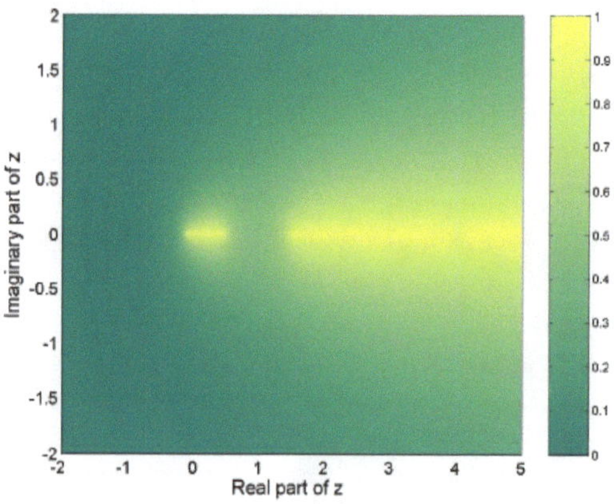

Fig. (**3**): **Case C:** Modulus of σ with respect to z.

Example 2. Let us briefly show how to obtain the constant coefficient case from the more general formula (11). The impedance for constant coefficients is given by

$$I(z) = -\sqrt{m\rho}\,\sqrt[+]{-z+\frac{V}{\rho}} \\ = -\sqrt[+]{m(V-\rho z)}. \tag{12}$$

All the eigenvalues of (9) are

$$\lambda_n = \frac{(\frac{n\pi}{S})^2 + mV}{m\rho}.$$

The eigenspace of λ_0 is the set of constant functions. For $n > 0$, the eigenvalue λ_n is degenerate. Its eigenspace is two-dimensional, spanned by $\cos(\pi x/S)$ and $\sin(\pi x/S)$. Notice that cos is even and sin is odd. Thus we have

$$a_n = \lambda_{n-1},\ n \geq 1, \quad \text{and} \quad b_n = \lambda_n,\ n \geq 1.$$

Since $a_{r+1} = b_r$ for any $r \geq 1$, the equation (11) yields the correct impedance expression

$$I = -\sqrt{m\rho}\,\sqrt[+]{-z+a_1} = -\sqrt[+]{m(V-\rho z)}.$$

2.3 Numerical Tests D and E

Let us consider another two numerical tests:

Case D: $m(x) = \rho(x) = 1,$

$$V(x) = \sum_{n=-\infty}^{+\infty} e^{-16(x-\pi/2-n\pi)^2},$$

Case E: $m(x) = 1,$

$$V(x) = 0, \quad \rho(x) = 1+\cos(2x)/5.$$

Case D corresponds to the Schrödinger equation with a periodic Gaussian potential, cf. Fig. (**4**), and Case E could arise from a second order hyperbolic wave equation in a periodic medium.

Figs. (**5**) and (**6**) present the impedance function $I(z)$ when z is very close to the real axis. It can be clearly seen that the impedance turns out to be either real or purely imaginary. Those real intervals with purely imaginary impedance are exactly those values of z for which the ODE (3) has no nontrivial L^2-solution. Recall that this statement does not rely on the symmetry property of the coefficients (4). In the engineering literature these intervals are called *pass bands*, while their complementary intervals are called *stop bands*. This notion of 'pass' and 'stop' refers to allowing and preventing the existence of traveling wave solutions.

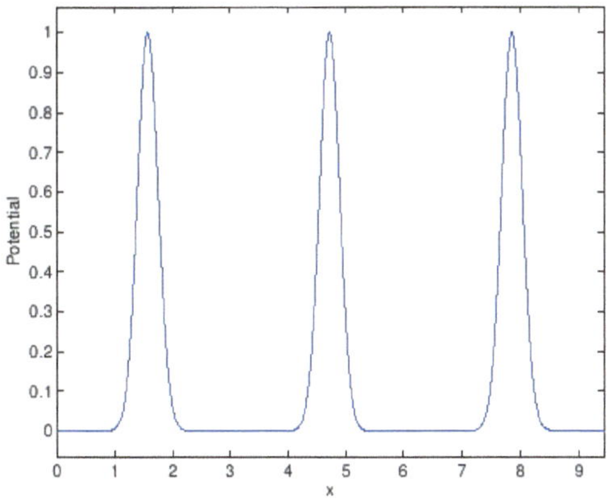

Fig. (**4**): Periodic Gaussian potential function $V(x) = \sum_{n=-\infty}^{+\infty} e^{-16(x-\pi/2-n\pi)^2}$.

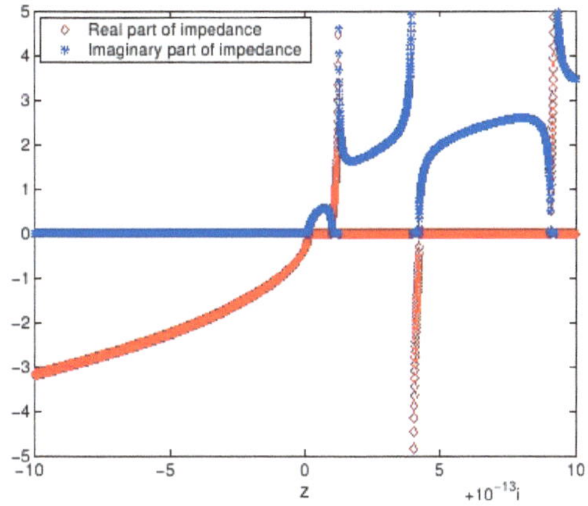

Fig. (**5**): **Case D:** Impedance $I(z)$ for the Schrödinger equation with a periodic Gaussian potential $V(x) = \sum_{n=-\infty}^{+\infty} e^{-16(x-\pi/2-n\pi)^2}$.

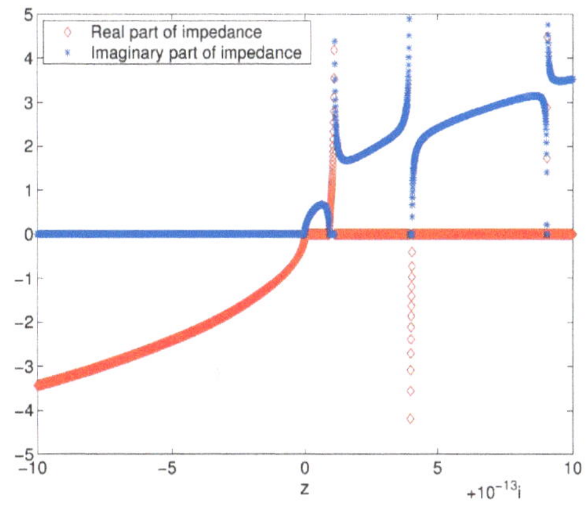

Fig. (**6**): **Case E:** Impedance plot for $m = 1$, $V = 0$ and $\rho = 1+\cos(2x)/5$.

Let us note that the impedance $I(z)$ becomes much more complicated as z approaches the real axis if one of the coefficient functions $m(x)$, $V(x)$ and $\rho(x)$ is not centrally symmetric, cf. (4).
Furthermore, we emphasize that the eigenvalues a_r and b_r can be computed with a high-accuracy solver for the characteristic problem (9). The first few eigenvalues are listed in Tables 1 and 2 with 6 digits. We observe that the relative difference between a_{r+1} and b_r decays quickly for increasing index r.

Table 1: **Case D:** The first several eigenvalues of (9) with $m(x) = \rho(x) = 1$ and $V(x) = \sum_{n=-\infty}^{+\infty} e^{-16(x-\pi/2-n\pi)^2}$.

r	a_{r+1}	b_r	r	a_{r+1}	b_r
0	1.30811(-1)		7	4.91344(1)	4.91486(1)
1	1.00842(0)	1.26431(0)	8	6.41442(1)	6.41386(1)
2	4.25428(0)	4.03081(0)	9	8.11403(1)	8.11423(1)
3	9.06010(0)	9.22586(0)	10	1.00142(2)	1.00141(2)
4	1.61965(1)	1.60886(1)	11	1.21141(2)	1.21141(2)
5	2.51111(1)	2.51730(1)	12	1.44141(2)	1.44141(2)
6	3.61574(1)	3.61260(1)	13	1.69141(2)	1.69141(2)

Table 2: **Case E:** The first few eigenvalues of (9), where $m(x) = 1$, $V(x) = 0$ and $\rho(x) = 1 + \cos(2x)/5$. Notice that $a_1 = 0$.

r	a_{r+1}	b_r	r	a_{r+1}	b_r
1	9.08164(-1)	1.10938	7	4.92536(1)	4.92537(1)
2	4.06748	3.98676	8	6.43296(1)	6.43296(1)
3	9.04010	9.06316	9	8.14157(1)	8.14157(1)
4	1.60896(1)	1.60838(1)	10	1.00512(2)	1.00512(2)
5	2.51315(1)	2.51328(1)	11	1.21618(2)	1.21618(2)
6	3.61880(1)	3.61877(1)	12	1.44735(2)	1.44735(2)

If the coefficient functions $m(x)$ and $\rho(x)$ are constant and $V(x) = 2q\cos(2x)$ with $q > 0$, then the general ODE (3) is reduced to the well-known *Mathieu's equation* [9, 51]. In this case, we obtain

$$a_1 < b_1 < a_2 < b_2 < a_3 < b_3 < \dots.$$

However, in general this property does not hold, and we can only expect the following

$$a_1 < \min(a_2, b_1) \le \max(a_2, b_1)) \\ < \min(a_3, b_2) \le \max(a_3, b_2) < \dots.$$

Note that the stop bands are characterized as

$$(-\infty, a_1),\ (\min(a_2, b_1), \max(a_2, b_1)), \\ (\min(a_3, b_2), \max(a_3, b_2)),\ \dots$$

and the pass bands are given by

$$(a_1, \min(a_2, b_1)),\ (\max(a_2, b_1), \min(a_3, b_2)), \\ (\max(a_3, b_2), \min(a_4, b_3)),\ \dots$$

Now let us consider the expression (11) with the infinite product limited to R factors:

$$I_R(z) = -\sqrt{m(0)\rho(0)}\,\sqrt[+]{-z+a_1}\cdot \\ \cdot \prod_{r=1}^{R} \frac{\sqrt[+]{-z+a_{r+1}}}{\sqrt[+]{-z+b_r}}, \quad \operatorname{Im} z \neq 0. \quad (13)$$

Figs. (**7**) and (**8**) show the maximum errors between the impedance $I(z)$ and $I_R(z)$ on 4001 equidistant points on three segments of the upper half complex plane. We detect that these errors become very small with increasing R. This observation has also been made for many other numerical tests.

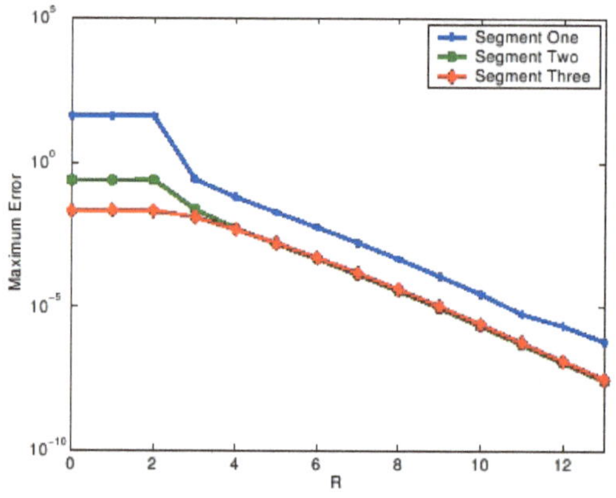

Fig. (**7**): **Case D:** Maximum error between the impedance $I(z)$ and $I_R(z)$. Segment One: $[-10, 10] + 10^{-13}i$. Segment Two: $[-10, 10] + i$. Segment Three: $[-10, 10] + 10i$.

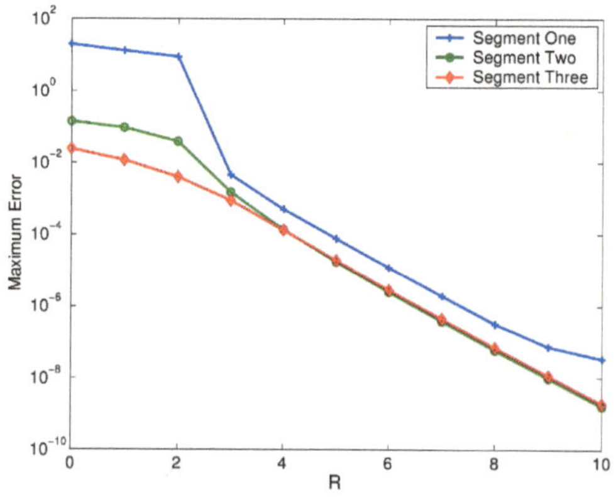

Fig. (**8**): **Case E:** Maximum error between the impedance $I(z)$ and $I_R(z)$. Segment One: $[-10, 10] + 10^{-13}i$. Segment Two: $[-10, 10] + i$. Segment Three: $[-10, 10] + 10i$.

It is thus reasonable to conjecture that the limit of $I_R(z)$ as R tends to $+\infty$ is the impedance $I(z)$, i.e. the formula (11) states the correct impedance expression.

If $z = z_0$ is a real number, then the impedance expression (11) might not be well-defined. If z_0 lies in one of the stop bands, we already know that

$$\lim_{\varepsilon \to 0^+} \operatorname{Im} I(z_0 + \varepsilon) = 0.$$

Due to the symmetry property of the impedance, i.e. $I(\bar{z}) = \overline{I(z)}$, we can define

$$I(z_0) = \lim_{\varepsilon \to 0^+} I(z_0 \pm \varepsilon).$$

Hence the impedance expression (11) still can be considered valid. If z_0 lies in one of the pass bands, the ODE (3) has no nontrivial bounded L^2-solution. In this case, we have to specify what kind of solution is really what we are seeking for. The impedance of this solution is thus the one-sided limit of $I(z_0 + \varepsilon)$ as either $\varepsilon \to 0^+$ or $\varepsilon \to 0^-$. In most cases, this choice can be made naturally under physical considerations.

Let us finally remark that the impedance formulation was proven very recently by Zhang and Zheng [76].

3. BOUND STATES FOR THE SCHRÖDINGER OPERATOR

As a first application of the impedance expression (11), we consider the following *bound state problem for the Schrödinger operator*:

Find an energy $E \in \mathbb{R}$ and a nontrivial real function $u \in L^2(\mathbb{R})$, such that

$$-\frac{d^2u}{dx^2} + V(x)u = Eu, \quad x \in \mathbb{R}, \tag{14}$$

where

$$V(x) = \begin{cases} 2 + 2\cos(\pi x), & |x| > 1, \\ 0, & |x| < 1. \end{cases}$$

The potential function $V(x)$ is periodic in $\mathbb{R} \backslash (-1, 1)$. In order to ensure that the solution u has a bounded L^2-norm, the energy E must be valued in the stop bands. The first few eigenvalues of the characteristic problem (9) with $m(x) = \rho(x) = 1$ and $V(x) = 2 - 2\cos(\pi x)$ (NOT $V(x) = 2 + 2\cos(\pi x)$) are listed in Table 3.

The first three stop bands are given by

$$(-\infty, 1.80087),\ (3.41926, 5.41414),\ (11.8359, 12.0349).$$

Table 3: The first few eigenvalues of (9) with $m(x) = \rho(x) = 1$ and $V = 2 - 2\cos(\pi x)$.

r	a_{r+1}	b_r	r	a_{r+1}	b_r
0	1.80087		3	2.42294(1)	2.42345(1)
1	3.41926	5.41414	4	4.14920(1)	4.14919(1)
2	1.20349(1)	1.18359(1)	5	6.36935(1)	6.36935(1)

If E is a bound state energy, then it must be an eigenvalue of the following *nonlinear characteristic problem*:

Find an energy $E \in \mathbb{R}$ and a nontrivial real function $u \in L^2(-1, 1)$, such that

$$-\frac{d^2u}{dx^2} + V(x)u = Eu, \quad x \in (-1, 1), \tag{15a}$$

$$-\frac{du}{dx}(-1) = I(E)u(-1), \tag{15b}$$

$$\frac{du}{dx}(1) = I(E)u(1). \tag{15c}$$

A direct discretization of the above problem (15) leads to a very complicated nonlinear algebraic equation with respect to E, and its solvability is not completely clear. Actually, the problem (15) is equivalent to the following *fixed point problem.*

For a given energy E we can solve the *linear characteristic problem*:

Find a function $\Phi(E) \in \mathbb{R}$ and a nontrivial real function $u \in L^2(-1, 1)$, such that the following boundary value problem holds

$$-u_{xx} + V(x)u = \Phi(E)u, \quad x \in (-1, 1), \tag{16a}$$

$$-\frac{du}{dx}(-1) = I(E)u(-1), \tag{16b}$$

$$\frac{du}{dx}(1) = I(E)u(1). \tag{16c}$$

The bound state energy thus satisfies $E = \Phi(E)$, i.e. E is a fixed point of the function $\Phi(E)$. Notice that $\Phi(E)$ is a multi-valued function and hence a series of bound states are expected.

Fig. (**9**) shows the first three branches of $\Phi(E)$ being restricted to $[-8, 15]$. The time-harmonic Schrödinger equation is discretized by 50 eighth-order finite elements in $[-1, 1]$. $I(E)$ is approximated by $I_{14}(E)$, which is equal to $I(E)$ within machine precision if $|E| < 20$. Three bound states exist in this energy range.

By performing the Newton-Steffenson iterations, the energies are found to be $E_0 = 0.642647$, $E_1 = 4.88651$ and $E_2 = 12.0164$. Our computations show that these values do not change within 6 digits by refining the finite element mesh.

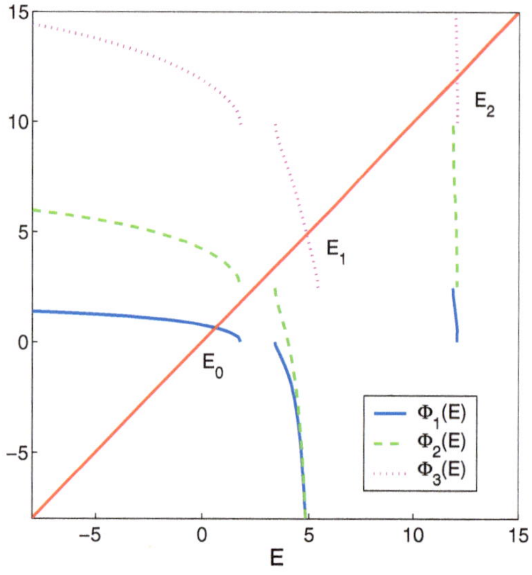

Fig. (**9**): The first three branches of $\Phi(E)$ being restricted to $[-8,15]$: $E_0 = 6.42647(-1)$. $E_1 = 4.88651$. $E_2 = 1.20164(1)$.

Next, the bound state wave functions (that are not normalized) are plotted in the Fig. (**10**). We observe in Fig. (**10**) that the ground state E_0 is well-localized, while the second excited bound state E_2 is greatly delocalized.
This demonstrates the advantage of the artificial boundary method and especially our ABCs (16b)–(16c), since a direct domain truncation method necessitates a very large computational domain to ensure the approximating accuracy of the wave function.

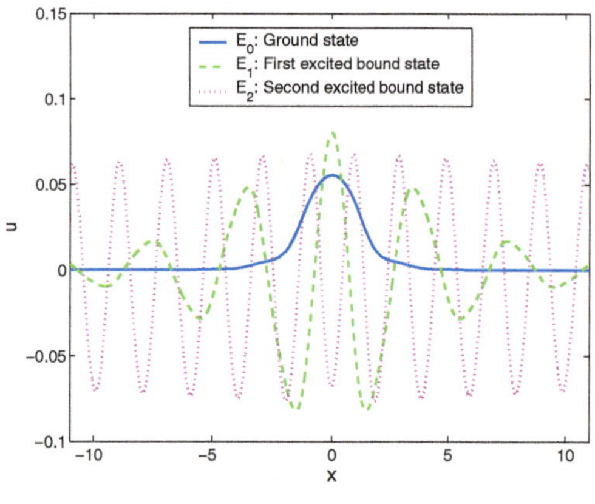

Fig. (**10**): The ground state E_0 and the first two excited bound states E_1, E_2.

4. EXACT ARTIFICIAL BOUNDARY CONDITIONS FOR TIME-DEPENDENT PROBLEMS

Based on the fundamental impedance expression (11), exact artificial boundary conditions can be derived for many time-dependent periodic structure problems, e.g., the Schrödinger equation (SE)

$$i\rho(x)\frac{\partial u}{\partial t} + \frac{\partial}{\partial x}\left(\frac{1}{m(x)}\frac{\partial u}{\partial x}\right) = V(x)u,$$

the diffusion equation (DE)

$$\rho(x)\frac{\partial u}{\partial t} = \frac{\partial}{\partial x}\left(\frac{1}{m(x)}\frac{\partial u}{\partial x}\right) - L(x)u,$$

and the second order hyperbolic equation (HE)

$$\frac{\partial}{\partial x}\left(\frac{1}{m(x)}\frac{\partial u}{\partial x}\right) - L(x)u = \rho(x)\frac{\partial^2 u}{\partial t^2}.$$

Here, the coefficients $V(x)$, $\rho(x)$, $m(x)$ and $L(x)$ are supposed to be centrally symmetric periodic functions at infinity. Moreover, $\rho(x)$ and $m(x)$ are positive, and $L(x)$ is nonnegative. Now the free parameter involved during the derivation of the impedance operator for stationary problems plays the role of the Laplace variable s. The impedances for these three equations are given by

$$I_{\mathrm{SE}}(is) = -\sqrt{m(0)\rho(0)}\,\sqrt[+]{-is+a_1}\cdot \cdot\prod_{r=1}^{+\infty}\frac{\sqrt[+]{-is+a_{r+1}}}{\sqrt[+]{-is+b_r}}, \quad (17)$$

and

$$I_{\mathrm{DE}}(-s) = -\sqrt{m(0)\rho(0)}\,\sqrt[+]{s+a_1}\cdot \cdot\prod_{r=1}^{+\infty}\frac{\sqrt[+]{s+a_{r+1}}}{\sqrt[+]{s+b_r}}, \quad (18)$$

and

$$I_{\mathrm{HE}}(-s^2) = -\sqrt{m(0)\rho(0)}\,\sqrt[+]{s^2+a_1}\cdot \cdot\prod_{r=1}^{+\infty}\frac{\sqrt[+]{s^2+a_{r+1}}}{\sqrt[+]{s^2+b_r}}. \quad (19)$$

In equations (17)-(19) the variable s with $\mathrm{Re}\, s > 0$ denotes the free argument in the Laplace domain. Notice that due to our assumption, all coefficients a_r and b_r in (18) and (19) are nonnegative and thus the formulas (18), (19) are well-defined. The numerical solution to the Schrödinger equation in conjunction with the ABC (17) has been investigated in [78]. Similar techniques can be used for the diffusion equation with the ABC (18) with minor modifications.

4.1 A second order hyperbolic equation in 2D

We consider the propagation of electromagnetic waves in a waveguide with cavity, cf. the schematic map Fig. (**11**). For a TM polarized electromagnetic wave, the electric field E is governed by the equation

$$\frac{\partial^2 E}{\partial x^2}+\frac{\partial^2 E}{\partial z^2}-\frac{\varepsilon(x,z)}{c^2}\frac{\partial^2 E}{\partial t^2}=0. \tag{20}$$

The relative dielectric permittivity ε, depending only on x after the artificial boundary, is supposed to be periodic. We assume that this waveguide is enclosed with a perfect conductor and hence we have a homogeneous Dirichlet boundary condition $E=0$ on the physical boundary.

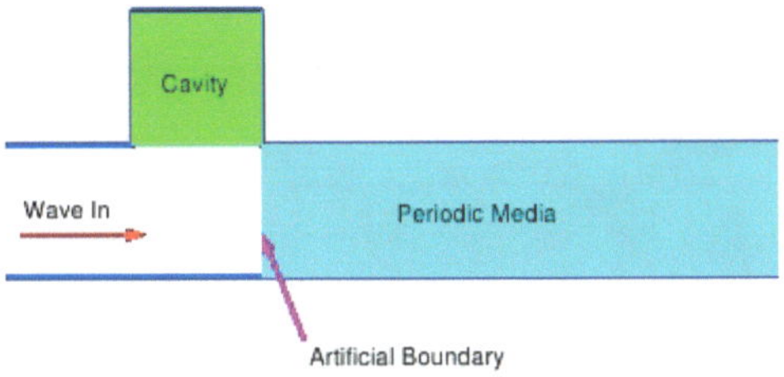

Fig. (**11**): Schematic of a waveguide with cavity.

On the semi-infinite slab region $[0,+\infty)\times[0,1]$, the characteristic decomposition can be applied with respect to the z variable. The eigenvalues are given by $n^2\pi^2$ and the eigenfunctions are $\sin(n\pi z)$, $n\geq 1$. An exact ABC in the frequency domain is thus set up as

$$\hat{E}_x^n(0,s)=-\frac{\sqrt{\varepsilon(0)}}{c}\sqrt[+]{s^2+a_1^n}\cdot\prod_{r=1}^{\infty}\frac{\sqrt[+]{s^2+a_{r+1}^n}}{\sqrt[+]{s^2+b_r^n}}\hat{E}^n(0,s),\quad n\geq 1. \tag{21}$$

Here, $\hat{E}^n(x,s)$ denotes the n-th mode of $\hat{E}(x,z,s)$ in the z-direction defined as

$$\hat{E}^n(x,s)=2\int_0^1\hat{E}(x,z,s)\sin(n\pi z)dz,\quad x\geq 0,\ n\geq 1.$$

$\hat{E}(x,z,s)$ is determined by $\hat{E}^n(x,s)$ as

$$\hat{E}(x,z,s)=\sum_{n=1}^{+\infty}\hat{E}^n(x,s)\sin(n\pi z),\quad x\geq 0.$$

The constants a_r^n and b_r^n in (21) are the eigenvalues of the characteristic problem (9) with the coefficients $m(x)=1$, $V(x)=n^2\pi^2$ and $\rho(x)=\varepsilon(x)/c^2$.

By setting

$$\hat{w}_k^n(s)=\prod_{r=k}^{\infty}\frac{\sqrt[+]{s^2+a_{r+1}^n}}{\sqrt[+]{s^2+b_r^n}}\hat{E}^n(0,s),\quad k\geq 1,\ n\geq 1,$$

we get the recursion relation

$$\sqrt[+]{s^2+b_k^n}\,\hat{w}_k^n(s)=\sqrt[+]{s^2+a_{k+1}^n}\,\hat{w}_{k+1}^n(s),$$

$k\geq 1$, $n\geq 1$ and (21) reads

$$\hat{E}_x^n(0,s)=-\frac{\sqrt{\varepsilon(0)}}{c}\sqrt[+]{s^2+a_1^n}\,\hat{w}_1^n(s), \tag{22}$$

$n\geq 1$. Returning to the physical domain yields

$$\frac{dw_k^n}{dt}=\frac{dw_{k+1}^n}{dt}+\frac{\sqrt{a_{k+1}^n}J_1(\sqrt{a_{k+1}^n}t)}{t}*w_{k+1}^n-\frac{\sqrt{b_k^n}J_1(\sqrt{b_k^n}t)}{t}*w_k^n,\ k\geq 1,\ n\geq 0,$$

and from (22) we obtain

$$\begin{aligned}\frac{\partial E^n}{\partial x}(0,t)&=-\frac{\sqrt{\varepsilon(0)}}{c}\Big(\frac{dw_1^n}{dt}+\frac{\sqrt{a_1^n}J_1(\sqrt{a_1^n}t)}{t}*w_1^n\Big)\\&=-\frac{\sqrt{\varepsilon(0)}}{c}\Big(\frac{\partial E^n}{\partial t}(0,t)\\&\quad+\sum_{k=0}^{+\infty}\frac{\sqrt{a_{k+1}^n}J_1(\sqrt{a_{k+1}^n}t)}{t}*w_{k+1}^n\\&\quad-\sum_{k=1}^{+\infty}\frac{\sqrt{b_k^n}J_1(\sqrt{b_k^n}t)}{t}*w_k^n\Big).\end{aligned} \tag{23}$$

Here, $*$ denotes a convolution with respect to the time variable t and J_1 is the Bessel function of first order. In a real implementation the infinite summation in (23) has to be truncated by only keeping the first K_n terms:

$$\begin{aligned}\frac{\partial E^n}{\partial x}(0,t)=-\frac{\sqrt{\varepsilon(0)}}{c}\Big(&\frac{\partial E^n}{\partial t}(0,t)\\&+\sum_{k=0}^{K_n}\frac{\sqrt{a_{k+1}^n}J_1(\sqrt{a_{k+1}^n}t)}{t}*w_{k+1}^n\\&-\sum_{k=1}^{K_n}\frac{\sqrt{b_k^n}J_1(\sqrt{b_k^n}t)}{t}*w_k^n\Big),\end{aligned} \tag{24}$$

and

$$w_{K_n+1}^n(t)=E^n(0,t).$$

If we want to resolve the n-th mode in the z-direction, we typically set $K_n\geq 0$. In order to ensure the approximating accuracy of the ABC, K_n should be increased for larger values of n. Of course, if we are not interested in the n-th mode at all, we only need to set $K_n = -1$. In the next numerical example, we set $K_n = 10$ for any $n = 0, 1 \ldots, N$, and $K_n = -1$ for any $n = N+1, \ldots$, where N denotes the number of modes in the z-direction we want to resolve.

4.2 Numerical Example

We now study the wave field generated by a *periodic disturbance* at the left physical boundary

$$E(-2,z,t) = \sin(\pi z) \sum_{n=0}^{+\infty} e^{-160(t-(n+0.5))^2},\ z \in (0,1).$$

The wave speed is set to 1, and the dielectric permittivity ε is set to be

$$\varepsilon(x,z) = \begin{cases} 1 & ,x < 0, \\ 1.2 - 0.2\cos(2\pi x) & ,x > 0. \end{cases}$$

We limit our computational time interval to $[0,6]$. Due to the finite wave propagation speed (at most 1), we can compute a *reference solution* E_{ref} in a large domain $(-2,4) \times (0,1) \cup (-1,0) \times (1,2)$ with small mesh sizes $\Delta x = \Delta z = 0.00125$ and $\Delta t = 0.000625$. The leap-frog central difference scheme is employed in all the computations. We use the standard *fast evaluation technique* proposed by Alpert, Greengard and Hagstrom [2] (cf. also [77]) for the convolution operations involved in the ABC (24). The poles and weights are taken from the web page of Hagstrom. The relative L^2-error is defined as

$$\frac{||E_{\text{ref}}(\cdot,\cdot,t) - E_{\text{num}}(\cdot,\cdot,t)||_{L^2}}{||E_{\text{ref}}(\cdot,\cdot,6)||_{L^2}},$$

where E_{ref} stands for the reference solution, while E_{num} denotes the numerical solution.

In Figs. (**12**) and (**13**) we compare the numerical solutions with the reference solutions at two different time steps $t = 3$ and $t = 3$. No difference can be observed with eyes.

In Fig. (**14**) we depict the errors when different number of modes in the z-direction are used. The accuracy of the numerical solutions is greatly improved for large number of modes.

The error evolution with respect to the time t is shown in Fig. (**15**). At the initial stage, the wave does not reach the artificial boundary, thus the ABC has no influence on the numerical solutions. The error arises completely from the interior discretization. After a critical time point (almost $t = 2.5$), the artificial boundary condition comes into effect.

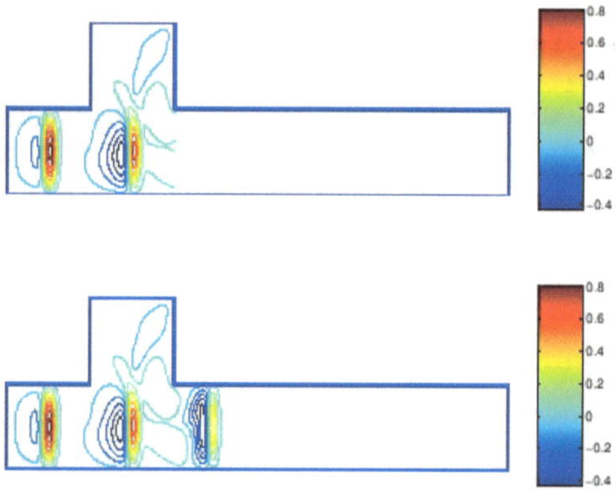

Fig. (**12**): Solutions at time $t = 3$. The number of modes is 10. The reference solution is obtained by taking $\Delta x = \Delta z = 0.00125$ and $\Delta t = 0.000625$.

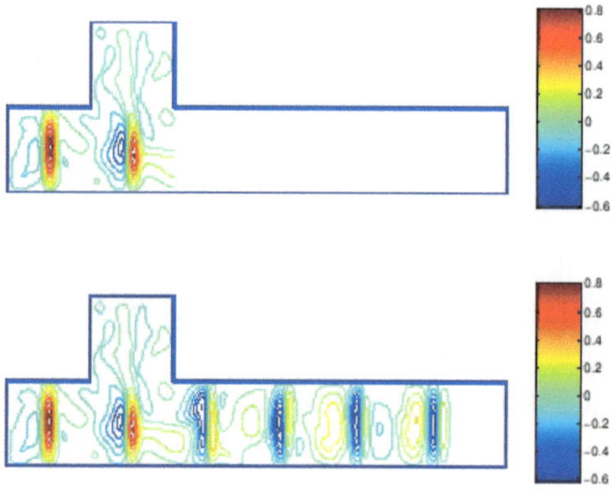

Fig. (**13**): Solutions at time $t = 6$. The number of modes is 10. The reference solution is obtained by taking $\Delta x = \Delta z = 0.00125$ and $\Delta t = 0.000625$.

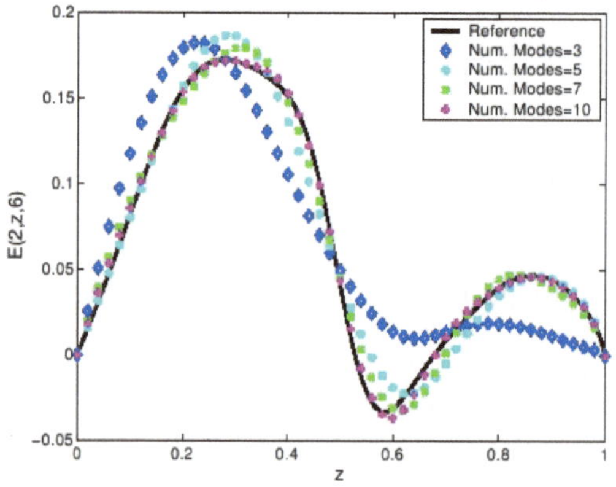

Fig. (**14**): Errors at time $t = 6$. $\Delta x = \Delta z = 0.02$. $\Delta t = 0.01$. The reference solution is obtained by taking $\Delta x = \Delta z = 0.00125$ and $\Delta t = 0.000625$. The line is $x = 0$.

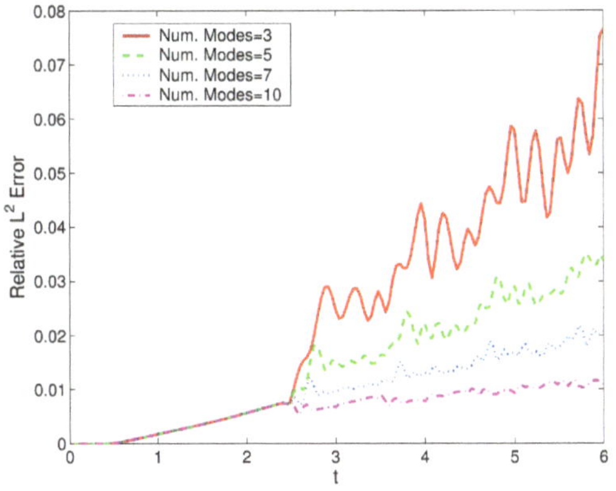

Fig. (**15**): Relative L^2 error. $\Delta x = \Delta z = 0.005$. $\Delta t = 0.0025$. The reference solution is obtained by taking $\Delta x = \Delta z = 0.00125$ and $\Delta t = 0.000625$.

We see that if enough number of modes are used, the error from the approximate boundary condition is nearly on the same level of interior discretization, which means the ABC is sufficiently accurate in this parameter regime.

Finally, we analyzed numerically in Fig. (**16**) the convergence rate of the relative L^2-errors at $t = 6$. Data-fitting reveals that the errors decay with an order of 1.851 in the parameter range $\Delta t \in [\frac{0.02}{7}, 0.01]$, when the number of modes in the z-direction is set to 10.

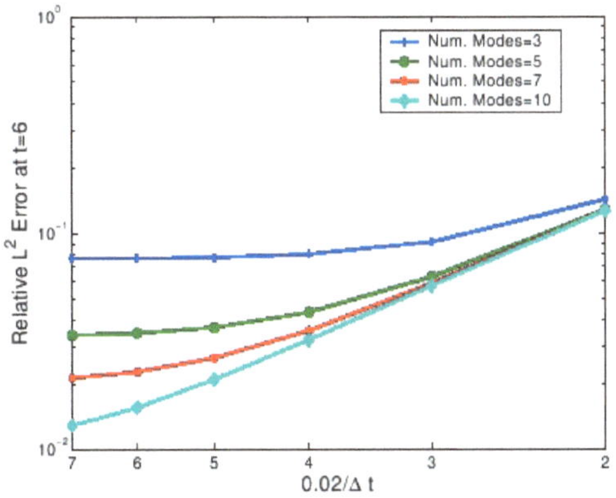

Fig. (**16**): Relative L^2 error. $\Delta x = \Delta z = 2\Delta t$. The reference solution is obtained by taking $\Delta x = \Delta z = 0.00125$ and $\Delta t = 0.000625$.

5. BOUNDARY MAPPINGS FOR PERIODIC ARRAYS

Let us consider the Helmholtz equation

$$-\Delta u(\mathbf{x}) + (V - z)u(\mathbf{x}) = f(\mathbf{x}). \qquad (25)$$

without source term, i.e., $V(\mathbf{x}) \equiv 0$ and $f(\mathbf{x}) \equiv 0$, on an array that is periodic in one direction, i.e. consisting of N identical cells as illustrated in Fig. (**17**).

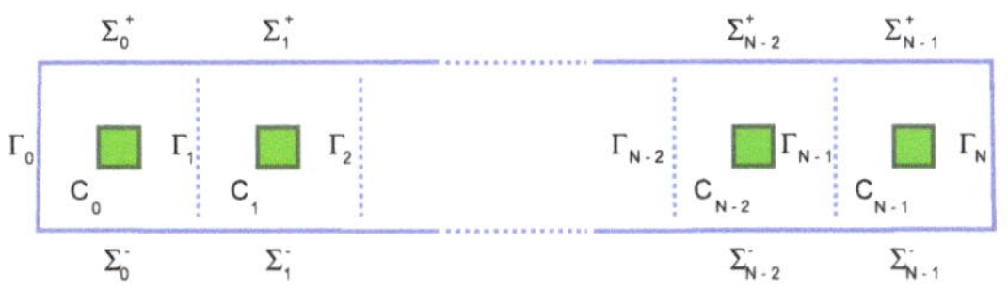

Fig. (**17**): Schematic view of a periodic array consisting of N cells.

We suppose that *appropriate* homogeneous linear boundary conditions are specified at the upper and lower, and the interior (if existing) boundaries, and these boundary conditions have the same periodicity consistent with that of the periodic structure. Here, "appropriate" means that these boundary conditions do not influence the well-posedness of the interior Helmholtz equation.
We define two *Sommerfeld mappings* of u as

$$\mathscr{G}_u^x = (\partial_x + \sqrt[+]{z})u, \qquad \mathscr{F}_u^x = (-\partial_x + \sqrt[+]{z})u.$$

It was proven in [20] that for given boundary data $\mathscr{F}_u^x$ on Γ_i and $\mathscr{G}_u^x$ on Γ_{i+k}, the Helmholtz equation (25), together with the boundary conditions on the upper and lower, and the interior (if existing) boundaries, is well-posed on the domain $\cup_{l=i}^{i+k-1} C_l$:

Lemma 3 (Lemma A.1. [20])**.** *The Helmholtz equation (25) is uniquely solvable in $\cup_{l=i}^{i+k-1} C_l$ for any $f_i \in (H^{1/2}(\Gamma_i))'$, $g_k \in (H^{1/2}(\Gamma_k))'$ and any $z \in \mathbb{Z}\backslash\{0\}$*

$$\begin{aligned}
&-\Delta u(\mathbf{x}) + zn^2 u(\mathbf{x}) = 0, \quad \mathbf{x} \in \Omega_{i,k} \equiv \cup_{l=i}^{i+k-1} C_l,\\
&\partial_y u(\mathbf{x}) = 0, \quad \mathbf{x} \in \cup_{i=0}^{N-1} \Sigma_i^{\pm},\\
&-\partial_x u(\mathbf{x}) + \sqrt[+]{z}\, u(\mathbf{x}) = f_i, \quad \mathbf{x} \in \Gamma_i,\\
&\partial_x u(\mathbf{x}) + \sqrt[+]{z}\, u(\mathbf{x}) = g_k, \quad \mathbf{x} \in \Gamma_k,
\end{aligned}$$

*where $n \in C^1(\cup_{l \in \mathbb{Z}} C_l)$, C_l and $\Sigma_l^{\pm}$ are defined as in Fig. (**17**).*

This implies that there exist four linear operators $\mathscr{A}_k$, $\mathscr{B}_k$, $\mathscr{C}_k$ and $\mathscr{D}_k$ satisfying

$$\begin{aligned}
\mathscr{G}_u^x|_{\Gamma_i} &= \mathscr{A}_k \mathscr{F}_u^x|_{\Gamma_i} + \mathscr{B}_k \mathscr{G}_u^x|_{\Gamma_{i+k}},\\
\mathscr{F}_u^x|_{\Gamma_{i+k}} &= \mathscr{C}_k \mathscr{F}_u^x|_{\Gamma_i} + \mathscr{D}_k \mathscr{G}_u^x|_{\Gamma_{i+k}}.
\end{aligned} \qquad (26)$$

Numerically, these operators can be derived by an appropriate spatial discretization of the domain $\cup_{l=0}^{k-1} C_l$. But if k is big, a large number of unknowns would get involved, which leads to a high computational effort. Our task is now to design an efficient and robust algorithm for evaluating these operators.

5.1 The recursive doubling method

Suppose for $k \in \{m,n\}$, the four linear operators $\mathscr{A}_k$, $\mathscr{B}_k$, $\mathscr{C}_k$ and $\mathscr{D}_k$ have already been obtained. From (26) we obtain

$$\mathscr{G}_u^x|_{\Gamma_i} = \mathscr{A}_m(\mathscr{C}_n\mathscr{F}_u^x|_{\Gamma_{i-n}} + \mathscr{D}_n\mathscr{G}_u^x|_{\Gamma_i}) + \mathscr{B}_m\mathscr{G}_u^x|_{\Gamma_{i+m}},$$
$$\mathscr{F}_u^x|_{\Gamma_i} = \mathscr{C}_n\mathscr{F}_u^x|_{\Gamma_{i-n}} + \mathscr{D}_n(\mathscr{A}_m\mathscr{F}_u^x|_{\Gamma_i} + \mathscr{B}_m\mathscr{G}_u^x|_{\Gamma_{i+m}}).$$

It is easy to prove that $I - \mathscr{A}_m\mathscr{D}_n$ and $I - \mathscr{D}_n\mathscr{A}_m$ (I denotes the identity operator) are invertible and thus

$$\begin{aligned} \mathscr{G}_u^x|_{\Gamma_i} &= (I-\mathscr{A}_m\mathscr{D}_n)^{-1}\mathscr{A}_m\mathscr{C}_n\mathscr{F}_u^x|_{\Gamma_{i-n}} \\ &\quad + (I-\mathscr{A}_m\mathscr{D}_n)^{-1}\mathscr{B}_m\mathscr{G}_u^x|_{\Gamma_{i+m}}, \\ \mathscr{F}_u^x|_{\Gamma_i} &= (I-\mathscr{D}_n\mathscr{A}_m)^{-1}\mathscr{C}_n\mathscr{F}_u^x|_{\Gamma_{i-n}} \\ &\quad + (I-\mathscr{D}_n\mathscr{A}_m)^{-1}\mathscr{D}_n\mathscr{B}_m\mathscr{G}_u^x|_{\Gamma_{i+m}}. \end{aligned} \tag{27}$$

Substituting the above expressions into (26) gives

$$\begin{aligned} \mathscr{G}_u^x|_{\Gamma_{i-n}} &= [\mathscr{A}_n + \mathscr{B}_n(I-\mathscr{A}_m\mathscr{D}_n)^{-1}\mathscr{A}_m\mathscr{C}_n]\mathscr{F}_u^x|_{\Gamma_{i-n}} \\ &\quad + \mathscr{B}_n(I-\mathscr{A}_m\mathscr{D}_n)^{-1}\mathscr{B}_m\mathscr{G}_u^x|_{\Gamma_{i+m}}, \\ \mathscr{F}_u^x|_{\Gamma_{i+m}} &= \mathscr{C}_m(I-\mathscr{D}_n\mathscr{A}_m)^{-1}\mathscr{C}_n\mathscr{F}_u^x|_{\Gamma_{i-n}} + [\mathscr{D}_m \\ &\quad + \mathscr{C}_m(I-\mathscr{D}_n\mathscr{A}_m)^{-1}\mathscr{D}_n\mathscr{B}_m]\mathscr{G}_u^x|_{\Gamma_{i+m}}, \end{aligned}$$

which imply the relations

$$\begin{aligned} \mathscr{A}_{m+n} &= \mathscr{A}_n + \mathscr{B}_n(I-\mathscr{A}_m\mathscr{D}_n)^{-1}\mathscr{A}_m\mathscr{C}_n, \\ \mathscr{B}_{m+n} &= \mathscr{B}_n(I-\mathscr{A}_m\mathscr{D}_n)^{-1}\mathscr{B}_m, \\ \mathscr{C}_{m+n} &= \mathscr{C}_m(I-\mathscr{D}_n\mathscr{A}_m)^{-1}\mathscr{C}_n, \\ \mathscr{D}_{m+n} &= \mathscr{D}_m + \mathscr{C}_m(I-\mathscr{D}_n\mathscr{A}_m)^{-1}\mathscr{D}_n\mathscr{B}_m. \end{aligned} \tag{28}$$

Hence, for any fixed cell number N, the operators $\mathscr{A}_N$, $\mathscr{B}_N$, $\mathscr{C}_N$, and $\mathscr{D}_N$ can be obtained by the following steps:

1. Derive $\mathscr{A}_1, \mathscr{B}_1, \mathscr{C}_1$, and $\mathscr{D}_1$ by the *cell analysis*. If $N = 1$, it is done.

2. Write the number N into *binary form* $(j_L \cdots j_0)_2$, with $L = [\log_2 N]$ and $j_L = 1$.

3. Use the doubling relations (28) L times by setting $m = n = 2^{k-1}$ to get $\mathscr{A}_{2^k}$, $\mathscr{B}_{2^k}$, $\mathscr{C}_{2^k}$, and $\mathscr{D}_{2^k}$ for $k = 1,\dots,L$.

4. For $l = L-1,\dots,0$, if $j_l \neq 0$, then use (28) by setting $m = (j_L \cdots j_{l+1}0\cdots 0)_2$ and $n = 2^l$ to obtain $\mathscr{A}_{(j_L\cdots j_l 0\cdots 0)_2}$, $\mathscr{B}_{(j_L\cdots j_l 0\cdots 0)_2}$, $\mathscr{C}_{(j_L\cdots j_l 0\cdots 0)_2}$ and $\mathscr{D}_{(j_L\cdots j_l 0\cdots 0)_2}$.

This procedure uses (28) at most $2[\log_2 N]$ times. Given the boundary data $\mathscr{F}_u^x|_{\Gamma_0}$ and $\mathscr{G}_u^x|_{\Gamma_N}$, in some cases it is necessary to obtain other data in a subdomain of $\cup_{l=0}^{N-1} C_l$, for example, $\mathscr{F}_u^y|_{\Sigma^-}$ and $\mathscr{G}_u^y|_{\Sigma^-}$ where $\Sigma^- = \cup_{i=0}^{N-1}\Sigma_i^-$. We need only to compute all $\mathscr{F}_u^x|_{\Gamma_i}$ and $\mathscr{G}_u^x|_{\Gamma_{i+1}}$ since for each i they completely determine the function u restricted to C_i. If N happens to be a power of 2, say $N = 2^L$, this can be achieved efficiently with the following algorithm: For $p = L,\dots,1$ and $k = 0,\dots,2^{L-p}-1$, compute $\mathscr{G}_u^x|_{\Gamma_{k2^p+2^{p-1}}}$ and $\mathscr{F}_u^x|_{\Gamma_{k2^p+2^{p-1}}}$ using (27) by setting $i = k2^p + 2^{p-1}$ and $n = m = 2^{p-1}$.

For a general cell number N, we proceed in the following way:

1. Write N into binary form $(j_L \cdots j_0)_2$, with $L = [\log_2 N]$ and $j_L = 1$;

2. For $l = 0,\dots,L$, if $j_l \neq 0$, compute $\mathscr{A}_k$, $\mathscr{B}_k$, $\mathscr{C}_k$, and $\mathscr{D}_k$ for $k = (j_l \cdots j_0)_2$ and $k = N - (j_l \cdots j_0)_2$, and use (27) by replacing i,n with $(j_l \cdots j_0)_2$ and m with $N - (j_l \cdots j_0)_2$ to derive $\mathscr{G}_u^x|_{\Gamma_{(j_l\cdots j_0)_2}}$ and $\mathscr{F}_u^x|_{\Gamma_{(j_l\cdots j_0)_2}}$. Then use the algorithm above for a power of 2 to derive $\mathscr{G}_u^x|_{\Gamma_i}$ and $\mathscr{F}_u^x|_{\Gamma_i}$ for any $i = (j_k \cdots j_0)_2 + 1, \cdots, (j_l \cdots j_0)_2 - 1$, where k is the largest number satisfying $k < l$ and $j_k \neq 0$.

For any $i = 1,\dots,N-1$, the above algorithm uses (28) at most $2[\log_2 N]$ times and (27) at most $[\log_2 N]$ times. After all $\mathscr{F}_u^x|_{\Gamma_i}$ and $\mathscr{G}_u^x|_{\Gamma_i}$ are derived, $\mathscr{F}_u^y|_{\Sigma_i^-}$ and $\mathscr{G}_u^y|_{\Sigma_i^-}$ are then obtained by the cell analysis. The final results can be written into the following form

$$\mathscr{G}_u^y|_{\Sigma^-} = (\mathscr{F} \to \mathscr{G})\mathscr{F}_u^x|_{\Gamma_0} + (\mathscr{G} \to \mathscr{G})\mathscr{G}_u^x|_{\Gamma_N},$$
$$\mathscr{F}_u^y|_{\Sigma^-} = (\mathscr{F} \to \mathscr{F})\mathscr{F}_u^x|_{\Gamma_0} + (\mathscr{G} \to \mathscr{F})\mathscr{G}_u^x|_{\Gamma_N}.$$

Here $(\mathscr{F} \to \mathscr{G})$, $(\mathscr{G} \to \mathscr{G})$, $(\mathscr{F} \to \mathscr{F})$ and $(\mathscr{G} \to \mathscr{F})$ are four linear operators defined in suitable distributional spaces.

Remark. If the boundary condition on Γ_N is given as a *Sommerfeld-to-Sommerfeld* (StS) mapping

$$\mathscr{G}_u^x|_{\Gamma_N} = \mathscr{E}_N\mathscr{F}_u^x|_{\Gamma_N} + S_N, \tag{29}$$

where $\mathscr{E}_N$ is a linear operator and S_N is a function defined on Γ_N, we have

$$\begin{aligned} \mathscr{G}_u^x|_{\Gamma_N} &= \mathscr{E}_N(I-\mathscr{D}_N\mathscr{E}_N)^{-1}\mathscr{C}_N\mathscr{F}_u^x|_{\Gamma_0} \\ &\quad + [I + \mathscr{E}_N(I-\mathscr{D}_N\mathscr{E}_N)^{-1}\mathscr{D}_N]S_N, \end{aligned}$$

and

$$\mathscr{G}_u^x|_{\Gamma_0} = [\mathscr{A}_N + \mathscr{B}_N \mathscr{E}_N (I - \mathscr{D}_N \mathscr{E}_N)^{-1} \mathscr{C}_N] \mathscr{F}_u^x|_{\Gamma_0} + [\mathscr{B}_N + \mathscr{B}_N \mathscr{E}_N (I - \mathscr{D}_N \mathscr{E}_N)^{-1} \mathscr{D}_N] S_N. \quad (30)$$

The invertibility of $I - \mathscr{D}_N \mathscr{E}_N$ is obvious if the periodic array problem is well-posed with the StS boundary mapping (29) on Γ_N. This expression (30) yields an exact StS mapping at the leftmost boundary Γ_0. Furthermore, if the Dirichlet-to-Neumann (DtN) mapping is well-defined on Γ_0, it can be derived straightforwardly from (30).

Remark. Recently, Yuan and Lu [73] proposed an analogous technique for deriving the exact DtN mapping. In their cell analysis, instead of using Sommerfeld data on Γ_i and Γ_{i+1}, they used Dirichlet data to determine Neumann data. A problem will appear if $-z$ happens to be one of the eigenvalues of the operator $-\Delta$ on $\cup_{k=i}^{i+2^J-1} C_k$ for some J with homogeneous Dirichlet boundary conditions specified on Γ_i and Γ_{i+2^J}, since in this case, the Dirichlet-to-Neumann (DtN) mapping does not exist at all. One might argue that the probability for this to happen is very small, but if the total number of periodic cells is large, the eigenvalues of $-\Delta$ with Dirichlet boundary conditions are very dense in the pass bands. This implies that if there are some eigenvalues of $-\Delta$ very close but never equal to s, though the DtN mapping exists, it is very ill-conditioned.

6. APPLICATION TO WAVEGUIDES

Here we present a first application of the proposed technique to waveguide problems. Consider the Helmholtz equation in the waveguide shown in Fig. (**18**). The domain between Γ_1 and Γ_2 consists of four periodic cells. Each cell has a size of 1×2 with a hole of 0.5×1 in the center. The domain between Γ_3 and Γ_4 also contains four periodic cells. Each cell has a size of 1×1 with a hole of 0.5×0.5. These two periodic structures are joined with a junction region between Γ_2 and Γ_3. The domains left to Γ_1 and right to Γ_4 are homogeneous.

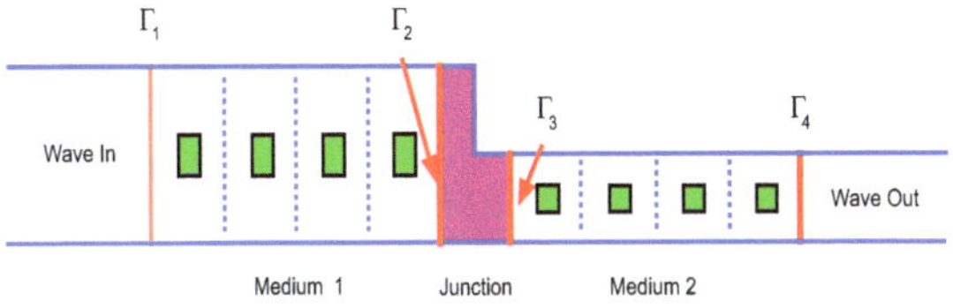

Fig. (**18**): Schematic of a model waveguide. Two waveguides with different periodic materials are joined with a junction zone between Γ_2 and Γ_3.

The governing equation is the Helmholtz equation (25) without source term $f(\mathbf{x})$ and $z(\mathbf{x}) \equiv -k^2$, i.e.

$$\Delta u + k^2 u = 0, \quad (31)$$

where $k > 0$ is the real wave number. Zero Dirichlet data is specified on the interior boundaries, and zero Neumann data on the top and bottom boundaries. A plane wave $u_0(x,y) = e^{-ikx}$ is traveling in the waveguide from the left side. It is well-known that the *disturbance part* $u - u_0$ satisfies the left-going boundary condition on Γ_1, i.e.,

$$\frac{\partial}{\partial x}(u - u_0) = \sqrt[+]{-\partial_y^2 - k^2}\,(u - u_0), \quad (x,y) \in \Gamma_1,$$

or equivalently in the form of StS mapping,

$$\mathscr{F}_u^x = \frac{ik - \sqrt[+]{-\partial_y^2 - k^2}}{ik + \sqrt[+]{-\partial_y^2 - k^2}} \mathscr{G}_u^x + 2iku_0, \quad (x,y) \in \Gamma_1. \quad (32)$$

The wave function u satisfies the right-going boundary condition on Γ_4, i.e.,

$$\frac{\partial u}{\partial x} = -\sqrt[+]{-\partial_y^2 - k^2}\,u, \quad (x,y) \in \Gamma_4,$$

or equivalently,

$$\mathscr{G}_u^x = \frac{ik - \sqrt[+]{-\partial_y^2 - k^2}}{ik + \sqrt[+]{-\partial_y^2 - k^2}} \mathscr{F}_u^x, \quad (x,y) \in \Gamma_4. \quad (33)$$

Now by using the technique in the last section, we could derive the StS mapping on Γ_2 and Γ_3. The wave function is then resolved by solving the Helmholtz equation *only* in the junction region between Γ_2 and Γ_3.

6.1 Band Structure Diagrams

To understand the typical wave behaviour in periodic waveguides we must consider the band structure diagrams of the characteristic equation $-\Delta u = \lambda u$ restricted to a single periodic cell. As assumed, the top and bottom boundary conditions are homogeneous Neumann, and the interior boundary condition is homogeneous Dirichlet. The boundary conditions at the left and right boundaries of the single cell are *pseudoperiodic*, namely,

$$u_{\text{right}} = e^{i\theta} u_{\text{left}}, \qquad \partial_x u_{\text{right}} = e^{i\theta} \partial_x u_{\text{left}},$$

where the parameter θ lies in the interval $[0, 2\pi]$. For each value of θ, there exists a sequence of

real eigenvalues λ that are shown in the following *band structure diagrams*. These eigenvalues, also regarded as *discrete energies*, correspond to a series of Bloch waves which could travel through the waveguides without damping.
Figs. (**19**) and (**20**) show these band structure diagrams for the two periodic structures to the left and to the right.

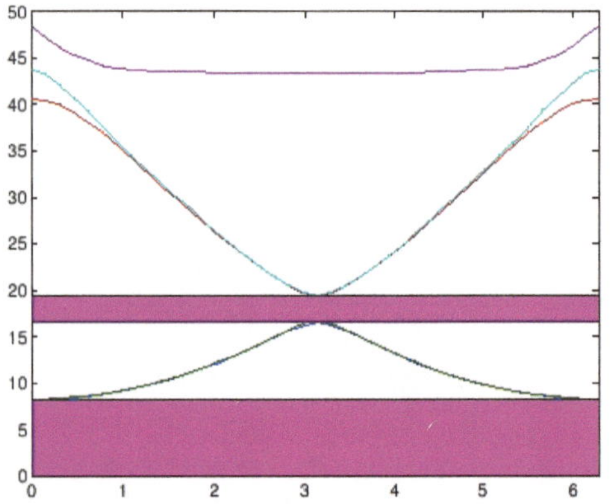

Fig. (**19**): Band structure with stop bands for the left periodic structure. The first two stop bands are the intervals $(-\infty, 8.27_{\pm 0.01})$ and $(16.69_{\pm 0.01}, 19.49_{\pm 0.01})$.

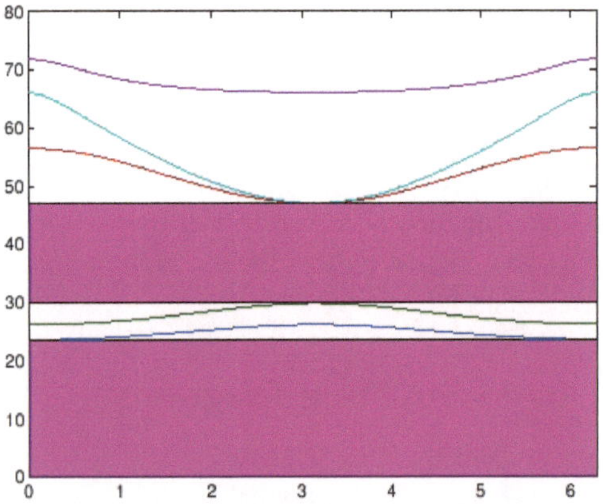

Fig. (**20**): Band structure with stop bands for the right periodic structure. The first two stop bands are the intervals $(-\infty, 23.61_{\pm 0.01})$ and $(29.85_{\pm 0.01}, 47.10 \pm 0.01)$.

The results are obtained by an eighth-order finite element discretization using the step sizes $\Delta x = \Delta y = 0.125$. For the left periodic structure between Γ_1 and Γ_2, the first two stop bands are $(-\infty, 8.27_{\pm 0.01})$ and $(16.69_{\pm 0.01}, 19.49_{\pm 0.01})$, while for the right periodic structure between Γ_3 and Γ_4, they are $(-\infty, 23.61_{\pm 0.01})$ and $(29.85_{\pm 0.01}, 47.10_{\pm 0.01})$. The first eigenvalue of the Dirichlet boundary value problem for the left periodic structure is $19.49_{\pm 0.01}$, while the first eigenvalue is $47.10_{\pm 0.01}$ for the right periodic structure.

We consider in the sequel *five cases*: $k = \sqrt{8}$, $k = \sqrt{19.49}$, $k = 6$, $k = \sqrt{47.10}$ and $k = 8$.

1. $k^2 = 8$ lies in stop bands of both two structures.
2. $k^2 = 19.49$ is the first eigenvalue of the Dirichlet boundary value problem for the left periodic structure.
3. $k^2 = 36$ lies in pass bands of both two structures.
4. $k^2 = 47.10$ is the first eigenvalue of the Dirichlet boundary value problem of the right periodic structure.
5. $k^2 = 64$ lies in pass bands of both two structures.

We point out the fact that the cases $k = \sqrt{19.49}$ and $k = \sqrt{47.10}$ cannot be solved with Yuan and Lu's method [73]. Figs. (**21**)-(**25**) show the real part of the wave function for the five chosen wave numbers. Again an eighth-order finite element code was used in the computation with the step sizes $\Delta x = \Delta y = 0.125$.

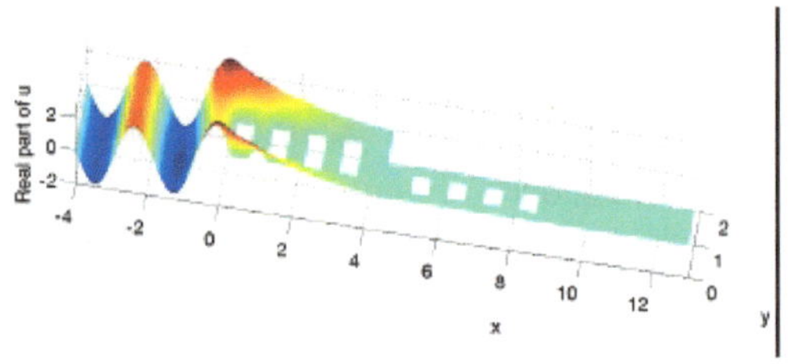

Fig. (**21**): Real part of the wave function for $k = \sqrt{8}$.

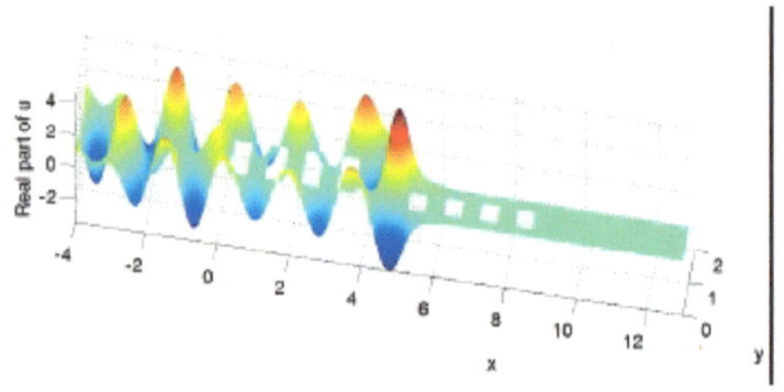

Fig. (**22**): Real part of the wave function for $k = \sqrt{19.49}$.

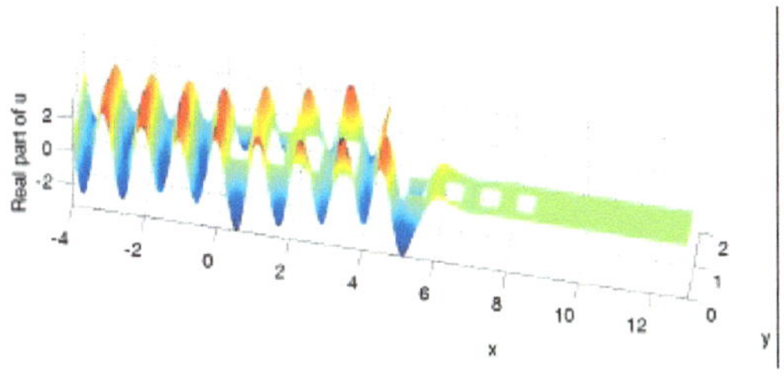

Fig. (**23**): Real part of the wave function for $k = 6$.

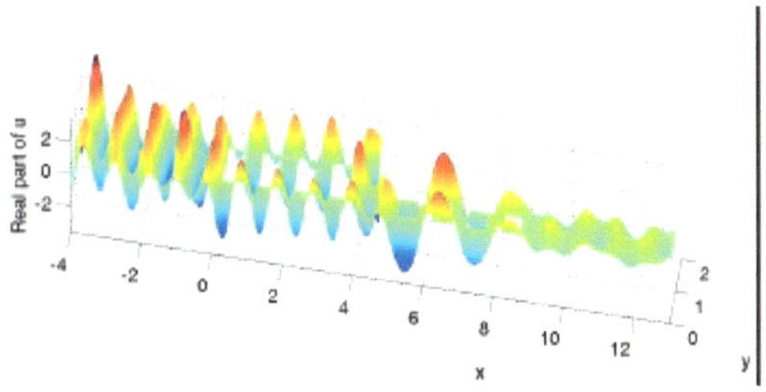

Fig. (**24**): Real part of the wave function for $k = \sqrt{47.10}$.

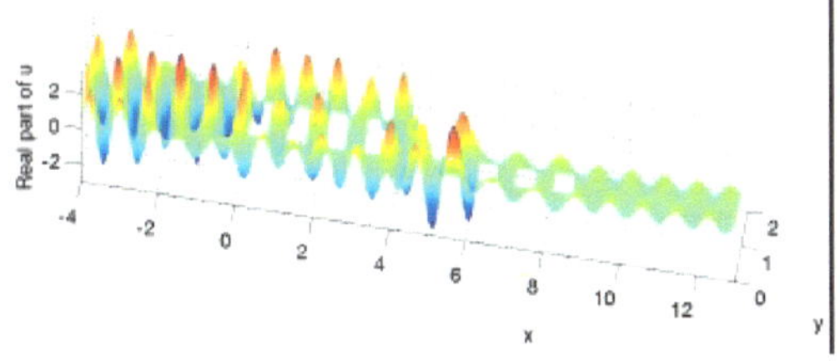

Fig. (**25**): Real part of the wave function for $k = 8$.

7. EXACT StS MAPPING FOR SEMI-INFINITE PERIODIC PROBLEMS

In many cases the exact StS (thus DtN or DtS) mapping is necessary to handle semi-infinite periodic array problems properly, see Fig. (**26**). Recently, Joly, Li and Fliss [40] presented a Newton-type method for the Helmholtz equation when z has a nonzero imaginary part. In this case any outgoing wave decays to zero exponentially fast at infinity. In the previous section, we have proposed a fast algorithm within $O(\log_2 N)$ operations for computing the exact StS mapping

$$g_0 = \mathscr{A}_N f_0 + \mathscr{B}_N g_N.$$

If the solution decays in one periodic cell with a factor of σ, by setting $N = \lceil -\frac{\ln \varepsilon}{\sigma} \rceil$, it is hopeful that $\mathscr{A}_N$ gives an approximation of the exact StS mapping on Γ_0 with an error of $O(\varepsilon)$. Here ε denotes the machine precision.

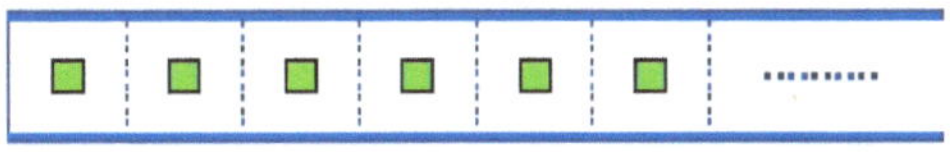

Fig. (**26**): Schematic view of a semi-infinite periodic array. Each cell has a size of 1×1, and a hole of 0.5×0.5 lies in the center.

It turns out that if $\operatorname{Im} z \neq 0$, or z is real but in the stop bands, the operator $\mathscr{A}_N$ converges with an exponential rate to the exact StS operator. In Fig. (**27**), we plot the relative errors of $\mathscr{A}_N$ w.r.t. $\mathscr{A}_{\text{ref}}$, which is obtained by setting $N = 1024$. In the computation we set $\Delta x = \Delta y = 0.125$ and use an eighth-order finite element method, thus in the discrete level A_N is expressed with a 65-by-65 matrix. Recall that $k^2 = 23, 31$ are in stop bands, and $k^2 = 25, 50$ in pass bands, cf. Fig. (**20**).

As a conclusion, using the doubling procedure of Section 5.1 at most $J = \lceil \log_2 \lceil -\frac{\ln \varepsilon}{\sigma} \rceil \rceil$ times gives the exact StS boundary mapping at the leftmost boundary up to machine precision. Our technique presents a very fast evaluation of the exact StS mapping.

If z lies in the stop bands, some traveling Floquet modes would appear, and the above argument ceases to hold. To obtain a well-posed PDE problem, we have to specify the outgoing waves and incoming waves. In a recent work of Joly et al. [40] a method is proposed to resolve this problem. However, we will not discuss this issue in this chapter.

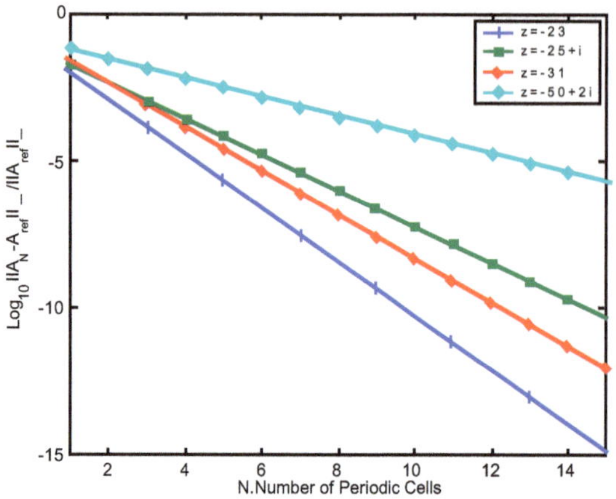

Fig. (**27**): Convergence of the StS mapping.

7.1 Application to the time-dependent Schrödinger equation

As an application, we consider the linear time-dependent Schrödinger equation

$$iu_t + u_{xx} = Vu, \quad x \in \mathbb{R}. \tag{34}$$

The initial data $u_0(x)$ is chosen as

$$u_0(x) = \exp(-x^2 + ik_0 x)$$

and the potential function V is set to

$$V(x) = \sum_{n\in\mathbb{Z}, n\neq 0} V_0 \exp\big(-(x-10n)^2\big).$$

In the Laplace domain the Schrödinger equation (34) is transformed into

$$-\hat{u}_{xx} + (V - is)\hat{u} = -iu_0, \quad x \in \mathbb{R}, \tag{35}$$

where Re $s > 0$ and $\hat{u}$ denotes the Laplace transformation of u defined by

$$\hat{u}(x,s) = \int_0^{+\infty} u(x,t)e^{-st}dt.$$

The function u_0 is well-supported in the interval $[-5,5]$. Outside of $[-5,5]$, the potential function V can be considered periodic with a period of 10. For any fixed s, the equation (35) can be solved in $[-5,5]$ with a high-order spatial discretization method. Here we use M eighth-order finite elements, which include $8M+1$ grid points. The StS boundary conditions at $x = \pm 5$ are derived by the method presented in the beginning of this section. The same number of grid points are used in the discrete periodic cell analysis.

The inverse Laplace transformation is evaluated numerically as

$$\begin{aligned} u(x,t) &= \frac{1}{2\pi i}\int_{\gamma-i\infty}^{\gamma+i\infty} e^{st}\hat{u}(x,s)\,ds \\ &\approx \frac{1}{2\pi}\int_{-f_{\max}}^{f_{\max}} \chi(f)e^{(\gamma+if)t}\hat{u}(x,\gamma+if)\,df, \end{aligned} \tag{36}$$

and the integral is further approximated by the middle-point rule. Several parameters need to be tuned: the damping factor γ, the cutoff frequency f_{max}, and the number of quadrature points N_f. In principle, the bigger is γ, the smoother is the function $\hat{u}$, thus the number of quadrature points can be made smaller. But to guarantee stability γ cannot be too large. This is typically because there is an exponential factor $e^{\gamma t}$ involved in the integral. The *cutoff frequency* f_{max} depends on the regularity of the solution. The smoother is u, the smaller is f_{max}. We leave open the theoretical investigation on the optimal choice of these parameters in this chapter. For the considered model problem when $k_0 = 2$ and $M = 16$, we set

$$f_{max} = 200, \quad \gamma = 1, \quad N_f = 1024,$$

and the *filtering function* χ as

$$\chi(f) = \exp\left(-(1.2f/f_{max})^{20}\right).$$

If $V_0 = 0$, the exact solution is

$$u(x,t) = \sqrt{\frac{i}{-4t+i}}\exp\left(\frac{-ix^2 - k_0 x + k_0^2 t}{-4t+i}\right)..$$

The relative L^2-errors in the computational region $[-5,5]$ are listed in Table 4 at different time points. We observe that in this time regime the relative errors are very small. If $V_0 \neq 0$, the analytical exact solution is in general not available. In Fig. (**28**) we illustrate the solution at different time points for $V_0 = 10$. The dashed blue line shows the potential function scaled by $1/V_0$.

8. NUMERICAL SIMULATION OF THE 2D SCHRÖDINGER EQUATION

Here we consider the following two-dimensional time-dependent Schrödinger equation

$$iu_t + u_{xx} + u_{yy} = Vu, \ \forall(x,y) \in \mathbb{R}^2, \ \forall t > 0, \tag{37}$$

$$u(x,y,0) = u_0(x,y), \ \forall(x,y) \in \mathbb{R}^2, \tag{38}$$

$$u(x,y,t) \to 0, \ r = \sqrt{x^2+y^2} \to +\infty, \ \forall t > 0. \tag{39}$$

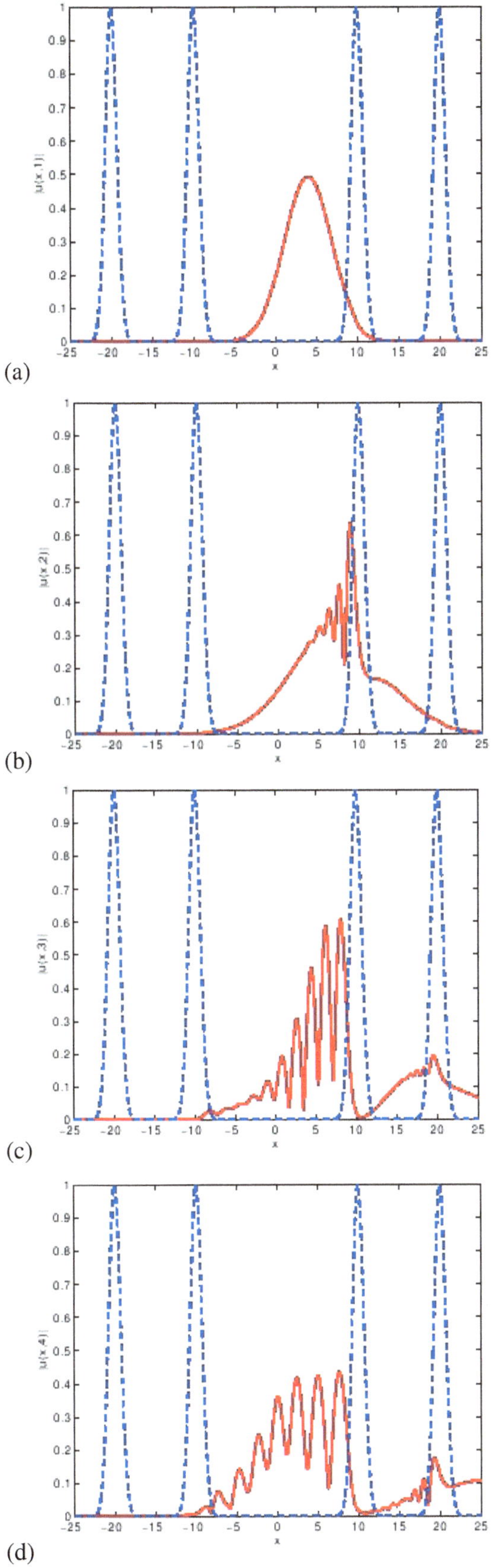

Fig. (**28**): Evolution of Gaussian packet in a periodic potential. (a) $t = 1$. (b) $t = 2$. (c) $t = 3$. (d) $t = 4$.

Table 4: Relative L^2-errors in $[-5,5]$ at different time points for $V_0 = 0$.

Time Point	Relative L^2-Error
1.0	1.83(-8)
1.5	2.12(-8)
2.0	2.66(-8)
2.5	3.14(-8)
3.0	3.56(-8)
3.5	3.92(-8)

The time evolution of the Gaussian wave packet is presented in Fig. (**29**). The potential function $V = V(x,y)$ is bi-periodic with a periodicity of 1×1 and a defect exists in the center of this periodic structure. The initial data u_0 is assumed locally supported, say in the defect cell.

The definition domain of the above problem is unbounded, and as a first step we could truncate the domain by introducing a rectangular artificial boundary and on it imposing the periodic boundary condition.

This treatment is justified if the time interval of simulation is finite and the number of cells enclosed by the artificial boundary is sufficiently large.

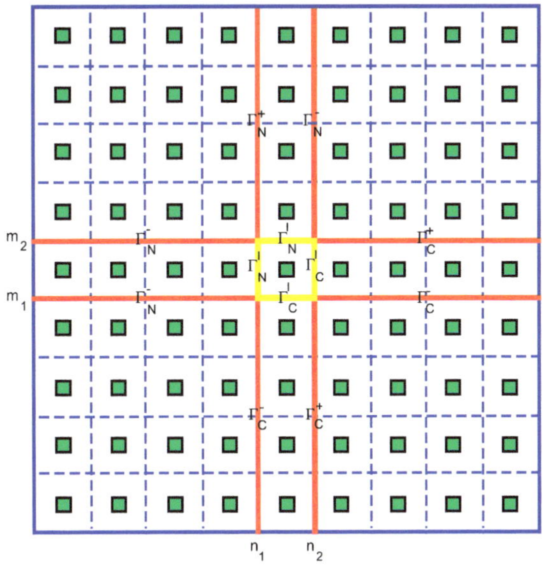

Fig. (**29**): A bi-periodic potential function with a defect in the center.

The next step is to find a suitable numerical scheme to resolve the wave field. Our basic idea is analogous to that in the last section for handling the one-dimensional Schrödinger equation with periodic potentials at infinity.

We first go to the frequency domain by solving the Helmholtz equation

$$-u_{xx}-u_{yy}+(V-is)u=-iu_0 \tag{40}$$

with a series of complex parameters s, and then perform the inverse Laplace transformation with a frequency filter. Notice that in (40) we use the same notation u to represent its Laplace-transformed function. This is mainly for the brevity of notations used in the following of this section. Of course we do not indent to solve the equation (40) on the whole truncated domain, since a large number of unknowns would still get involved. Instead, we try to find in the following subsection an accurate boundary condition on the defect cell boundary $\Gamma_E^i\cup\Gamma_S^i\cup\Gamma_W^i\cup\Gamma_N^i$, and perform computation only on the defect cell.

8.1 The Boundary Condition on Defect Cell Boundary

Let us first consider the equation (40) on the geometry shown in Fig. (**30**). Suppose periodic boundary conditions are specified on Σ_0 and Σ_M. Set $\Gamma_W=\cup_{k=0}^{M-1}\Gamma_{W,k}$ and $\Gamma_E=\cup_{k=0}^{M-1}\Gamma_{E,k}$. In the y-direction, we have M periodic layers.

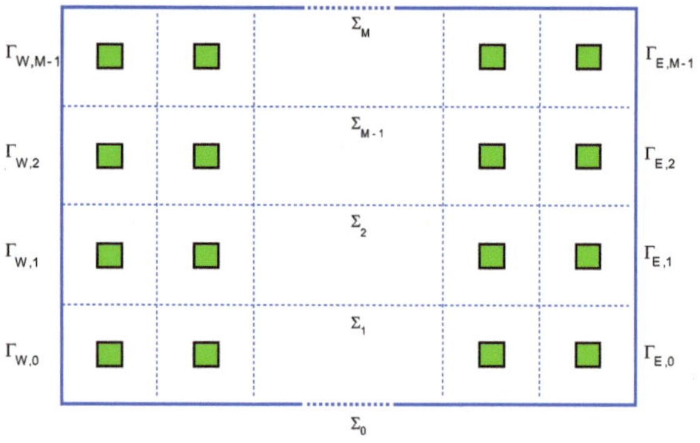

Fig. (**30**): Schematic view of a bi-periodic structure with periodic boundary conditions on Σ_0 and Σ_M.

We define the *discrete Fourier transformation* in the y-direction as

$$\hat{u}_k(x,y)=\sum_{m=0}^{M-1}u(x,y+mL)\,\omega^{km},\quad \omega=e^{-2i\pi/M},$$

$k=0,1,\ldots,M-1$. The inverse transformation is given as

$$u(x,y+mL)=\frac{1}{M}\sum_{k=0}^{M-1}\hat{u}_k(x,y)\,\omega^{-km}.$$

It is straightforward to verify that

$$\hat{u}_k(x,y+L)=\omega^{-k}\hat{u}_k(x,y).$$

Thus the problem on the domain shown in Fig. (**30**) can be reduced to M periodic array problems with pseudo-periodic boundary conditions on Σ_0 and Σ_1. By the analysis in the first section, we get

$$\mathscr{G}^x_{\hat{u}_k}|_{\Gamma_{W,0}}=\hat{\mathscr{A}}_k\mathscr{F}^x_{\hat{u}_k}|_{\Gamma_{W,0}}+\hat{\mathscr{B}}_k\mathscr{G}^x_{\hat{u}_k}|_{\Gamma_{E,0}},$$
$$\mathscr{F}^x_{\hat{u}_k}|_{\Gamma_{E,0}}=\hat{\mathscr{C}}_k\mathscr{F}^x_{\hat{u}_k}|_{\Gamma_{W,0}}+\hat{\mathscr{D}}_k\mathscr{G}^x_{\hat{u}_k}|_{\Gamma_{E,0}},$$

and

$$\mathscr{G}^y_{\hat{u}_k}|_{\Sigma_0}=\widehat{(\mathscr{F}\to\mathscr{G})}_k\mathscr{F}^x_{\hat{u}_k}|_{\Gamma_{W,0}}+\widehat{(\mathscr{G}\to\mathscr{G})}_k\mathscr{G}^x_{\hat{u}_k}|_{\Gamma_{E,0}},$$
$$\mathscr{F}^y_{\hat{u}_k}|_{\Sigma_0}=\widehat{(\mathscr{F}\to\mathscr{F})}_k\mathscr{F}^x_{\hat{u}_k}|_{\Gamma_{W,0}}+\widehat{(\mathscr{G}\to\mathscr{F})}_k\mathscr{G}^x_{\hat{u}_k}|_{\Gamma_{E,0}}.$$

Then going back to the variable u of (40) yields

$$\mathscr{G}^x_u|_{\Gamma_{W,m}}=\mathscr{A}^m\mathscr{F}^x_u|_{\Gamma_W}+\mathscr{B}^m\mathscr{G}^x_u|_{\Gamma_E},$$
$$\mathscr{F}^x_u|_{\Gamma_{E,m}}=\mathscr{C}^m\mathscr{F}^x_u|_{\Gamma_W}+\mathscr{D}^m\mathscr{G}^x_u|_{\Gamma_E},$$

and

$$\mathscr{G}^y_u|_{\Sigma_m}=(\mathscr{F}\to\mathscr{G})_m\mathscr{F}^x_u|_{\Gamma_W}+(\mathscr{G}\to\mathscr{G})_m\mathscr{G}^x_u|_{\Gamma_E},$$
$$\mathscr{F}^y_u|_{\Sigma_m}=(\mathscr{F}\to\mathscr{F})_m\mathscr{F}^x_u|_{\Gamma_W}+(\mathscr{G}\to\mathscr{F})_m\mathscr{G}^x_u|_{\Gamma_E},$$

where

$$\mathscr{A}^m\mathscr{F}^x_u|_{\Gamma_W}=\frac{1}{M}\sum_{n=0}^{M-1}\left[\sum_{k=0}^{M-1}\hat{\mathscr{A}}_k\omega^{k(n-m)}\right]\mathscr{F}^x_u|_{\Gamma_{W,n}},$$
$$\mathscr{B}^m\mathscr{G}^x_u|_{\Gamma_E}=\frac{1}{M}\sum_{n=0}^{M-1}\left[\sum_{k=0}^{M-1}\hat{\mathscr{B}}_k\omega^{k(n-m)}\right]\mathscr{G}^x_u|_{\Gamma_{E,n}},$$
$$\mathscr{C}^m\mathscr{F}^x_u|_{\Gamma_W}=\frac{1}{M}\sum_{n=0}^{M-1}\left[\sum_{k=0}^{M-1}\hat{\mathscr{C}}_k\omega^{k(n-m)}\right]\mathscr{F}^x_u|_{\Gamma_{W,n}},$$
$$\mathscr{D}^m\mathscr{G}^x_u|_{\Gamma_E}=\frac{1}{M}\sum_{n=0}^{M-1}\left[\sum_{k=0}^{M-1}\hat{\mathscr{D}}_k\omega^{k(n-m)}\right]\mathscr{G}^x_u|_{\Gamma_{E,n}},$$

and

$$(\mathscr{F}\to\mathscr{G})_m\mathscr{F}^x_u|_{\Gamma_W}=\frac{1}{M}\sum_{n=0}^{M-1}\left[\sum_{k=0}^{M-1}\widehat{(\mathscr{F}\to\mathscr{G})}_k\omega^{k(n-m)}\right]\mathscr{F}^x_u|_{\Gamma_{W,n}},$$

$$(\mathscr{G} \to \mathscr{G})_m \mathscr{G}_u^x|_{\Gamma_E} = \frac{1}{M}\sum_{n=0}^{M-1}\left[\sum_{k=0}^{M-1}(\widehat{\mathscr{G} \to \mathscr{G}})_k \omega^{k(n-m)}\right]\mathscr{G}_u^x|_{\Gamma_{E,n}},$$

$$(\mathscr{F} \to \mathscr{F})_m \mathscr{F}_u^x|_{\Gamma_W} = \frac{1}{M}\sum_{n=0}^{M-1}\left[\sum_{k=0}^{M-1}(\widehat{\mathscr{F} \to \mathscr{F}})_k \omega^{k(n-m)}\right]\mathscr{F}_u^x|_{\Gamma_{W,n}},$$

$$(\mathscr{G} \to \mathscr{F})_m \mathscr{G}_u^x|_{\Gamma_E} = \frac{1}{M}\sum_{n=0}^{M-1}\left[\sum_{k=0}^{M-1}(\widehat{\mathscr{G} \to \mathscr{F}})_k \omega^{k(n-m)}\right]\mathscr{G}_u^x|_{\Gamma_{E,n}}.$$

Note that the above operators can be evaluated efficiently by FFT.

Now come back to the equation (40) on the geometry shown in Fig. (**29**). Since periodic boundary conditions are specified on the boundary of the truncated domain, applying the above analysis we have

$$\begin{aligned}
\mathscr{G}_u^y|_{\Gamma_E^- \cup \Gamma_W^+} &= (\mathscr{F} \to \mathscr{G})_{m_1}^H \mathscr{F}_u^x|_{\Gamma_S^+ \cup \Gamma_E^i \cup \Gamma_N^-} \\
&\quad + (\mathscr{G} \to \mathscr{G})_{m_1}^H \mathscr{G}_u^x|_{\Gamma_S^- \cup \Gamma_W^i \cup \Gamma_N^+}, \\
\mathscr{F}_u^y|_{\Gamma_E^+ \cup \Gamma_W^-} &= (\mathscr{F} \to \mathscr{F})_{m_2}^H \mathscr{F}_u^x|_{\Gamma_S^+ \cup \Gamma_E^i \cup \Gamma_N^-} \\
&\quad + (\mathscr{G} \to \mathscr{F})_{m_2}^H \mathscr{G}_u^x|_{\Gamma_S^- \cup \Gamma_W^i \cup \Gamma_N^+}, \\
\mathscr{G}_u^x|_{\Gamma_N^+ \cup \Gamma_S^-} &= (\mathscr{F} \to \mathscr{G})_{n_1}^V \mathscr{F}_u^y|_{\Gamma_W^- \cup \Gamma_N^i \cup \Gamma_E^+} \\
&\quad + (\mathscr{G} \to \mathscr{G})_{n_1}^V \mathscr{G}_u^y|_{\Gamma_W^+ \cup \Gamma_S^i \cup \Gamma_E^-}, \\
\mathscr{F}_u^x|_{\Gamma_N^- \cup \Gamma_S^+} &= (\mathscr{F} \to \mathscr{F})_{n_2}^V \mathscr{F}_u^y|_{\Gamma_W^- \cup \Gamma_N^i \cup \Gamma_E^+} \\
&\quad + (\mathscr{G} \to \mathscr{G})_{n_2}^V \mathscr{G}_u^y|_{\Gamma_W^+ \cup \Gamma_S^i \cup \Gamma_E^-},
\end{aligned} \tag{41}$$

and

$$\begin{aligned}
\mathscr{G}_u^x|_{\Gamma_E^i} &= \mathscr{A}_{m_1}^H \mathscr{F}_u^x|_{\Gamma_S^+ \cup \Gamma_E^i \cup \Gamma_N^-} + \mathscr{B}_{m_1}^H \mathscr{G}_u^x|_{\Gamma_S^- \cup \Gamma_W^i \cup \Gamma_N^+}, \\
\mathscr{F}_u^x|_{\Gamma_W^i} &= \mathscr{C}_{m_1}^H \mathscr{F}_u^x|_{\Gamma_S^+ \cup \Gamma_E^i \cup \Gamma_N^-} + \mathscr{D}_{m_1}^H \mathscr{G}_u^x|_{\Gamma_S^- \cup \Gamma_W^i \cup \Gamma_N^+}, \\
\mathscr{G}_u^y|_{\Gamma_N^i} &= \mathscr{A}_{n_1}^V \mathscr{F}_u^y|_{\Gamma_W^- \cup \Gamma_N^i \cup \Gamma_E^+} + \mathscr{B}_{n_1}^V \mathscr{G}_u^y|_{\Gamma_W^+ \cup \Gamma_S^i \cup \Gamma_E^-}, \\
\mathscr{F}_u^y|_{\Gamma_S^i} &= \mathscr{C}_{n_1}^V \mathscr{F}_u^y|_{\Gamma_W^- \cup \Gamma_N^i \cup \Gamma_E^+} + \mathscr{D}_{n_1}^V \mathscr{G}_u^y|_{\Gamma_W^+ \cup \Gamma_S^i \cup \Gamma_E^-}.
\end{aligned} \tag{42}$$

Here we use the superscripts H and V to distinguish those operators in two different directions. Given $\mathscr{F}_u^x|_{\Gamma_E^i}$, $\mathscr{G}_u^x|_{\Gamma_W^i}$, $\mathscr{F}_u^y|_{\Gamma_N^i}$ and $\mathscr{G}_u^x|_{\Gamma_S^i}$, in principle $\mathscr{G}_u^y|_{\Gamma_E^- \cup \Gamma_W^+}$, $\mathscr{F}_u^y|_{\Gamma_E^+ \cup \Gamma_W^-}$, $\mathscr{G}_u^x|_{\Gamma_N^+ \cup \Gamma_S^-}$ and $\mathscr{F}_u^x|_{\Gamma_N^- \cup \Gamma_S^+}$ can be determined by the operator equations (41). Thus then (42) implicitly define an StS mapping from $\mathscr{F}_u^x|_{\Gamma_E^i}$, $\mathscr{G}_u^x|_{\Gamma_W^i}$, $\mathscr{F}_u^y|_{\Gamma_N^i}$ and $\mathscr{G}_u^x|_{\Gamma_S^i}$, to $\mathscr{G}_u^x|_{\Gamma_E^i}$, $\mathscr{F}_u^x|_{\Gamma_W^i}$, $\mathscr{G}_u^y|_{\Gamma_N^i}$ and $\mathscr{F}_u^x|_{\Gamma_S^i}$. A DtN mapping can be further derived on the boundary of the defect cell, and the computation can now be performed solely on the defect cell.

Unlike the periodic array problems which are periodic only in one direction, the derivation of StS mapping becomes much more complicated. On the discrete level we need to solve a linear system with unknowns $\mathscr{G}_u^y|_{\Gamma_E^- \cup \Gamma_W^+}$, $\mathscr{F}_u^y|_{\Gamma_E^+ \cup \Gamma_W^-}$, $\mathscr{G}_u^x|_{\Gamma_N^+ \cup \Gamma_S^-}$ and $\mathscr{F}_u^x|_{\Gamma_N^- \cup \Gamma_S^+}$. This operation is still much time-consuming. However, if the size of domain is enlarged, the number of unknowns is only increased *linearly* for two-dimensional problems.

8.2 A Numerical Example

The initial function is

$$u_0(x,y) = \exp(-100x^2 - 100y^2 + 20xi).$$

The potential function is

$$V(x,y) = \sum_{m,n\in\mathbb{Z}} V_0 \exp(-100(x-m)^2 - 100(y-n)^2) - V_0\exp(-100x^2 - 100y^2).$$

The periodic cell is of size 1×1, and the origin is located in the center of the defect cell $[-0.5, 0.5] \times [-0.5, 0.5]$. The whole computational domain contains $9 \times 9 = 81$ periodic cells. We set the cut-off frequency as 40000, and use the middle-point quadrature rule to approximate the integral (36). The number of quadrature points is 1024. Each cell is discretized into $8 \times 8 = 64$ eighth-order finite elements. In Table 5 we list the relative L^2-errors at different time points when $V_0 = 0$. The reference solution is obtained by the spectral method with same grid points. We see that in this time regime, the errors are always less than 0.02 percent. In Figs. (**31**)-(**33**), we show several snapshots for the modulus of wave functions when $V_0 = 0$, and in Figs. (**34**)-(**36**) for the potential $V_0 = 4000$. Note that only 9 cells including the defect cell are shown in those figures.

Table 5: Relative L^2-errors in $[-0.5, 0.5]^2$ at different time points for $V_0 = 0$.

Time Point ($\times 0.0125$)	Relative L^2-Error
1.0	4.05(-5)
1.5	6.21(-5)
2.0	8.65(-5)
2.5	1.19(-4)
3.0	1.51(-4)

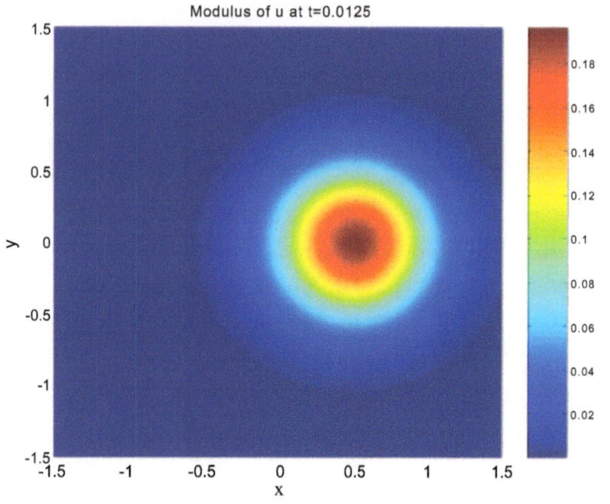

Fig. (**31**): $V_0 = 0$.

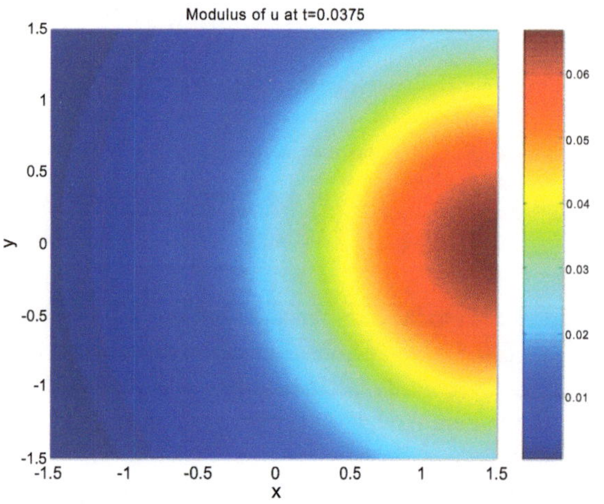

Fig. (**32**): $V_0 = 0$.

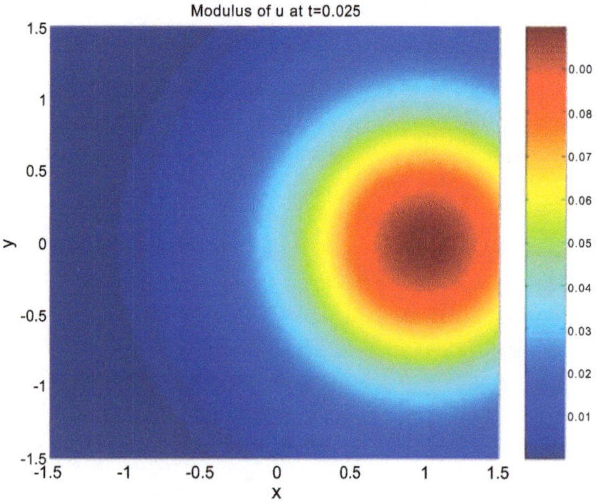

Fig. (**33**): $V_0 = 0$.

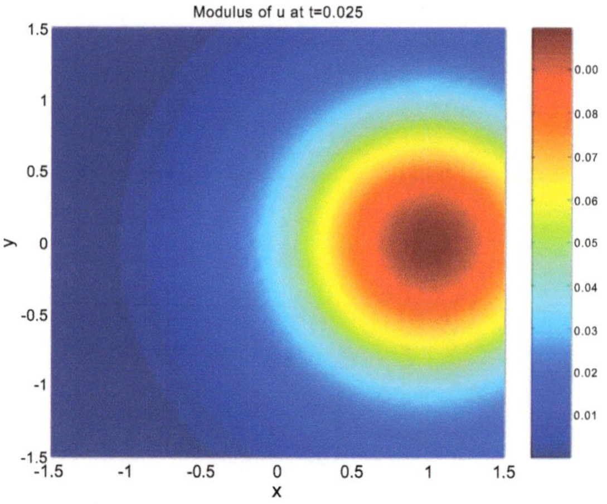

Fig. (**34**): $V_0 = 4000$.

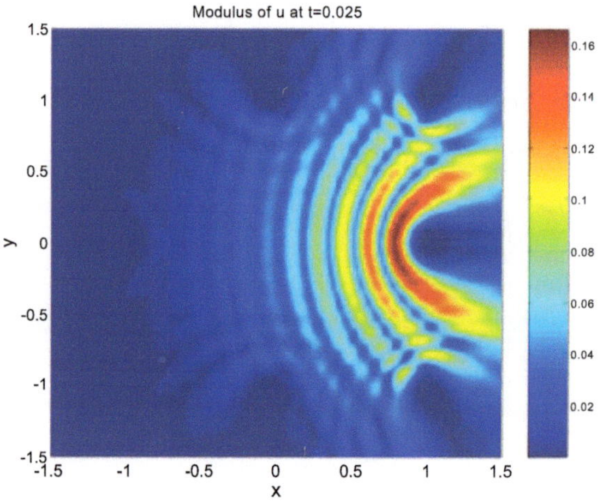

Fig. (**35**): $V_0 = 4000$.

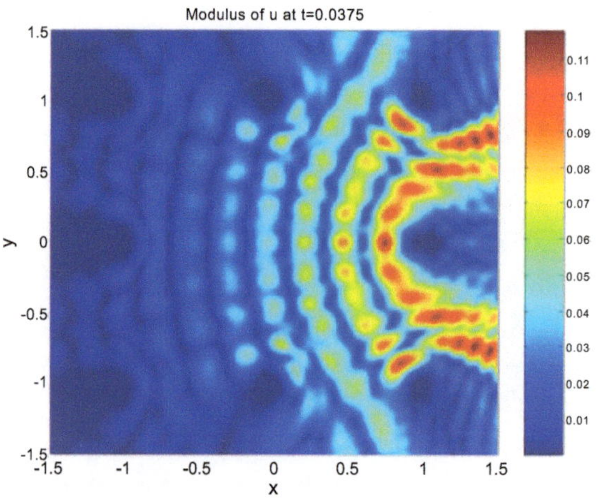

Fig. (**36**): $V_0 = 4000$.

9. A MODEL PROBLEM

We consider a closed waveguide consisting of an infinite number of identical cells, see Fig. (**37**). There C_j denotes the j-th periodic cell, and Γ_j the j-th cell boundary. The governing wave equation is the *Helmholtz equation*

$$\Delta u + k^2 n^2 u = 0, \quad (x,y) \in \Omega = \cup_{j=1}^{+\infty} C_j, \qquad (43)$$

where k denotes the reference wave number, and $n = n(x,y)$ is the refraction index function. On each cell boundary Γ_j we define two *Sommerfeld data* associated with the function u as

$$f_j(u) = (\partial_x + ik)u|_{\Gamma_j}, \quad g_j(u) = (\partial_x - ik)u|_{\Gamma_j}, \quad (44)$$

where i denotes the imaginary unit. To clarify the physical meaning of these two data, let us first return to the one-dimensional constant coefficient Helmholtz equation

$$u_{xx} + k^2 u = 0.$$

Two linearly independent solutions are $e^{\pm ikx}$. As a common convention, e^{ikx} represents a wave traveling to the right, and e^{-ikx} to the left. An easy computation yields

$$(\partial_x + ik)e^{ikx} = 2ike^{ikx}, \quad (\partial_x - ik)e^{ikx} = 0,$$

and

$$(\partial_x + ik)e^{-ikx} = 0, \quad (\partial_x - ik)e^{-ikx} = -2ike^{-ikx}.$$

These expressions above imply that the operator $\partial_x + ik$ eliminates the left-going wave while the operator $\partial_x - ik$ eliminates the right-going wave. Thus, the functions f_j and g_j in (44) contain some information about the right-going and left-going waves respectively. They are further referred to as *incoming* or *outgoing* relying on the location of Γ_j with respect to (w.r.t.) the concerned part of the domain. For example, w.r.t. C_j, f_j is incoming and g_j is outgoing, but w.r.t. C_{j-1}, f_j is outgoing and g_j is incoming.

The boundary conditions on the top, bottom and interior (if existing) boundaries could be either Neumann or Dirichlet, or any combination, but they need to be consistent with the geometry periodicity. Moreover, these boundary conditions should guarantee the well-posedness of the Helmholtz equation (43) on the union of any finite number of periodic cells, say $\cup_{j=0}^{N-1} C_j$, if the incoming Sommerfeld data are prescribed on its left and right boundaries, say Γ_0 and Γ_N.

We remark that these restrictions are in fact very mild thanks to the Holmgren uniqueness theorem [35, Section 5.3]. In the sequel, if not specified otherwise, we assume homogeneous Neumann boundary conditions at the top and bottom boundaries.

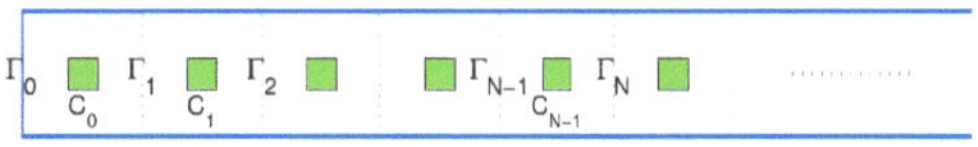

Fig. (**37**): Schematic of a semi-infinite periodic array. C_j denotes the j-th periodic cell. Γ_j is the left cell boundary of C_j and the right cell boundary of C_{j-1} (for $j \geq 1$).

9.1 The periodic Arrays

Three different periodic arrays (PA) will be considered in this chapter, and we will refer to them as PA-One, PA-Two and PA-Three. All of them consist of periodic cells with size of 1×1. More details are given below.

- **PA-One**. Homogeneous waveguide. $n = 1$.
- **PA-Two**. A hole of size 0.5×0.5 is located in the center of every periodic cell. Zero Dirichlet boundary condition is applied at the hole boundary. $n = 1$.
- **PA-Three**. Rectangular waveguide. $n(x,y) = 1 + 0.5\cos(2\pi x)\sin(2\pi y)$.

To explore the wave property in a periodic array, it is usually helpful to consider the dispersion diagram of the characteristic equation $-\Delta u = En^2 u$, restricted to a single periodic cell, say C_0. The boundary conditions at the left and right boundaries are *pseudoperiodic*, namely,

$$u|_{\Gamma_1} = e^{i\theta} u|_{\Gamma_0}, \qquad u_x|_{\Gamma_1} = e^{i\theta} u_x|_{\Gamma_0},$$

where the parameter θ is valued in $[0, 2\pi)$. For each θ, there exists a sequence of real eigenvalues E, usually called *energies*. All energies E w.r.t. θ then compose the dispersion diagram. The dispersion relation for PA-One, the homogeneous waveguide, can be obtained analytically as

$$E_{jm} = j^2\pi^2 + (\theta + 2\pi m)^2.$$

This multi-valued function is plotted in Fig. (**38**). For PA-Two and PA-Three, no analytical expressions of dispersion relation are available, and a spatial discretization method has to be employed.

We use the eighth-order FEM method with mesh sizes $\Delta x = \Delta y = 0.125$ for all the numerical tests reported in this chapter.

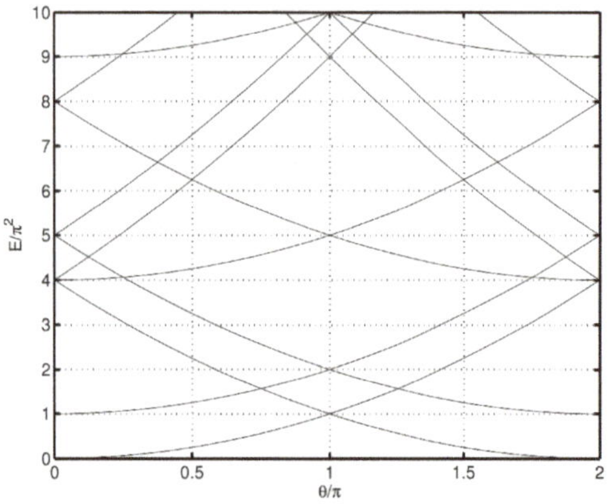

Fig. (**38**): Dispersion diagram of PA-One, an homogeneous waveguide.

The dispersion diagrams for PA-Two and PA-Three are shown in Figs. (**39**)-(**40**). A significant phenomena could be observed that unlike the homogeneous waveguide, there are some bands of energy values in the dispersion diagrams of PA-Two and PA-Three that could not be reached for any parameter θ.

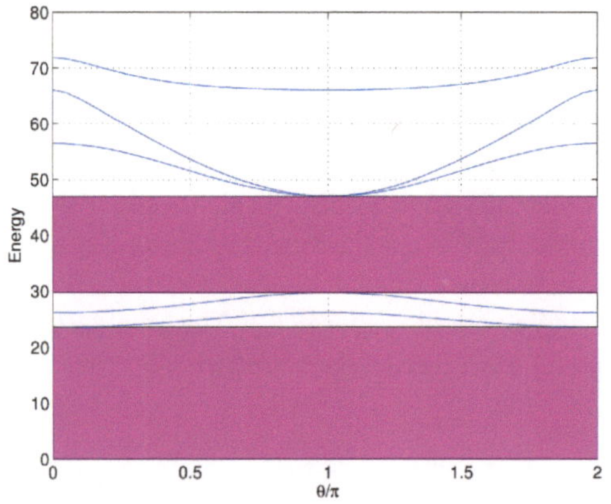

Fig. (**39**): Dispersion diagram of PA-Two. The first two stop bands are $(0, 23.61_{\pm 0.01})$ and $(29.85_{\pm 0.01}, 47.10_{\pm 0.01})$.

Physically, waves with energy (here k^2) in these bands could not propagate in the medium. Right in this context, they are usually referred to as *stop bands* in the literature. In fact, it is exactly this remarkable property which makes the periodic structures extremely useful, for example, they could be elaborately designed to act as some kind of frequency selecting modules in the microwave and optical engineering.

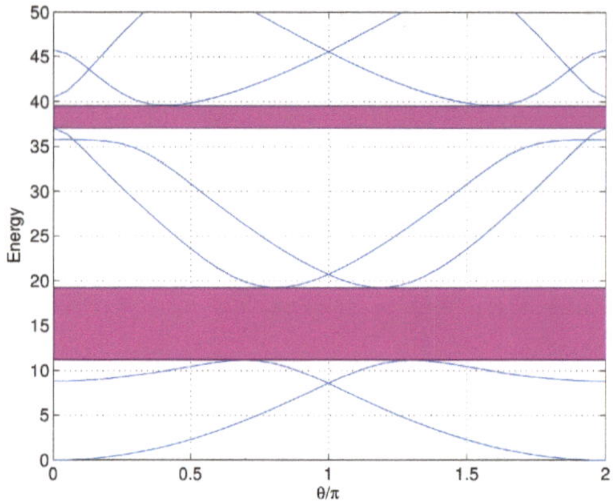

Fig. (**40**): Dispersion diagram of PA-Three.The first two stop bands are $(11.20_{\pm 0.01}, 19.29_{\pm 0.01})$ and $(37.08_{\pm 0.01}, 39.58_{\pm 0.01})$.

This work is aimed at developing an efficient method for deriving an exact boundary mapping of semi-infinite periodic arrays for *any real wavenumber k*.

10. THE LIMITING ABSORPTION PRINCIPLE

The first problem we are facing is how to guarantee the well-posedness of the Helmholtz equation (43), which naturally arises due to the absence of a radiation-like condition at infinity. Although the constant coefficient case with separable geometries is well solved, this problem is not trivial at all and largely remains open for the variable coefficient Helmholtz equation.

There are at least three methods of possibly deriving a unique solution of the Helmholtz equation in unbounded domains: asymptotic radiation condition, limiting absorption principle and limiting amplitude principle [67]. In this chapter we employ the *limiting absorption principle* (LABP). The LABP is said to hold at $k > 0$ if and only if for any $f_0(u) \in L^2(\Gamma_0)$ (take $f_0(u)$ as a unity), the solution $u^\varepsilon \in H^1(\Omega)$ of the following *damped Helmholtz equation*

$$\Delta u^\varepsilon + (k^2 + i\varepsilon) n^2 u^\varepsilon = 0 \qquad (45)$$

with the boundary condition

$$f_0(u^\varepsilon) = f_0(u),$$

converges to a unique solution $u \in H^1_{loc}(\Omega)$ of the Helmholtz equation (43), and the outgoing Sommerfeld datum $g_0(u^\varepsilon) = \mathscr{A}^\varepsilon_{\inf} f_0(u^\varepsilon)$ also converges to the unique function $g_0(u)$. This makes it possible to define a *Sommerfeld-to-Sommerfeld* (StS) mapping $\mathscr{A}_{\text{inf}}$ as the limit of $\mathscr{A}^{\varepsilon}_{\text{inf}}$, which maps $f_0(u)$ to $g_0(u)$, namely,

$$g_0(u) = \mathscr{A}_{\text{inf}} f_0(u).$$

Let us start considering PA-One first. In this case the separation of variables method is available. We set

$$u^{\varepsilon} = \sum_{n=0}^{+\infty} u^{\varepsilon,n} \cos(n\pi y)$$

and

$$f_0(u) = \sum_{n=0}^{+\infty} f_0(u^n) \cos(n\pi y),$$

$$g_0(u^{\varepsilon}) = \sum_{n=0}^{+\infty} g_0(u^{\varepsilon,n}) \cos(n\pi y).$$

Then (45) is transformed into a sequence of ODE problems:

$$u^{\varepsilon,n}_{xx} + (k^2 + i\varepsilon - n^2\pi^2) u^{\varepsilon,n}_{xx} = 0,$$
$$f_0(u^{\varepsilon,n}) = f_0(u^n), \quad \forall n = 0, 1, \ldots.$$

The bounded solutions of the above problems are

$$u^{\varepsilon,n} = \frac{f_0(u^n)}{i\sqrt{k^2 + i\varepsilon - n^2\pi^2} + ik} e^{i\sqrt{k^2+i\varepsilon-n^2\pi^2}x}.$$

Hence, we have

$$g_0(u^{\varepsilon,n}) = \frac{i\sqrt{k^2 + i\varepsilon - n^2\pi^2} - ik}{i\sqrt{k^2 + i\varepsilon - n^2\pi^2} + ik} f_0(u^n),$$

and

$$g_0(u^n) \overset{def}{=} \lim_{\varepsilon \to 0} g_0(u^{\varepsilon,n}) = \frac{i\sqrt{k^2 - n^2\pi^2} - ik}{i\sqrt{k^2 - n^2\pi^2} + ik} f_0(u^n). \tag{46}$$

Besides, it is straightforward to verify that

$$g_0(u^{\varepsilon,n}) = g_0(u^n) + \begin{cases} \frac{2\sqrt{i\varepsilon} f_0(u^n)}{k} + O(\varepsilon), & k = n\pi, \\ \frac{ik\varepsilon f_0(u^n)}{(\sqrt{k^2-n^2\pi^2}+k)^2\sqrt{k^2-n^2\pi^2}} + O(\varepsilon^2), & k \neq n\pi. \end{cases} \tag{47}$$

The expression (47) states that the convergence rate of $g_0(u^{\varepsilon})$ to

$$g_0(u) = \sum_{n=0}^{+\infty} g_0(u^n) \cos(n\pi y)$$

is of first order with respect to ε if k is unequal to any $n\pi$ with $n \geq 0$. If k is equal to some $n_0\pi$, which implies the resonance of the n_0-th mode in the y-direction, the convergence rate would degenerate to half order. But the LABP holds independent of the wavenumber k.

Based on the above analysis, we conjecture that, under some mild restrictions on the geometry and the refraction index function, the LABP holds for every $k > 0$ for more general semi-infinite periodic arrays. Some numerical evidences will be reported in the end of this section.

The LABP itself suggests a method for deriving the exact StS mapping on the left boundary Γ_0: first compute the exact StS mapping of the problem (45) for a given ε, denoted by $\mathscr{A}^{\varepsilon}_{\text{inf}}$, and then let ε tend to zero. In [20] the authors proposed a fast evaluation method for the exact StS mapping of the damped Helmholtz equation (45). The basic idea is as follows. For any $N > 0$, the damped Helmholtz equation (45) is well-posed on the domain $\cup_{j=0}^{N-1} C_j$, with the incoming Sommerfeld data f_0^{ε} and g_N^{ε} prescribed at the boundaries Γ_0 and Γ_N. Thus there are four linear scattering operators $\mathscr{A}^{\varepsilon}_N$, $\mathscr{B}^{\varepsilon}_N$, $\mathbb{C}^{\varepsilon}_N$ and $\mathscr{D}^{\varepsilon}_N$ satisfying

$$g_0^{\varepsilon} = \mathscr{A}^{\varepsilon}_N f_0^{\varepsilon} + \mathscr{B}^{\varepsilon}_N g_N^{\varepsilon}, \qquad f_N^{\varepsilon} = \mathbb{C}^{\varepsilon}_N f_0^{\varepsilon} + \mathscr{D}^{\varepsilon}_N g_N^{\varepsilon}.$$

Since g_N^{ε} goes to zero exponentially fast as N tends to infinity, it is reasonable to expect that $\mathscr{A}^{\varepsilon}_N$ converges and the limit is just the exact StS mapping $\mathscr{A}^{\varepsilon}_{\text{inf}}$. Note that the *fast doubling procedure* and the involved scattering operators are explained previously in Section

In Fig. (**41**) we plot the relative errors of the scattering operators $\mathscr{A}^{\varepsilon}_N$ compared to the reference operator $\mathscr{A}^{\varepsilon}_{\text{ref}}$, which is obtained by using the doubling technique 20 times, i.e., $N = 2^{20}$. Since FEM is used, the scattering operators are approximated by matrices of rank 65×65. We could see that the doubling technique really leads to an efficient algorithm. Also notice that when k^2 lies in the stop bands, for example $k^2 = 23, 31$, $\mathscr{A}_N$ itself converges as N goes to infinity. This implies that when k^2 is in the stop bands, we could derive the StS mapping directly without considering the LABP.

Next we explain how to let ε tend to zero. In light of the expression (47), if the resonance does not occur, the exact StS mapping $\mathscr{A}_{\text{inf}}$ is expected to bear an asymptotic expansion like

$$\mathscr{A}^{\varepsilon}_{\text{inf}} = \mathscr{A}_{\text{inf}} + \varepsilon \mathscr{A}^{(1)}_{\text{inf}} + \varepsilon^2 \mathscr{A}^{(2)}_{\text{inf}} + \cdots. \tag{48}$$

Thus in most cases, the convergence rate of the LABP is of first order. This observation is supported by the numerical evidences shown in Fig. (**42**). Note

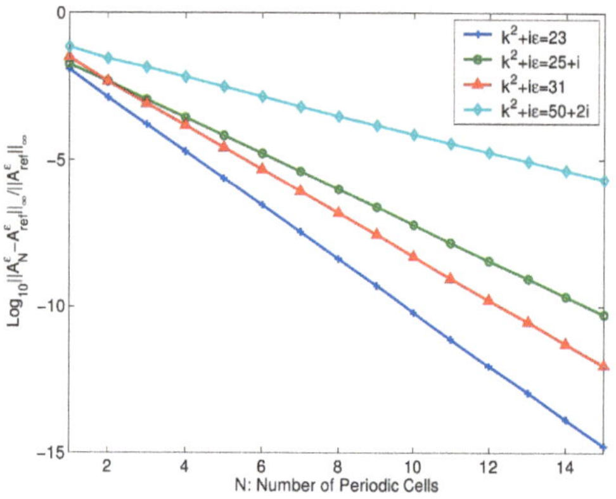

Fig. (**41**): Relative errors of $\mathscr{A}_N^\varepsilon$ to the reference StS mapping $\mathscr{A}_{\text{ref}}^\varepsilon$, which is obtained by setting $N = 2^{20}$.

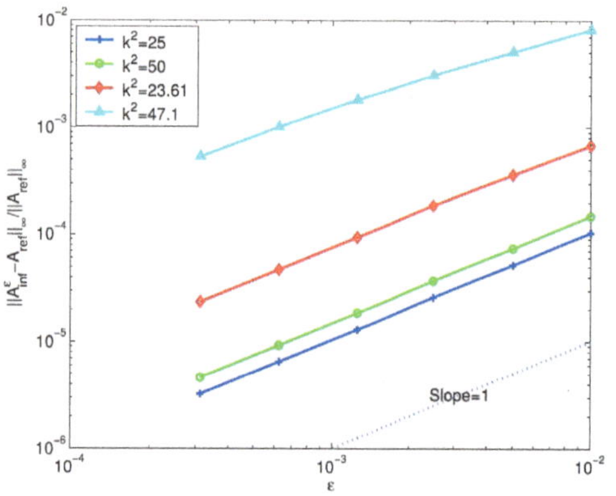

Fig. (**42**): The reference operator $\mathscr{A}_{\text{ref}}$ is obtained by setting $\varepsilon = 10^{-7}$.

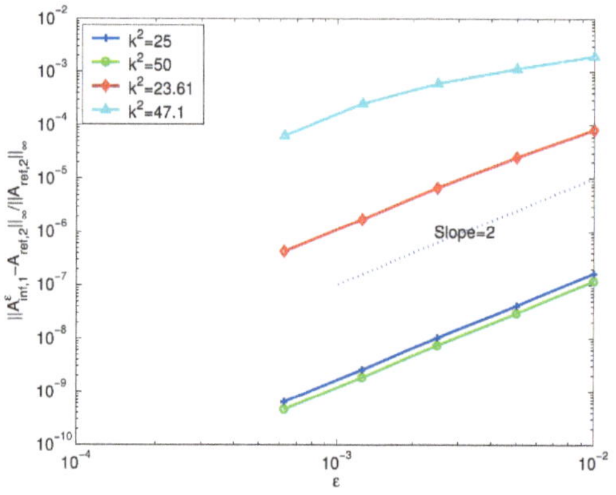

Fig. (**43**): The reference matrix $\mathscr{A}_{\text{ref},2}$ is obtained by using extrapolation technique twice with $\varepsilon_0 = 0.00125$, i.e., $\mathscr{A}_{\text{ref},2} = \mathscr{A}_{\text{inf}}^{\varepsilon_0}/3 - 2\mathscr{A}_{\text{inf}}^{\varepsilon_0/2} + 8\mathscr{A}_{\text{inf}}^{\varepsilon_0/4}/3$. $\mathscr{A}_{inf,1}^\varepsilon = -\mathscr{A}_{\text{inf}}^\varepsilon + 2\mathscr{A}_{\text{inf}}^{\varepsilon/2}$ is obtained by using extrapolation technique once.

that the convergence rate could be improved by standard extrapolation techniques. In Fig. (**43**) we show the errors of the StS operators extrapolated once to the reference operator, which is obtained by using extrapolations twice and setting a small damping parameter $\varepsilon_0 = 0.00125$. We could see that the accuracy is greatly improved, and second order rate can be clearly observed. We should also notice that if k is close to a resonance wave number, for example $k^2 = 23.61,\ 47.1$, the asymptotic convergence rate could only manifest for sufficiently small damping parameters.

11. ASYMPTOTIC BEHAVIOUR OF AN LABP SOLUTION

The last section showed that if k is not a resonance wave number, the extrapolation technique could yield very accurate solution. Obviously this algorithm needs to evaluate the scattering operators for a sequence of ε, and this turns out to be computationally quite expensive. Besides, though the chance of k being a resonance wave number is very rare, if k is close to a resonance wave number, the extrapolation method could not present very accurate result. In this section we will develop a new method by directly using the scattering operators for the undamped Helmholtz equation.

Recall from the last section that when k^2 lies in the stop bands, the exact StS mapping could be computed by the doubling technique without using the LABP. This is due to the fact that the solution lies in $L^2(\Omega)$, and thus it decays exponentially fast at infinity. If k^2 lies in the pass bands (complementary energy intervals of stop bands), in general an LABP solution cannot be expected to decay. Our basic idea is to separate those traveling (not-decaying) waves and evanescent (decaying) waves, and handle them by different means.

First let us introduce some notations. Suppose u and v are two solutions of the Helmholtz equation (43). Define the *co-related energy flux* of u and v as

$$\begin{aligned}\mathscr{E}(u,v) &= -2ik\left[(u_x,v)_{\Gamma_j} - (u,v_x)_{\Gamma_j}\right]\\ &= (f(u),f(v))_{\Gamma_j} - (g(u),g(v))_{\Gamma_j}.\end{aligned}$$

Besides, the energy flux of u is defined as $\mathscr{E}(u,u)$, which is also equal to

$$\mathscr{E}(u,u) = 4k\,\mathrm{Im}\int_{\Gamma_j} u_x\bar{u}\,dy.$$

We should remark that the co-related energy flux does not rely on the choice of Γ_j. Moreover, $\mathscr{E}(\cdot,\cdot)$ defines a sesquilinear form.

A nontrivial solution u of the Helmholtz equation (43) or (45) is regarded as a *Bloch wave* associated with the *Floquet multiplier* $\alpha \in \mathbb{C}$ if it satisfies the following two conditions

$$u|_{\Gamma_{i+1}} = \alpha u|_{\Gamma_i}, \quad u_x|_{\Gamma_{i+1}} = \alpha u_x|_{\Gamma_i}, \quad \forall i = 0, 1, \ldots.$$

We denote by $\mathbb{F}$ the set of all Floquet factors. A Bloch wave is referred to as evanescent, traveling, or anti-evanescent if the associated Floquet multiplier α satisfies $|\alpha| < 1$, $|\alpha| = 1$, or $|\alpha| > 1$. If $|\alpha| = 1$, we refer to α as a *unitary Floquet multiplier*. The set of unitary Floquet multipliers is denoted by $\mathbb{UF}$. Note that the Floquet factor cannot be zero due to the mentioned Holmgren uniqueness theorem. For any $\alpha \in \mathbb{F}$, all associated Bloch waves together with zero function form a linear space. This space, denoted by $\mathbb{E}_\alpha$, is called an *(α-periodic) eigenfunction space*. Here we list a couple of propositions about the Floquet theory from [42].

Proposition 4. *If $\alpha \in \mathbb{F}$, then $1/\alpha \in \mathbb{F}$ either.*

Proposition 5. *$\mathbb{UF}$ is a finite set. For any $\alpha \in \mathbb{UF}$, $N_\alpha = \dim \mathbb{E}_\alpha < +\infty$.*

Proposition 6. *Given two Floquet multipliers α_j and α_k, and two functions $\varphi_j \in \mathbb{E}_{\alpha_j}$ and $\varphi_k \in \mathbb{E}_{\alpha_k}$. If $\alpha_j \alpha_k^* \neq 1$, then $\mathscr{E}(\varphi_i, \varphi_j) = 0$.*

Proposition 7. *If u is an LABP solution, then the energy flux of u is nonnegative.*

Obviously, an LABP solution u cannot include the anti-evanescent Bloch waves, thus asymptotically, u is a combination of traveling Bloch waves. It is known that not every traveling Bloch wave is an LABP solution. We need to pick out those compatible with the LABP. To get some insight, let us consider the homogeneous waveguide problem.
Suppose $k = \pi$. Then the traveling Bloch wave space is given by

$$\mathrm{Span}\{e^{-i\pi x}, e^{i\pi x}, \cos(\pi y)\}.$$

If the x-period L is set as a non-integer positive number, then we get three unitary Floquet multipliers: $e^{-i\pi L}$ associated with $\mathrm{Span}\{e^{-i\pi x}\}$, $e^{i\pi L}$ with $\mathrm{Span}\{e^{i\pi x}\}$ and, 1 with $\mathrm{Span}\{\cos(\pi y)\}$. Since

$$\begin{aligned}\mathscr{E}(e^{-i\pi x}, e^{-i\pi x}) &= 4\pi \,\mathrm{Im} \int_0^1 (-i\pi e^{-i\pi x}) e^{i\pi x} dy\Big|_{x=0} \\ &= -4\pi^2,\end{aligned}$$

and an LABP solution has a nonnegative energy flux, $e^{-i\pi x}$ is thus not admissible. Comparatively, we have

$$\begin{aligned}\mathscr{E}(e^{i\pi x}, e^{i\pi x}) &= 4\pi \,\mathrm{Im} \int_0^1 (i\pi e^{i\pi x}) e^{-i\pi x} dy\Big|_{x=0} \\ &= 4\pi^2,\end{aligned}$$

and

$$\begin{aligned}\mathscr{E}(\cos(\pi y), \cos(\pi y)) &= 4\pi \,\mathrm{Im} \int_0^1 (0) e^{i\pi x} dy\Big|_{x=0} \\ &= 0.\end{aligned}$$

The problem appears when L is taken as an integer. For example, let us take $L = 1$. In this case there are two unitary Floquet multipliers 1 and -1, namely,

$$\begin{aligned}\alpha_1 &= -1 \longleftrightarrow \mathbb{E}_{\alpha_1} = \mathrm{Span}\{e^{-i\pi x},\ e^{i\pi x}\}, \\ \alpha_2 &= \ \ 1 \longleftrightarrow \mathbb{E}_{\alpha_2} = \mathrm{Span}\{\cos(\pi y)\}.\end{aligned}$$

$\mathbb{E}_{\alpha_2}$ represents a resonance space, and two-dimensional space $\mathbb{E}_{\alpha_1}$ contains both the left-going and right-going traveling waves. The problem is how to classify these two kind of waves. One may say the energy principle could still work, since obviously the Bloch wave $e^{i\pi x}$ is outgoing, and $e^{-i\pi x}$ is incoming. But the question is that $\mathbb{E}_{\alpha_1}$ may have different basis representation, for example,

$$\begin{aligned}\mathbb{E}_{\alpha_1} &= \mathrm{Span}\{e^{-i\pi x} + 2e^{i\pi x},\ e^{-i\pi x} + 3e^{i\pi x}\} \\ &= \mathrm{Span}\{e^{i\pi x} + 2e^{-i\pi x},\ e^{i\pi x} + 3e^{-i\pi x}\}.\end{aligned}$$

For the first representation, both basis functions are right-going, and for the second, both are left-going. However, generally we could not distinguish an LABP outgoing traveling wave only through its energy flux.
The above problem becomes even more severe if we take $L = 2$. In this case there exists only one unitary Floquet multiplier

$$\alpha = 1 \longleftrightarrow \mathbb{E}_\alpha = \mathrm{Span}\{e^{-i\pi x},\ e^{i\pi x}, \cos(\pi y)\}.$$

It is not hard to find different basis representations for $\mathbb{E}_\alpha$, which have completely different signs of energy flux. As a conclusion, if α is a unitary Floquet multiplier and the associated eigenfunction space $\mathbb{E}_\alpha$ is multi-dimensional, we have to resort to other criterion to determine the LABP right-going Bloch waves.
Let us remark here that for a three-dimensional waveguide problem, the chance for $\mathbb{E}_\alpha$ being multi-dimensional is absolutely not rare, though it seems true for two-dimensional waveguide problems.

Suppose $\alpha \in \mathbb{UF}$, and $\{\varphi_j\}_{j=1}^{N_\alpha}$ constitute a set of basis functions of $\mathbb{E}_\alpha$, *orthonormal* w.r.t. the n^2-weighted inner product $(\cdot,\cdot)_{n^2}$ defined as

$$(\varphi_j, \varphi_k)_{n^2} = \int_{C_0} n^2 \varphi_j \bar{\varphi}_k \, dy.$$

We define the *energy flux matrix* $M = (m_{jk})$ as

$$m_{jk} = \mathscr{E}(\varphi_j, \varphi_k), \qquad \forall j,k = 1,2,\cdots,N_\alpha.$$

It is easy to verify that M is a Hermitian matrix, which implies the existence of a unitary matrix U, such that

$$U^\top M \bar{U} = \Lambda = \mathrm{diag}(\lambda_1, \lambda_2, \ldots, \lambda_{N_\alpha}),$$

where λ_j are real eigenvalues of M ordered by

$$\lambda_1 \geq \lambda_2 \geq \cdots \geq \lambda_{m_1} > 0 = \lambda_{m_1+1} = \ldots$$
$$\cdots = \lambda_{m_2} = 0 > \lambda_{m_2+1} \geq \cdots \geq \lambda_{N_\alpha}.$$

We introduce a new set of basis function $\{\psi_j\}_{j=1}^{N_\alpha}$ as

$$(\psi_1, \ldots, \psi_{N_\alpha}) = (\varphi_1, \ldots, \varphi_{N_\alpha})U,$$

which will be referred to as a *canonical set of basis functions* of $\mathbb{E}_\alpha$. Now we could separate $\mathbb{E}_\alpha$ into three parts, i.e.,

$$\mathbb{E}_\alpha = \mathbb{R}_\alpha \oplus \mathbb{S}_\alpha \oplus \mathbb{L}_\alpha,$$

with

$$\mathbb{R}_\alpha = \mathrm{Span}\{\psi_1, \cdots, \psi_{m_1}\},$$
$$\mathbb{S}_\alpha = \mathrm{Span}\{\psi_{m_1+1}, \cdots, \psi_{m_2}\},$$
$$\mathbb{L}_\alpha = \mathrm{Span}\{\psi_{m_2+1}, \cdots, \psi_{N_\alpha}\}.$$

Proposition 8. *For any $\alpha \in \mathbb{E}_\alpha$, $\{\lambda_j\}_{j=1}^{N_\alpha}$ are invariant quantities, and $\mathbb{R}$, $\mathbb{S}$ and $\mathbb{L}$ are invariant subspaces of $\mathbb{E}_\alpha$. Besides, for any $\varphi_1 \in \mathbb{R}_\alpha$, $\varphi_2 \in \mathbb{S}_\alpha$, $\varphi_3 \in \mathbb{L}_\alpha$, we have*

$$\mathscr{E}(\varphi_1, \varphi_1) > 0, \quad \mathscr{E}(\varphi_2, \varphi_2) = 0, \quad \mathscr{E}(\varphi_3, \varphi_3) < 0.$$

For the homogeneous waveguide problem, it is straightforward to verify that $\mathbb{R}_\alpha$ is the admissible LABP Bloch wave space with positive energy flux. $\mathbb{S}_\alpha$ is the resonance wave space, which is also admissible to the LABP. Note that if $\mathbb{S}_\alpha$ is excluded from the asymptotic solution space, the Helmholtz equation would loose solvability for some incoming Sommerfeld data f_0.
Based on these facts, for a general semi-infinite periodic array, we make the following conjecture.

Conjecture 1. Suppose $\alpha_1, \ldots, \alpha_M$ are all unitary Floquet multipliers, and $\varphi_1^{\alpha_j}, \ldots, \varphi_{N_{\alpha_j}}^{\alpha_j}$ constitute a set of orthonormal basis functions of $\mathbb{R}_{\alpha_j} \oplus \mathbb{S}_{\alpha_j}$. Then asymptotically, any LABP solution u lies in the space

$$\mathrm{Span}\{\varphi_k^{\alpha_j} | j = 1, \ldots, M, k = 1, \ldots, N_{\alpha_j}\}. \tag{49}$$

Although we have no proof of this conjecture yet, its validity is strongly supported by the numerical tests given in the next section. Let us remark here that according to Proposition 6, $\{\varphi_k^{\alpha_j}\}_{j=1,k=1}^{M,N_{\alpha_j}}$ in fact constitute a set of basis functions of the LABP right-going Bloch wave space.

12. EVALUATION OF THE EXACT StS MAPPING

Based on Conjecture 1, we know when N is large, asymptotically,

$$f_N(u) \approx \sum_{j=1}^{M} \sum_{k=1}^{N_{\alpha_j}} t_k^j f_0(\varphi_k^{\alpha_j}),$$

$$g_N(u) \approx \sum_{j=1}^{M} \sum_{k=1}^{N_{\alpha_j}} t_k^j g_0(\varphi_k^{\alpha_j}).$$

Or in an abbreviated vector form,

$$f_N(u) \approx FT, \qquad g_N(u) \approx GT, \tag{50}$$

where

$$F = (F_1, \cdots, F_M), \quad G = (G_1, \cdots, G_M),$$
$$T = (T_1, \cdots, T_M)^\top$$

with

$$F_j = (f_0(\varphi_1^{\alpha_j}), \cdots, f_0(\varphi_{N_{\alpha_j}}^{\alpha_j})),$$
$$G_j = (g_0(\varphi_1^{\alpha_j}), \cdots, g_0(\varphi_{N_{\alpha_j}}^{\alpha_j})), \tag{51}$$
$$T_j = (t_1^{\alpha_j}, \cdots, t_{N_{\alpha_j}}^{\alpha_j}),$$

Recall that

$$g_0(u) = \mathscr{A}_N f_0(u) + \mathscr{B}_N g_N(u),$$
$$f_N(u) = \mathbb{C}_N f_0(u) + \mathscr{D}_N g_N(u).$$

Using (50) T could be derived by the least square method as

$$T \approx (F - \mathscr{D}_N G)^{-1} \mathbb{C}_N f_0(u). \tag{52}$$

Here, $^{-1}$ denotes the pseudo-inverse operator. We then have

$$g_0(u) = \mathscr{A}_N f_0(u) + \mathscr{B}_N g_N(u) \approx (\mathscr{A}_N + \mathscr{B}_N G(F - \mathscr{D}_N G)^{-1}\mathbb{C}_N) f_0(u),$$

which means that by putting

$$\tilde{\mathscr{A}_N} = \mathscr{A}_N + \mathscr{B}_N G(F - \mathscr{D}_N G)^{-1}\mathbb{C}_N,$$

the limit of $\tilde{\mathscr{A}_N}$ would give the exact StS mapping $\mathscr{A}_{\inf}$ on the left boundary Γ_0.

The key step to implement the above algorithm is to derive a canonical set of basis functions for all unitary Floquet multipliers, i.e. we need to compute the functions F_j and G_j defined in (51). This objective can be achieved by the following steps:

1. Solve the generalized eigenvalue problem

$$\begin{pmatrix} -\mathscr{A}_1 & I \\ -\mathbb{C}_1 & 0 \end{pmatrix} \begin{pmatrix} f_0 \\ g_0 \end{pmatrix} = \alpha \begin{pmatrix} 0 & \mathscr{B}_1 \\ -I & \mathscr{D}_1 \end{pmatrix} \begin{pmatrix} f_0 \\ g_0 \end{pmatrix}$$

   to obtain all (different) unitary Floquet multipliers $\{\alpha_j\}_{j=1}^M$ and their associated generalized eigenfunctions $(f_{0,k}^{\alpha_j}, g_{0,k}^{\alpha_j})$, $k = 1, \cdots, N_{\alpha_j}$.

2. If $\mathbb{E}_{\alpha_j}$ is one-dimensional, i.e. $N_{\alpha_j} = 1$, compute the energy flux of the eigenfunction φ_1^j associated with the Sommerfeld data $(f_{0,1}^{\alpha_j}, g_{0,1}^{\alpha_j})$ by

$$\mathscr{E}(\varphi_1^j, \varphi_1^j) = (f_{0,1}^{\alpha_j}, f_{0,1}^{\alpha_j})_{\Gamma_0} - (g_{0,1}^{\alpha_j}, g_{0,1}^{\alpha_j})_{\Gamma_0}.$$

   If and only if $\mathscr{E}(\varphi_1^j, \varphi_1^j) \geq 0$, then φ_1^j is an admissible LABP traveling Bloch wave, i.e., $F_j = (f_{0,1}^{\alpha_j})$, $G_j = (g_{0,1}^{\alpha_j})$. Otherwise, $F_j = G_j = \emptyset$.

3. If $\mathbb{E}_{\alpha_j}$ is multi-dimensional, i.e., $N_{\alpha_j} > 1$, derive a set of orthonormal eigenfunctions $\{\varphi_k^{\alpha_j}\}_{k=1}^{N_{\alpha_j}}$ of the following problem

$$\Delta u + k^2 n^2 u = 0,$$
$$u|_{\Gamma_1} = \alpha_j u|_{\Gamma_0}, \quad u_x|_{\Gamma_1} = \alpha_j u_x|_{\Gamma_0}.$$

   Compute the associated Sommerfeld data $\{f_0(\varphi_k^{\alpha_j})\}$ and $\{g_0(\varphi_k^{\alpha_j})\}$. Compute the energy matrix $M = (m_{kl})$ with

$$m_{kl} = (f_0(\varphi_k^{\alpha_j}), f_0(\varphi_l^{\alpha_j}))_{\Gamma_0} - (g_0(\varphi_k^{\alpha_j}), g_0(\varphi_l^{\alpha_j}))_{\Gamma_0},$$

   for all $k, l = 1, \cdots, N_{\alpha_j}$. Find a unitary matrix $U = (u_{lk})$ to diagonalize M, such that

$$U^\top M\bar{U} = \Lambda = \text{diag}(\lambda_1, \lambda_2, \ldots, \lambda_{N_{\alpha_j}}),$$

   where λ_j are real eigenvalues of M ordered by

$$\lambda_1 \geq \lambda_2 \geq \cdots \geq \lambda_{m_1} > 0 = \lambda_{m_1+1} = \ldots \cdots = \lambda_{m_2} = 0 > \lambda_{m_2+1} \geq \cdots \geq \lambda_{N_{\alpha_j}}.$$

   Set $F_j = (F_j^1, \cdots, F_j^{m_2})$ and $G_j = (G_j^1, \cdots, G_j^{m_2})$ with

$$F_j^k = \sum_{l=1}^{N_{\alpha_j}} f_0(\varphi_l^{\alpha_j}) u_{lk}, \qquad G_j^k = \sum_{l=1}^{N_{\alpha_j}} g_0(\varphi_l^{\alpha_j}) u_{lk},$$

   for all $k = 1, \cdots, m_2$.

4. Finally, set $F = (F_1, \cdots, F_N)$ and $G = (G_1, \cdots, G_N)$.

In the following we will report our numerical tests. For simplicity, we refer to the StS mapping derived with the LABP as LABP-StS, and the StS mapping based on the asymptotic expansion of the traveling Bloch waves as ASYM-StS. First we consider the PA-One. In this case the analytical StS mapping is available. For the n-th mode in the y-direction, the exact StS mapping is given as in (46). The computed StS mapping, no matter which method is employed, is diagonalizable. In Table 6 we list the errors of ASYM-StS. We see generally the asymptotic method presents very accurate results except on the resonance wave number. For example, if $k = \pi$, the first y-mode is resonant.

Table 6: Errors of Direct computation.

	$k=\pi$	$k=\frac{5\pi}{4}$	$k=\sqrt{2}\pi$	$k=\sqrt{3}\pi$
$n=0$	1.50(-9)	4.60(-9)	7.02(-12)	5.91(-13)
$n=1$	7.58(-6)	1.78(-9)	1.07(-9)	9.44(-13)
$n=2$	2.13(-12)	3.52(-12)	1.31(-11)	3.23(-12)
$n=3$	5.44(-13)	8.74(-13)	2.80(-12)	5.40(-13)
$n=4$	2.28(-13)	3.24(-13)	1.00(-12)	2.10(-13)

In Table 7 we list the errors of the LABP-StS. They are derived with two times of extrapolation. We see that except at the resonance wave numbers, this method presents the results at least of the same quality of those derived by the asymptotic method. But when resonance occurs, the extrapolation technique is only of little use. In order to obtain high accuracy, one has to make the damping parameter very small, but this probably implies a numerical stability problem.

For the other two periodic structures PA-Two and PA-Three, no analytical expression is available on the exact StS mapping. We compare the numerical solutions by two different methods. From Table 8-9, we could conclude in principle these two methods

Table 7: $\varepsilon = 0.00125$. Extrapolation.

	$k=\pi$	$k=\frac{5\pi}{4}$	$k=\sqrt{2}\pi$	$k=\sqrt{3}\pi$
$n=0$	5.03(-9)	3.53(-12)	6.91(-12)	1.07(-12)
$n=1$	5.68(-3)	7.30(-12)	1.52(-8)	1.49(-12)
$n=2$	2.26(-12)	3.43(-12)	1.26(-11)	3.40(-12)
$n=3$	7.51(-13)	8.22(-13)	2.99(-12)	5.82(-13)
$n=4$	2.39(-13)	2.74(-13)	1.06(-12)	2.37(-13)

bring the same results. When k is away from the resonance wave number, these two methods present the results of same quality. But their difference becomes big when k approaches the resonance wave number. Considering the results for the homogeneous waveguide problem, we thus believe at the resonance wave numbers, the asymptotic method presents better solution.

Table 8: $\varepsilon = 0.00125$. Comparison. PA-Two.

	Relative error
$k^2 = 25$	1.31(-12)
$k^2 = 50$	3.26(-12)
$k^2 = 23.61$	3.89(-8)
$k^2 = 47.1$	6.76(-5)

Table 9: $\varepsilon = 0.00125$. Comparison. PA-Three.

	Relative error
$k^2 = 5$	9.58(-13)
$k^2 = 25$	9.26(-13)
$k^2 = 11.20$	7.16(-9)
$k^2 = 19.29$	6.23(-10)

CONCLUSIONS

In this chapter we have generalized a recent result of Zheng [78] and derived an exact Dirichlet-to-Neumann artificial boundary condition for general problems with periodic structures at infinity. We considered in detail the bound state problem for the Schrödinger operator and a second order hyperbolic equation in two space dimensions. The proof of this new kernel expression for the artificial boundary condition was presented recently by Zhang and Zheng [76].

Secondly, we introduced a fast evaluation method of the Sommerfeld-to-Sommerfeld (StS) mapping for periodic structure problems. Our proposed strategy is an improvement of the recently developed recursive doubling process by Yuan and Lu for the evaluation of Dirichlet-to-Neumann maps. We presented numerical results for the Helmholtz equation and the time-dependent Schrödinger equation in one and two space dimensions with periodic structures including cases where the method of Yuan and Lu fails.

In the last part of this chapter we considered the Helmholtz equation in the semi-infinite periodic array in this paper. Since no radiation-like boundary condition is specified at infinity, the Helmholtz equation is in general not well-posed. To solve this problem we employed the limiting absorption principle. We have proposed a new algorithm which combines the doubling procedure of the second part and the extrapolation technique to obtain high-accuracy approximation to the exact StS mappings. Considering the computational complexity, we present another method which uses the asymptotic behavior of a limiting absorption principle solution. Though we could not prove, the validity of this method is strongly supported by our numerical evidences.

We believe that these results can be generalized to the derivation of fully discrete artificial boundary conditions in the spirit of [15] for periodic potential problems. These boundary conditions are directly derived for the numerical scheme. Another very challenging task would be the extension of the present work to multi-dimensional problems with periodic structures. Furthermore, our ideas can be extended to more complicated wave-like equations, such as Maxwell's equations and elastic wave equations. Besides, many relevant theoretical problems are left open in this chapter.

REFERENCES

[1] Abboud T. Electromagnetic waves in periodic media. in: Kleinman R, Angell T, Colton D, Santosa F, Stakgold I (Eds.). Proceedings of the 2nd International Conference on Mathematical and Numerical Aspects of Wave Propagation, Newark, DE, 1993, SIAM, Philadelphia, 1993: 1-9.

[2] Alpert B, Greengard L, Hagstrom T. Nonreflecting boundary conditions for the time-dependent wave equation. J Comput Phys 2002; 180: 270-296.

[3] Antoine X, Besse C. Unconditionally stable discretization schemes of non-reflecting boundary conditions for the one-dimensional Schrödinger equation. J Comput Phys 2003; 181: 157-175.

[4] Antoine X, Besse C, Mouysset V. Numerical schemes for the simulation of the two-dimensional Schrödinger equation using non-reflecting boundary conditions. Math Comp 2004; 73: 1779-1799.

[5] Antoine X, Besse C, Descombes S. Artificial boundary conditions for one-dimensional cubic nonlinear Schrödinger equations. SIAM J Numer Anal 2006; 43: 2272-2293.

[6] Antoine A, Arnold A, Besse C, Ehrhardt M, Schädle A. A review of transparent and artificial boundary conditions techniques for linear and nonlinear Schrödinger equations. Commun Comput Phys 2008; 4: 729-796. (open-access article)

[7] Antoine X, Besse C, Szeftel J. Towards accurate artificial boundary conditions for nonlinear PDEs through examples. CUBO A Math J 2009; 11: 29-48.

[8] Antoine X, Besse C, Klein P. Absorbing boundary conditions for the one-dimensional Schrödinger equation with an exterior repulsive potential. J Comput Phys 2009; 228: 312-335.

[9] Arscott FM. Periodic differential equations. Pergamon Press, Oxford, 1964.
[10] Barth M, Benson O. Manipulation of dielectric particles using photonic crystal cavities. Appl Phys Lett 2006; 89: 253114.
[11] Bastard G. Wave mechanics applied to semiconductor heterostructures. les éditions de physique, Les Ulis Cedex, France, 1988.
[12] Bienstman P, Baets R. Optical modelling of photonic crystals and VCSELs using eigenmode expansion and perfectly matched layers. Opt Quant Electron 2001; 33: 327-341.
[13] Braga AMB, Hermann G. Floquet waves in anisotropic periodically layered composites. J Acoust Soc Am 1992; 91: 1211-1227.
[14] Brillouin L. Wave propagation in periodic structures. Dover Publications, New york, 1953.
[15] Ehrhardt M. Discrete Artificial Boundary Conditions. Dissertation, TU Berlin, 2001.
[16] Ehrhardt M, Arnold A. Discrete transparent boundary conditions for the Schrödinger equation. Riv Mat Univ Parma 2001; 6: 57-108.
[17] Ehrhardt M, Mickens RE. Solutions to the discrete Airy equation: Application to parabolic equation calculations. J Comput Appl Math 2004; 172: 183-206.
[18] Ehrhardt M. Discrete transparent boundary conditions for Schrödinger-type equations for non-compactly supported initial data. Appl Numer Math 2008; 58: 660-673.
[19] Ehrhardt M, Zheng C. Exact artificial boundary conditions for problems with periodic structures. J Comput Phys 2008; 227: 6877-6894.
[20] Ehrhardt M, Han H, Zheng C. Numerical simulation of waves in periodic structures. Commun Comput Phys 2009; 5: 849-870.
[21] Ehrhardt M, Sun J, Zheng C. Evaluation of scattering operators for semi-infinite periodic arrays. Commun Math Sci 2009; 7: 347-364.
[22] Ehrhardt M, Zheng C. Implementing exact absorbing boundary condition for the linear one-dimensional Schrödinger problem with variable potential by Titchmarsh-Weyl theory. Preprint No. 1426, WIAS Berlin, July 2009,
[23] Fliss S, Joly P. Exact boundary conditions for time-harmonic wave propagation in locally perturbed periodic media. Appl Numer Math 2009; 59: 2155-2178
[24] Foulkes WMC, Mitas L, Needs RJ, Rajagopal G. Quantum Monte Carlo simulations of solids. Rev Mod Phys 2001; 73: 33-83.
[25] Fox C, Oleinik V, Pavlov B. A Dirichlet-to-Neumann map approach to resonance gaps and bands of periodic networks. In: N. Chernov, Y. Karpeshina, I. W. Knowles, R. T. Lewis and R. Weikard (Eds.), Recent advances in differential equations and mathematical physics Contemp Math 412, American Mathematical Society, Providence, RI, 2006: 151-169.
[26] Givoli D. Non-reflecting boundary conditions. J Comput Phys 1991; 94: 1-29.
[27] Griffiths DJ, Steinke CA. Waves in locally periodic media. Am J Phys 2001; 69: 137-154.
[28] Guo SP, Albin S. Numerical techniques for excitation and analysis of defect modes in photonic crystals. Opt Express 2003; 11: 1080-1089.
[29] Hagstrom T. Radiation boundary conditions for the numerical simulation of waves. Acta Numer 1999; 8: 47-106.
[30] Han H. The artificial boundary method-numerical method of partial differential equations on unbounded domains. in: Frontiers and Propests of Contemporary Applied Mathematics. Li T, Zheng P (Eds.) Higher Education Press, World Scientific, 2005, pp. 33-58.
[31] Han Z, Forsberg E, He S. Surface plasmon Bragg gratings formend in metal-insulator-metal waveguides. IEEE Photonics Tech Lett 2007; 19: 91-93.
[32] Harari I, Patlashenko I, Givoli D. Dirichlet-to-Neumann maps for unbounded wave guides. J Comput Phys 1998; 143: 200-223.
[33] Helfert SF, Pregla R. Efficient analysis of periodic structures. J Lightwave Technol 1998; 16:1694-1702.
[34] Ho PL, Lu YY. A bidirectional beam propagation method for periodic waveguides. IEEE Photonics Tech Lett 2002; 14: 325-327.
[35] Hörmander L. Linear partial differential operators, 4th Printing, Grundlehren der mathematischen Wissenschaften 116, Springer-Verlag, 1969.
[36] Jacobsen J. Analytical, numerical, and experimental investigation of guided waves on a periodically strip-loaded dielectric slab. IEEE Trans Antennas Prop 1970; 18: 379-388.
[37] Hu Z,Lu YY. Efficient analysis of photonic crystal devices by Dirichlet-to-Neumann maps. Opt Express 2008; 16: 17383-17399.
[38] Hu Z, Lu YY. Efficient numerical method for analyzing coupling structures of photonic crystal waveguides. IEEE Photonics Tech Lett 2009; 23: 1737-1739.
[39] Johnson SG, Joannopoulos JD. Photonic crystals : the road from theory to practice. Kluwer Academic Publishers, 2002.
[40] Joly P, Li JR, Fliss S. Exact Boundary Conditions for Periodic Waveguides Containing a Local Perturbation. Commun Comput Phys 2006; 1: 945-973.
[41] Keller JB, Givoli D. Exact non-reflecting boundary conditions. J Comput Phys 1989; 82: 172-192.
[42] Kuchment P. Floquet theory for partial differential equations. Operator Theory: Advances and Applications 60. Birkhäuser Verlag, Basel, 1993.
[43] Kuchment P. The mathematics of photonic crystals. Chapter 7 in: Mathematical modeling in optical science. Frontiers in applied mathematics 22. SIAM, Philadelphia, 2001.
[44] Li S, Lu YY. Computing photonic crystal defect modes by Dirichlet-to-Neumann maps. Opt Express 2007; 15: 14454-14466.
[45] Little BE, Haus HA. A variational coupled-mode theory for periodic waveguides. IEEE J Quantum Elect 1995; 31: 2258-2264.
[46] Y.Y. Lu, Computing Dirichlet-to-Neumann maps for numerical simulation of photonic crystal structures, Proc Nat Inst Math Sci 2008; 3: 65-70.
[47] Ludwig A, Leviatan Y. Analysis of bandgap characteristics of two-dimensional periodic structures by using the source-model technique. J Opt Soc Am A 2003; 20: 1553-1562.
[48] Papadakis JS. Impedance formulation of the bottom boundary condition for the parabolic equation model in underwater acoustics. NORDA Parabolic Equation Workshop, NORDA Tech Note 143, 1982.
[49] Potel C, Gatignol P, De Belleval JF. Energetic criterion for the radiation of floquet waves in infinite anisotropic periodically multilayered media. Acustica-Acta Acustica 2001; 87: 340-351.
[50] Reed M, Simon B. Methods of modern mathematical physics II: Fourier analysis, self-adjointness, Academic Press, San Diego 1975.
[51] Richards JA. Analysis of Periodically Time-Varying Systems. Springer-Verlag, 1983.
[52] Rodríguez-Esquerre VF, Koshiba M, Hernández-Figueroa HE. Finite-element analysis of photonic crystal cavities: Time and frequency domains. J Lightwave Tech 2005; 23: 1514-1521.
[53] Sakoda K. Optical Properties of Photonic Crystals. Springer-Verlag, Berlin, 2001.
[54] Schmidt F. Solution of Interior-Exterior Helmholtz-Type Problems Based on the Pole Condition Concept: Theory and Algorithms. Habilitation thesis, Free University Berlin, 2002.
[55] Schulenberger JR, C.H. Wilcox CH. The limiting absorption principle and spectral theory for steady-state wave propagation in inhomogeneous anisotropic media. Arch Ration Mech Anal 1971; 41: 46-65.
[56] Semenikhin IA, Pavlov BS, Ryzhii VI Plasma waves in two-dimensional electron channels: propagation and trapped modes. Preprint No. NI07028-AGA of the Isaac Newton Institute for Mathematical Sciences, 2007.
[57] Shipman S, Volkov D. Guided modes in periodic slabs: existence and nonexistence. SIAM J Appl Math 2007; 67: 687-713.
[58] Sjöberg D. Analysis of large finite periodic structures using infinite periodicity methods. Technical Report TEAT-7143, Lund Institute of Technology, Sweden, 2006.
[59] Smith DR, Dalichaouch R, Kroll N, Schultz S, McCall SL, Platzman PM. Photonic band structure and defects in one and two dimensions. J Opt Soc Am B 1993; 10: 314-321.
[60] Smith DR, Pendry JB, Wiltshire MCK. Metamaterials and Negative Refractive Index. Science 2004; 305: 788-792.
[61] Søndergard T, Bozhevolnyi SI, Boltasseva A. Theoretical analysis of ridge gratings for long-range surface plasmon polaritons. Phys Rev B 2006; 73: 045320.
[62] Soussi S. Convergence of the supercell method for defect modes calculations in photonic crystals. SIAM J Numer Anal 2005; 43: 1175-1201.
[63] Tausch J, Butler J. Floquet Multipliers of periodic Waveguides via Dirichlet-to-Neumann Maps. J Comput Phys 2000; 159: 90-102.
[64] Tausch J, Butler J. Efficient Analysis of Periodic Dielectric Waveguides using Dirichlet-to-Neumann Maps. J Opt Soc Amer A 2002; 19: 1120-1128.
[65] Tsynkov SV. Numerical solution of problems on unbounded domains. A review. Appl Numer Math 1998; 27: 465-532.
[66] Vouvakis MN, Kezhong Z, Lee JF. Finite-element analysis of infinite periodic structures with nonmatching triangulations. IEEE Trans Magn 2006; 42: 691-694.
[67] Vainberg BR. Principles of radiation, limit absorption and limit amplitude in the general theory of partial differential equations. Russ Math Surv 1966; 21: 115-193.

[68] Wacker A. Semiconductor Superlattices: A model system for nonlinear transport. Phys Rep 2002; 357: 1-111.
[69] Wu Y, Lu YY. Dirichlet-to-Neumann map method for analyzing crossed arrays of circular cylinders. J Opt Soc Am B 2009; 26: 1442-1449.
[70] Yariv A, Yeh P. Optical Waves in Crystals – Propagation and Control of Laser Radiation. Wiley Series in Pure and Applied Optics, Wiley, 2002.
[71] Yuan L, Lu YY. An efficient bidirectional propagation method based on Dirichlet-to-Neumann maps. IEEE Photonics Tech Lett 2006; 18: 1967-1969.
[72] Yuan L, Lu YY. Dirichlet-to-Neumann map method for second harmonic generation in piecewise uniform waveguides. J Opt Soc Am B 2007; 24: 2287-2293.
[73] Yuan L, Lu YY. A Recursive Doubling Dirichlet-to-Neumann Map Method for Periodic Waveguides. J Lightwave Technol 2007; 25: 3649-3656.
[74] Yuan J, Lu YY, Antoine X. Modeling photonic crystals by boundary integral equations and Dirichlet-to-Neumann maps. J Comput Phys 2008; 227: 4617-3629.
[75] Yuan L, Lu YY. An efficient numerical method for optical waveguides with holes. J Lightwave Technology 2009; 27: 2557-2562.
[76] Zhang M, Zheng C. Closed form impedance expression for periodic Schrödinger operators with symmetric coefficient functions. to be submitted to: Frontiers of Mathematics in China, 2009.
[77] Zheng C. Approximation, stability and fast evaluation of an exact artificial boundary condition for the one-dimensional heat equation. J Comput Math 2007; 25: 730-745.
[78] Zheng C. An exact boundary condition for the Schrödinger equation with sinusoidal potentials at infinity. Commun Comput Phys 2007; 3: 641-658.

CHAPTER 7

Negative Refraction Based Applications in Artificial Periodic Media

O. Vanbésien

Institut d'Electronique, de Microélectronique et de Nanotechnologie (IEMN – UMR CNRS 8520) - Université des Sciences et Technologies de Lille - Avenue Poincaré BP 60069 - 59652 Villeneuve d'Ascq Cedex – France (email: Olivier.Vanbesien@iemn.univ-lille1.fr)

Abstract: In this chapter, we will address the potential applications of negative refraction in artificial periodic media. Different approaches will be considered depending on the targeted wavelength operation from microwaves down to optics. First, physical concepts to create such an abnormal propagation regime will be described: (i) negative permittivity and negative permeability engineering in patterned metallic or metallo-dielectric structures namely metamaterials and (ii) band structure engineering in full dielectric structures namely photonic crystals. Second, practical examples of real devices will be given, starting from negative refraction evidence in a two dimensional prism in microwaves, backward wave propagation in a periodically loaded transmission line to end with subwavelength focusing by a photonic crystal slab for optical waves. In the third part, other exciting properties beyond negative refraction will be evoked with the design of hyperlenses and cloaking devices.

INTRODUCTION

Artificial Periodic Media (APM) receive a strong interest from various scientific communities since they represent a quasi infinite field of new problems for physicists and/or mathematicians and of potential original applications for engineers [1]. Quite general concepts can be established and applied to the whole wavelength spectrum from microwaves down to optics including terahertz, far-, mid- and near-infrared. If electromagnetic waves (including optical waves) are thoroughly investigated (the photonic approach), more recently strong efforts have also been devoted to acoustic waves (the phononic approach). In this chapter, we will restrict ourselves to the photonic approach. We will explore the new wave propagation properties afforded by these APM by assessing the strong correlation between the length scale at which the APM is structured (a) and the targeted wavelength of operation (λ). To this aim, we will examine precisely how and when it will be possible to describe the APM by classical effective parameters as permittivity (ε), permeability (μ), refraction index (n) or surface impedance (z). Using these concepts, two classes of materials could be defined, metamaterials and electromagnetic or photonic crystals whose operating regimes obey $a/\lambda << 1$ and $a \sim \lambda/2- \lambda/4$ respectively. In both cases, ultra-refraction or negative refraction regimes can be identified even if underlying physical principles will differ. As one goal of this chapter is to describe some real prototypes for different characteristic operating wavelengths, we will see that the choice of one class of material or the other will be mainly imposed by the technological feasibility of the device. Indeed, as wavelength decreased down to a few hundreds of nanometers for the visible, material structuring has to follow the same trend. This implies that one has to control fabrication at nanometer scale and one has to face numerous challenges to reach such objectives. Moreover the choice of constitutive materials and of their properties in the targeted wavelength domain (metals or dielectrics including semiconductors) will direct the design choice.

One of the main motivations of this research domain is that it is expected that this tailoring of parameters will permit to go beyond classical wave properties. As specific examples, let us mention the break of the Rayleigh limit of $\lambda/2n$ for lenses resolution or the possibility to continuously move from a positive to a negative value of refractive index passing through the zero value but with a non-zero wave group velocity. Also expected is a dramatic miniaturization of devices owing to the possibility to bend light over distances much shorter than the wavelength without additional losses due to radiation. Dream or future reality, we will see that all these effects based on the dispersive properties of complex arrangements will have to face trade-offs in terms of losses or possible bandwidths depending strongly on the primary choices of constitutive elements or cells made to build the APM.

This chapter will be organized as follows. First, some basic concepts to define abnormal propagation (ultra- and negative refraction) regimes in APM will be given.

Matthias Ehrhardt (ED)

Physical properties will be detailed: (i) negative permittivity and negative permeability engineering in patterned metallic or metallo-dielectric structures namely metamaterials and (ii) band structure engineering in full dielectric structures namely photonic crystals. In a second part, negative refraction based devices will be shown, starting from negative refraction evidence in a two dimensional prism in microwaves, backward wave propagation in a periodically loaded transmission line to end with subwavelength focusing by a photonic crystal slab for optical waves. For each different approach, constraints for fabrication will be addressed as well as pre-requisites for measurement techniques which have to be included at the design level. In a third part, it will be theoretically shown that other exciting properties beyond negative refraction can be obtained if the material patterning is performed in the three dimensions of space by a full engineering of effective parameter tensors. This will allow, sometimes with rough approximations, the design of new functionalities as hyperlenses and cloaking devices.

PHYSICAL APPROACHES FOR NEGATIVE REFRACTION

Basic concepts

For the electromagnetic or optical wave propagation studies in the linear regime, it is very convenient to describe the propagation medium by effective parameters. They can be described as tensors if the studied material possesses spatial dependent properties or by just one constant value if full isotropy can be assumed. In this latter case, that we will consider in the two next sections, to solve the wave propagation equation (namely, the Helmholtz equation deduced from Maxwell's equations), a couple of parameter is sufficient: the effective permittivity (ε) and the effective permeability (μ). In optics, more commonly used are the refractive index (n) and the surface impedance (z). These two sets of parameters are relied by:

$$n^2 = \epsilon\mu \,;\; z^2 = \varepsilon/\mu \qquad \text{(Eq. 1)}$$

In general, as real materials present losses, all these parameters are complex numbers and the passage from one set to another is not straightforward and special care has to be taken to extract the correct parameters from the different mathematical branches arising from the square extraction.
Veselago [2], in the late sixties, has explored what could be the properties of materials with simultaneous negative real parts of ε and μ. He stated that in this case, the negative real part of the refractive index became also negative whereas the real part of surface impedance remained positive. Using this assumption and generalizing the "Snell-Descartes" law to any value of refraction index, Veselago has proposed the flat lens concept : a simple slab of material can focus behind the slab a point source located in front of the slab owing to simple geometrical conditions as illustrated in Fig. (1a). Moreover special properties can be obtained if n = -1 and z = 1, in other terms if $\varepsilon = \mu = -1$ simultaneously, the anti-vacuum like situation. In this case, particular excitation of surface modes is able to feed the image also with the evanescent components of the harmonic wave decomposition leading in an "idealized" world to perfect resolution (Fig (1c)). This concept of superlens, often related to the notion of evanescent wave "amplification" (Fig. (1b)) has provoked many debates not definitely closed up to now [3-10]. It can be mentioned that this proposition to use negative refraction was not limited to focusing. Veselago has also investigated the reversal of the Doppler and the Cerenkov effects. Let's see now how these theoretical concepts have become reality.

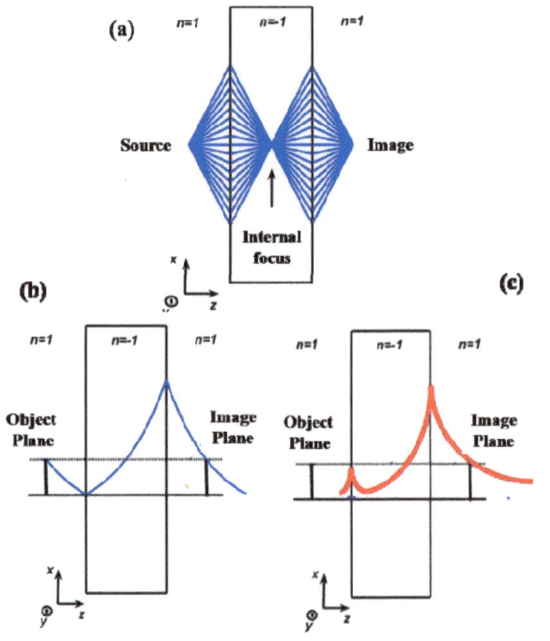

Fig. (1). Flat lens concept based on a negative refraction index homogeneous material: (a) propagating waves, (b) evanescent wave amplification, (c) surface mode excitation.

The metamaterial approach: permittivity and permeability engineering

Veselago's proposal was exciting, but at that time no natural or fabricated object was able to show these properties. In the late nineties, Pendry [11] has proposed to use some metallic resonators called split ring resonators to mimic a negative permeability medium. He based his demonstration on a classical Lorentz approach to describe the permeability evolution versus frequency, which become negative over a limited frequency bandwidth. A network of such resonators associated under certain conditions with a diluted array of metallic wires [12] exhibiting below their equivalent plasma frequency a negative permittivity leads theoretically to an equivalent medium with a real but negative refractive index where waves can propagate. Smith et al [4] first bring this concept to reality with the measurement of a significant transmission level in a prototype operating in microwaves [13-15]. This first impressive demonstration was either an indirect way to evidence the existence of negative refraction. Insertion loss is not sufficient to prove the effect. One can imagine that when mixed the two networks interfere thus modifying their intrinsic properties. In other words, this material was also labeled as a "left-handed" material since in this operating regime magnetic field, electric field and wave vector forms an indirect trihedron. In such regime, phase velocity and group velocity are of opposite sign. In addition since energy always flows in the direction imposed by user, it results in a backward wave propagation regime.

Drude model:

$$\varepsilon = 1 - \frac{\omega_p^2}{\omega^2 + i\gamma\omega}$$

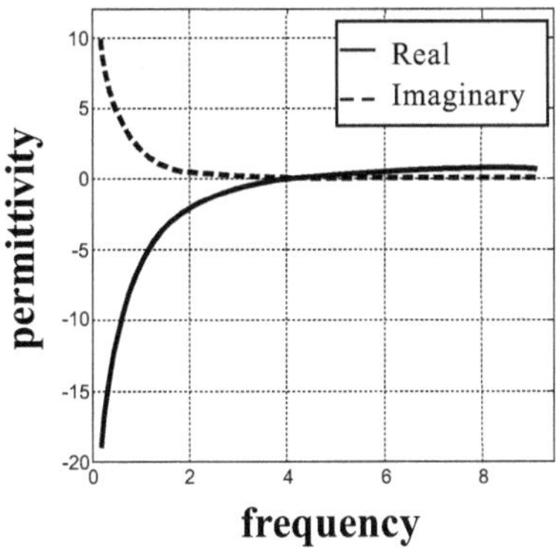

Fig. (2). Drude model for the permittivity of a fictitious metallic wire array with $\omega_p = 5$ and $\gamma = 0.5$.

At first order, a first model for such a metamaterial can be done by considering separately both phenomena: the wire array and the metallic resonator. For the former whose electrical activity is exploited, the permittivity can be described using a classical Drude model (see Fig. (2)) in which the plasma frequency is fixed owing to the geometrical parameters of the array (period, wire thickness…) leading to a reduced plasma frequency characteristic of the "dilution" of the metal. By varying parameters values comprised between the bulk metal one (thousand of terahertz) and thin wire networks, low values (a few gigahertz) can be obtained. As shown in Fig. (2), below the plasma or equivalent plasma frequency, the equivalent permittivity is negative. One can note that this engineering is necessary if one wants to work reasonable negative values since the real value drops rapidly below the plasma frequency.

To be more precise, an imaginary part can be added (γ) to take losses into account.

For the permeability, to traduce the resonant effect, a Lorentz model (see Fig. (3)) is assumed. Here, two frequencies are used: the first describes the resonance itself and depends on the resonator geometry whereas the second one points the transition between a negative and a positive value (tending asymptotically towards 1) after the resonance. Let us mention that this magnetic activity is obtained with classical metals without any active magnetic material. Conversely to ε, the negative region is here bounded between these two frequencies as shown in Fig. (3). Here again, the choice of pertinent geometrical parameters allows to reach a wide range of frequencies and losses can be added. Let's notice that to increase frequency operation, one have to reduce resonator dimensions with a mid infrared limit for nanometric structures.

Lorentz model :

$$\mu = 1 - \frac{\omega_p^2 - \omega_0^2}{\omega^2 - \omega_0^2 + i\gamma\omega}$$

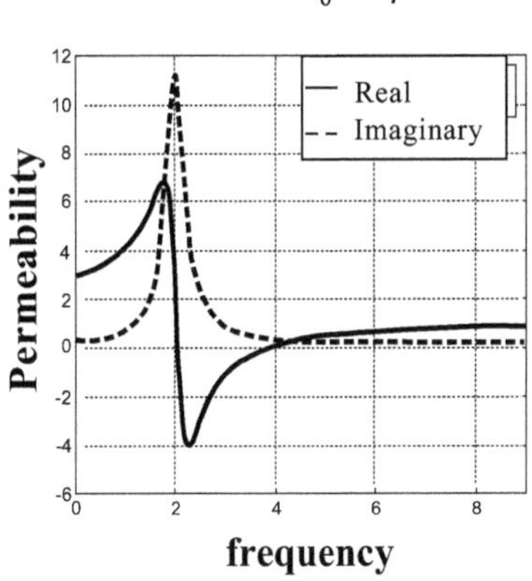

Fig. (3). Lorentz model for the permeability of a fictitious metallic resonator with $\omega_0 = 3$, $\omega_p = 5$ and $\gamma = 0.5$.

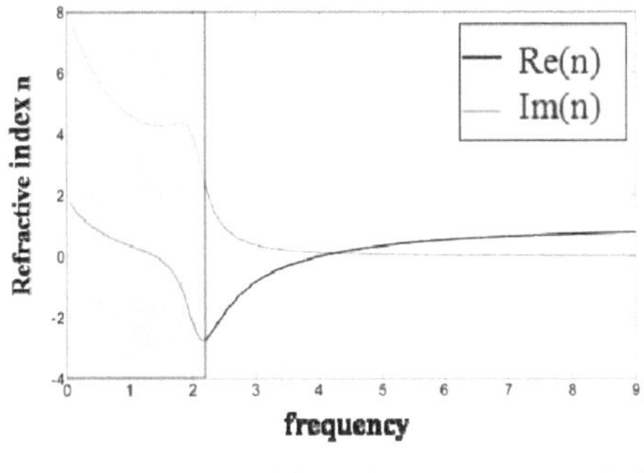

Fig. (4). Negative index (left-handed) metamaterial obtained by mixing metallic wire (Drude model of Fig. (3)) and metallic resonator (Lorentz model of Fig. (4)) arrays.

When ε and μ are designed separately, one has to hope that by mixing both structures a superimposition of the electric and magnetic properties can be obtained without significant cross related effects. Fortunately, if care is given to the design to reach simultaneously reasonable values of ε and μ in a given bandwidth and to constitutive "particle" position with respect to EM field excitation, it can work and a recovery of transmission corresponding to a negative n can be evidenced as shown in Fig. (4).

(a)

(b)

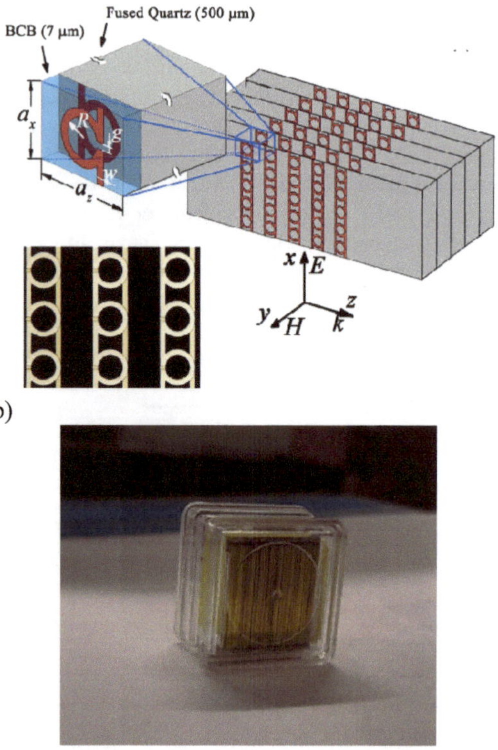

Fig. (5). Fig. (5). (a) Basic cell for a W-type particle array aimed at operating as a left-handed material above 100 GHz. (b) Typical block of metamaterial built using the W-type particle.

However, to fully characterize such a new material, the transmission information is quite poor and do not assess solely the left-handed character of the propagation. Other tools need to be developed. Even if it is not a necessary condition, APM are often investigated and considered as periodically patterned structures. Thus, a basic cell describing the structure can be defined and dispersion characteristics assuming an infinite medium (band structure) can be investigated. Different strategies can be envisaged depending on the intrinsic nature of the basic cell. If many tools (even commercial) exist for dielectric materials that we will use for photonic crystals, less has been done to analyze metal based structures.

To illustrate this point, let's turn to an Ω-type metamaterial. Such structures have been widely investigated in microwaves [16-23], here we will detail a design aimed to operate around 100 GHz. The idea is here to design a left-handed structure with only one type of metallic particles which exhibit negative ε and μ. Indeed, the geometrical shape of the Ω letter can fulfill the requisites for negative refraction: the round shape for the magnetic resonator and the arms for the electric activity. Such a basic cell is given in Fig. (5.a). In fact, to define the latter, two metallic Ω particles deposited back to back are used on both sides of a dielectric substrate to alleviate bi-anisotropic effect.

The goal is now to retrieve the complex propagation constant of the infinite medium created by the arrangement of these particles in the three dimensions of space. The 3D reality is often complex and in general, one chooses a given configuration in terms of geometry or wave polarization to extract a set of effective parameters usable for applications. Once the propagation direction is chosen, the scattering matrix (S) which gives the reflection and transmission coefficients under matched conditions, can be calculated. This can be done using a full three dimensional solution of the Maxwell equations (typically, a finite element method is used as the one commercialized by Ansoft corporation named HFSS). Proper boundary conditions are imposed in the plane perpendicular to the propagation direction to mimic an infinite array of particles. Then, the S matrix is converted in the corresponding chain matrix (or ABCD matrix) as follows:

$$\begin{cases} A = \dfrac{(1+S_{11})(1-S_{22}) + S_{12}S_{21}}{2S_{21}} \\ B = z_0\dfrac{(1+S_{11})(1+S_{22}) - S_{12}S_{21}}{2S_{21}} \\ C = \dfrac{1}{z_0}\dfrac{(1-S_{11})(1-S_{22}) - S_{12}S_{21}}{2S_{21}} \\ D = \dfrac{(1-S_{11})(1+S_{22}) + S_{12}S_{21}}{2S_{21}} \end{cases} \quad \text{(Eq. 2)}$$

z_0 is the impedance of reference used to calculate the S matrix. Let us mention that for passive symmetric structures $S_{11} = S_{22}$ and $S_{12} = S_{21}$.

If the device is treated as a homogeneous monomode propagation medium, the chain matrix can be written:

$$M = \begin{pmatrix} ch(\gamma a) & z sh(\gamma a) \\ \dfrac{1}{z} sh(\gamma a) & ch(\gamma a) \end{pmatrix} \quad \text{(Eq. 3)}$$

where a represents the length of the cell, z the cell impedance and $\gamma = \alpha + j\beta$, the complex propagation constant.

Using A, one can write:

$$\alpha = \frac{1}{a}\ln\left|A \mp \sqrt{A^2-1}\right|$$

$$\beta a = \text{Phase}\left(A \mp \sqrt{A^2-1}\right) + 2k\pi, k \in \mathbb{Z} \quad \text{(Eq. 4)}$$

Fig. (6) gives the dispersion characteristics obtained for the structure of Fig. (5a). On this figure, the signs of group velocity ($v_g = d\omega/d\beta$) and phase velocity ($v_\phi = \omega/\beta$) are reported depending on the sign of βa. Two or three frequency domains can be defined depending on geometrical parameters. For low frequencies, a left-handed band characterized by v_ϕ and v_g of opposite sign, is obtained. At high frequencies, the opposite right-handed regime is obtained (same sign for v_ϕ and v_g). In between, a gap where propagation is forbidden can be obtained. The gap bandwidth depends on geometry and can be cancelled leading to what is called a "balanced composite" behavior. In this case, the structure can be continuously frequency tuned from a negative to a positive refraction index (as in Fig.6). More interesting in terms of potential applications, the zero value can be obtained with a non-zero group velocity, a particular regime solely obtained in such artificial periodic media.

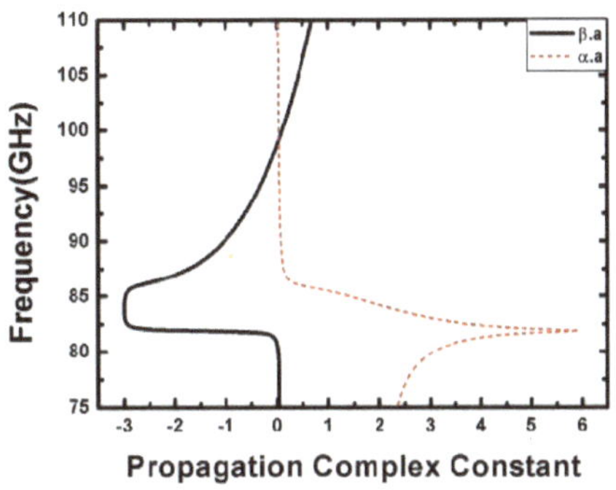

Fig. (6). Dispersion characteristic obtained for the Ω-type based metamaterial.

The chain matrix can also be used to retrieve the effective parameters of the structure, that is to say the couple (ε_{eff}, μ_{eff}) and/or the couple (n, z) previously defined. Note that these deduced values are valid as long as the device is used under the same restrictions assumed to calculate the S matrix. They rarely represent in these metamaterials an isotropic parameter which can be used under any configuration of incidence and polarization.
The information concerning the index can be found in the complex propagation constant:

$$n = \frac{c\gamma}{j\omega} \qquad \text{(Eq. 5)}$$

and the impedance is deduced using :

$$z = \mp\sqrt{\frac{B}{C}} \qquad \text{(Eq. 6)}$$

Finally:

$$\varepsilon_{eff} = \frac{n}{z} \text{ and } \mu_{eff} = nz \qquad \text{(Eq. 7)}$$

All the previous parameters are complex numbers, and the extraction task is not straightforward. Indeed, many branches can be calculated and one has to choose the physically based ones. As example, all imaginary parts must remain negative to traduce propagation losses… Indeed, no amplification can be found for a medium constituted of passive individual particles.

(a)

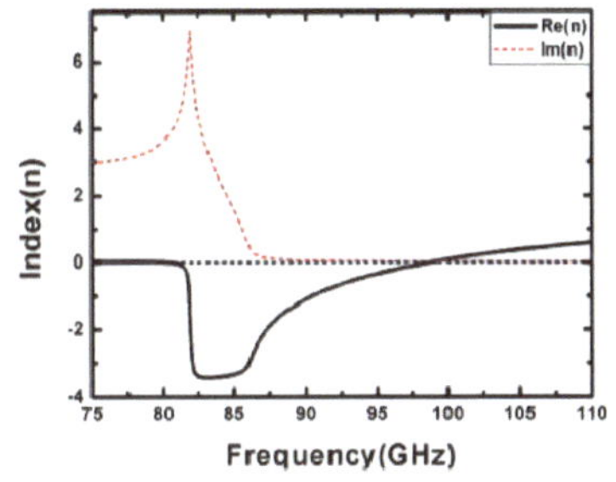

(b)

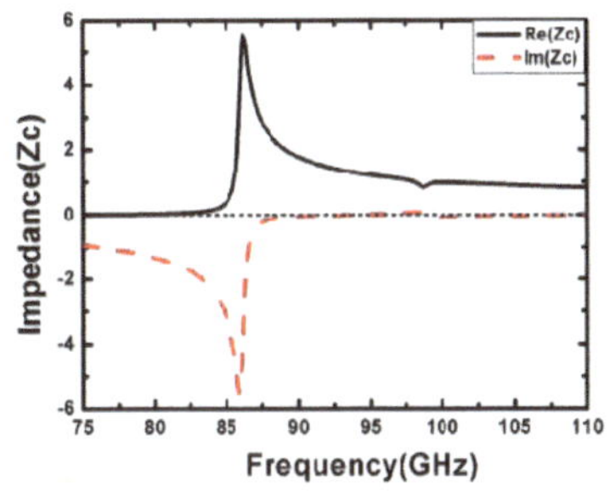

(c)

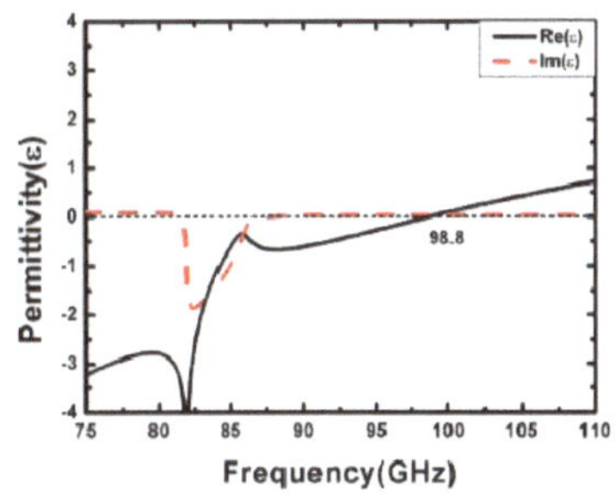

(d)

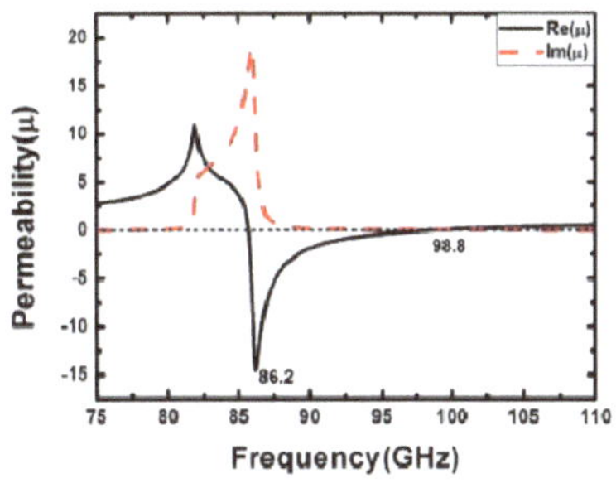

Fig. (7). Effective complex parameters for the Ω-type based metamaterial under normal incidence: (a) n: refractive index; (b) z: surface impedance; (c) ε_{eff}: effective permittivity; (d) μ_{eff} : effective permeability

Fig. (7) illustrates such a retrieval for the structure depicted in Fig. (5) The different regimes observed in Fig. (6) are now expressed in terms of effective parameters which can be engineered "as will" as a function of the targeted application. Positive or negative values for the different parameters can be reached and losses can be estimated as a function of the constitutive materials. Problems related to impedance matching to ensure an efficient power transfer between the metamaterial and its environment can also be addressed. As stated in introduction, let us recall that all the procedures developed above require monomode propagation and that the wavelength is large compared to the material patterning, at least one order of magnitude.

The photonic crystal approach: band structure engineering

Left handedness is not limited to metallo-dielectric based devices. It has been shown that full dielectric crystals can also under certain operating conditions exhibit negative refraction, even if the patterning scale is not large compared to the wavelength as it is the case for metamaterials.

Photonic crystals (PCs) have been preliminary designed in the 1980's for their stop band properties. Many applications for nanophotonics have then been addressed using 2D and 3D crystals including high-Q cavities for lasers, linear defects for wave guiding or multiplexing and for surface wave extinction in radiating systems [24-37].

More recently, attention has been paid to the dispersion properties of such crystals [1,38], especially in two dimensions, in the pass-bands. In two dimensions, such a crystal can be constituted of dielectric inclusions periodically arranged in air or by an array of air holes patterned in a dielectric matrix. If limited to two dielectric species, a limited set of periodical geometries in 2 dimensions is available which can be labeled using the crystallographic tools (Bravais lattices) as square, triangular, hexagonal lattices… Degree of freedom can be found here in the shapes that take the air or dielectric inclusions.

To establish the basic concepts of negative refraction in 2D-PCs, let us consider a triangular array of circular air holes patterned in a dielectric medium. Such a choice is guided by its compatibility with technological equipments required to fabricate such lattices to operate at optical wavelengths, domain where metallo-dielectric metamaterials are difficult to find both for technological and physical (losses) reasons.

In Fig. (8), a schematic of the structure is given, with the associated reciprocal lattice (which defines the Brillouin zone and the principal symmetry points). Two main parameters can be tuned to fix the propagation properties, namely the lattice period (labeled a) and the filling factor (labeled f.f, which represents the proportion of air in the dielectric matrix in our case).

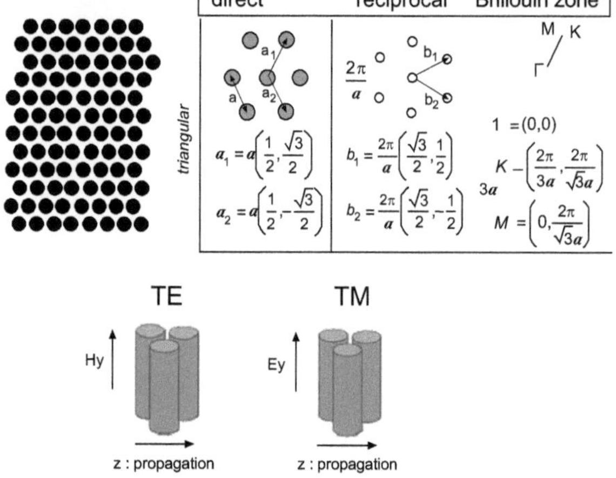

Fig. (8). Schematic of the 2D triangular photonic crystal and its reciprocal lattice. TE and TM polarizations are also defined.

A plane wave method (BandSolve© from Rsoft) allows us to calculate the dispersion curves of the periodic lattice. Two polarizations of the wave, TE (transverse electric) and TM (transverse magnetic) can be defined as shown in Fig. (8). Fig. (9) illustrate a typical result of such band diagram for the main crystallographic directions as a function of a/λ where a is the period of the triangular lattice and λ the free space wavelength. This calculation is done for a f.f of 50 %.

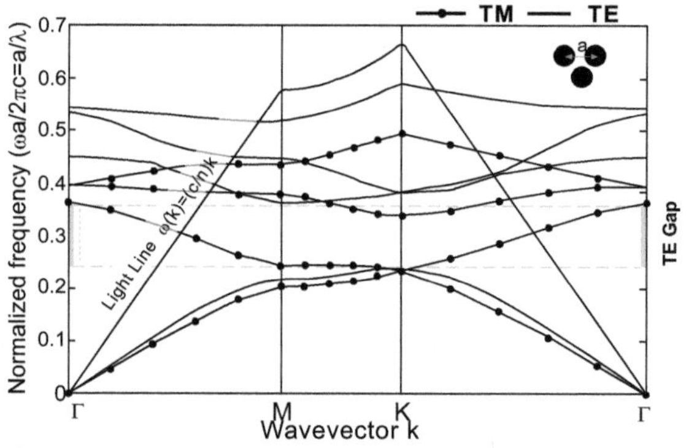

Fig. (9). Band diagram for the 2D triangular photonic crystal for the main crystallographic directions. Also plot the light line.

As long as the first band is concerned (small a/λ), the band structure is quite equivalent for TE and TM polarizations. It is a classical right-handed behavior with a wave vector increasing as λ is decreasing. For higher values of a/λ, TE and TM behaviors appear strongly different. For TE waves, a band gap is obtained whereas a second band with a reverse slope occurs for TM waves. Above this domain (> 0.38 a/λ), TE and TM waves coexist again with a rather complex distribution.

These results can be also interpreted in terms of refractive index. Indeed, it is well known that the index refraction of a matter can be related to the phase velocity of light in the matter compares to the light velocity in the vacuum. It is also known that phase velocity can be deduced from the dispersion curve ($v_\phi = \omega/k$). To describe the matter by a unique constant at a given wavelength, isotropic conditions must be obtained. That is to say that the phase velocity is solely given by the modulus of the wave vector and is independent of its direction. This can be checked by tracing the equi-frequency plots related to a specific band of the band diagram. To illustrate this point, Fig.(10) gives such contours for the first two TM bands of the diagram of Fig.(9).

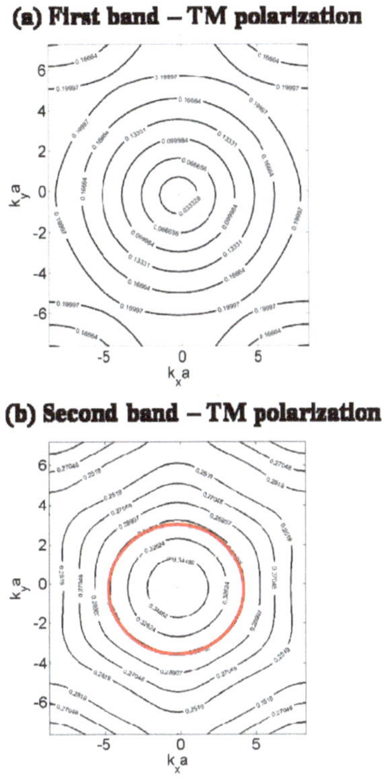

Fig. (10). Equi-frequency contours for the first two TM bands of Fig.(9) in the first Brillouin zone defined by $0 < k_x a < 8\pi/3$ *and* $0 < k_z a < 4\pi/\sqrt{3}$. Frequencies are normalized in a/λ. The red circle represents the 2D light cone.

At first glance, one can note that isotropy can be found in both bands and that index of refraction can be defined in both cases. To help, the light line corresponding to ω/k is also plotted. It represents an effective index of 1. That is to say that if this line crosses one band for a given wavelength and that a photonic stat exists, an effective index with a modulus of 1 can be extracted if isotropy is verified for this wavelength. This does not occur for the first band. It should occur if a line ω/nk is plotted with $|n| > 1$. This means that for the first band only index of refraction larger than 1 can be obtained.

For the second band, the situation is a little bit more complex. A crossing with the light line is possible. But at this point, we can observe that phase velocity ($v_\phi = \omega/k$) and group velocity ($v_g = d\omega/dk$) are of opposite sign. For a real system which combines different propagation media (air and PC for example), it appears natural to keep as positive reference the direction by which energy is transmitted, that is to say the one given by the group velocity. In other words, his implies that of a right handed medium is kept as a reference, the medium where v_ϕ and v_g are of opposite sign corresponds to a reversal of the phase velocity. As a consequence, the deduced refractive index has to be considered negative. In our case, the crossing between the light line and the second band gives n = -1. If the same reasoning is pursued on this second band, we are able to define negative refractive index as long as equi-frequency contours are circular in the second band leading to values in modulus smaller and greater than one.

Does this physically sound? The answer could be yes and no. Indeed, a smaller than one index value in this case would mean to operate above the light line. In a real 3D prototype obviously of finite dimension, this means high losses, in other terms a high coupling of the PC propagating mode with radiating ones in air. Here, to exploit such a regime, one has to imagine how to avoid such leaky modes, probably by surrounding the PC with a lossless environment to promote in-plane propagation.

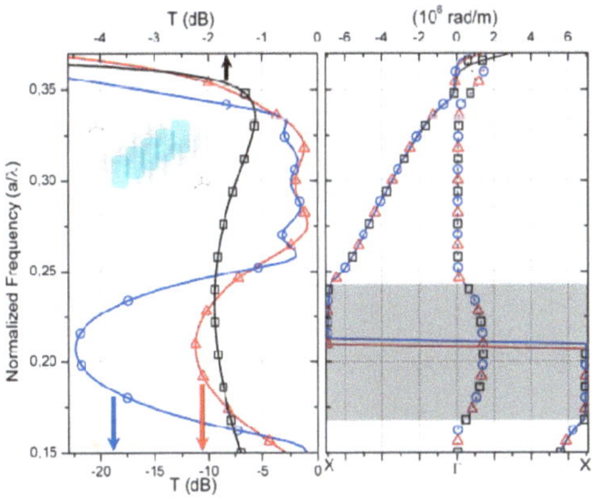

Fig. (11). Frequency dependence of transmission and of α (dotted lines) and β (solid lines) for one (black □), three (red Δ) and five (blue ○) cells, along the Γ-X direction. The grayed out frequency band is the first forbidden gap.

As Fig.(9) is considered, one can find another wavelength domain where negative refraction can be mimicked without any negative index. To this aim, one has to tilt the lattice and work around K point of the Brillouin zone at the top of the first band. In this case a/λ is close to 0.2 and can be considered largely smaller than one. If phase and group velocities are determined around this point, one can encounter cases where they show opposite directions. A PC slab can also focus under these conditions a divergent point source. The major drawback is that it can only operate under very near field conditions (much lower than λ) whatever the PC slab thickness!

Turning back to the first regime around a/λ ~ 0.3 where an negative effective index can be defined, one can ask ourselves if such a condition is sufficient to claim that a PC slab can operate as a superlens. Let us remain that superlensing in air was defined as $\varepsilon = \mu = -1$ simultaneously, leading to n = -1 and z = 1. Here, using band structures, calculated for a periodic lattice (which implies infinite dimensions), the question of impedance cannot be addressed directly. To reach this information, one has to use a finite slab in one direction and calculate the transmission coefficient of an incoming wave to deduce by de-embedding the impedance value as done with the method described above for metamaterials [39].

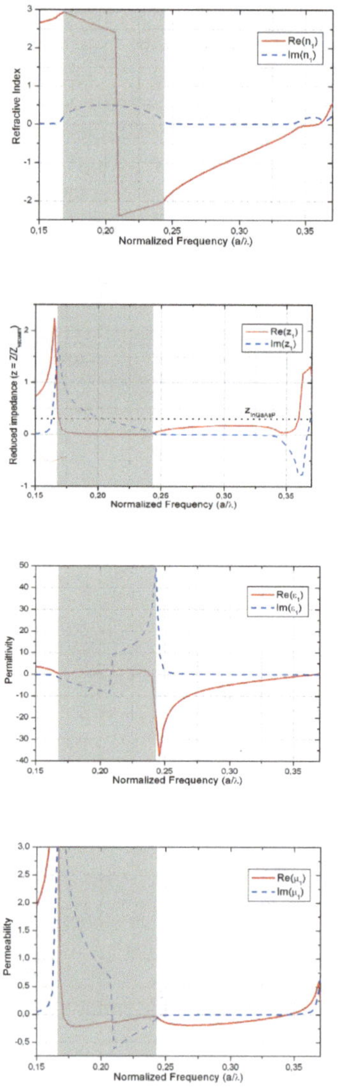

(a)

(b)

(c)

(d)

Fig. (12). Effective parameters for the square photonic crystal as a function of the normalized frequency: (a) refractive index, (b) reduced impedance, (c) permittivity and (d) permeability. The gray areas represent the forbidden gap.

For the range of a/λ and f.f considered here, the main difficulty is to avoid extra resonances in the transmission due to the Fabry-Perot effect which develops in the cavity formed by the crystal surrounded by air. Such a result is obtained when the reference port access for the simulation are reference to an impedance as close as possible to the true medium impedance.

To illustrate this point, such a procedure has been carried on a PC consisting of a square lattice of air holes etched in a semiconductor. Fig.(11) shows the transmission coefficient for different slab thicknesses and the α and β values deduced along Γ-X direction of the square Brillouin zone of the crystal. One can observe that even with one row of holes, α and β are correctly estimated and corresponds to the dispersion characteristics of the 2D crystal.

From these values, n and z can be extracted as well as ε and μ. All results are given in Fig.(12). As expected, n is negative and close to -1 around a/λ = 0.3. In the meantime, z is positive and limited to a value lower than 0.3 which corresponds roughly to the impedance of the semiconductor matrix. It never reaches one. As a consequence, impedance matching with air is not possible with the present structure. Moreover, as confirmed by the deduced values of permittivity and permeability, superlensing with $\varepsilon = \mu = -1$ is also impossible. In general, in modulus, ε is greater than one and μ smaller than one, leading to a n value which can reach 1 but with an associated impedance lower than one.

Let us remain that this calculation has been conducted under normal incidence that is to say in the Γ-X direction of the reciprocal lattice. What happen in other directions? Results shown that impedance matching is not directly possible and that ε and μ vary not only with frequency but also as a function of incidence angle.

All these primary conclusions show that the use of full dielectric photonic crystals for negative refraction applications as lenses will be a tricky task to reach the extraordinary properties expected by the theory as subwavelength resolution for example. Nevertheless, in the framework of nanophotonics and microlenses, no real solution appears better than the others nowadays and PCs remain attractive. In the next section, we will report on optimization methods dedicated to specific prototypes for different regions of the frequency spectrum which deal with theory to match with reality.

NEGATIVE REFRACTION BASED APPLICATIONS

To illustrate negative refraction regime or backward wave propagation, three different examples in terms of application and frequency domain have been chosen. First, negative refraction effect using a prism built using a Ω-type based metamaterial will be evidenced experimentally at microwaves. Second, backward wave propagation will be studied at Terahertz frequencies on a periodically loaded transmission line interpreted in terms of negative permittivity and negative permeability. Third, lens operation at a wavelength of 1.55 μm using an optimized n = -1 photonic crystal slab characterized by scanning near field optical microscopy will be addressed. All these examples will show that specific design or technological approaches are necessary to address particular wavelength domains. If concepts are quite equivalent throughout the frequency spectrum, fabrication and characterization facilities required are strongly dependent on wavelength and become more and more expensive as dimensions decrease with wavelength with the use of the most advanced tools of nanotechnology.

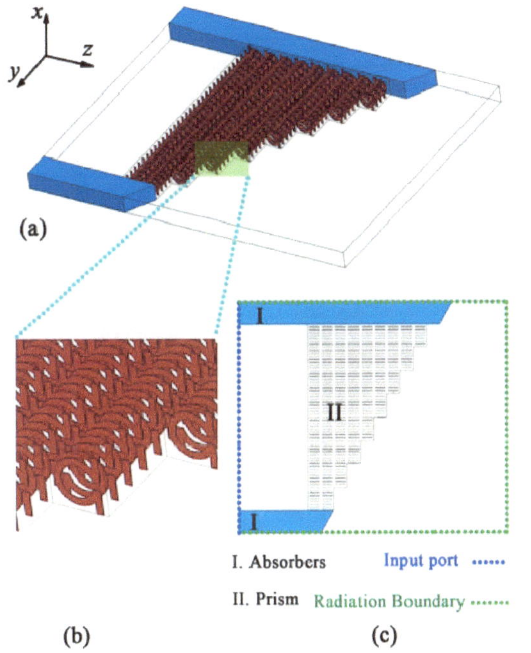

Fig. (13). Illustration of the prism configuration for refraction simulation. (a) 3D view of the prism model which consists of eight stairs; (b) Close up view of one step which consists of four cells along *y* direction and one cell along *x* direction; (c) Top view of the prism model indicating the boundaries.

2D Ω-type metamaterial based prism at microwaves

Starting from the Ω-type lattice defined in the first part of this chapter (see Fig.(5)), a prism has been designed to evidence the negative refraction by assessing Snell-Descartes law [40]. All dimensions have been scaled to shift the "zero effective index" [41-43] regime down to X and Ku frequency bands (between 8 and 18 GHz) where both fabrication and characterization techniques are facilitated. Let us recall that, as the structure is anisotropic, a negative value of n can only be obtained in one direction of propagation z and characterized in the (y,z) plane as shown in Fig.(13). More details about the design are given in Annex A.

To operate properly, the structure has to be illuminated by a plane wave under normal incidence. Reflection can occur at the first interface but no refraction meaning that the wave is injected within the structure in z direction. To study refraction, a tilted interface is designed at output with a mean angle of ~30°. In reality this can only be achieved, as shown in Fig.(5b), with a stair-like configuration over a finite number of basic cells. To be sure that such a rough interface is able to show an equivalent refraction effect as a smooth one, two sets of simulation will be performed: first, by considering an homogeneous medium characterized by its tensors of permittivity and permeability extracted from band structure and numerical transmission experiment for various wave polarizations and second by ab-initio simulations of the full designed microstructure [44-47].

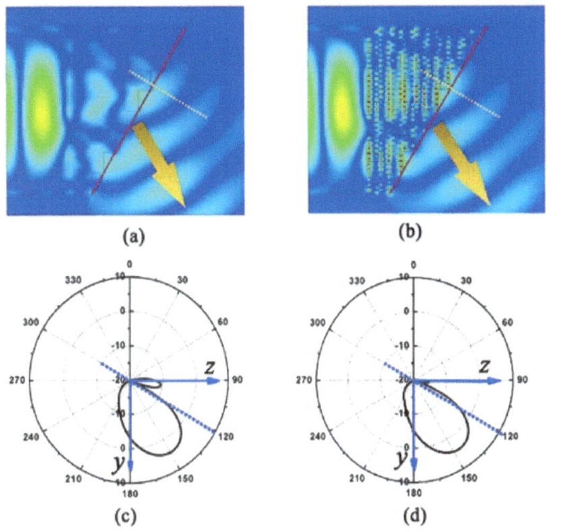

Fig. (14). Numerical evidence of negative refraction at 10.9 GHz. Electric field intensity of (a) homogenous and (b) micro-structured prisms at the same input phase; refractive angle plotted in far field for (c) homogenous and (d) micro-structured prisms.

Frequency (GHz)	10.9	12.8	15
n	-1.0052	-0.038	0.777
ε_x	-3.49+0.13i	-0.17+0.04i	4.90+0.57i
μ_y	-0.28+0.08i	-0.0082+0.0002i	0.1232+0.05i

TABLE 1 : EFFECTIVE PARAMETERS OF THE Ω-TYPE BASED METAMATERIAL

Fig.(14), Fig.(15) and Fig.(16) give the simulation results of the electric field intensity for three different frequencies, 10.9 GHz, 12.8 GHz and 15 GHz respectively. The respective theoretical extracted values of n (real), ε_x and μ_y (complex) of interest for the studied polarization are synthesized in Table (1).

To analyze precisely the refraction angle, far-field patterns are used. These plots analyse the region of space where electric field propagates referenced to the full space solid angle. In the three cases, the normal to the interface is shown for clarity and corresponds on the far field plot to the angle of 120°.

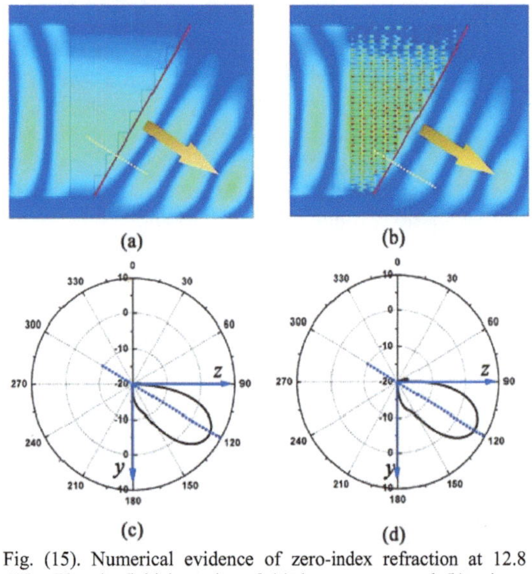

Fig. (15). Numerical evidence of zero-index refraction at 12.8 GHz. Electric field intensity of (a) homogenous and (b) micro-structured prisms at the same input phase; refractive angle plotted in far field for (c) homogenous and (d) micro-structured prisms

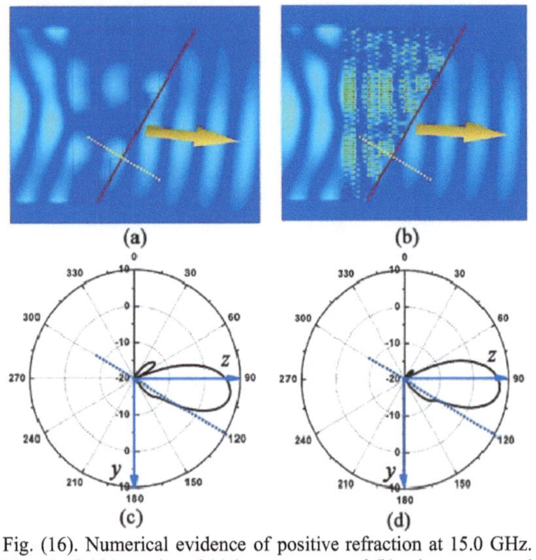

Fig. (16). Numerical evidence of positive refraction at 15.0 GHz. Electric field intensity of (a) homogenous and (b) micro-structured prisms at the same input phase; refractive angle plotted in far field for (c) homogenous and (d) micro-structured prisms

The similarity of the two simulations is impressive, in terms of field intensity, and assesses the powerfulness of homogenization techniques as long as the medium patterning remains much smaller than the operating wavelength.

Let us notice that if n = -1 is obtained around 11 GHz, it does not mean automatically $\varepsilon = \mu = -1$, values "a priori" required for super-focusing. In this case, impedance matching with air is not realized and reflection exists at the input interface of the prism.

The prism sample, composed of eleven steps rather than eight for the simulation prototype, is shown in Fig.(17a). The prism sample was fabricated using a Printed Circuit Board technology. In practice, 0.8 mm thick epoxy substrates coated with a 35 μm copper layers ($\sigma = 5.8\ 10^7$ S/m) were assembled according to the layout shown in Fig.(13). Owing to the large dimension of the completed prism-type prototype, we preferred to use a chemical etching technique by Ion Peroxide rather than micro milling as in our previous works on microwave metamaterial. The number of unit cells of each layer along the propagation direction at the first and last steps was increased from five to fifteen, respectively. Each layer consists of 18 unit cells along x-direction with a periodicity of 3.9 mm.

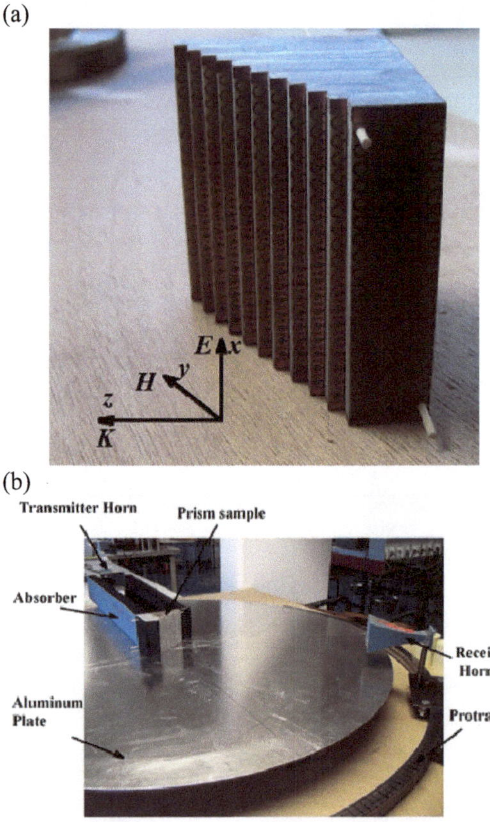

Fig. (17) (a) Photograph of a prism–type sample; (b) Photo of the angle resolved measurement setup (the upper metal aluminium plate was removed for clarity).

The prism refraction experimental setup is shown in Fig.(17b). Two absorbing layers (Eccosorb AN-75) are used in the input section for shaping the incident beam with a quasi-plane phase front. Absorber layers were also placed around the prototype for avoiding cavity effects which could result from the reflected beams. The angle resolved measurements of the transmitted field are carried out by means of a rotating detector whose angular position is recorded via a homemade large scale circular protractor with a step of 1.25°, as shown in Fig.(17b). With respect to the similar scattering chambers which can be found in the literature, the emission as well as the detection of electromagnetic wave was performed via two horn antennas rather than by using waveguides flanges. For covering all the bands, two sets of horn pairs were used (Nadar 640 for operating in the X band and Nadar 639 for covering the Ku band). The two horns were placed at the periphery of the 96-cm-diameter aluminium plates in order to satisfy the far field condition. The refraction measurement was performed by a HP 85107A Vector Network Analyzer.

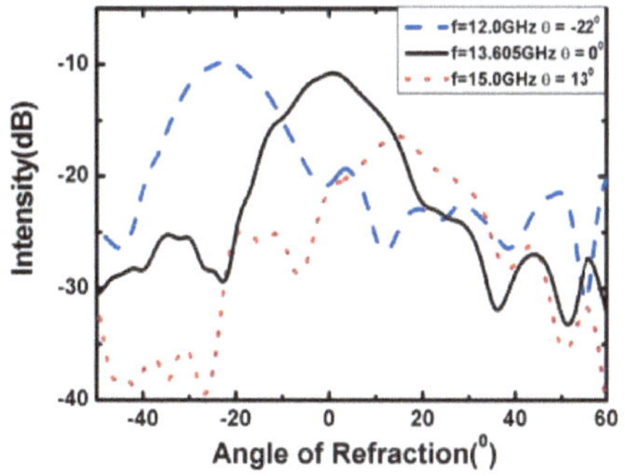

Fig. (18) Angle dependence of the detected signal at three characteristic frequencies: at the negative index band @ 12.0 GHz (dashed line), zero index @ 13.6 GHz (solid line) and positive index @ 15 GHz (dotted line).

Fig.(18) shows the angle dependence of the detected signal, which was measured for three representative frequencies of the negative index band (@ 12.0 GHz), zero index characteristic frequency (@ 13.6 GHz) and the positive index band (@ 15.0 GHz). For positive and zero indices, the maximum transmitted signal levels are quite comparable around -10 dB. As it can be calculated using the parameter extraction techniques, the impedance matching to the embedding air medium is more favourable for the lower frequencies of the spectrum. On the contrary, there is a bigger mismatch in the upper part which explains the weaker level of the transmitted wave. The same behaviour can be noted for the scaled prototype at 100 GHz on Fig.(7b) with an impedance decreasing below 1 (air reference) as frequency is increasing below and above n = 0 regime. This shows clearly the invariance of concepts with frequency if the scaling in dimensions is properly done. This remark can be tempered since differences can appear when constitutive material frequency dependent properties are considered. For example, dielectric or metallic losses which naturally increase with frequency can rapidly alter the metamaterial dispersion characteristics if not enough care is taken during the design process.

By applying the Snell-Descartes law ($n_1\sin\theta_1 = n_2\sin\theta_2$) generalized to any index value, it is possible to retrieve experimentally the frequency dependence of the effective refractive index. In the equation of the refraction, the incident angle is the angle of the tilted interface ($\theta_1 = 30.5°$) and the refraction angle (θ_2) is drawn from the maximum of transmitted power at each measured frequency. n_2 is here 1 and so n_1, the real value of the metamaterial index can be deduced. Experimentally, an angle of -22° is obtained at 12 GHz, 0° at 13.6 GHz and 13° at 15 GHz. An excellent agreement between the simulation and experimental results can be obtained when the permittivity of epoxy glass is assumed to be 3.6 rather than 4.0 used for the design and the simulations.

A periodically loaded backward transmission line at terahertz frequencies

As we will see in the next two parts, technological challenges become of prime importance as operating wavelengths and thus characteristic dimensions are decreased. For this second illustration of negative refraction, we will return to a quite old subject concerning backward travelling waves revisited using the metamaterial approach and transposed to terahertz frequencies. To study such waves a periodically loaded transmission line will be designed, fabricated and characterized [48]. A simple ideal model in terms of lumped circuit elements as inductances and capacitances will be used to assess the theoretical approach.

Let's start by considering the model of an ideal transmission line as shown in Fig. (19a). A classical line is right-handed, can be considered as a low-pass filter and modeled in the lossless case using a series inductance and a parallel capacitance. Such circuit describes the unit cell repeated periodically to represent the line. Ideally, as shown in Fig.(19b), a left-handed line should be represented by the dual circuit, a high pass-filter, with a series capacitance and a parallel inductance. In reality, such a device doesn't exist and we face by construction a composite character represented by the circuit of Fig.(19c) that we will detail in the following [49-50]

The composite is described using four elements, two in series (L'_R and C'_L) and two in parallel (C'_R and L'_L) [51-52]. We thus can define, by unit length, an impedance Z' and an admittance Y' as the following:

$$Z' = j\left(\omega L'_R - \frac{1}{\omega C'_L}\right) \qquad \text{(Eq. 8)}$$

$$Y' = j\left(\omega C'_R - \frac{1}{\omega L'_L}\right) \qquad \text{(Eq. 9)}$$

The propagation constant of the line is defined by:

$$k = \alpha + j\beta = \sqrt{Z'Y'} \quad \text{(Eq. 10)}$$

Then in the case of a lossless homogeneous line [53], the dispersion relationship can be written as:

$$\beta(\omega) = \mp \sqrt{\omega^2 L'_R C'_R + \frac{1}{\omega^2 L'_L C'_L} - \left(\frac{L'_R}{L'_L} + \frac{C'_R}{C'_L}\right)} \quad \text{(Eq.11)}$$

As a function of ω, if β is real a transmission band exists. Conversely, if β is purely imaginary we face a forbidden band (see Fig.(20))

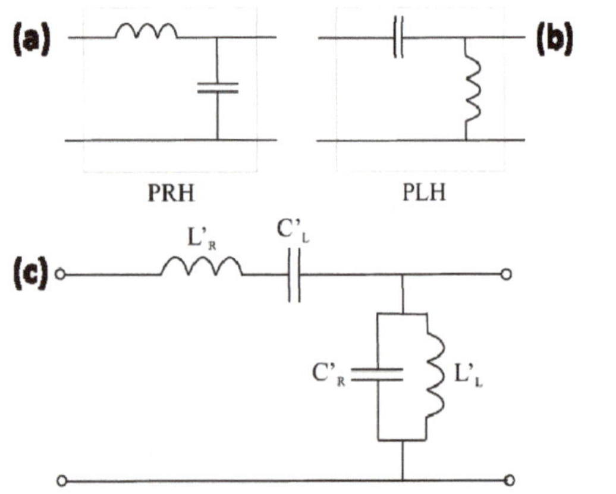

Fig. (19) Electrical model for a left-handed/right handed transmission line: (a) perfect right-handed line, (b) perfect left-handed line, (c) composite structure

The forbidden band is characteristic of unbalanced transmission lines. In other words, this means that the right-handed impedance term is not matched to the left-handed one. This results to the appearance of ω_1 and ω_2 on the dispersion diagram.

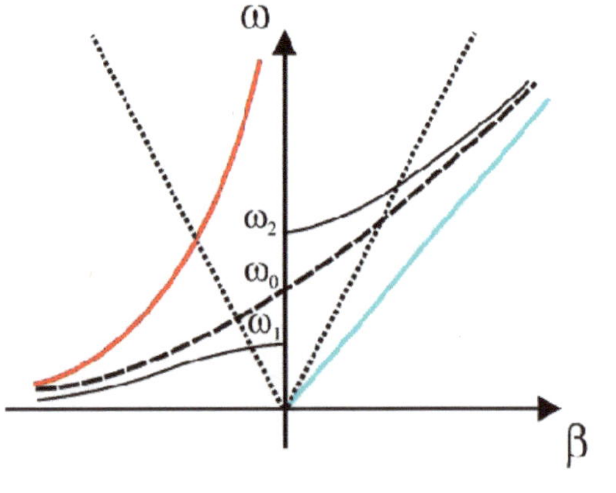

Fig. (20) Dispersion diagram (full black lines) for the transmission line given in Fig. (19). Also given in dotted bold lines the case of a balanced composite line (no gap around ω_0). For understanding, the dispersion curve for an ideal right handed line is given in blue and for an ideal left-handed line is given in red. The light line is then plotted in dotted thin black line.

On the ω-β diagram, it is possible to extract phase (v_ϕ) and group (v_g) velocities:

$$v_g = \frac{\partial \omega}{\partial \beta} \text{ and } v_\phi = \frac{\omega}{\beta} \quad \text{(Eq. 12)}$$

Below ω_1, $v_\phi v_g < 0$ whereas > 0 above ω_2. This means that the two velocities are anti-parallel for the first band (backward waves – negative refractive index) and parallel for the second band (forward waves – positive refractive index). The forbidden band (gap) in between can be tuned depending on the respective values of the lumped elements and can be cancelled when the following equality is verified [54]:

$$L'_R C'_L = L'_L C'_R \quad \text{(Eq. 13)}$$

This "balanced" case gives rise to a singular operating mode with a zero phase velocity (infinite guided wavelength) and associated to a non-zero group velocity. It corresponds to a continuous evolution of the refractive index from negative, zero to positive values.

From these equations, effective parameters can be extracted as shown in the beginning of this chapter: n, z, ε and μ. For example, ε and μ can be related to the lumped circuit elements by:

$$\begin{cases} \mu = L'_R - \dfrac{1}{\omega^2 C'_L} \\ \varepsilon = C'_R - \dfrac{1}{\omega^2 L'_L} \end{cases} \quad \text{(Eq. 14)}$$

Let us mention that the model used here is very convenient to understand the basic principle of such composite lines. But, it is too simple... To go further in the analysis, one has to take into account that the described cells in solely the basic cell of a periodically loaded transmission line. Indeed, the left-handed character can only be given by lumped elements and to obtain a significant effect, they have to be repeated along the main transmission medium. In this case, the infinite character of the two bands of Fig.(20) is no longer valid and a real multi-band structure (over the Brillouin zone and β values bounded by π/Λ – Λ : spatial period of the series capacitance and shunt inductance inserted in the right-handed line) has to be considered.

To obtain a quantitative description, full electromagnetic simulations are required and extracted scattering parameters of the designed lines can be extracted to reconstruct the band diagram by the methods described earlier.

In short, the design steps of a composite line are the following: First, lumped elements values are calculated in order to fix the main frequency domains that is to say the left-handed band, the gap (if desires) and the right-handed band. This is done with the help of the "simplified" band diagram obtained from the line theory. Second, these values are used to size the right-handed line, the series capacitance and the shunt inductance. Third, a full wave simulation is carried out to check the real band diagram of the periodically loaded (composite) transmission line. If needed, corrections are made to be in accordance with the primary objectives.

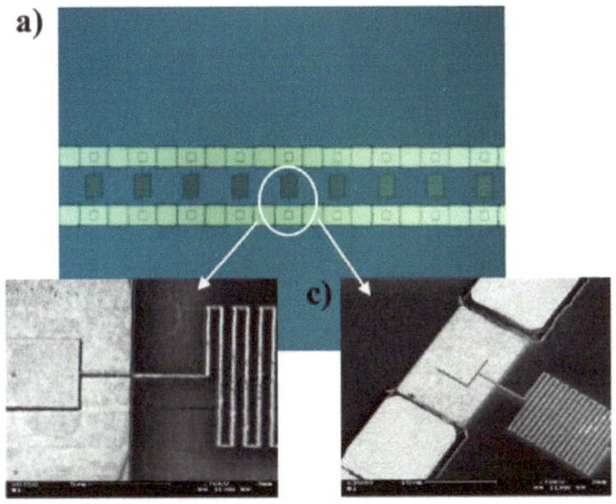

Fig. (21) Example of a composite left handed/right handed line for a left handed operating regime around 300GHz. The structure is based on a classical coplanar strip right handed line (CPS) whereas the left handed properties are afforded by the plot capacitance inserted in series with the line and the meander inductance in parallel.

To illustrate the approach, Fig.(21) shows a CPS (coplanar strip) right handed line periodically loaded with series capacitances and shunt inductances. The goal is to obtain an effective refractive index negative around 300 GHz (ω_1). A CPS line is constituted of two thin metallic conductors deposited on a semiconductor substrate. A forward single mode propagating regime can be obtained from zero to high frequencies before the apparition of higher order modes mainly in the substrate. Moreover this configuration is favourable to the periodic insertion of shunt inductances (meander metallic lines between the two conductors) and series capacitance (plate Si_3N_4 dielectric layer between two metallic contacts). See details on Fig.(21). To give orders of magnitude, such a design corresponds to values of 35.7 pF/m and 870 nH/m for the capacitance and inductance distributes values of the right-handed line. For the "left-handed" lumped elements, it corresponds to a series capacitance of 6.2 fF and a shunt inductance of 38 pH.

From the technological point of view, this gives typically plot capacitances of 25 μm^2. For the shunt inductance the metallic wire is 200 nm thick for a width of 200 nm and a length of about 200 µm (unfolded). To fabricate the prototype of Fig.(21), one have to use advanced tools of micro- and nano-electronics such as electron lithography (for the smallest details) or optical lithography combined with reactive ion etching to define accurately the plot capacitances. If fabrication represents a first challenge to operate at frequencies of a few hundred GHz, characterization is a second one, far more difficult to handle than network analysis in microwaves. Here, to analyse the lines and to evidence the phase properties of the backward waves, electro-optic sampling techniques [55-56] will be used.

Briefly, the device will be illuminated by a short pulse whose spectral extent includes the frequency range investigated. To reach THz frequencies an optical pulse (femtosecond laser) will be used. Using an appropriate conversion, the pulse can be electrically transmitted through the line and collected at the output using the reciprocal conversion. This pump-probe experiment is quite classical and details of operation can be found in the literature[57-58]. This time dependent signal which contains in general a lot of information (incident wave, reflected wave, echoes....) can be numerically treated and transposed in the frequency domain by Fourier transform. Fig.(22a) illustrates schematically the basic principle.

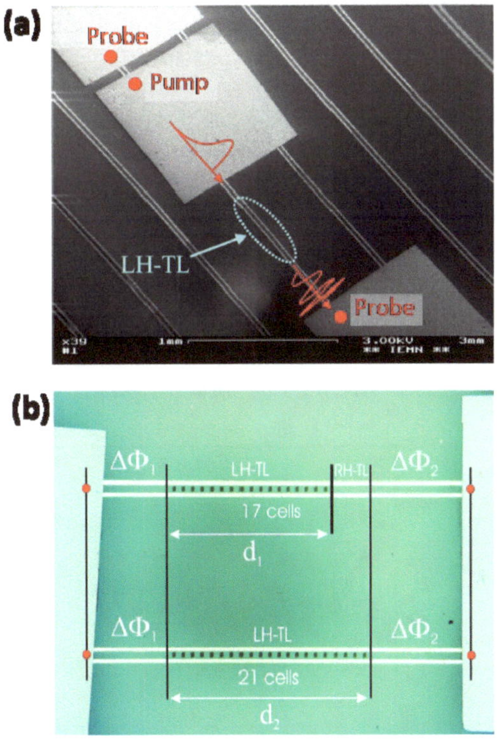

Fig. (22) Snapshot of two lines used for the measurement by electro-optic sampling. The two lines are of different lengths (17 and 21 cells). The phase difference extracted permits to derive the effective index of refraction of the line. (a) full view of the of the pump-probe experiment. (b) zoom view on the two lines.

In our case, the aim is to characterize our left-handed line in terms of negative refractive index. It can be shown that this parameter can be easily extracted from the measurement of two lines of different lengths, as illustrated in Fig.(22b) [59]. By taking into account the difference between the two configurations ($\Delta L = d_2 - d_1$), the index of the left-handed line (n_{LH}) can be related at a given frequency to the phase difference ($\Delta\Phi$) of the transmission coefficient by:

$$n_{LH} = -\frac{c}{f\Delta L}\frac{\Delta\Phi}{2\pi} + n_{RH} \qquad \text{(Eq. 15)}$$

where n_{RH} represents the effective index of the right-handed CPS line.

The results of such a procedure are shown Fig.(23) and Fig.(24). Fig.(23) gives the two time dependant signals through the 17-cell and 21-cell lines, and one can note that oscillations in advance (Δt) for the longer line compared to the shorter one indicating the left-handed character of the propagation. Transposed in the frequency space, the n_{LH} index can be extracted and it is found negative between 260 and 320 GHz with a minimum of -0.8 at 280 GHz. Note that this frequency band was the one targeted by the design.

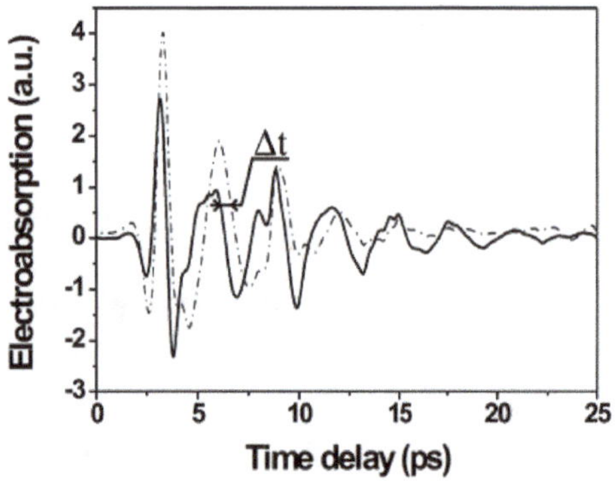

Fig. (23) Time evolution of the transmitted signals by the two lines of different lengths. The Δt value can be related to the phase advance of the signal issued from the longest line.

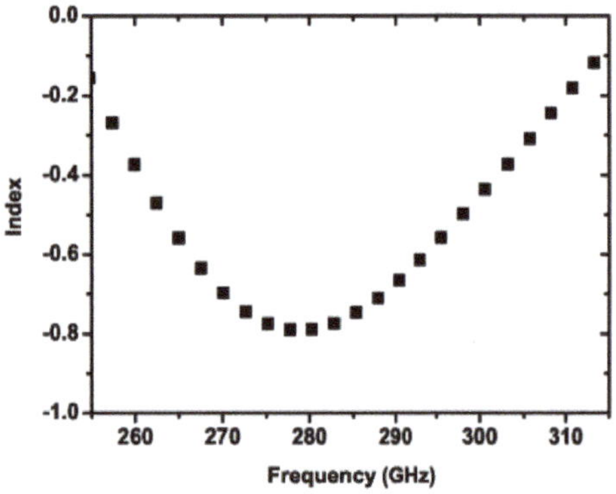

Fig. (24) Extraction of the effective index as a function of frequency extracted from Fig. (23).

This second example of applications illustrates most of the challenges that we face to design and characterize metamaterials to evidence negative refraction effects as wavelength is decreasing. In the next part, we will address smaller wavelengths, typically around 1 µm, to reach the optical domain. As previously mentioned, the concept of metamaterial which implies a structuring much shorter than the wavelength becomes very difficult to satisfy. That is the reason why photonic crystal will be considered even if we will have to deal with a mix of diffraction and refraction...

A n = -1 photonic crystal based flat lens

At optical wavelengths, the use of metallic particles is a real challenge since it is well known that losses increase dramatically at these frequencies of a few hundred of Terahertz [60]. This is the main reason why researchers have tried to explore the properties of full dielectric based materials, namely photonic crystals. Here again, two approaches can be envisaged. As in the previous examples, one can try to modulate the effective parameters of the artificial medium patterned. The main difficulty concerns the "artificial magnetism" and to this aim, some authors have tried to use Mie resonances than can be observed in pillars of material with very high intrinsic values of permittivity. In such case, a negative permeability can be mimicked in a small frequency band whose value depends on the pillar diameter. However, to reach optical wavelengths, here again nano-patterning is required. Unfortunately, the maturity of technologies for such materials (ferro-electrics for example) is too weak to envisage at short terms applications in nanophotonics.

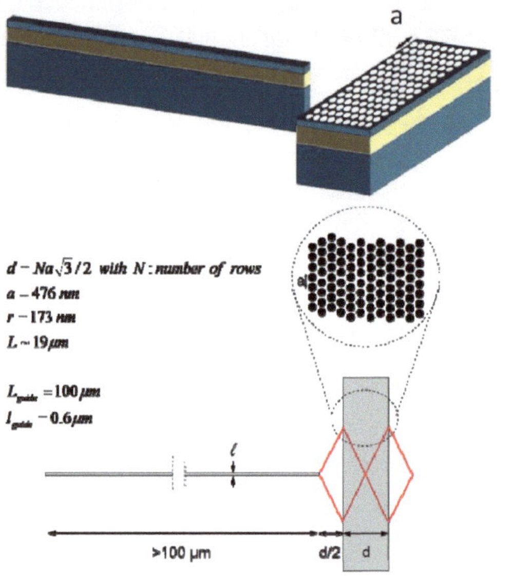

Fig. (25) 2D and 3D views of the lens prototype fabricated at IEMN along with characteristic dimensions to operate at 1.55 µm.

The second approach consists to use photonic crystals whose properties have been described earlier. For the latter, semiconductor based technologies appear more mature and the fabrication of prototypes exploiting negative refraction is possible [61-67].

To illustrate the operation of a n = -1 flat lens, we make use of the 2D photonic crystal describe in Fig.(8) and Fig.(9) [68]. Additional details concerning the optimization of the 2D geometry for a finite structure are given in Annex B.

As reality is tridimensional, one has to confine the light in the third dimension (labeled vertical direction) since the PC only is in a 2D plane (labeled lateral plane). Fig.(25) gives schematic 2D and 3D views of the lens prototype.

(a)

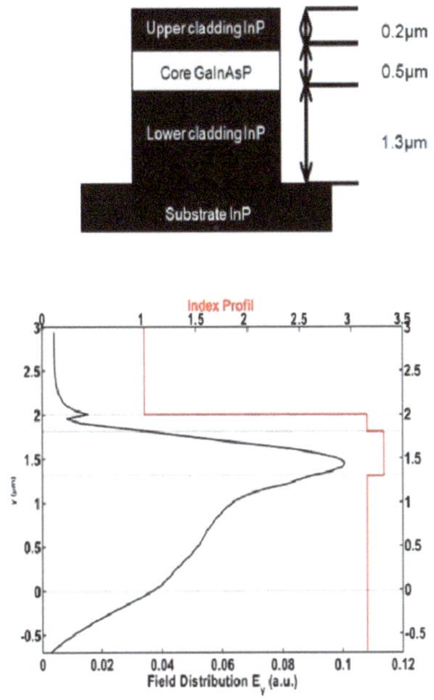

(b)

Fig. (26) (a) Sketch of the semiconductor heterostructure aimed to confine the light in the third direction. (b) Index profile along with the field distribution of the confined mode in the InGaAsP layer.

The device fabrication makes use of the most advanced techniques of nanoelectronics available in IEMN. The general scheme to fabricate the lens consists of three major stages: (i) the growth of the epilayer by molecular beam epitaxy (MBE), (ii) the nanoscale patterning by e-beam lithography and (iii) the fabrication of the hole array along with the ridge waveguide by inductively coupled plasma (ICP) etching. (see Fig. (26)).

The epilayer consists of a 500 nm-$Ga_{0.27}In_{0.73}As_{0.6}P_{0.4}$ grown by Gas source MBE system on an InP substrate and covered by a 200-nm InP capping layer. Here, lateral monomode propagation is expected in the wavelength range of interest. The high index contrast between InP and air permits one to limit dramatically the spreading of optical mode above the quaternary layer. At the opposite, on the substrate side, special care has been paid during ridge guide or lens fabrication to deeply etch the InP substrate in order to confine the evanescent tail of the optical mode.

The hole nanopatterning is performed by e-beam lithography (LEICA 5000 equipment – voltage 50 kV – 32 doses for correction of proximity effects). Since the following step is a highly anisotropic deep etching, it requires a mask which can resist to etching conditions. To this aim, a negative resist (HSQ Hydrogen Silses Quioxane - removal on non-exposed regions) will be used to define in one step the ridge waveguide and the hole array (filling factor close to 50 % so no extra time for lithography). The mask obtained shows an excellent regularity with very well defined circular holes and a periodicity maintained on a large area.

The following step is the ridge guide and hole array definition by ICP etching. Using $Cl_2/H_2/CH_4$, a ratio close to 10:1 has been obtained and a depth higher than 2 µm has been achieved using HSQ as a mask. High anisotropy with a good surface quality has been obtained as show in Fig. (27). The very last step consists in cleaving the structure at the end of the ridge waveguide in order to be able to inject light for the SNOM (Scanning Near field Optical Microscopy) experiment.

Fig. (27) Prototype of a 13 rows flat lens fabricated at IEMN.

Fig. (28) and (29) give experimental results on a 21 rows flat lens obtained by SNOM (Scanning Near field Optical Microscopy) [69-70]. The measurement consists in collecting the optical power coupled to a thin probe above the prototype at a fixed controlled distance. Usually, a SNOM is used to collect the transverse evanescent tail of an optical mode confined in a device. Here, we extent its use to the collection of optical waves travelling in open space. In these successive top views of Fig. (28), one can note on the left sides of the figures the initial source with a high intensity at the waveguide end. Also, it can be observed as function of wavelength that a clear spot appears behind the lens on the right side of the image between 1.53 and 1.54 µm. Complex patterns also exist due to the interferences which develop between the lens and the probe [71].

The "best" result in terms of focusing is given in Fig.(29) and corresponds to a wavelength of $\lambda_0 = 1.525$ µm. The resolution of the lens can be estimated to $0.8\lambda_0$ which is a result very close to the optimal theoretical value predicted for such a structure 0.6 λ_0. This result doesn't overcome the classical Rayleigh limit of 0.5 λ_0 but it represents a first step toward this objective not yet experimentally reached by any proposal based on classical or negative refraction for such optical wavelengths. It has to be mentioned that if subwavelength resolution still needs to be investigated [72-75], the power level transmitted through the lens needs also to be enhanced. In other terms, the matching between the lens and the surrounding air is not realized and reflections at input are detrimental. In practice, only 1/1000^{e} of the incident power is transmitted.

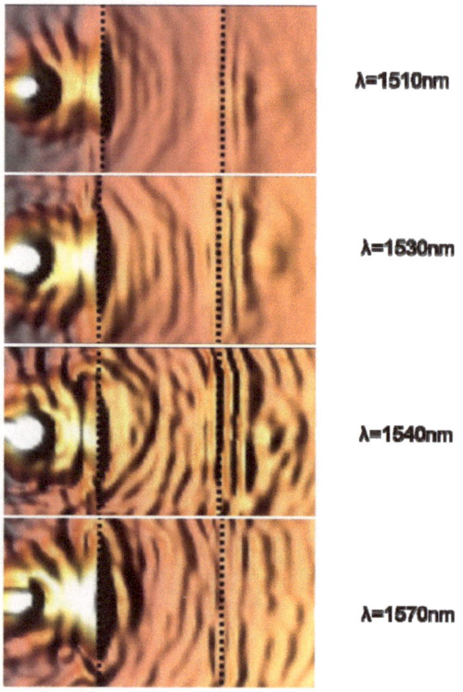

Fig. (28) Optical measurement as a function of wavelength of a 21 rows flat lens. (SNOM measurements are carried out at ICB / University of Bourgogne, Dijon, France).

To understand clearly the origin of losses, three dimensional simulations of the whole prototype have been performed. Let us mention that such simulations cannot be used for optimisation or design purposes. Indeed, such simulations require huge resources both in terms of memory and time computing. These calculations help us to have a deeper insight of the physical processes involved in the focusing process. They can also help us to understand the wavelength shift measured 0.025 μm with the best result at 1.525 μ whereas the design was done for 1.55 μm.

Fig.(30) illustrates such a 3D calculation. The result is a cut view equivalent to the measured image of Fig.(29) at 1.525 μm. To compare theory and experiment, ray tracing is recalled in both cases. Most of the physical processes involved are well described except the interference fringes especially around the second interface which results from tip/lens interaction. The spot position (taken numerically when the optical field is the narrower) and the reflection at input fit satisfactorily.

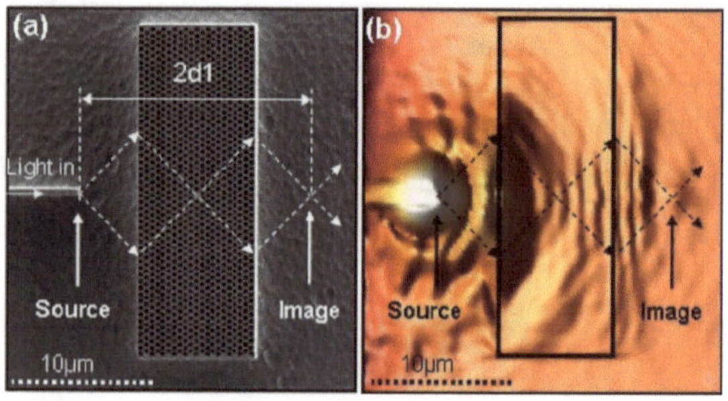

Fig. (29) Optical measurement a 21 rows flat lens at 1.525 μm where lens resolution is optimal and can be estimated around 0.8λ_0 (λ_0: wavelength in air (SNOM measurements are carried out at ICB / University of Bourgogne, Dijon, France).

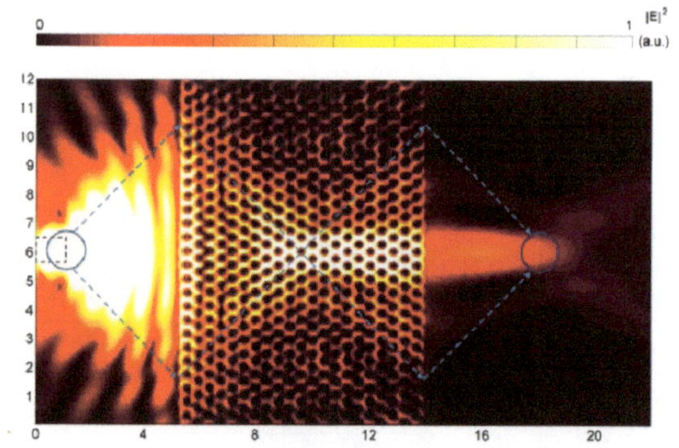

Fig. (30) Top view of a three dimensional FDTD simulation of the prototype at λ = 1.525 μm.

Fig.(31) gives a similar calculation but at 1.55 μm, the initially targeted wavelength in the two dimensional design process. Fig.(31a) is a side view whereas Fig.(31b) is a top view as Fig.(30). The PC parameters used corresponds to the 2D design of Fig.(25) but here the vertical heterostructure is simulated along with holes of finite depth. It appears clearly on Fig(31b) that the effective refractive index of the lens is not equal to -1. Indeed the constriction of the field within the lens is shifted to the right hand side of the lens and not exactly located at the middle expected by the design. This shift means that the effective index is negative and greater than one in modulus. On the band structure it corresponds to a smaller a/λ value below the light line. It also means than -1 should be obtained with a greater a/λ value, that is to say a smaller wavelength, result consistent with the one observed experimentally and calculated in Fig.(30). Fig.(31a) learns us more about insertion losses in our device. In the central part above the PC, one can note that the optical wave is strictly localized in the heterostructure. In consequence, radiation losses above the crystal are negligible, a result in accordance to the fact that this operation point stands below the light line of the real 2.5D crystal. However, within the crystal one can note some losses toward the substrate under holes depth. The major loss component can be found at lens input. A first origin comes from the waveguide itself with a non-negligible backward reflection in the guide at the emitting point. In second, the reflection at lens input, impedance mismatch between the air and the PC, with the result of radiation above the lens and energy storing between the guide and the first lens interface.

In a last stage, these 3D simulations illustrates that the definition of the focus point behind the lens is not as trivial as in two dimensions. Light is diffracted up and down by the second interface and this effect spreads severely the spot image. This is especially true since optical power level is small at output. On Fig.(31b), the spot location is not as clearly defined as in Fig.(30). But let us remain that a perfect spot is supposed to exist only for n = -1 if surrounded with air and that aberrations appear in any other case.

(a)

(b)

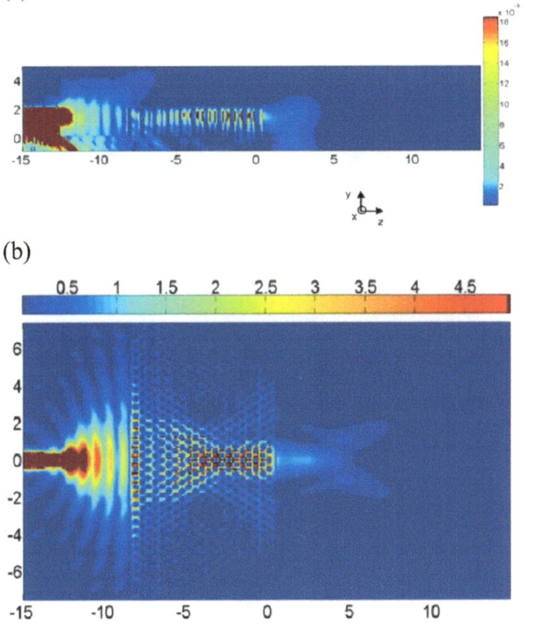

Fig. (31) Three dimensional FDTD simulation of the prototype at $\lambda = 1.55$ µm. (a) side cut; (b) top view.

The lens experiment is a perfect illustration of the difficulty encountered in reality with a prototype patterned at nanometer scale even very carefully designed. Two dimensional approaches are fundamental for fast design but suffer from approximations that can modify significantly the expected real 3D results. Negative refraction represents only one aspect of these new guiding properties afforded by APM. In the last part of this chapter, some new perspectives afforded by the association of several photonic crystals or by full 3D material effective parameter engineering will be presented.BEYOND NEGATIVE REFRACTION: A 3D MATERIAL PARAMETER ENGINEERING

The extraordinary refraction properties illustrated up to now in a propagation line, a prism or a lens can be enriched when combined to other operating regimes of PC or when a local control of the effective parameters is considered. This leads in general to the concept of "cloaking devices" where one searches to hide a region of space to any incoming wave (concept of invisibility) or "hyperlenses" for high resolution focusing purposes. These properties are mainly obtained by a full 3D engineering of the permittivity and permeability tensors using transformation optics techniques (or conformal mapping...). This local control of the effective parameters goes beyond the general scope of this chapter since the artificial medium is no longer be periodic in these cases. Nevertheless, they represent an extension of the concepts described up to now and the potential applications appear so exciting and promising that they can be ignored [76-88]...

A photonic crystal based cloaking device

Below, we illustrate an alternative route to the definition of a cloaking device at optical wavelengths [89]. The concept is to "rebuild" behind the acting device an incident Gaussian modulated plane wave whereas no light can penetrate in the core of the device. The proposal is based on the association of two photonic lattices operating in different regimes, namely stop-band and negative refraction. The operation principle relies on the abrupt changes of direction in wave propagation afforded by negative refraction. Device dimensions can be easily adapted to the desired hidden area as well as to the incident plane wave spatial extent. Conversely to the conformal mapping approaches, real "invisibility" is not reached in the sense that the device has no cylindrical or spherical symmetry. However, it could appear as an original approach for compact optical devices for future integrated nanophotonics including 3D architectures where the need for optical "via-holes" could help complex circuit designers.

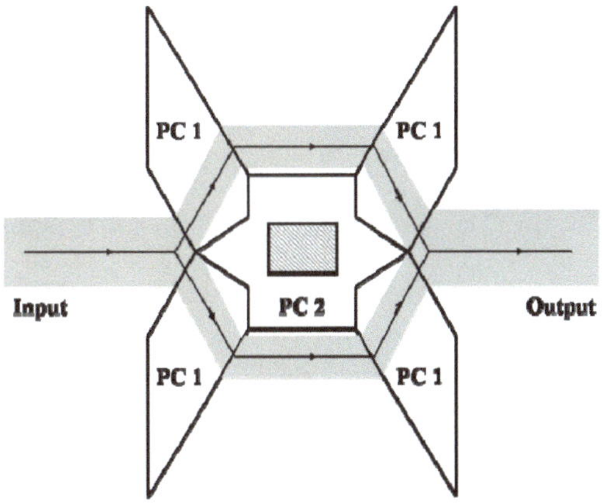

Fig.(32): Schematic of the "butterfly" optical device for cloaking operation

This leads to the "butterfly" optical device whose schematic is given in Fig.(32). For the wavelength range investigated, a first photonic crystal labeled PC1 will be used in the negative refraction regime encountered within the second pass-band, and a second one labeled PC2 will be used in stop-band regime. The hidden region appears hatched. The gray region illustrates the optical path from input to output. The incident optical wave is here polarized with an electric field perpendicular to the propagation plane (parallel to the air holes for a 3D structure).

The V-shape of PC1 on left side allows splitting the incident wave in two equal streams with a low reflection level whereas the second interface permits to turn the two in-phase streams in the initial propagation direction. On the right side, the symmetric design reciprocally regenerates the incident optical wave from the two separated streams. PC2 is inserted in the core of the device to protect the hidden zone from especially the straightforward component of the incident optical wave (it is the main role of the two front and back PC2 tips). Surrounding the cloaked zone, the lattice thickness is adjusted to reach negligible field magnitude in the central region.

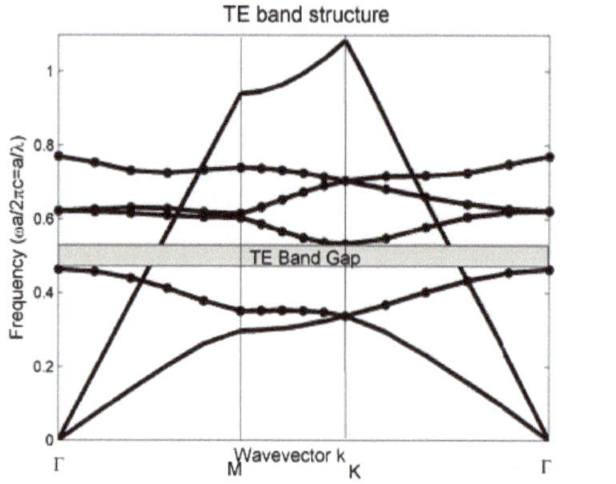

Fig.(33): Band structure of PC2 (period : 775 nm, hole diameter : 736 nm) showing the full band gap around $\lambda = 1.55$ µm.

For the basic cloaking device, the PCs characteristics are the following. PC1 is the photonic crystal used previously for the flat lens designed to operate with n = -1 at $\lambda = 1.55$ µm. The V-shape used to split the incident stream is formed along triangle sides.

For PC2, the period is 775 nm for a hole diameter of 736 nm. The corresponding dispersion diagram is given Fig.(33). This high filling factor in air is necessary to reach a full band gap regime at 1.55 µm, corresponding here to $a/\lambda = 0.5$.

Fig.(34) illustrates the operating principle at 1.55 µm with the amplitude (a) and the phase (b) of the electric field. A Gaussian shaped continuous plane wave (attested by the regular phase fronts) is injected in air on the left hand side just in front of the V shaped interface air/PC1. Owing to the negative index close to –1 of PC1 at 1.55 µm, the incident stream is equally shared in two parts propagating within PC 1. The presence of larger holes (triangular tip - PC 2 in forbidden band regime) in the incident propagation direction at the V corner permits one to avoid straightforward wave propagation to protect to cloaked zone. A second negative refraction occurs which redirects the two half-streams in the incident propagation direction along the cloaked zone again protected by PC2. To retrieve the incident propagation direction both interfaces of PC 1 needs to be parallel. On the other side the incident wave is reconstructed using a reciprocal design with again two successive negative refraction effects. The phase information confirms that regular phase fronts characteristic of a continuous plane wave are recovered at the output, the same regularity is observed in air between the two pair of wings of the "butterfly" device. Since protected by PC2 the core of the device is free of electromagnetic energy whereas the four parts of PC 1 allows to split and to reconstruct in phase the incident stream.

(a)

(b)

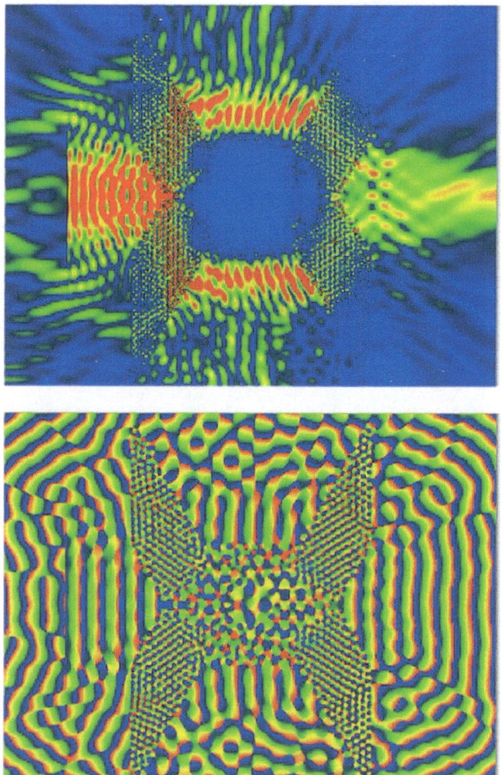

Fig.(34): Cloaking device simulation at 1. 55 µm, (a) electric field amplitude (linear scale), (b) electric field phase

The cloaking operation is efficient as long as the index remains close to -1 and as PC 2 presents a full gap to the optical wave. In this case, we obtain a 50 nm bandwidth between 1.54 µm and 1.59 µm. Let us mention that the equivalent refractive index variation is of about 10% around –1 and that if –1 is optimal it is not an absolute prerequisite.

To analyze deeply the wave reconstruction, we present a spatial analysis (Fig.35) of the transmitted wave at different positions behind the cloaked area. The Gaussian shape is mostly preserved but the transmitted signal is slightly attenuated. Looking also at time evolution of the optical wave, one can measure the phase regularity of the transmitted signal. As it appears in Fig.(34b) the phase is quasi-independent of the lateral shift. Only a slight deviation of a few degrees is observed at 6 microns from the center but it corresponds to a wave magnitude lower than -25 dB. It can be mentioned that the spectral purity behind the device remains good and that propagation is monomode throughout the device. The plane nature of the wave behind the device cannot be distinguished from the incident wave which was the primary goal of the device.

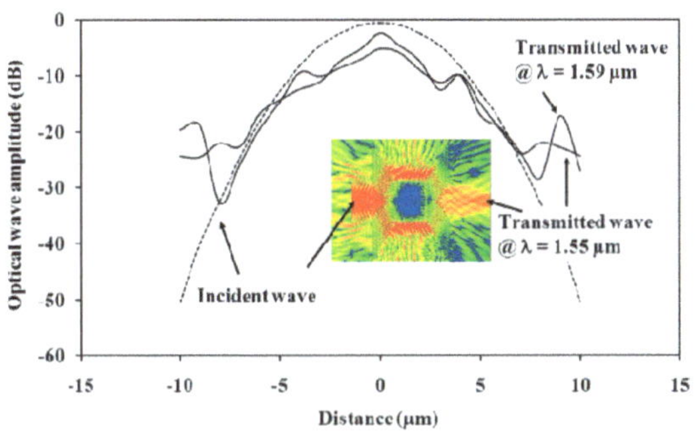

Fig.(35): Amplitude analysis of the incident and transmitted wave at different wavelengths

As mentioned in introduction, this "butterfly" optical device is not a full cloaking device since it has no cylindrical or spherical symmetry and only works for a unidirectional incident plane wave. Its major advantage stems in its principle which is quite simple. Only two different PC's are needed and no refined space index (permittivity or permeability) modulation is required compared to the different proposals made to reach full 2D cloaking (see next part). It is also intrinsically different of devices based on the decoupling and mixing of a unique guided wave as, for example, a planar Mach-Zender interferometer. Here, the incident signal is geometrically split in two sub-signals which are added at device output. The signal amplitude is unchanged throughout the device, only its spatial extent. If the example is given for a Gaussian transverse modulated wave, it can be extended to any waveform in free space as long as the wave interacts with the V-shaped area at input. No symmetry with respect to the main propagation direction is required. Moreover the size of the hidden zone can be enlarged at will (and can become much longer than the wavelength) as soon as the geometrical rules given above are respected and that propagation losses in PC1 remain at acceptable levels. Lastly, let us emphasize that the lateral device size can be designed as a function of the incident monochromatic wave space extent. Indeed, as propagation takes place in free space or in bulk PC's, no dispersion induced by mode superposition as in PC based guided structures exists when large multimode waveguides are used. We thus achieve a high degree of versatility for integrated nanophotonics.

Transformation optics for a full dielectric electromagnetic cloak and a metal-dielectric planar hyper-lens

In this last part, we will develop briefly a complementary approach for artificial media loosing the concept used up to now of periodicity. Conformal transformation of electromagnetic domains has been proposed as an exciting approach to control the flow of propagating waves. It enables the design of objects displaying unprecedented functionalities with the requirement of space gradients and anisotropy of the constitutive parameters (permittivity and permeability). Approaches have been proposed in photonic crystals using a precise engineering of the refraction index, positive or negative, to bend light over very short distances (smaller than the wavelength) so that it was possible to create artificial "mirages". The structures proposed below represent a further step of effective parameter engineering following the seminal works of Pendry [78] and Leonhardt [83] using optical conformal mapping techniques to reach cloaking and the exciting concept of invisibility.

First, let us consider a full dielectric electromagnetic cloak design top operate in the Terahertz region based on permeability engineering via Mie resonances of dielectric pillars [90].

Here, Mie theory is applied to engineer the magnetic plasma frequency of high-κ ferroelectric rods at THz frequencies[91]. Full-wave simulations coupled with a field-summation retrieval technique were employed to assess the effective parameters (permittivity ε_{eff} and permeability μ_{eff}) of ferroelectric $Ba_xSr_{1-x}TiO_3$ (BST) rods excited by an H-field along the rod axis. From the results of this study, a device that performs cloaking at terahertz frequencies was carefully built from individual cylinders. Here, the idea is to superimpose the electromagnetic response of cylinders with different radii in such a way that the micro-structured cloak shell displays a strong radial distribution in permeability and falls into a class of cylindrical objects derived from the conformal transformation theory.

The dielectric rods were excited by a plane wave along the y-axis, as the inset in Fig.(37a) depicts. The dispersion of the investigated material was neglected and the permittivity ε' was set to 200 with a dielectric loss tangent ε"/ ε' = 2.10-2 for all frequencies. This assumption relies on the fact that the presented micro-structured device operates, by definition, within a narrow frequency range where the bulk parameters can be assumed constant. Perfect magnetic conductors and perfect electric conductors were used for the lateral and top/bottom faces, respectively. Hence, the incident wave injected by the input port propagates through a unique rod along the propagation direction x and infinite set of rods in the y-z plane. Then, the scattering parameters (reflection S11 and transmission S21) were computed from the input/output ports to determine the Mie resonance frequency.

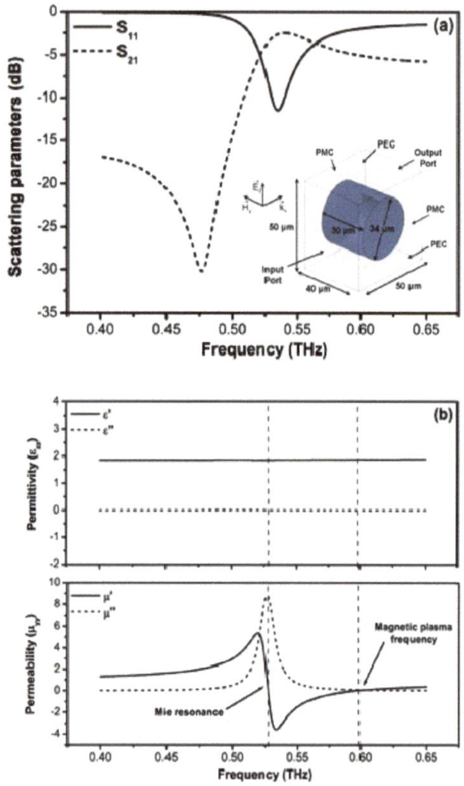

Fig.(36) (a) Scattering parameters of a micro-cut BST rod computed using the finite-element solver HFSS. The inset presents the computational cell and applied boundary conditions. The rod displays a permittivity function ε_{BST} = 200-5j (b) Top and bottom view present complex ε_{zz} and μ_{yy}, respectively, retrieved using a field-summation method. The Mie resonance occurs at 0.527 THz.

Fig.(37b) presents the frequency dependence on the reflection and transmission parameters of a plane wave impinging onto a BST cylinder with a 34 µm diameter and a height of 30 µm and for frequency values ranging from 0.36 to 0.66 THz. These frequencies are within the range where the rod electromagnetic behavior can be described by an effective medium. Here, the simulation domain was 50 µm x 40 µm x 50 µm. The 5 µm spacing between the top/bottom surfaces of the rod and the lateral faces insures that the incident wave does not impinge onto an infinitely long rod along the y axis. Although this aspect is not discussed in this work, note that the Mie resonance is rejected to lower frequencies with an infinite rod. For this example, the Mie resonance is predicted to occur at 0.527 THz; frequency at which the rod displays the most significant magnetic response. The complex effective permittivity and permeability were retrieved using a field-summation technique over the simulation domain and are presented in Fig.(36b). The data indicate that the real part of the permeability displays a Lorentz-like behavior around the Mie resonance frequency. In contrast, the real part of the permittivity remains almost constant with a value of 1.86 over the frequency range of interest. The imaginary part of the permittivity and permeability is ~1.10-5 and ~8.10-2, respectively. Finally, the data show that the magnetic plasma frequency at which the permeability vanishes is 0.597 THz. This information is critical because it enables the design of metamaterials with permeability values ranging from 0 to 1.

In order to tailor the magnetic activity of the dielectric rods, the simplest approach would be to shift upward or downward the Mie resonance frequency by changing the dielectric function of the material. However, this solution is technologically impractical since it would require mastering the rod stoichiometry with high accuracy. Another alternative would be to increase or decrease the path length that the displacement currents undergo at the resonance. Physically, this means adjusting the radius of the rod to explore μ_{yy} values between 0 and 1.

Now that a set of rod dimensions is known, one can turn to the definition of the cloak itself. In the case of an annular cloak defined by its inner radius a and outer radius b [see Fig.(37a)], the effective parameters of the cloak shell (permittivity and permeability) must be independently engineered to satisfy a set of equations derived from the conformal transformation theory. On the other hand, it was shown that this design burden can be overcome by using a reduced set of equations. This allows one parameter only (permittivity or permeability) to be varied with the cloak radius. In our case, a progressive variation in the permeability is obtained by positioning the Mie rods radially. Under this configuration, a transverse-electric (TE) polarized plane wave (H_r, H_θ, E_z in cylindrical coordinates) has to be employed to illuminate the magnetic cloak and the original set of equations can be reduced to:

$$\mu_r(r) = \left(\frac{r-a}{r}\right)^2 \quad \textbf{(Eq. 16)}$$

$$\mu_\theta(r) = 1 \quad \textbf{(Eq. 17)}$$

$$\varepsilon_z(r) = \left(\frac{b}{b-a}\right)^2 \quad \textbf{(Eq. 18)}$$

Here, the cloak consists of 7 levels of 30-µm-high BST rods with diameter values increasing from 34 to 40 µm by steps of 1 µm and 10 µm radial pitch separating each rod, as shown in Fig.(37a). Finally, the inner radius of the cloak was arbitrarily set to a = 280 µm and outer radius b = 560 µm to limit the size of the computational cell. Fig.(37b) presents the cloak effective parameters ($\mu_r(r)$, $\mu_\theta(r)$, and $\varepsilon_z(r)$) dependence on the normalized radius r/*a* calculated (i) for the investigated BST rods at 0.58 THz and (ii) from Eq.16. This operating frequency was selected by matching the discrete distribution in permeability from the rods with Eq.16. The data also show that the permittivity value is about half of the one computed from Eq.18. In the following, we will see that this permittivity mismatch does not fundamentally perturb the cloaking ability.

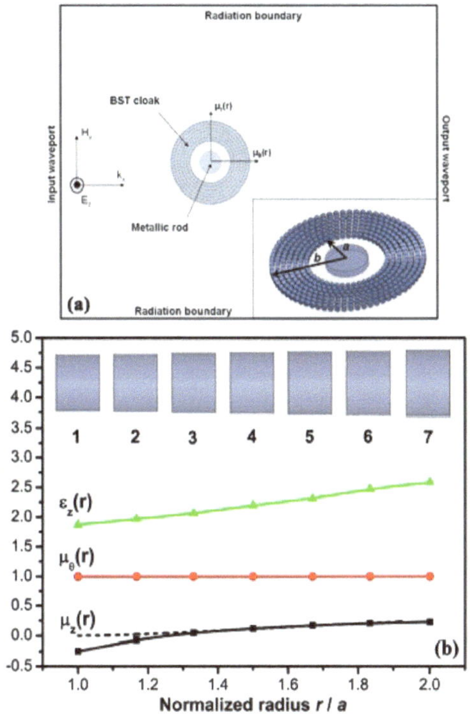

Fig.(37). (a) Top view of the computational domain used to simulate the discretized dielectric cloak. The inset presents a 3D rendering of the device which consists of 7 concentric rings of micro-cut BST rods. An ideal copper cylinder was placed at the center to assess the cloaking efficiency. (b) Dependence of the cloak effective parameters (μ_r, μ_θ, and ε_z) at 0.58 THz on the normalized radius r/a where a is the cloak inner radius and for radii values ranging from a to b = 2a. The dashed line plots the theoretical distribution of the radial permeability given by the Eq(14).

In order to achieve an efficient cloaking device, key points must be satisfied. First, the device itself must not be detected. Consequently, the field pattern should not present wave front distortions around the cloak nor shadow regions behind it. This particular point is critical because it is understandable that the electromagnetic signature of the device must not be revealed. Hence, the outer surface of the device must be well impedance matched with its surrounding environment, typically air, to minimize reflection.

We first studied the scattering of a plane wave at 0. 58 THz (λ= 517μm) for an uncloaked copper rod. The results of the E-field distribution achieved in this case are displayed in Fig.(38a). Strong distortions of the E-field are clearly observed behind and in front of the metal scatterer. Fig.(38b) shows the E-field z component for the metallic object now surrounded by the micro structured cloak. These results unambiguously demonstrate that cloaking is achieved at the targeted frequency. First, the E-field wave fronts are well reconstructed behind the cloak with minimal scattering. A high power transmission value of ~66% and low reflection value of ~13% were computed, thus indicating that ~20% of the total energy is either absorbed within the BST rods or reflected toward the radiation boundaries. Also few back-scattered ripples are observed at the front and side of the cloak.

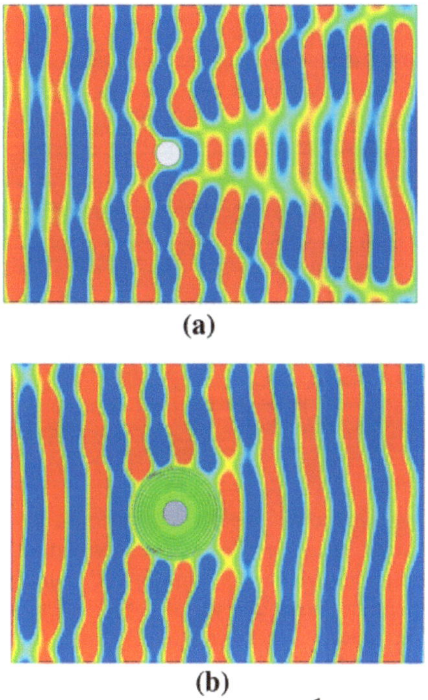

Fig. (38). Steady-state E_z pattern calculated at 0.58 THz for a copper rod without (a) and with a micro structured cloak (b). The plane of observation is located at mid-distance between the bottom and top face of the simulation domain. The wave fronts are well reconstructed behind the cloak without noticeable backscattering. The metallic particle placed at the center of the device is nearly "invisible" to a detector located at the output port.

The copper rod is effectively hidden and invisible for any incoming waves. Let us mention that this result is obtained using "realistic" parameters for the constitutive material, notably in terms of intrinsic losses. Thus, the cloaking effect is not perfect but indicates us quite precisely what could be observed in the real world. These results show the potential of this all-dielectric solution for cloaking applications at microwave and terahertz frequencies. Is it reasonable to envisage operation at optical wavelengths? Up to now, the question remains open, since BST permittivity tends to decrease rapidly as wavelengths reach the infrared region and that rod dimensions required for the engineering become too small to be technologically fabricated. Is there any other material with high permittivity that can be structured at nanometer scale? (Highly-doped) semiconductors are potential candidates…but no real demonstrations are available at the present time.

At last, let us consider another field of applications of transformation optics: hyperlenses [92-96]. In this section, we will use this technique to magnify near-field patterns. The main idea is to channel and magnify any near-field pattern in order to bring its fine details, originally carried by evanescent waves only, over the Abbe limit. Consequently, at the output of the hyperlens, those details are converted into propagating waves, which can then be probed in the far field by a conventional imaging system. The design procedure is as follows: let us consider a flat channeling device in the x y-plane; z is the direction of the magnetic field (TM polarization), whereas y is the main propagation direction. Consequently, we have to choose a very low value for ε_{xx}. In the following, we will use the lossless parameters (ε_{xx}, ε_{yy}, μ_{zz}) = (0.001, 2.5, 1). The device is positioned between y = a and y = b. The magnification factor is t. The coordinate transformation used in the procedure is:

$$\begin{cases} \mathbf{x}' = \left[\left(\frac{\mathbf{y}-\mathbf{a}}{\mathbf{b}-\mathbf{a}}\right)(\mathbf{t}-\mathbf{1})+\mathbf{1}\right]\mathbf{x} \\ \mathbf{y}' = \mathbf{y} \\ \mathbf{z}' = \mathbf{z} \end{cases} \qquad \textbf{(Eq. 19)}$$

This transformation is linear with respect to x and y. On the input interface (y = a), we have an identity between the original and the transformed coordinates. On the output interface (y = b), we 'stretch' the coordinates along the x-direction by a factor of t (x' = tx). It should be noted that such a transformation introduces an 'optical axis' since the translation along x is proportional to the distance to x = 0 at the input. However, at this point, it is not necessary to limit the device along the x-direction. By following the procedure by Schurig et al [79] for this linear transformation, we obtain the local expressions for both the 2×2 permittivity tensor and the out-of-plane permeability (in our 2D TM case) of the device equivalent to the transformed space. These expressions only depend on the position, the geometrical parameters (a, b, t) and the original material parameters (ε_{xx}, ε_{yy}, μ_{zz}).

Fig.(39) illustrates such a device. The input near-field pattern is simply two very narrow ($\lambda_0/60$) sources of magnetic field centered at + and − $\lambda_0/20$ and positioned directly at the bottom interface of the lens. The rest of the bottom boundary is a perfect magnetic conductor. Above the lens is placed an air layer. The top and side boundaries are perfectly matched layers. The wavelength in air was chosen to be 7.5 times larger than the thickness of the lens. Fig.(39) shows the resulting magnetic field map (a) an d the plot of the same field on the input and output boundaries of the lens (b). The distance between the two peaks is multiplied by 5 thanks to the lens. It should be noted that a significant spreading of the peaks can also be observed.

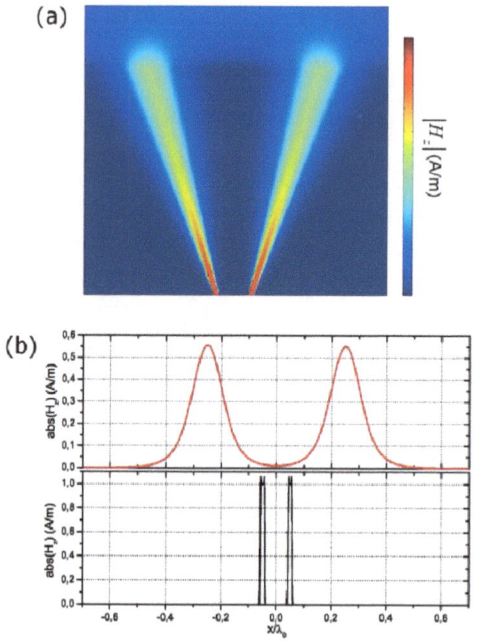

Fig.(39) Hyperlens operation : (a) Simulated magnetic field map inside the flat hyperlens and in the air above for an input pattern of two sub-wavelength sources. (b) Corresponding field plot for both interfaces of the lens.

Beyond the search for invisibility, which is an ultimate goal in the control of light, other applications could be addressed through transformation optics such as lenses. In particular, we show that hyperlens or high focusing devices can be designed by such a procedure. The main challenge in such design is to keep in mind the technological constraints depending on the targeted wavelength domain.

CONCLUSIONS

In this chapter, we have given a brief overview of the potential applications of negative refraction in artificial periodic media. We have seen that the concepts can rely on permittivity and permeability engineering in metamaterials or on photonic band structure engineering in photonic crystals. In both cases, whatever the choice of parameters, the goal is to obtain a double negative medium with ε and μ simultaneously negative or described by a negative index of refraction (and a positive surface impedance). The basic concepts are dimensionless and can be applied to any wavelength, starting from microwaves and centimeter wavelengths down to optics and sub-micrometer wavelengths. The challenge in each case is the fabrication since advance tools of nanoscience are required as wavelength decreases. Also, the intrinsic properties of the constitutive materials appear of prime importance as dimensions are lowered.

We believe that this new class of artificial materials with their "extraordinary" properties in terms of propagation, waveguiding or radiation, will bring original applications in the near future, from microwaves, terahertz region, far and mid-infrared and optics. Particularly promising is the domain of integrated nanophotonics, where applications based on sub-wavelength imaging by super- or hyper-lenses or cloaking are expected...

ACKNOWLEDGMENTS

All the results presented here come from the DOME group of IEMN (Opto- and Micro-Electronics Devices) headed by Prof. D. Lippens. Particular thanks to N. Fabre, X. Mélique, F. Zhang, T. Crépin, C. Croenne, D. Gaillot, G. Houzet, M. Hofman and E. Lheurette. Have also contributed to these works: S. Potet, J. Carbonell, T. Decoopman, S. Fasquel, M. Perrin and J. Danglot. Optical measurements have been carried out at Institut Carnot de Bourgogne in Dijon. Thanks to F. de Fornel, B. Cluzel and L. Lalouat.

ANNEX A: Design rules for the Ω-type metamaterial (from ref. [40])

Fig.(A1a) shows a schematic of the basic cell while Fig.(A1b) illustrates the interconnection arrangement along the *x* axis, transverse to the propagation direction along *z*. The basic unit cell consists of two back-to-back omega metal inclusions. The core of the pattern is responsible for a magnetic dipole through current loops via an H field excitation along the *y* axis. It is imperative to satisfy this H-field polarization condition in order to achieve a pure magnetic response namely B= -μH [97]. The two arms of the omega pattern are responsible for the electric response when the electric field is polarized along the *x* axis. One can compare the omega patterns with the combination of split rings resonators (the central current loops) and wires (the arms of omega patterns). The back-to-back configuration suppresses the magneto-electric response according to the basic ideas developed in broad side coupled split ring resonators [98].

The fact of combining in one motif, the magnetic and electric responses means that there exists an interdependence of phenomena. However, as for distinct SRR and wire array, a first optimization of the electromagnetic properties can be carried out by considering separately the dispersion of the magnetic and of the electric dipoles.

The frequency dependence of the effective permeability is governed by the two characteristic frequencies of a Lorentz-like response, namely the resonant frequency of the current loop (ω_0) and the magnetic plasma frequency (ω_{mp}). The latter corresponds to the transition frequency between negative and positive values of μ_{eff} in the upper part of the spectrum [11]. Many experimental and theoretical works have addressed the derivation of the resonant frequencies of C- or U-shaped micro-resonators either on the basis of full-wave analysis or by a lumped element approach. The latter is particularly useful in an optimization procedure notably through adjustments of the permittivity and thickness of the substrate and of the width of the metal strips which tune the coupling capacitance between the opposite omega motifs.

For the current loops, the magnetic angular resonance frequency ω_0 can be predicted by a LC resonance, in which:

$$\mathbf{L} = \mu_0 \frac{\pi r^2}{d} \qquad (\mathbf{Eq. A1})$$

L is the inductance of the current loops and:

$$\mathbf{C} = \frac{1}{2}\frac{\varepsilon_r \varepsilon_0 w}{d}\pi\left(\frac{2r + w}{2} - \frac{g}{2}\right) \qquad (\mathbf{Eq. A2})$$

C is the total capacitance between the broadside coupled split ring resonators [21] (here ε_0 and ε_r are the permittivity of vacuum and substrate, respectively. μ_0 is the permeability of vacuum, *w* is the wire width, r is the inner radius, *g* is the gap of omega pattern and d is the thickness of the substrate). The so-called magnetic plasma frequency ω_{mp}, is directly related to the magnetic resonant frequency ω_o and the filling factor F:

$$\mathbf{F} = \frac{\pi r^2}{S} \qquad (\mathbf{Eq. A3})$$

the area ratio occupied by the interior ring with respect to the area of basic unit cell in the *x-z* plane (see Fig.(A1a)), according to the following approximation [11]:

$$\omega_{mp} = \frac{\omega_0}{\sqrt{1-F}} = \frac{1}{\sqrt{(1-F)LC}} \qquad (\mathbf{Eq. A4})$$
$$= \frac{1}{\sqrt{1-F}\sqrt{\left(\mu_0 \frac{\pi r^2}{d}\right)\frac{1}{2}\frac{\varepsilon_0\varepsilon_r w}{d}\left(\pi\frac{2r+w}{d} - \frac{g}{2}\right)}}$$

For the electrical response which can be compared to that of an array of infinite metal wire, the effective permittivity follows a Drude-like model. Under this assumption, the frequency dependence of the permittivity exhibits negative values below the plasma frequency ω_p, which can be considered as a characteristic frequency. An estimate of the ω_p can be found in the seminal work of Pendry [12].

$$\omega_p^2 = \frac{2\pi c_0^2}{a^2 \ln(a/r)} \qquad \text{(Eq. A5)}$$

where a and r are the periodicity of wire array and radius of wire, respectively. The cross section of the wire, considered here, is rectangular instead of circular as in Pendry's model. However, the plasma frequency has an analogous dependence as a function of the wire width (w), i.e. increasing with w [99].

As outlined in the text, a balanced condition can be achieved when $\omega_{mp} = \omega_p$. To this aim, the degrees of freedom for a given permittivity of a substrate are the width of the metallization and the substrate thickness. As aforementioned, the dielectric and magnetic responses are not strictly independent and a resonant feature is superimposed onto the frequency dependence of the electric response. This spurious response is however weak and the simple rule which insists on saying that the effective permittivity is negative below the plasma frequency and positive above still hold.

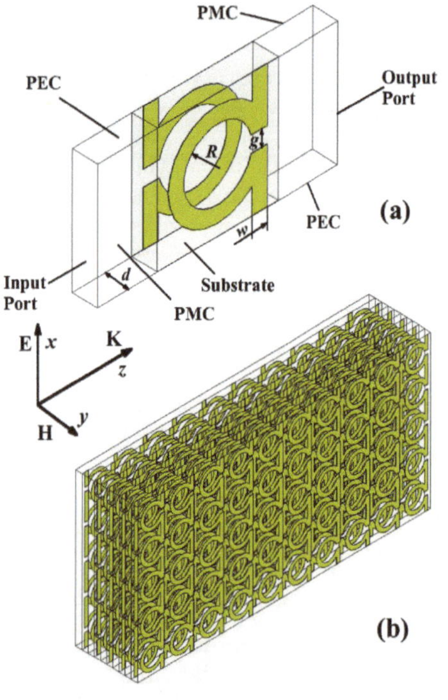

Fig. (A1) (a). Schematic of omega unit cell. (b) Schematic of omega pattern array with an edge illumination by a plane wave. The dimension parameters are as follows: $r = 1.1$ mm, $w = g = 0.5$ mm, $d = 0.8$ mm. Both the periodicities along x and z directions are 3.9mm. The substrate is chosen as epoxy glass with an assumption of relative permittivity of 4.0 ($tan\delta = 0.02$).

In practice, the fine tuning of the balanced condition between ω_{mp} and ω_p was performed by full wave analysis (Commercial package, High Frequency Structure Simulator, HFSS simulation code from Ansoft) of the electromagnetic behaviour of the basic cell by applying the Bloch-Floquet theorem. The boundary conditions are displayed in Fig. (A1a). Perfect electrical conductors are placed along the x direction while perfect magnetic conductors are imposed along the y direction. A plane wave with the electric field polarized along x was assumed to illuminate along the z direction. Two ports according to Fig.(A1a) allow the calculation of the scattering parameters.

ANNEX B: Optimization procedure for the design of a n = -1 flat lens using a 2D photonic crystal (from ref. [68])

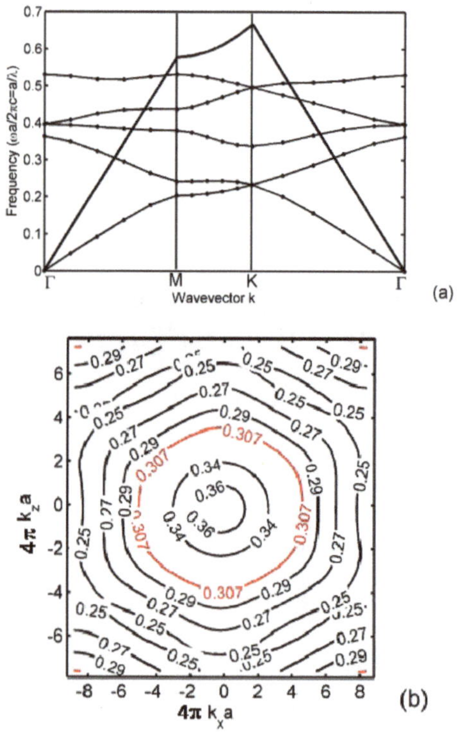

Fig. (B1) : (a) 2D Band structure for the triangular lattice with period a = 476 nm and air hole diameter d = 350 nm (filling factor of 0.38) – TM mode (E field parallel to the holes) - The index value of the patterned semiconductor is 3.32 ; (b) 2D isofrequency plot for the second band where negative refraction index can be defined (red line: n = -1 circle).

Our approach is based on a photonic crystal constituted of air holes etched in a semiconductor heterostructure. Calculations are performed throughout the paper using the commercial codes BandSolve (for band structures) and Fullwave (FDTD - lens experiment) by RSoft. For the two dimensional simulation, an effective index of 3.32 is assumed corresponding to an InP/GaInAsP/InP based double heterojunction. The desired operating wavelength is 1.55 μm. To obtain a negative refractive index equal to –1 at this wavelength in second band of the crystal, we use a triangular lattice of air holes to promote isotropy with a periodicity a = 476 nm and a hole diameter d = 350 nm (filling factor : 38 %). Fig.(B1) gives the corresponding TM band structure (E field parallel to the holes) along with the light line. The second band stands from a/λ = 0.24 to 0.36 (wavelength range 1.25 - 1.98 μm). As shown in Fig.(B1b) this second band is almost isotropic (circular iso-frequency plots around Γ point) up to high wave vector values. Let us remain at this point that working with n = -1 (crossing between the band and the light line) is a necessary condition to obtain focusing of an incident point source with a slab but is not sufficient to obtain the "superlens". Indeed, surface impedance matching between air and the crystal is required to minimize reflection at input.

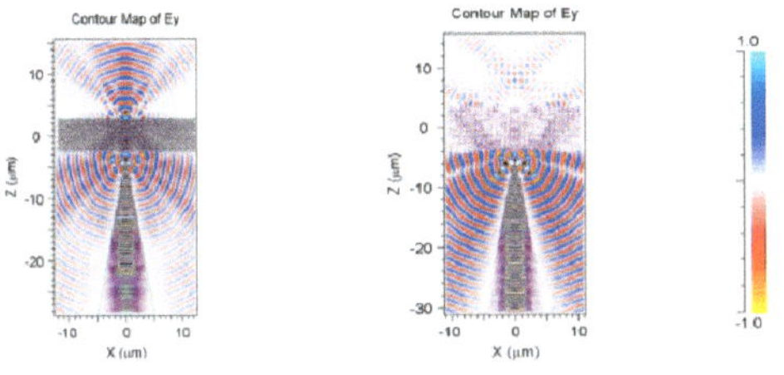

Fig. (B2) : Lens experiment using a triangular photonic crystal slab: ΓM oriented (left side) and ΓK oriented (right side)

To illustrate this, Fig.(B2a) and (B2b) show a virtual prototyping of such a lens fed by an optical tapered waveguide ended by a small ($\lambda/10$) diffracting hole to mimic a point source. For the left hand side of Fig.(B2) the crystal is ΓM oriented (injection perpendicular to the triangle sides) whereas ΓK orientation (injection parallel to the triangle sides) is used for the right side of Fig.(B2). Even if the dispersion curve is isotropic for this frequency and an effective refraction index can be deduced, the lens behavior is strongly dependent on incidence. The focusing is clearly apparent in the first case, but has almost disappeared for the second. Moreover the signal intensity behind the lens is strongly attenuated since reflection at input is huge. In Fig.(B2a), a clear focus is present issued from the lens and optical rays can be superimposed as in classical geometrical optics. The extension along propagation axis of the focus point is very small which indicates a refractive index value very close to –1 in this case. Using Gauss approximation of small angles and the distance ratios between focus points inside and outside the lens leads to an estimated value of – 0.95. At the opposite, the width of the focus point behind the lens is larger than the incident electric field pattern revealing the fact that subwavelength resolution is not reached, and consequently that effective permittivity and permeability are not close to –1. Also, reflection is important in input and needs to be optimized. Such behavior has already been observed and studied in the literature [100-101] and will be analyzed more deeply later. As a first indication, let us mention that the lens thickness is different in ΓM and ΓK directions since an equivalent number of air hole rows has been assumed in both cases but distance between two successive rows is different (equal to $a\sqrt{3}/2$ in Fig.(B2a) and equal to a in Fig.(B2(b)). In terms of cavities, the two lenses possess distinct Fabry-Perot modes. Even if n = -1 in both cases, we face a quasi-matched case in terms of impedance with low reflection and high transmission in Fig.(B2a) but the impedance mismatch prevails in Fig.(B2b) with opposite performances. This first all-angle interpretation needs now to be refined.

Indeed, at this point, it appears clearly that the lens properties depend strongly on the incidence angle of the optical wave and thus on the first interface. More than this first interface, we will see that the thickness and the number of crystal periods used in the main propagation direction are also of prime importance. To analyze all these effects, the waveguide is suppressed at input and we will use a simple Gaussian pulse excitation. Its width is chosen so that transmission and reflection spectra on both sides of the lens can be calculated by fast Fourier transform over the second pass band of the photonic crystal and around 1.55 μm, the targeted wavelength of operation. For the study, the incident angle between the pulse direction of propagation and the normal to the lens will be varied. Finally, special attention will be paid to the position and size of the input and output ports of our FDTD simulation in order to recover the totality of the incident optical signal so that the sum of transmitted power and reflected power is equal to the incident power carried by the pulse. For sake of simplicity, we will focus our attention in the following on a six row lens to evidence the main limitations and performances of such lenses.

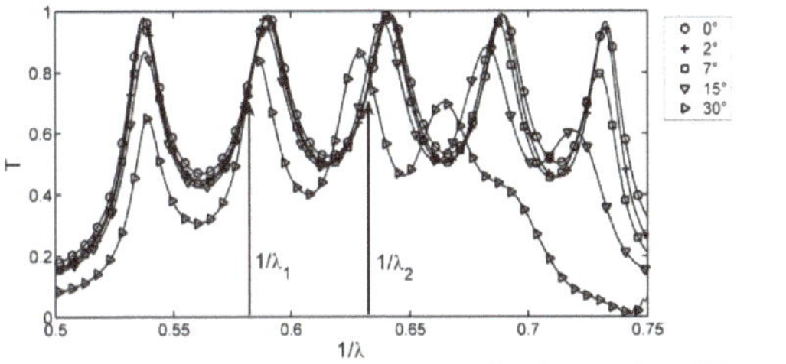

Fig. (B3) : Transmission through a 6 row flat lens under different incidence angles (0°, 2°, 7°, 15°, 30°)

Fig.(B3) gives the transmission spectra obtained for a six row lens for incidence angles varying from 0° to 30° as a function of $1/\lambda$ (1.55 μm corresponds to 0.645 μm^{-1}). As mentioned in Ref [62], special care has to be devoted to the interface cut to promote the focusing properties. The row width is defined as $a\sqrt{3}/2$ with the air hole located at the center. This ensures a perfect symmetry along the direction perpendicular to the triangle sides if an odd number of rows are used. However it has been checked that our optimization procedure is valid whatever the number of rows, odd or even. The fact that air holes are staggered does not affect focusing properties. Under normal incidence (here labeled 0°), five transmission peaks are clearly visible, each very close to unity. In between, the transmission decreases rapidly and can be less than 40 % even if the wavelength stands within the second band. This impedance mismatch between air and the lens, responsible for a quite high reflection coefficient is not only induced by the first interface as often claimed but the result of the cavity formed by the crystal embedded with "air". The number of peaks is equal to the number of air hole rows minus one within the propagation direction. This effect is very well known for any heterogeneous periodical medium [102-103]. If air hole rows are considered as a specific medium with n_1 index and the semiconductor in between as a medium n_2 index, this effect corresponds to the Fabry-Perot resonances induced by the n_1/n_2 alternate layers. It can be mentioned that the low levels of transmission obtained between peaks evolve very weakly with the number of rows even if the number of peaks increases. Conversely, their position within the band depends of the number of rows. This means that if an operating wavelength is targeted, care has to be taken that the thickness of the slab allows the presence of a maximum in transmission at this wavelength. Here, the structure is designed to give a maximum under normal incidence at 1.55 μm for that given number of rows.

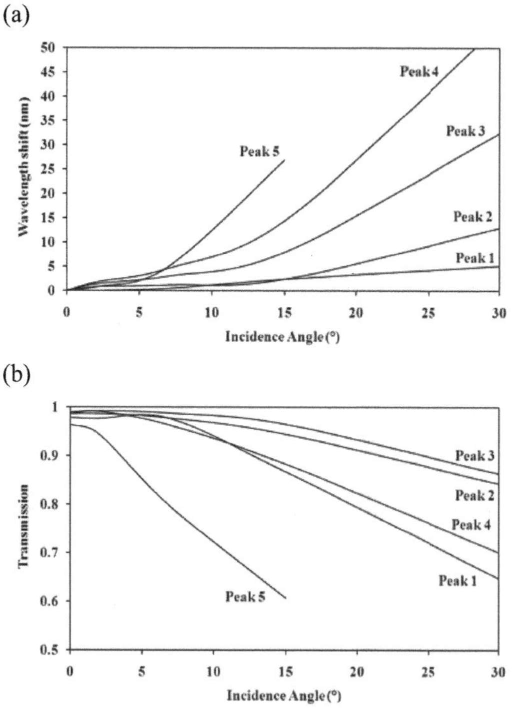

Fig. (B4) : (a) Wavelength shifts versus incidence of maximum transmission peaks for a 6 row photonic crystal based slab in negative refraction regime; (b) Maximum transmission values associated.

This first precaution is not enough to ensure good performances for the lens. Indeed, to image a point source even in near field we face a circular wavefront at the first interface with an extension which is function of the source-lens separation. This means a multi-incidence beam interacting with the lens. Obviously, the matching (unitary transmission) is not preserved for any incidence, as exemplified by the additional spectra reported in Fig.(B3). This effect was already present in Fig.(B2) for the two extreme incidence angles (0° - ΓM and 30° - ΓK). But this effect has to be considered whatever the incidence angle value. As this angle increases, one can note a shift to higher wavelengths of the respective transmission peaks along with a decrease in amplitude. This figure shows clearly the sensitivity to incidence angle is more pronounced as λ decreases experiencing deeper the interspace between hole rows. These evolutions are summarized in Fig.(B4) where wavelength shifts are reported versus incidence angle. For the higher wavelength, the peak position is insensitive (less than 5 nm) to reasonable (in the sense of geometrical optics Gauss approximation) incidence angles, whereas for the shorter wavelength the peak almost disappears for an incidence of 30°. For the intermediate peaks, the shifts are respectively 13 nm, 32 nm and 55 nm for peaks 2, 3 and 4 between 0° and 30°. Looking at Fig.(B4b), it is clear that peaks 2 and 3 leads to high transmission levels for all incidence angles lower than 30°, whereas peaks 1 and 4 decrease faster as soon as angles higher than 10° are under concern. This appears as a strong restriction for good focusing properties. Moreover, the ripple between the peaks has to be correctly evaluated for a specific operating wavelength. As a matter of example, even if the maximum of transmission remains higher than 80 %, one can note that the wavelength associated to the maximum under normal incidence induces a transmission drop to less than 50 % at this same wavelength for the incidence of 30°. Different theories exist to optimize and decrease such ripples, notably in "filtering" theory. In general, it consists in considering the structure under study as a series of cascaded cells and in modifying the characteristic properties of each cell (but preserving lens symmetry about its middle in the propagation direction) to reach flat band conditions. Such approaches have also been applied to semiconductor superlattices in nanoelectronics. Concerning photonic crystal slabs, a first attempt has been presented recently with the proposal of a shape modification of the first row of air holes, replacing circles by ovoid forms, to minimize input reflection [65, 104]. The associated results are very promising in terms of matching but more has to be done to reach good focusing performances in terms of associated negative refractive index. We believe that this hole shape engineering should not be limited to the first row but the slab has to be considered entirely to reach such flat band conditions so that it could work efficiently for various incidence angles. As shown in this last work, such engineering is particularly tricky in terms of fabrication due to the precision needed. An alternate way is to incorporate some extra material by atomic layer deposition, as proposed in [102], in order to modify locally the structural parameters.

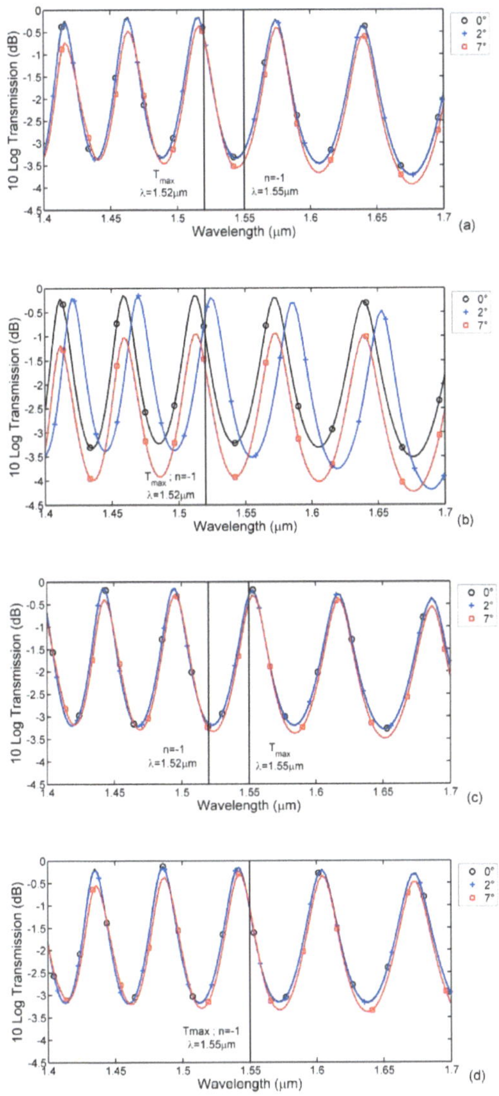

Fig. (B5) : Transmission spectra for a twelve row photonic crystal slab for various incidence angles and geometry
(a) a = 476 nm ; d = 350 nm
(b) a = 476 nm ; d = 366 nm
(c) a = 469 nm ; d = 350 nm
(d) a = 469 nm ; d = 332 nm

The best results can be obtained at the two particular wavelengths labeled λ_1 and λ_2 in Fig.(B3) for which high transmission levels (higher than 80 %) can be obtained whatever the angle of incidence but with the detriment of a negative index shift around –1. These two values λ_1 and λ_2 are equal to 1.71 μm and 1.58 μm. By considering the iso-frequency plots of Fig.(B1b), we obtain respectively negative refractive index of –1.52 for λ_1 and –1.08 for λ_2 . However, neither n = -1 nor λ = 1.55 μm requirements are fulfilled. Or, as mentioned by several authors, optimal performances for the focusing required n = -1 if fed by an optical wave in air.

In the following, an original optimization procedure will be followed which does not need any shape engineering. Our goal is to achieve simultaneously a maximum of transmission and n = -1 by design. Here, the search for maximal transmission does not mean surface impedance matching since for the second band of such a photonic crystal the impedance is close to the dielectric impedance value that is to say about 0.3 rather than the expected 1. The intrinsic nature of high transmission is the geometrical dependence on the number of rows of the lens. Concerning the photonic crystal itself, two parameters can be adjusted, the lattice period and the filling factor (hole diameter).

The first dimension to fix is the lens thickness desired. Then as a function of a typical lattice period around 450-500 nm and a filling factor about 30 to 50 % chosen to obtain n = -1 at 1.55 μm to work in second band, the number of rows can be deduced. For this given number of rows, transmission spectra as function of wave incidence can be calculated (as the one of figure 3 for 6 rows or the one of Fig.(B5a) for 12 rows). Here for clarity, incidences up to 7° are plotted but the study remains strictly valid up to 15°. If "lucky", coincidence between a middle-band transmission peak (or high values whatever incidence) and n = -1 at 1.55 μm can be directly obtained, then the lens is quasi-optimized. If not, two routes can be followed as illustrated in Fig.(B5b) and (B5c) depending on prerequisites.

As shown Fig.(B5a) for a lattice period of 476 nm and hole diameter of 350 nm, transmission is minimum (less than –3 dB) at 1.55 μm where n = -1. If we decide to choose the 6th transmission peak , high values of transmission (higher than –1 dB up to – 15° incidence angle) can be obtained at λ = 1.522 μm. Here the idea is to shift the n = -1 value to this wavelength by changing the filling factor. Using band calculation procedure, this goal can be reached for a hole diameter of 366 nm instead of 350 nm with the same period. As shown Fig.(B5b), the new transmission spectra shows that the maximum peak coincides with the n = -1 wavelength. It can be mentioned here that the filling factor change induces also slight changes in the transmission spectra, notably as various incidence angles are explored. But these changes are of second order, and if needed a refinement in filling factor can be operated to reach more accuracy. This first route shows that it is possible to reach a high transmission level along with n = -1, but the operating wavelength is dependent on the transmission peak chosen and on the lens number of rows.

If λ = 1.55 μm, or any other wavelength, is a strong prerequisite, one have to add at least one step in the optimization procedure. Once again we can start from

Fig.(B5a). Let us modify now, for a fixed hole diameter, the lattice period in order to shift one transmission peak to 1.55 µm for that given number of rows. This is illustrated in Fig.(B5c) with a lattice period of 469 nm where the 7th peak coincides now with the targeted wavelength 1.55 µm. Looking at the band structure, now n = -1 corresponds to λ = 1.52 µm. As a final step, the filling factor can now be modulated to shift the n = -1 value back to 1.55 µm. This is obtained with a hole diameter of 332 nm. The procedure can be repeated when small additional adjustments are required. At the end the lens is optimized with n = -1 and high transmission levels at 1.55 µm for a given lens thickness, which was the pursued goal (Fig.(B5d)).

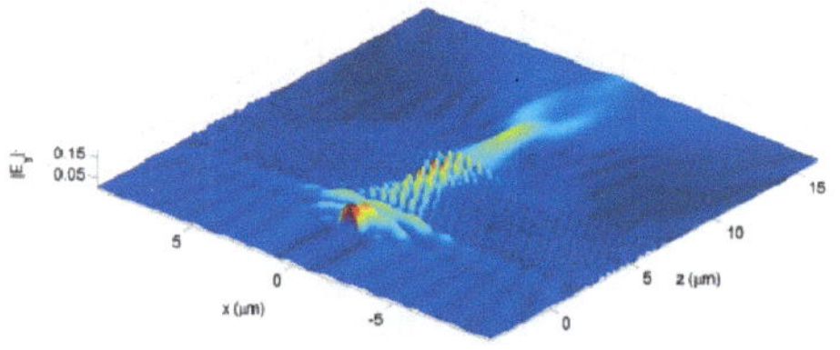

Fig. (B6) : 3D map of the electric field for the optimized lens (a = 469 nm ; d = 332 nm) operating at 1.55 µm.

Fig.(B6) gives a 3D field map of the lens operation for the optimized structure. Reflection appears minimized owing to the optimization of the transmission which reaches unity and thus improves the impedance matching. A clear focus spot appears also behind the lens. The source-lens distance is 2.23 µm for a 4.87 µm thick lens. The spot position at 9.70 µm confirms the value of n very close to -1. It has a full width at half maximum (FWHM) of about 1 µm, which corresponds to 2λ/3. Strictly speaking, we are far from a super-lens operating regime. Let us mention that such a possibility of an ideal lens with PC's structured at a scale comparable to the wavelength is still questionable [9, 10]. In the lower part of the spectrum, a resolution of λ/5 with a photonic crystal has been deduced from experiments [105]. However, if subwavelength resolution is of major concern, it has to be mentioned that such a flat lens has no optical axis, that is to say that the image is invariant by any translation parallel to the lens interface. This property only afforded by negative refraction justifies solely such lens optimization.

REFERENCES

Metamaterials and photonic crystals, ultra- and negative refraction, super- and hyper- lenses, cloaking devices are part of a research field in intense activity since a decade. The establishment of a relevant bibliography is a tricky task since one can find easily thousands of papers dealing with these exciting topics. Obviously, the hundred of articles referenced below is not exhaustive. Most of the seminal papers are cited as well as those directly related to the thematics and prototypes presented in this chapter. Many other papers deserve to be quoted, sorry for those unintentionally forgotten...

[1] Perrin M, Fasquel S, Decoopman T, Mélique X, Vanbésien O, Lheurette E, Lippens D. Left-handed electromagnetism obtained via nanostructured metamaterials: comparison with that from microstructured photonic crystals. J Opt A: Pure Appl Opt 2005; 7(2): S3.

[2] Veselago VG. The electrodynamics of substances with simultaneously negative of ε and μ. Sov Phys-Usp 1968; 10: 509-14.

[3] Pendry JB. Negative refraction makes a perfect lens. Phys Rev Lett 2000; 85(18): 3966-69.

[4] Smith DR, Kroll N. Negative refraction Index in Left-handed materials. Phys Rev Lett 2000; 85(14): 2933-36.

[5] Ben-Aryeh Y. Nonclassical high resolution effects produced by evanescent waves. J Opt B: Quantum Semiclassical Opt 2003; 5(6): S557.

[6] Pendry JB, Anantha Ramakrishna S. Near-field lenses in two dimensions. J Phys Condens Matter 2002; 14(36): 8463.

[7] Pendry JB, Anantha Ramakrishna S. Focusing light using negative refraction. J Phys Condens Matter 2003; 15(37): 6345.

[8] Garcia N, Nieto-Vesperinas M. Left-handed materials do not make a perfect lens. Phys Rev Lett 2002; 88(20): 207403.

[9] Maystre D, Enoch S. Perfect lenses made with left-handed materials: Alice's mirror? J Opt Soc Am A 2004; 21(1): 122-31.

[10] Efros AL, Li C. Electrodynamics of left-handed medium. Solid State Phenomena 2007; 121-3:1065-68.

[11] Pendry JB, Holden AJ, Robbins JD, Stewart WJ. Magnetism from conductors and enhanced non linear phenomena. IEEE Trans Microwave Theory Tech 1999; 47(11): 2075-84.

[12] Pendry JB, Holden AJ, Stewart WJ, Youngs I. Extremely low frequency plasmons in metallic microstructures. Phys Rev Lett 1996; 76: 4773-76.

[13] Shelby RA, Smith DR, Schultz S. Experimental verification of a negative index refraction. Science 2001; 292 : 77-9.

[14] Parazolli CG, Greegor RB, Li K, Koltenbach BEC, Tanielian M. Experimental verification and simulation of negative index of refraction using Snell's law. Phys Rev Lett 2003; 90 : 107401.

[15] Greegor RB, Parazolli CG, Li K, Koltenbach BEC, Tanielian M. Experimental determination and numerical simulation of the properties of negative index of refraction materials. Opt Express 2003; 11: 688-95.

[16] Lheurette E., Vanbésien O, Lippens D. Double negative media using interconnected omega type metallic particles. Microwave Opt Technol Lett 2007; 49 :84-9.

[17] Lheurette E, Houzet G, Carbonell J, Fuli Zhang, Vanbésien O, Lippens D. Omega type balanced composite negative refractive index materials. IEEE Trans Antennas Propag 2008; 56(11): 3462-69.

[18] Saadoun MMI, Engheta N. A reciprocal phase shifter using novel pseudochiral or Ω medium. Microwave Opt Technol Lett 1992; 5: 184-8.

[19] Saadoun MMI, Engheta N. Theoretical study of the electromagnetic properties of nolocal Ω medium. Prog. Electromagnetic Res 1994; PIER9: 351-97.

[20] Simovski CR, He S. Frequency range and explicit expressions for negative permittivity and permeability for an isotropic medium formed by a lattice of perfectly conducting Ω particles. Phys Lett A 2003; 311: 254-63.

[21] Huangfu J, Ran L, Chen H, Zhang X, Chen K, Grzegorczyk T, Kong JA. Experimental confirmation of negative refractive index of a metamaterial composed of Ω-like metallic patterns. Appl Phys Lett 2004; 84: 1537-39.

[22] Ran L, Huangfu J, Chen H, Li Y, Zhang X, Chen K, Kong JA. Microwave solid-state left-handed material with a broad bandwidth and an ultralow loss. Phys Rev B: Condens. Matter 2004; 70: 073102.

[23] Ran L, Huangfu J, Chen H, Zhang X, Cheng K, Grzegorczyk TM, Kong JA. Experimental study on several left-handed metamaterials. Progr Electromagnetic Res 2005; PIER51: 249-79.

[24] Yablonovitch E. Inhibited spontaneous emission in Solid State Physics and Electronics. Phys Rev Lett 1987; 58(20): 2059-62.

[25] John S. Strong localization of photons in certain disordered dielectric superlattices. Phys Rev Lett 1987; 58(23) : 2486-89.

[26] Joannopoulos JD, Meade RD, Winn JN. Photonic crystal: molding the flow of light. 1995; Princeton, NJ, Princeton University Press

[27] Yablonovitch E, Gmitter TJ, Leung KM. Photonic band structures ;The face-centered cubic case employing non spherical atoms. Phys Rev Lett 1991; 67(17): 2295-98.

[28] Wijnhoven J, Vos WL. Preparation of photonic crystals made of air spheres in Titania. Science 1998; 281: 802-4.

[29] Cassagne D, Reynolds A, Jouanin C. Modelling of 3D photonic crystals based on opals. Opt Quantum Electron 2000; 32(6-8): 923-33.

[30] Noda S, Tomoda K, Yamamoto N, Chutinan A. Full three dimensional photonic band gap crystals at near-infrared wavelengths. Science 2000; 289: 604-6.

[31] Tanabe T, Notomi N, Kuramochi E. Measurement of ultra-high-Q photonic crystal nanocavity using single-sideband frequency modulator. Electron Lett 2007; 43(3): 187-8.

[32] Talneau A, Le Gouezigou L, Bouadma N, Kafesaki M, Soukoulis CM, Agio M. Photonic crystal ultrashort bends with improved transmission and low reflection at 1.55 µm. Appl Phys Lett 2002; 80(4): 547-9.

[33] Louvion N, Gérard D, Mouette J, de Fornel F, Seassal C, Letartre X, Rahmani A, Callard S. Local observation and spectroscopy of optical modes in an active photonic crystal microcavity. Phys Rev Lett 2005; 94(11): 113907.

[34] Smith CJM, Benisty H, Olivier S, Rattier M, Weisbuch C, Krauss TF, De la Rue RM, Houdré R, Oesterle U. Low-loss channel waveguides with two-dimensional photonic crystal boundaries. Appl Phys Lett 2000; 77(18): 2813-15.

[35] Fasquel S, Mélique X, Vanbésien O, Lippens D. Wave channeling in compact multiport waveguides patterned in electromagnetic crystals. Superlattices Microstruct 2002; 32(2-3) : 145-51.

[36] Danglot J, Carbonell J, Fernandez M, Vanbésien O, Lippens D. Modal analysis of guided structures patterned in a metallic photonic crystal. Appl Phys Lett 1998; 73: 2712-14.

[37] Fasquel S, Mélique X, Lippens D, Vanbésien O. Three dimensional calculation of propagation losses in photonic crystal waveguides. Opt Commun 2005; 246(1-3) : 91-6.

[38] Fabre N, Fasquel S, Legrand C, Mélique X, Muller M, François M, Vanbésien O, Lippens D. Towards focusing using photonic crystal lens. Opto-electron Rev 2006; 14 : 225.

[39] Croënne C, Fabre N, Gaillot DP, Vanbésien O, Lippens D. Bloch impedance in negative index photonic crystals. Phys Rev B: Condens Matter 2008; 77: 125333.

[40] Fuli Zhang, Potet S, Carbonell J, Lheurette E, Vanbésien O, Zhao X, Lippens D. Negative-zero-positive refractive index in a prism-like Omega-type metamaterial. IEEE Trans Microwave Theory Tech 2008; 56(11): 2566-73.

[41] Enoch S, Tayeb G, Sabouroux P, Guérin N, Vincent P. A metamaterial for directive emission. Phys Rev Lett 2002; 89: 213902.

[42] Ziolkowski RW. Propagation in and scattering from a matched metamaterial having a zero index of refraction. Phys. Rev. E: Stat Phys Plasmas Fluids Relat Interdisciplin Top 2004; 70: 046608.

[43] Alu A, Silveirinha MG, Salandrino A, Engheta N. Epsilon near zero metamaterials and electromagnetic sources: tailoring the radiation. Phys Rev B: Condens Matter 2007; 75: 155410.

[44] Zhang F, Potet S, Carbonell J, Lheurette E, Vanbésien O, Zhao X, Lippens D. Application of omega-type and related metamaterials for beam steering in X-Ku bands. Proc. 37th European Microwave Conference Munich Germany 2007; 909-12.

[45] Smith DR, Schultz S, Markos P, Soukoulis CM. Determination of effective permittivity and permeability of metamaterials from reflection and transmission coefficients. Phys Rev B: Condens Matter 2002; 65: 195104.

[46] Chen X, Grzegorczyk TM, Wu B-I, Pacheco J, Kong JA. Robust method to retrieve the constitutive effective parameters of metamaterials. Phys Rev E: Stat Phys Plasmas Fluids Relat Interdisciplin Top 2004; 70: 016608.

[47] Smith DR, Rye PM, Mock JJ, Vier DC, Starr AF. Enhanced diffraction from a grating on the surface of a negative index metamaterial. Phys Rev Lett 2004; 93: 137405.

[48] Crépin T, Lampin JF, Decoopman T, Mélique X, Desplanque L, Lippens D. Experimental evidence of backward waves on terahertz left-handed transmission lines. Appl Phys Lett 2005, 87: 104105.

[49] Grbic A, Eleftheriades GV. Experimental verification of backward wave radiation from a negative refractive index metamaterial. J Appl Phys 2002; 92(10): 5930-35.

[50] Liu L, Caloz C, Chang CC, Itoh T. Forward coupling phenomena between artificial left-handed transmission lines. J Appl Phys 2002; 92(9): 5560-65.

[51] Caloz C, Itoh T. Positive/negative refractive index anisotropic 2-D metamaterials. IEEE Microwave Wireless Compon Lett 2003; 13(12): 547-9.

[52] Caloz C, Itoh T. Transmission line approach of left-handed (LH) materials and microstrip implementation of an artificial LH transmission line. IEEE Trans Antennas Propag 2004; 52(5): 1159-66.

[53] Caloz C, Itoh T. Metamaterials for high-frequency electronics. Proc. IEEE 2005; 93(10): 1744-52.

[54] Caloz C, Sanada A, Itoh T. A novel composite right-/left-handed coupled-line directional coupler with arbitrary coupling level and broad bandwidth. IEEE Trans Microwave Theory Techn. 2004; 52(3): 980-92.

[55] Auston DH. Picosecond optoelectronic switching and gating in Silicon. Appl Phys Lett 1975; 26(3): 101-3.

[56] Valdmanis JA, Mourou G, Gabel CW. Picosecond electro-optic sampling system. Appl Phys Lett 1982; 41: 211-3.

[57] Lampin JF, Desplanque L, Mollot F. Detection of picoseconds electrical pulses using Franz-Keldysh effect. Appl Phys Lett 2001; 78(26): 4103-05.

[58] Desplanque L, Lampin JF, Mollot F. Generation and detection of terahertz pulses using post-process bonding of low temperature grown GaAs and AlGaAs. Appl Phys Lett 2004; 84(12): 2049-51.

[59] Siddiqui OF, Mojahedi M, Eleftheriades GV. Periodically loaded transmission line with effective negative refractive index and negative group velocity. IEEE Trans Antennas Propag 2003; 51(10): 2619-25.

[60] Dolling G, Wegener M, Soukoulis CM, Linden S. Negative index metamaterial at 780 nm wavelength. Opt Lett 2007; 32(1): 53-5.

[61] Aydin K, Bulu I, Ozbay E. Subwavelength resolution with a negative index metamaterial superlens. Appl Phys Lett 2007; 90(25): 254102.

[62] Decoopman T, Tayeb G, Enoch S, Maystre D, Gralak B. Photonic crystal lens: from negative refraction and negative index to negative permittivity and permeability. Phys Rev Lett 2006; 97(7): 073905.

[63] Berrier A, Mulot M, Swillo M, Qiu M, Thylen L, Talneau A, Anand S. Negative refraction at infrared wavelengths in a two dimensional photonic crystal. Phys Rev Lett 2004; 93(7): 073902.

[64] Schonbrun E, Yamashita T, Park W, Summers CJ. Negative index imaging by an index-matched photonic crystal slab. Phys Rev B: Condens Matter 2006; 73(19): 195117.

[65] Matsumoto T, Eom KS, Baba T. Focusing of light by negative refraction in a photonic crystal slab superlens on silicon-on-insulator substrate. Opt Lett 2006; 31(18): 2786-88.

[66] Lu Z, Miao B, Hodson TR, Lin C, Muralowski JA, Prather DW. Negative refraction imaging in a hybrid photonic crystal device at near-infrared frequencies. Opt Express 2007; 15(3): 1286-91.

[67] Moussa R, Foteinopoulou S, Zhang L, Tuttle G, Guven K, Ozbay E, Soukoulis CM. Negative refraction and superlens behavior in a two dimensional photonic crystal. Phys Rev B: Condens Matter 2005; 71(8): 085106.

[68] Fabre N, Mélique X, Lippens D, Vanbésien O. Optimized focusing properties of photonic crystal slabs. Opt Commun 2008; 281(13): 3571-77.
[69] Cluzel B, Gérard D, Picard E, Charvolin T, de Fornel F, Hadji E. Subwavelength imaging of field confinement in a waveguide integrated photonic crystal cavity. J Appl Phys 2005; 98(8): 086109.
[70] Kramper P, Kafesaki M, Soukoulis CM, Birner A, Muller F, Gosele U, Wehrspohn RB, Mlynek J, Sandoghdar V. Near-field visualization of light confinement in a photonic crystal microresonator. Opt Lett 2004; 29(2): 174-6.
[71] Fabre N, Lalouat L, Cluzel B, Mélique X, Lippens D, de Fornel F, Vanbesien O. Optical near-field microscopy of light focusing through a photonic crystal flat lens. Phys Rev Lett 2008; 101: 073901.
[72] Luo C, Johnson SG, Joannopoulos JD, Subwavelength imaging in photonic crystals. Phys Rev B: Condens Matter 2003; 68(4): 045115.
[73] Stockman MI. Nanofocusing of optical energy in tapered plasmonic waveguides. Phys Rev Lett 2004; 93(13): 137404.
[74] Verhagen E, Polman A, Kuipers LK. Nanofocusing in laterally tapered plasmonic waveguides. Opt Express 2008; 16(1): 45-7.
[75] Robinson JT, Manolatou C, Chen L, Lipson M. Ultrasmall mode volumes in dielectric optical microcavities. Phys Rev Lett 2005; 95(14): 143901.
[76] Centeno E, Cassagne D. Graded Photonic crystals. Opt Lett 2005; 30: 2278-81.
[77] Shalaev VM. Optical negative index metamaterials. Nat Photonics 2007; 1: 41-8.
[78] Pendry JB, Schurig D, Smith DR. Controlling electromagnetic fields. Science 2006; 312: 1780-82.
[79] Schurig D, Pendry JB, Smith DR. Calculation of material properties and ray tracing in transformation media. Opt Express 2006; 14(21): 9794-804
[80] Cummer SA, Popa BI, Schurig D, Smith DR, Pendry JB. Full wave simulations of electromagnetic cloaking structures. Phys Rev E: 2006; 74: 036621.
[81] Schurig D, Mock JJ, Justice BJ, Cummer SA, Pendry JB, Starr AF, Smith DR. Metamaterial electromagnetic cloak at microwave frequencies. Science 2006; 314: 977-80.
[82] Cai W, Chettiar UK, Kildishev AV, Shalaev VM. Optical cloaking with non-magnetic metamaterials. Nat Photonics, 2007; 1: 224-7.
[83] Leonhardt U. Optical Conformal Mapping. Science 2006; 312: 1777-80.
[84] Li J, Pendry JB. Hiding under the carpet: a new strategy for cloaking. Phys Rev Lett 2008; 101(20): 203901.
[85] Magnus F, Wood B, Moore J, Morrison K, Perkins G, Fyson J, Wiltshire MCK, Caplin D, Cohen LF, Pendry JB. A d.c magnetic metamaterial. Nat Mater 2008; 7(4): 295-7.
[86] Demetriadou A, Pendry JB. Taming spatial dispersion in wire metamaterial. J Phys Condens Matter 2008; 20(29): 295222.
[87] Shvets G, Trendafilov S, Pendry JB, Sarychev A. Guiding focusing and sensing on the subwavelength scale using metallic wire arrays. Phys Rev Lett 2007; 99(5): 053903.
[88] Zolla F, Guenneau S, Nicolet A, Pendry JB. Electromagnetic analysis of cylindrical invisibility cloaks and mirage effects. Opt Lett 2007; 32(9): 1069-71.
[89] Vanbesien O, Fabre N, Mélique X, Lippens D. Photonic crystal based cloaking device at optical wavelengths. Appl Opt 2008; 47(10): 1358-62.
[90] Peng L, Chen LRH, Zhang H., Kong JA, Grzegorczyk TM. Experimental Observation of Left-Handed Behavior in an Array of Standard Dielectric Resonators. Phys Rev Lett 2007; 98: 157403.
[91] Gaillot DP, Croenne C, Lippens D. An all-dielectric route for terahertz cloaking. Opt Express 2008; 16(6) 3986-92.
[92] Gaillot DP, Croenne C, Zhang F, Lippens D. Transformation optics for the full dielectric electromagnetic cloak and metal-dielectric planar hyperlens. New J Phys 2008; 10: 115039.
[93] Salandrino A, Engheta N. Far-field sub diffraction optical microscopy using metamaterial crystals: theory and simulations. Phys Rev B: Condens Matter 2006; 74(7): 075103.
[94] Jacob Z, Alekseyev LV, Narimanov E. Optical hyperlens: far field imaging beyond the diffraction limit. Opt Express 2006; 14(18): 8247-56.
[95] Liu Z, Xiong Y, Sun C, Zhang X. Far-field optical hyperlens magnifying sub-diffraction limited objects. Science 2007; 315: 1686.
[96] Lee H, Liu Z, Xiong Y, Sun C, Zhang X. Development of optical hyperlens for imaging below the diffraction limit. Opt Express 2007; 15(24): 15886-91.
[97] Lippens D. Metamaterials and infra-red applications. Comptes Rendus Physique 2008; 9: 184-96.
[98] Marqués R, Mesa F, Martel J, Medina F. Comparative analysis of edge- and broadsided-coupled split ring resonators for metamaterial design-theory and experiments. IEEE Trans Antennas Propag 2003; 51(10): 2572-81.
[99] Bulu I, Caglayan H, Ozbay E. Designing materials with desired electromagnetic properties. Microwave Opt Technol Lett 2006; 48: 2611-15.
[100] Ruan Z, Qiu M, Xiao S, He S, Thylen L. Coupling between plane waves and Bloch waves in photonic crystals with negative refraction. Phys Rev B: Condens Matter 2005; 71(4): 045111.
[101] Vanbésien O, Leroux H, Lippens D. Maximally flat transmission windows in finite superlattices. Solid State Electron 1992; 35(5): 665-9.
[102] Carbonell J, Vanbésien O, Lippens D. Electric field patterns in finite two dimensional wire photonic lattices. Superlattices Microstruct 1997; 22(4) : 597-605.
[103] Graugnard E, Gaillot DP, Nudham SN, Neff CW, Yamashita T, Summers CJ. Photonic band tuning in two-dimensional photonic crystal slab waveguides by atomic layer deposition. Appl Phys Lett 2006; 89: 181108.
[104] Baba T, Ohsaki D. Interfaces of photonic crystals for high efficiency light transmission. J J Appl Phys 2001; 40(10): 5920-24.
[105] Cubukcu E, Aydin K, Ozbay E, Foteinopolou S, Soukoulis CM. Subwavelength resolution in a two dimensional photonic crystal based superlens. Phys Rev Lett 2003; 91(20): 207401.

CHAPTER 8

Modeling of Metamaterials in Wave Propagation

G. Leugering, E. Rohan and F. Seifrt

Lehrstuhl für Angewandte Mathematik II, Universität Erlangen-Nürnberg, Germany.
New Technologies Research Center, Research Institute at University of West Bohemia, Plzen, Czech Republic.

Abstract: This chapter focuses on acoustic, electromagnetic, elastic and piezo-electric wave propagation through heterogenous layers. The motivation is provided by the demand for a better understanding of meta-materials and their possible construction. We stress the analogies between the mathematical treatment of phononic, photonic and elastic meta-materials. Moreover, we treat the cloaking problem in more detail from an analytical and simulation oriented point of view. The novelty in the approach presented here is with the interlinked homogenization- and optimization procedure.

1. INTRODUCTION

The terminology 'metamaterials' refers to 'beyond conventional material properties' and consequently those 'materials' typically are not found in nature. It comes as no surprise that research in this area, once the first examples became publicly known, has undergone an exponential growth. Metamaterials are most often man-made, are engineered materials with a wide range of applications. Starting in the area of micro-waves where one aims at cloaking objects from electromagnetic waves in the invisible frequency range, the ideas rather quickly inflicted researcher from optics for a variety of reasons. Superlenses allowing nanoscale imaging and nanophotolithography, couple light to the nanoscale yielding a family of negative-index-material(NIM)-based devices for nanophotonics, such as nanoscale antennae, resonators, lasers, switchers, waveguides and finally cloaking are just the most prominent fascinating fields. Nano-structured materials are characterized by 'ultra-fine microstructure'. There are at least two reasons why downscaling the size of a microstructure can drastically influence its properties. 'First, as grain size gets smaller, the proportion of atoms at grain boundaries or on surfaces increases rapidly. The other reason is related to the fact that many physical phenomena (such as dislocation generation, ferromagnetism, or quantum confinement effects) are governed by a characteristic length. As the physical scale of the material falls below this length, properties change radically'(see [44]).

Metamaterial properties, therefore, emerge under the controlled influence of microstructures. Inclusions on the nano-scale together with their material properties and their shape are to be designed in order to fulfill certain desired material properties, such as 'negative Poisson' ratio in elastic material foams, negative 'mass' and 'negative refraction indices' for the forming of band-gaps in acoustic and optical devices, respectively.

Thus given acoustic, elasto-dynamic, piezo-electric or electromagnetic wave propagation in a non-homogeneous medium and given a certain merit function describing the desired material-property or dynamic performance of the body involved, one wants to find e.g. the location, size, shape and material properties of small inclusions such that the merit function is increased towards an optimal material or performance. This, at the the first glance, sounds like the formulation of an ancient dream of man-kind. However, proper mathematical modelling, thorough mathematical analysis together with a model-based optimization and simulation can, when accompanied by experts in optics and engineering, lead to such metamaterial-concepts and finally to products.

Designing optimal microstructures can be seen from two aspects. Firstly, inclusions, their size, positions and properties are considered on a finite, say, nanoscale and are subject to shape, topology and material optimization. Secondly, such potential microstructures are seen from the macroscopic scale in form of some effective or averaged material. This brings in the notion and the theory of homogenization of microstructures. The interplay between homogenization and optimization becomes, thus, most prominent.

Besides the optimal design approach to metamaterial, in particular in the context of negative refraction indices, permittivities, permeabilities, there is another fascinating branch of research that concentrates on 'Transformation Optics', a notion pro-

Matthias Ehrhardt (ED)

moted by Pendry et.al. [27, 45] in optics and Greenleaf et.al. [16] in the more mathematically inclined literature. We refrain from attempting any recollection of major contribution to this field and refer to these survey articles ([27, 45, 16]) and the references therein. In order to be more specific and because in this contribution we will not dwell on this approach on any research level, we give a brief account of the underlying idea.

1.1 The Cloaking Problem and Metamaterials: Transformation Method

In order to keep matters as simple as possible, we consider the following classical problem

$$\begin{cases} \nabla \cdot \sigma \nabla u = 0, & \text{in } \Omega, \\ u = f, & \text{on } \partial\Omega. \end{cases} \tag{1}$$

We have the *Dirichlet-to-Neumann map* (DtN)

$$\Lambda_\sigma(f) := \nu \cdot \sigma \nabla u|_{\partial\Omega}. \tag{2}$$

Calderón's problem is to reconstruct σ from Λ_σ! For smooth and isotropic σ this is possible. Thus, in that case the Cauchy data $(f, \Lambda_\sigma(f))$ uniquely determine σ. Therefore, no cloaking is possible with smooth variations of the material! In the heterogeneous an-isotropic case, we may consider a diffeomorphism $F : \Omega \to \Omega$ with $F|_{\partial\Omega} = I$ and then make a change of variables $y = F(x)$ s.t. $u = v \circ F^{-1}$. The so-called push forward is defined as

$$\begin{aligned} (F_*\sigma)^{jk}(y) &:= \frac{1}{\det DF_{jk}} S^{jk}(x)|_{x=F^{-1}(y)} \\ S^{jk}(x) &:= \sum_{p,q=1}^{n} \frac{\partial F^j}{\partial x^p}(x) \frac{\partial F^k}{\partial x^q}(x) \sigma^{pq}(x). \end{aligned} \tag{3}$$

We notice that

$$\Lambda_\sigma = \Lambda_{F_*\sigma}, \tag{4}$$

where DF_{jk} denotes the Jacobi-matrix of F ($DF = \nabla F^T$). The idea behind is that the coefficients σ can be interpreted as a Riemann metric. Transformations into curvilinear coordinates are classic in mechanics, see e.g. Gurtin[17]. Thus, transformations into curvilinear coordinates correspond one-to-one with transformation between different materials. The construction of a transformation that allows for cloaking is as follows.

Denote $\hat{x} := \frac{x}{|x|}$, $\hat{y} := \frac{y}{|y|}$ and define the mapping $F : \mathbb{R}^3 \setminus \{0\} \to \mathbb{R}^3 \setminus \{B_a(0)\}$

$$x = F(y) := \begin{cases} x = x(y) = f(y) := g(|y|)\hat{y}, \\ \quad \text{for } 0 < |y| \le b, \\ x = x(y) := y, \text{ for } |y| > b, \end{cases} \tag{5}$$

where $B_r(x_0) := \{x \in \mathbb{R}^3 : |x - x_0| \le r\}$ and such that g satisfies: for a, b with $0 < a < b$, $g \in C^2([0,b])$, $g(0) = a$, $g(b) = b$ and $g'(\rho) > 0$, $\forall \rho \in [0,b]$ This transformation maps the punctuated three-space into a spherical ring with inner radius a and outer radius b, such that the exterior of the ball $B_b(0)$ is left unchanged. We consider the ball $K := B_a(0)$ as the cloaked object, the layer $\{x : a < |x| \le b\}$ as the cloaking layer and the union as the spherical cloak. The shape of the cloak can be arbitrary, however. Examples for spherical cloaks are $g(\rho) := \frac{b-a}{b}\rho + a$ (linear) or $g(\rho) := \left[1 - \frac{a}{b} + p(\rho - b)\right]\rho + a$ (quadratic)

We consider a similar construction as above, but now for many cloaked objects located at point $c_i, i = 1, \ldots, N$:

$$x = F(y) := \begin{cases} f(y) := c_i + g_i(|y - c_i|)(\hat{y} - c_i), \\ \quad \text{for } y \in B_{b_i}(c_i),\ i = 1, \ldots, N \\ y, \text{ for } y \in \mathbb{R}^3_0 \setminus \{\cup_{i=1}^N B_{b_i}(c_i) =: \tilde{\Omega}\}, \end{cases} \tag{6}$$

where the cloaked objects are now

$$K_i := \{x \in \mathbb{R}^3 : |x - c_i| \le a_i\}, i = 1, \ldots N \tag{7}$$

$K = \cup_{i=1}^N K_i$ is the entire cloaked object. The cloaked subregions are supposed to be separated:

$$\min \text{dist}\,(B_{b_i}(c_i), B_{b_j}(c_j)) > 0,\ \forall i \neq j,\ i, j = 1, \ldots, N \tag{8}$$

The domains of interest are now: $\Omega_0 := \mathbb{R}^3 \setminus \{c_1, \ldots, c_N\}$, $\Omega := \mathbb{R}^3 \setminus K$. $F(\cdot)$ is only piecewise smooth with singularities across ∂K.

$$DF(y)_{kl} = \begin{cases} \frac{g_j(|y-c_j|)}{|y-c_j|}\delta_{kl} + \left(\frac{g_j'(|y-c_j|)}{|y-c_j|^2} - \frac{g_j(|y-c_j|)}{|y-c_j|^3}\right) \cdot \\ \cdot (y - c_j)_k (y - c_j)_l, y \in B_{b_J}(c_j) \\ \delta_{kl}, y \in \tilde{\Omega} \end{cases} \tag{9}$$

We have the determinant $\Delta(y) = \det DF(y)$

$$\Delta(y) = \begin{cases} g_j'(|y - c_j|)\left(\frac{g_j(|y-c_j|)}{|y-c_j|}\right)^2, \\ \quad y \in B_{b_j}(c_j), j = 1, \ldots, N \\ 1, y \in \tilde{\Omega} \end{cases} \tag{10}$$

Obviously, $\sigma_* = F_*\sigma$ is degenerate along the boundary ∂K. In order to properly pose a self-adjoint extension of the corresponding Laplace(-Beltrami-)operator, we need to work in weighted spaces.
The idea above is extended to the phononic and the photonic situation. In particular treating the Maxwell system in its time-harmonic form the transformed system reads as

$$\nabla \times \boldsymbol{E} = \mathrm{j}k\mu(x)\boldsymbol{H}, \quad \nabla \times \boldsymbol{H} = -\mathrm{j}k\varepsilon(x)\boldsymbol{H} + \boldsymbol{J}_e \tag{11}$$

where ε, μ are given by:

$$\varepsilon = \frac{1}{\Delta(y)} D^T F \varepsilon_0 DF, \; \mu = \frac{1}{\Delta(y)} DF^T \mu_0 DF \tag{12}$$

The material matrices ε, μ are again degenerate at ∂K!
In order to obtain finite energy solutions to the Maxwell system, one needs to work in weighted spaces. For cloaking, one requires energy conservation. Introduce weighted scalar products

$$\begin{aligned}(\boldsymbol{E}^1, \boldsymbol{E}^2)_{\Omega,E} &:= \int_\Omega \boldsymbol{E}^1 \cdot \varepsilon \bar{\boldsymbol{E}}^2 dx, \; (\boldsymbol{H}^1, \boldsymbol{H}^2)_{\Omega,H} \\ &= \int_\Omega \boldsymbol{H}^1 \cdot \mu \bar{\boldsymbol{H}}^2 dx\end{aligned} \tag{13}$$

and require local energy conservation. To this end define the local energy for an open bounded subdomain $O \subset \Omega$

$$\int_\Omega \boldsymbol{E} \cdot \varepsilon \bar{\boldsymbol{E}} dx + \int_\Omega \boldsymbol{H} \cdot \mu \bar{\boldsymbol{H}} dx < \infty. \tag{14}$$

A solution satisfies the Maxwell system in the distributional sense and has finite local energy. One obtains **two** boundary (over-determined i.g.) conditions on ∂K

$$\begin{aligned}\boldsymbol{E} \times \boldsymbol{n} = 0, \; \boldsymbol{H} \times \boldsymbol{n} = 0, \text{ on } \partial K_+, \\ (\nabla \times \boldsymbol{E}) \cdot \boldsymbol{n} = 0, \; (\nabla \times \boldsymbol{H}) \cdot \boldsymbol{n} = 0, \text{ on } \partial K_-,\end{aligned} \tag{15}$$

This procedure of defining cloaking transformations is rather general and applies also to elliptic systems, 2-d and 3-d elasticity, elasto-dynamics and the time-dependent Maxwell equations. Thus, formally, from a purely mathematical point of view, the problem of cloaking can be regarded as analytically solved. The fundamental question however remains: How can the transformed material tensors be realized ?
Indeed, this problem is widely open. There is an approach to approximate the cloaking transforms by less singular mappings in particular by inflating a ball rather than a point to a ring-shaped domain. But still, the material could not be realized so far and further analysis is in order. On the positive side it is evident that even from the point of view of transformation optics the appearance of singular behaviour at the boundary of the region to be cloaked indicates that microstructures may genuinely occur. Indeed, a second approach [16] is based on a truncation of ε, μ to such tensors, say ε_R, μ_R that are uniformly (in x) bounded above and below. When $R \to 1$ they tend to ε, μ, respectively. It is shown in [16] that it is possible to match these tensors ε_R, μ_R by periodic microstructured material in the cloak in the homogenization limit. The result shows that utopian 'metamaterial' constructed by an approximation to exact cloaking can be 'realized' via homogenization of periodic microstructures within the cloaking region. This is a very encouraging result that needs to be further exploited.

1.2 Metamaterials via Homogenization

In this contribution we want to discuss the theme of object cloaking by 'homogenized metamaterials'. We are aiming at designing coating layers containing microstructure which are 'wrapped' around an object. The coated object may be subject to acoustic or electromagnetic incoming waves. We want to survey and present new results applying the method of homogenization and *at the same time* thin-domain approximation to such nano-structured layers. We investigate the resulting effective transmission condition and represent the cloaking problem as an optimization problem or a problem of exact controllability, the controls being shape, topology and material parameters for the inclusions constituting the microstructure.
In the context of mathematical modeling, there are many connections and analogies between acoustics and optics. Below we summarize some recent investigations on homogenization of periodically heterogeneous structures exposed to inciding acoustic, or electromagnetic waves. Namely the following issues are discussed:

- Phononic metamaterials which may exhibit negative effective mass for certain frequency ranges (the so called band-gaps).
- Homogenized 'acoustic sieve' problem; there the periodic perforation of a rigid layer (the obstacle) influences the acoustic impedance of the discontinuity interface.

- In analogy to the 'phononic' metamaterials, the 'photonic' ones may provide frequency-dependent magnetic permeability which may become even negative for some frequencies.
- As a central theme of this contribution is related to the cloaking problem, we discus the optical transmission on thin heterogeneous surface. The homogenization of such structure leads to a model resembling the homogenized acoustic sieve problem.

In all of the above cases combinations of 'classical' materials and geometrical arrangement of the heterogeneities gives rise to 'new' materials – *metamaterials* – characterized by their *effective properties* which makes their behaviour qualitatively different from any of the individual components. Especially the geometrical influence of materials' microstructures is challenging and inspires the *metamaterial optimal design*. We consider the cloaking problem formulated as the optimization problem parametrized by the *homogenized* metamaterial structure, i.e. by geometry of the heterogeneities distributed in the cloaking layer.
The optimization problem will also be considered in the context finite diameter material inclusion, thus without homogenization. For the interlacing of optimization and optimization and optimal control see Kogut and Leugering [20, 21, 22, 23]

1.3 Topology Optimization for the Cloaking Problem

Instead of transformation techniques and the method of optimizing micro-structures before or after homogenization one may look directly into material optimization of coated objects. Indeed, given a region to be cloaked by a layer with material inclusions or 'holes', one may want to use topology optimization and shape optimization in order to find such optimal 'micro-structures'. More precisely, the concept of material interpolation (SIMP) [5] can be used in order to detect material densities of a given class of materials around the object. Moreover, the concept of topological derivatives or topological sensitivities can be used to check as to whether at a given point in the cloaking region an inclusion should be considered. Once the location is detected a subsequent shape sensitivity analysis followed by shape variation will then assign the optimal shape of that inclusion. Variations of this theme will be discussed in this contribution.

2. HOMOGENIZATION FOR MODELING OF METAMATERIALS IN ACOUSTIC AND ELECTROMAGNETIC WAVE PROPAGATION

Homogenization of periodically heterogeneous structures is a well accepted mathematical tool which enables one to reduce significantly the complexity of modeling such structures. The complexity is due to "detailed geometry" associated with description of piecewise defined material coefficients (properties), which at the end may lead to an intractable numerical problem featured by millions of unknowns and huge data to be treated. "Averaging" of the material properties, based on the asymptotic analysis and the representative volume element (the representative periodic cell) leads to the "homogenized medium" described by the effective material parameters, so that the whole structure can be described with a few data.
In this section we demonstrate how the homogenization approach (see e.g. [1, 13, 14, 15, 41] for general references) can be used to approximate dispersion properties in strongly heterogeneous media. In the case of *phononic* and *photonic* materials, the dispersion (and thereby the possible occurrence of band gaps) is retained even in the homogenized medium, due to special scaling of material properties of one of the material components.

3. PHONONIC MATERIALS – ELASTIC AND PIEZOELECTRIC WAVES

We now consider an elastic medium formed by periodic structures involving very soft substructures. Thus, the material properties, being attributed to material constituents vary periodically with the local position. Throughout the text all the quantities varying with this microstructural periodicity are labeled with superscript $^{\varepsilon}$, where ε is the characteristic scale of the microstructure. Typically ε can be considered as the ratio between the microstructure size and the incident wave length.

3.1 Periodic Strongly Heterogeneous Material

The material properties are associated to the periodic geometrical decomposition which is now introduced. We consider an open bounded domain $\Omega \subset \mathbb{R}^3$ and the reference (unit) cell $Y =]0,1[^3$ with an embedded inclusion $\overline{Y_2} \subset Y$, whereby the matrix part is $Y_1 = Y \setminus \overline{Y_2}$. Let us note, that Y may be defined as a parallelepiped, the particular choice of the unit cube is just for ease of explanation. Using the reference cell we generate the decomposition of Ω as the union of all inclusions (which should not penetrate $\partial\Omega$), having the size $\approx \varepsilon$,

$$\Omega_2^\varepsilon = \bigcup_{k\in\mathbb{K}^\varepsilon} \varepsilon(Y_2+k)\,, \quad \text{where } \mathbb{K}^\varepsilon = \{k\in\mathbb{Z}|\ \varepsilon(k+\overline{Y_2})\subset\Omega\}\,, \tag{16}$$

whereas the perforated matrix is $\Omega_1^\varepsilon = \Omega\setminus\Omega_2^\varepsilon$. Also we introduce the interface $\Gamma^\varepsilon = \overline{\Omega_1^\varepsilon}\cap\overline{\Omega_2^\varepsilon}$, so that $\Omega = \Omega_1^\varepsilon\cup\Omega_2^\varepsilon\cup\Gamma^\varepsilon$.

Properties of a three dimensional body made of the elastic material are described by the elasticity tensor c_{ijkl}^ε, where $i,j,k = 1,2,\dots,3$. As usually we assume both major and minor symmetries of c_{ijkl}^ε ($c_{ijkl}^\varepsilon = c_{jikl}^\varepsilon = c_{klij}^\varepsilon$).

We assume that inclusions are occupied by a "very soft material" in the sense that the coefficients of *the elasticity tensor in the inclusions* are significantly smaller than those of the matrix compartment, however *the material density* is comparable in both the compartments. Such structures exhibit remarkable band gaps. Here, as an important feature of the modeling based on asymptotic analysis, the ε^2 scaling of elasticity coefficients in the inclusions appears. This *strong heterogeneity* in elasticity coefficients is related to the geometrical scale of the underlying microstructure (possibly another composite material involving "soft" and "hard" materials). The following ansatz is considered:

$$\rho^\varepsilon(x) = \begin{cases} \rho^1 & \text{in } \Omega_1^\varepsilon,\\ \rho^2 & \text{in } \Omega_2^\varepsilon,\end{cases} \qquad c_{ijkl}^\varepsilon(x) = \begin{cases} c_{ijkl}^1 & \text{in } \Omega_1^\varepsilon,\\ \varepsilon^2 c_{ijkl}^2 & \text{in } \Omega_2^\varepsilon.\end{cases} \tag{17}$$

3.1.1 Extension for Piezoelectric Materials

Properties of a three dimensional body made of the piezoelectric material are described by three tensors: the elasticity tensor c_{ijkl}^ε, the dielectric tensor d_{ij} and the piezoelectric coupling tensor g_{kij}^ε, where $i,j,k = 1,2,\dots,3$. The following additional symmetries hold: $d_{ij}^\varepsilon = d_{ji}^\varepsilon$ and $g_{kij}^\varepsilon = g_{kji}^\varepsilon$.

In analogy with the purely elastic case, the scaling of material coefficients by ε^2 is considered in Ω_2^ε, except of the density:

$$\rho^\varepsilon(x) = \begin{cases} \rho^1 & \text{in } \Omega_1^\varepsilon,\\ \rho^2 & \text{in } \Omega_2^\varepsilon,\end{cases} \quad c_{ijkl}^\varepsilon(x) = \begin{cases} c_{ijkl}^1 & \text{in } \Omega_1^\varepsilon,\\ \varepsilon^2 c_{ijkl}^2 & \text{in } \Omega_2^\varepsilon,\end{cases} \quad g_{kij}^\varepsilon(x) = \begin{cases} g_{kij}^1 & \text{in } \Omega_1^\varepsilon,\\ \varepsilon^2 g_{kij}^2 & \text{in } \Omega_2^\varepsilon,\end{cases} \quad d_{ij}^\varepsilon(x) = \begin{cases} d_{ij}^1 & \text{in } \Omega_1^\varepsilon,\\ \varepsilon^2 d_{ij}^2 & \text{in } \Omega_2^\varepsilon.\end{cases} \tag{18}$$

3.2 Modeling the Stationary Waves

We consider stationary wave propagation in the medium introduced above. Although the problem can be treated for a general case of boundary conditions, for simplicity we restrict the model to the description of clamped structures loaded by volume forces. Assuming a harmonic single-frequency volume forces,

$$\boldsymbol{F}(x,t) = \boldsymbol{f}(x)e^{i\omega t}\,, \tag{19}$$

where $\boldsymbol{f} = (f_i), i = 1,2,3$ is its local amplitude and ω is the frequency. We consider a dispersive displacement field with the local magnitude $\boldsymbol{u}^\varepsilon$

$$\boldsymbol{U}^\varepsilon(x,\omega,t) = \boldsymbol{u}^\varepsilon(x,\omega)e^{i\omega t}\,. \tag{20}$$

This allows us to study the steady periodic response of the medium, as characterized by displacement field $\boldsymbol{u}^\varepsilon$ which satisfies the following boundary value problem:

$$\begin{aligned} -\omega^2\rho^\varepsilon\boldsymbol{u}^\varepsilon - \mathrm{div}\sigma^\varepsilon &= \rho^\varepsilon\boldsymbol{f} \quad \text{in } \Omega,\\ \boldsymbol{u}^\varepsilon &= 0 \quad \text{on } \partial\Omega,\end{aligned} \tag{21}$$

where the stress tensor $\sigma^\varepsilon = (\sigma_{ij}^\varepsilon)$ is expressed in terms of the linearized strain tensor $\boldsymbol{e}^\varepsilon = (e_{ij}^\varepsilon)$ by the Hooke's law $\sigma_{ij}^\varepsilon = c_{ijkl}^\varepsilon e_{kl}(\boldsymbol{u}^\varepsilon)$. Problem (21) can be formulated in a weak form as follows: Find $\boldsymbol{u}^\varepsilon \in \mathbf{H}_0^1(\Omega)$ such that

$$\begin{aligned} -\omega^2\int_\Omega \rho^\varepsilon\boldsymbol{u}^\varepsilon\cdot\boldsymbol{v} + \int_\Omega c_{ijkl}^\varepsilon e_{kl}(\boldsymbol{u}^\varepsilon)e_{ij}(\boldsymbol{v}) &=\\ = \int_\Omega \boldsymbol{f}\cdot\boldsymbol{v} \quad &\text{for all } \boldsymbol{v}\in\mathbf{H}_0^1(\Omega)\,,\end{aligned} \tag{22}$$

where $\mathbf{H}_0^1(\Omega)$ is the standard Sobolev space of vectorial functions with square integrable generalized derivatives and with vanishing trace on $\partial\Omega$, as required by $(21)_2$. The weak problem formulation (22) is convenient for the asymptotic analysis using the two-scale convergence [1], or the unfolding method of homogenization [13].

3.2.1 Extension for Piezoelectric Materials

In addition, a synchronous harmonic excitation by volume charges with a single frequency ω can be considered $\tilde{q}(x,t) = q(x)e^{i\omega t}$, where q is the magnitude of the distributed volume charge. Accordingly, we should expect a dispersive piezoelectric field with magnitudes $(\boldsymbol{u}^\varepsilon, \varphi^\varepsilon)$

$$\tilde{\boldsymbol{u}}^\varepsilon(x,\omega,t) = \boldsymbol{u}^\varepsilon(x,\omega)e^{i\omega t} \,,$$
$$\tilde{\varphi}^\varepsilon(x,\omega,t) = \varphi^\varepsilon(x,\omega)e^{i\omega t} \,.$$

Then the periodic response of the medium is characterized by field $(\boldsymbol{u}^\varepsilon, \varphi^\varepsilon)$ which satisfies the following boundary value problem:

$$\begin{aligned} -\omega^2\rho^\varepsilon \boldsymbol{u}^\varepsilon - \mathrm{div}\sigma^\varepsilon &= \rho^\varepsilon \boldsymbol{f} \quad \text{in } \Omega, \\ -\mathrm{div}\boldsymbol{D}^\varepsilon &= q \quad \text{in } \Omega, \\ \boldsymbol{u}^\varepsilon &= 0 \quad \text{on } \partial\Omega, \\ \varphi^\varepsilon &= 0 \quad \text{on } \partial\Omega, \end{aligned} \tag{23}$$

where the stress tensor $\sigma^\varepsilon = (\sigma^\varepsilon_{ij})$ and the electric displacement $\boldsymbol{D}^\varepsilon$ are defined by constitutive laws

$$\begin{aligned} \sigma^\varepsilon_{ij} &= c^\varepsilon_{ijkl} e_{kl}(\boldsymbol{u}^\varepsilon) - g^\varepsilon_{kij}\partial_k \varphi^\varepsilon, \\ D^\varepsilon_k &= g^\varepsilon_{kij} e_{kl}(\boldsymbol{u}^\varepsilon) + d^\varepsilon_{kl}\partial_l \varphi^\varepsilon. \end{aligned} \tag{24}$$

The problem (23) can be weakly formulated as follows: Find $(\boldsymbol{u}^\varepsilon, \varphi^\varepsilon) \in \mathbf{H}^1_0(\Omega) \times H^1_0(\Omega)$ such that

$$\begin{aligned} -\omega^2 \int_\Omega \rho^\varepsilon \boldsymbol{u}^\varepsilon \cdot \boldsymbol{v} + \int_\Omega c^\varepsilon_{ijkl} e_{kl}(\boldsymbol{u}^\varepsilon) e_{ij}(\boldsymbol{v}) - \\ - \int_\Omega g^\varepsilon_{kij} e_{ij}(\boldsymbol{v}) \partial_k \varphi^\varepsilon = \int_\Omega \boldsymbol{f} \cdot \boldsymbol{v} \,, \\ \int_\Omega g^\varepsilon_{kij} e_{ij}(\boldsymbol{u}^\varepsilon) \partial_k \psi + \int_\Omega d_{kl} \partial_l \varphi^\varepsilon \partial_k \psi = \int_\Omega q\psi \,, \end{aligned} \tag{25}$$

for all $(\boldsymbol{v}, \psi) \in \mathbf{H}^1_0(\Omega) \times H^1_0(\Omega)$.

3.3 The Homogenized Model

Due to the *strong heterogeneity* in the elastic (and other piezoelectric) coefficients, the homogenized model exhibits dispersive behaviour; this phenomenon cannot be observed when standard two-scale homogenization procedure is applied to a medium without scale-dependent material parameters, as pointed out e.g. in [3]. In [4] the unfolding operator method of homogenization [13] was applied with the strong heterogeneity assumption (17), (18) We shall now record the resulting homogenized equations, as derived in [4], which describe the structure behaviour at the "macroscopic"scale. They involve the homogenized coefficients which depend on the characteristic responses at the "microscopic" scale.

Below it can be seen that the "frequency–dependent" mass coefficients are determined just by material properties of the inclusion and by the material density ρ^1 in the matrix, whereas the elasticity (and other piezoelectric) coefficients are related exclusively to the matrix material occupying the perforated domain.

For brevity in what follows we employ the following notations:

$$\begin{aligned} a_{Y_2}(\boldsymbol{u}, \boldsymbol{v}) &= \int_{Y_2} c^2_{ijkl} e^y_{kl}(\boldsymbol{u})\, e^y_{ij}(\boldsymbol{v}), \\ d_{Y_2}(\phi, \psi) &= \int_{Y_2} d^2_{kl} \partial^y_l \phi\, \partial^y_k \psi, \\ g_{Y_2}(\boldsymbol{u}, \psi) &= \int_{Y_2} g^2_{kij} e^y_{ij}(\boldsymbol{u})\, \partial^y_k \psi, \\ \rho_{Y_2}(\boldsymbol{u}, \boldsymbol{v}) &= \int_{Y_2} \rho^2 \boldsymbol{u} \cdot \boldsymbol{v}, \end{aligned} \tag{26}$$

whereby analogous notations are used when the integrations apply over Y_1.

3.3.1 Elastic Medium

Frequency–dependent homogenized mass involved in the macroscopic momentum equation are expressed in terms of eigenelements $(\lambda^r, \varphi^r) \in \mathbb{R} \times \mathbf{H}^1_0(Y_2)$, $r = 1, 2, \ldots$ of the elastic spectral problem which is imposed in inclusion Y_2 with $\varphi^r = 0$ on ∂Y_2:

$$\begin{aligned} \int_{Y_2} c^2_{ijkl} e^y_{kl}(\varphi^r)\, e^y_{ij}(\boldsymbol{v}) = \lambda^r \int_{Y_2} \rho^2 \varphi^r \cdot \boldsymbol{v} \;\forall \boldsymbol{v} \in \mathbf{H}^1_0(Y_2)\,, \\ \int_{Y_2} \rho^2 \varphi^r \cdot \varphi^s = \delta_{rs}\,. \end{aligned} \tag{27}$$

To simplify the notation we introduce the *eigenmomentum* $\boldsymbol{m}^r = (m^r_i)$,

$$\boldsymbol{m}^r = \int_{Y_2} \rho^2 \varphi^r. \tag{28}$$

The effective mass of the homogenized medium is represented by mass tensor $\boldsymbol{M}^* = (M^*_{ij})$, which is evaluated as

$$M^*_{ij}(\omega^2) = \frac{1}{|Y|} \int_Y \rho\, \delta_{ij} - \frac{1}{|Y|} \sum_{r \geq 1} \frac{\omega^2}{\omega^2 - \lambda^r} m^r_i m^r_j \,; \tag{29}$$

The elasticity coefficients are computed just using the same formula as for the perforated matrix do-

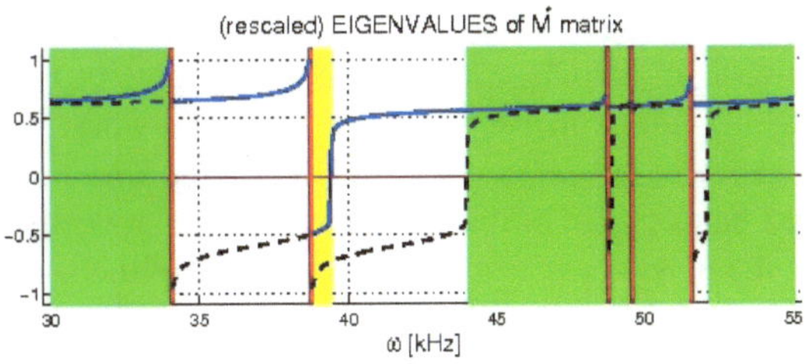

Fig. (**1**): Weak band gaps (white) and strong band gaps (yellow) computed for an elastic composite with L-shaped inclusions, the green bands are propagation zones.

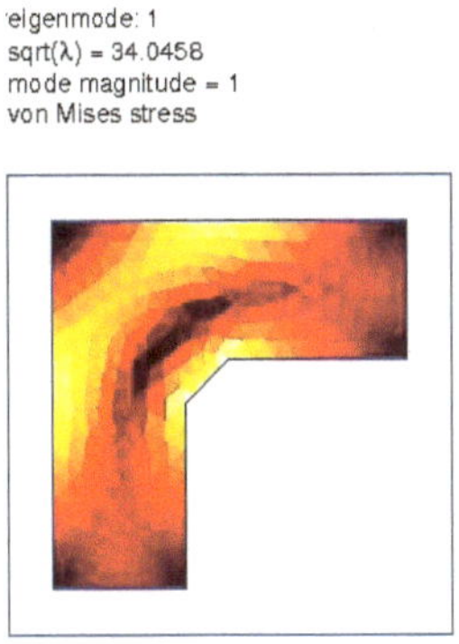

Fig. (**2**): The first eigenmode of the L-shaped clamped elastic inclusion.

main, thus being independent of the inclusions material:

$$C^*_{ijkl} = \frac{1}{|Y|}\int_{Y_1} c^1_{pqrs} e^y_{rs}(\boldsymbol{w}^{kl} + \Pi^{kl}) e_{pq}(\boldsymbol{w}^{ij} + \Pi^{ij}) \,, \tag{30}$$

where $\Pi^{kl} = (\Pi^{kl}_i) = (y_l \delta_{ik})$ and $\boldsymbol{w}^{kl} \in \mathbf{H}^1_\#(Y_1)$ are the corrector functions satisfying

$$\int_{Y_1} c^1_{pqrs} e^y_{rs}(\boldsymbol{w}^{kl} + \Pi^{kl}) e^y_{pq}(\boldsymbol{v}) = 0 \quad \forall \boldsymbol{v} \in \mathbf{H}^1_\#(Y_1) \,. \tag{31}$$

Above $\mathbf{H}^1_\#(Y_1)$ is the restriction of $\mathbf{H}^1(Y_1)$ to the Y-periodic functions (periodicity w.r.t. the homologous points on the opposite edges of ∂Y).
The *global (homogenized) equation* of the homogenized medium, here presented in its differential form, describes the macroscopic displacement field $\boldsymbol{u}$:

$$\omega^2 M^*_{ij}(\omega) u_j + \frac{\partial}{\partial x_j} C^*_{ijkl} e_{kl}(\boldsymbol{u}) = -M^*_{ij}(\omega) f_j \,, \tag{32}$$

Heterogeneous structures with finite scale of heterogeneities exhibit the frequency *band gaps* for certain frequency bands. In the *homogenized medium*, the wave propagation depends on the positivity of mass tensor $\boldsymbol{M}^*(\omega)$; this effect is explained below.

3.3.2 Piezoelectric Medium

In the piezoelectric medium, the spectral problem analogous to (27) comprises the additional constraint arising from electric charge conservation $(23)_2$: find eigenelements $[\lambda^r; (\varphi^r, p^r)]$, where $\varphi^r \in \mathbf{H}^1_0(Y_2)$ and $p^r \in H^1_0(Y_2)$, $r = 1, 2, \dots$, such that

$$\begin{aligned} a_{Y_2}(\varphi^r, \boldsymbol{v}) - g_{Y_2}(\boldsymbol{v}, p^r) &= \lambda^r \rho_{Y_2}(\varphi^r, \boldsymbol{v}) \\ &\quad \forall \boldsymbol{v} \in \mathbf{H}^1_0(Y_2), \\ g_{Y_2}(\varphi^r, \psi) + d_{Y_2}(p^r, \psi) &= 0 \quad \forall \psi \in H^1_0(Y_2), \end{aligned} \tag{33}$$

with the orthonormality condition imposed on eigenfunctions φ^r:

$$a_{Y_2}(\varphi^r, \varphi^s) + d_{Y_2}(p^r, p^s) = \lambda^r \rho_{Y_2}(\varphi^r, \varphi^s) \stackrel{!}{=} \lambda^r \delta_{rs}. \tag{34}$$

Moreover, if $q \not\equiv 0$ in $(23)_2$, then the following problem must be solved: find $\tilde{p} \in H^1_0(Y_2)$, the unique solution satisfying

$$d_{Y_2}(\tilde{p}, \psi) = \int_{Y_2} \psi \qquad \forall \psi \in H^1_0(Y_2) \,. \tag{35}$$

The homogenized mass $M^*_{ij}(\omega)$ is evaluated using the same formula (29), as in the elastic case. Further new coefficients $Q^*_i(\omega)$ are introduced using the solution of (35)

$$Q^*_i(\omega) = -\frac{1}{|Y|}\sum_{r\geq 1} \frac{\omega^2}{\omega^2 - \lambda^r} m^r_i g_{Y_2}(\varphi^r, \tilde{p}), \tag{36}$$

describing influence of the volume charge on the mechanical loading.
The *piezoelectric coefficients* of the homogenized medium are defined in terms of the corrector basis functions satisfying the microscopic auxiliary problems:

1. Find $(\chi^{ij}, \pi^{ij}) \in \mathbf{H}^1_\#(Y_1) \times H^1_\#(Y_1)$, $i,j = 1, \dots, 3$ such that (the notation corresponds to that introduced in (26))

$$\begin{cases} a_{Y_1}\left(\chi^{ij}+\Pi^{ij}, v\right)-g_{Y_1}\left(v, \pi^{ij}\right) & =0, \\ g_{Y_1}\left(\chi^{ij}+\Pi^{ij}, \psi\right)+d_{Y_1}\left(\pi^{ij}, \psi\right) & =0, \end{cases}$$
$$\forall v \in \mathbf{H}^1_\#(Y_1), \forall \psi \in H^1_\#(Y_1), \tag{37}$$

where $\Pi^{ij}=(\Pi^{ij}_k)=(y_j\delta_{ik})$;

2. Find $(\chi^k,\pi^k)\in \mathbf{H}^1_\#(Y_1)\times H^1_\#(Y_1)$, $i,j=1,\dots,3$ such that

$$\begin{cases} a_{Y_1}\left(\chi^{k}, v\right)-g_{Y_1}\left(v, \pi^{k}+\Pi^k\right) & =0, \\ g_{Y_1}\left(\chi^{k}, \psi\right)+d_{Y_1}\left(\pi^{k}+\Pi^k, \psi\right) & =0, \end{cases}$$
$$\forall v \in \mathbf{H}^1_\#(Y_1), \forall \psi \in H^1_\#(Y_1), \tag{38}$$

where $\Pi^k = y_k$.

Using the corrector basis functions just defined the homogenized coefficients are expressed, as follows:

$$\begin{aligned} C^*_{ijkl} &= \frac{1}{|Y|} a_{Y_1}\left(\chi^{kl}+\Pi^{kl}, \chi^{ij}+\Pi^{ij}\right)+ \\ &+\frac{1}{|Y|} d_{Y_1}\left(\pi^{kl}, \pi^{ij}\right), \\ D^*_{ki} &= \frac{1}{|Y|}\left[d_{Y_1}\left(\pi^k+\Pi^k, \pi^i+\Pi^i\right)+a_{Y_1}\left(\chi^k, \chi^i\right)\right], \\ G^*_{kij} &= \frac{1}{|Y|}\left[g_{Y_1}\left(\chi^{ij}+\Pi^{ij}, \Pi^k\right)+d_{Y_1}\left(\pi^{ij}, \Pi^k\right)\right]. \end{aligned} \tag{39}$$

The *global equation* describes the macroscopic field of displacements $\boldsymbol{u}$ and of electric potential φ

$$\begin{aligned} \omega^2 M^*_{ij}(\omega) u_j + \frac{\partial}{\partial x_j}\left(C^*_{ijkl} e_{kl}(\boldsymbol{u}) - G^*_{kij}\partial_k \varphi\right) &= \\ = -M^*_{ij}(\omega) - Q^*_i(\omega) q&, \\ \frac{\partial}{\partial x_k}\left(G^*_{kij} e_{ij}(\boldsymbol{u}) + D^*_{kl}\partial_l \varphi\right) = q&. \end{aligned} \tag{40}$$

Further related work on the sensitivity analysis can be found in [32, 34].

3.4 Band Gap Prediction

As the main advantage of the homogenized models (32) and (40), by analyzing the dependence $\omega \to \boldsymbol{M}^*(\omega)$ one can determine distribution of the band gaps; it was proved in [4] that there exist frequency intervals G^k, $k=1,2,\dots$ such that for $\omega \in G^k \subset]\lambda^k, \lambda^{k+1}[$ at least one eigenvalue of tensor $M^*_{ij}(\omega)$ is negative. Those intervals where all eigenvalues of M^*_{ij} are negative are called *strong*, or *full* band gaps. In the latter case the negative sign of the mass changes the hyperbolic type of the wave equation to the elliptic one, therefore any waves cannot propagate. In the "weak" bad gap situation only waves with certain polarization can propagate, as explained below.

The band gaps can be classified w.r.t. the polarization of waves which cannot propagate; the polarization is determined in terms of the eigenvectors of $M^*_{ij}(\omega)$. Given a frequency ω, there are three cases to be distinguished according to the signs of eigenvalues $\gamma^r(\omega)$, $r=1,2,3$ (in 3D), which determines the "positivity, or negativity" of the mass:

1. **propagation zone** – All eigenvalues of $M^*_{ij}(\omega)$ are positive: then homogenized model (32), or (40) admits wave propagation without any restriction of the wave polarization;

2. **strong band gap** – All eigenvalues of $M^*_{ij}(\omega)$ are negative: then homogenized model (32), or (40) does *not* admit any wave propagation;

3. **weak band gap** – Tensor $M^*_{ij}(\omega)$ is indefinite, i.e. there is at least one negative and one positive eigenvalue: then propagation is possible only for waves polarized in a manifold determined by eigenvectors associated with positive eigenvalues. In this case the notion of wave propagation has a local character, since the "desired wave polarization" may depend on the local position in Ω.

In Fig. (**1**) we introduce a graphical illustration of the band gaps analyzed for an *elastic* material with L-shaped inclusions (its eigenmode fig. (**2**)). Whenever inclusions (considered in 2D) are symmetric w.r.t. more than 1 axis of symmetry, only strong band gaps exist, see Fig. (**3**). This may not be the case for *piezoelectric materials*; in Fig. (**4**) we illustrate dispersion curves and the weak band gaps obtained for a homogenized piezoelectric composite with circular inclusions.

Usually the band gaps are identified from the *dispersion* diagrams. For the homogenized model the dispersion of guided plane waves is analyzed in the standard way, using the following ansatz:

$$\begin{aligned} \boldsymbol{u}(x,t) &= \bar{\boldsymbol{u}}\, e^{-\mathrm{j}(\omega t - x_j \kappa_j)}, \\ \varphi(x,t) &= \bar{\varphi}\, e^{-\mathrm{j}(\omega t - x_j \kappa_j)}, \end{aligned} \tag{41}$$

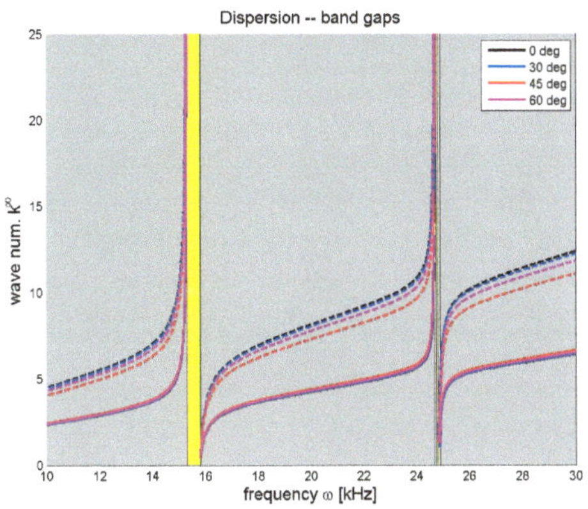

Fig. (**3**): Dispersion curves for guided waves in composites with circular inclusions: elastic material, only strong band gaps. Different angles of wave incidence displayed by different colours.

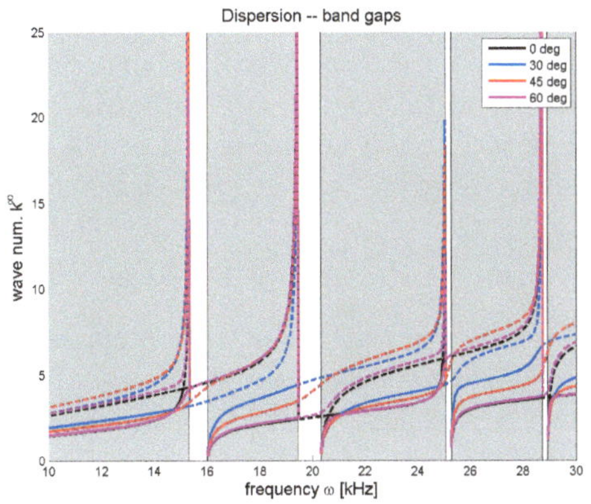

Fig. (**4**): Dispersion curves for piezoelectric material.

where $\bar{\boldsymbol{u}}$ is the displacement polarization vector (the wave amplitude), $\bar{\varphi}$ is the electric potential amplitude, $\kappa_j = n_j \varkappa$, $|\boldsymbol{n}| = 1$, i.e. $\boldsymbol{n}$ is the incidence direction, and $\varkappa$ is the wave number. The dispersion analysis consists in computing nonlinear dependencies $\bar{\boldsymbol{u}} = \bar{\boldsymbol{u}}(\omega)$ and $\varkappa = \varkappa(\omega)$. For this one substitutes (41) into the homogenized model (40); on introducing projections of the homogenized tensors into the direction of the wave propagation,

$$\Gamma_{ik} = C^*_{ijkl} n_j n_l \,,\; \gamma_i = G^*_{kij} n_j n_k \,,\; \zeta = D^*_{kl} n_l n_k \,, \quad (42)$$

and substituting in (40), we obtain

$$\begin{aligned} -\omega^2 M^*_{ij}(\omega^2)\bar{u}_j + \varkappa^2\left(\Gamma_{ik}\bar{u}_k - \gamma_i\bar{\varphi}\right) &= 0\,, \\ \varkappa^2\left(\gamma_k\bar{u}_k + \zeta\bar{\varphi}\right) &= 0\,. \end{aligned} \quad (43)$$

In (43) we can eliminate $\bar{\varphi}$ (assuming $\varkappa^2 \neq 0$), thus the dispersion analysis reduces to the "standard elastic case" where the acoustic tensor is modified, thus

$$\begin{aligned} -\omega^2 M^*_{ij}(\omega^2)\bar{u}_j + \varkappa^2 H_{ik}\bar{u}_k = 0\,, \\ \text{where} \quad H_{ik} = \Gamma_{ik} + \gamma_i\gamma_k/\zeta \end{aligned} \quad (44)$$

is analyzed as follows

- for all $\omega \in [\omega^a, \omega^b]$ and $\omega \notin \{\lambda^r\}_r$ compute eigenelements $(\eta^\beta, \boldsymbol{w}^\beta)$:

$$\omega^2 M^*_{ij}(\omega^2) w^\beta_j = \eta^\beta H_{ik} w^\beta_k \,, \quad \beta = 1,2,3\,; \quad (45)$$

- if $\eta^\beta > 0$, then $\varkappa^\beta = \sqrt{\eta^\beta}$,
- else ω falls in an *acoustic gap*, wave number is not defined.

In heterogeneous media *in general* the polarizations of the two waves (outside the band gaps) are *not mutually orthogonal*, which follows easily from the fact that $\{\boldsymbol{w}^\beta\}_\beta$ are $\boldsymbol{M}^*(\omega^2)$–orthogonal.

Moreover, in the presence of the piezoelectric coupling, which introduces another source of anisotropy, the standard orthogonality is lost even for heterogeneous materials with "symmetric inclusions" (circle,hexagon, etc.), in contrast with elastic structures where these designs preserve the standard orthogonality.

More details on the band gap properties and their relationship to the dispersion of guided waves were discussed in [35, 30, 10]. The sensitivity analysis for the optimization problem was discussed in [31, 32, 34, 33].

4. ACOUSTIC TRANSMISSION ON PERFORATED INTERFACES

In this section we present an example which illustrates, how homogenization can be employed to describe acoustic transmission between two halfspaces separated by an interface that establishes a microstructure. The detailed analysis was presented in [38].

We consider the acoustic medium occupying domain Ω^G which is subdivided by perforated plane Γ_0 in two disjoint subdomains Ω^+ and Ω^-, so that $\Omega^G = \Omega^+ \cup \Omega^- \cup \Gamma_0$, see Fig. (7). Denoting by p the acoustic pressure field in $\Omega^+ \cup \Omega^-$, in a case of no convection flow, the acoustic waves in Ω^G are described by the following equations (ω is the frequency of the incident wave),

$$\begin{aligned} c^2\nabla^2 p + \omega^2 p = 0 \quad & \text{in } \Omega^- \cup \Omega^+ , \\ + \text{ boundary conditions} \quad & \text{on } \partial\Omega^G , \end{aligned} \tag{46}$$

supplemented by the transmission conditions on interface Γ_0 — these present *the key issue of this section.* The boundary conditions on Γ_0 will be specified later on. Let p^+ and p^- be the traces of p on $\partial\Omega^+ \cap \Gamma_0$ and on $\partial\Omega^- \cap \Gamma_0$, respectively.

The standard treatment of the acoustic transmission on a sieve-like perforation Γ_0 results in the relationship between jump $p^+ - p^-$ and normal derivatives $\frac{\partial p^+}{\partial n^+} = -\frac{\partial p^-}{\partial n^-}$,

$$\begin{aligned} \frac{\partial p^+}{\partial n^+} &= -\mathrm{j}\frac{\omega\rho}{Z}(p^+ - p^-), \\ \frac{\partial p^-}{\partial n^-} &= -\mathrm{j}\frac{\omega\rho}{Z}(p^- - p^+), \end{aligned} \tag{47}$$

where n^+ and n^- are the outward unit normals to Ω^+ and Ω^-, respectively, ω is the frequency, ρ is the density and Z is the *transmission impedance*. This quantity incorporates many physical aspects of the transmission, namely the geometry – the design of the perforation. In [38] a homogenized transmission conditions were proposed which describe the acoustic impedance of the interface characterized by a periodically perforated obstacle embedded in a layer of thickness δ. In Figure (5) we illustrate such a layer Ω_δ embedded in $\Omega^G = \Omega_\delta^+ \cup \Omega_\delta^+ \cup \Omega_\delta \cup \Gamma_\delta^\pm$.

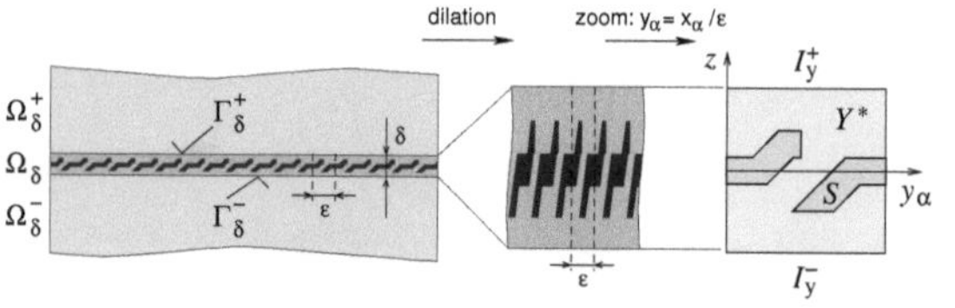

$$\Omega_\delta^\varepsilon \xrightarrow{\text{dilatation}} \Omega^\varepsilon \xrightarrow{\text{unfolding}} Y^*$$

Fig. (5): Left: global problem imposed in entire domain Ω^G before homogenization of the layer Ω_δ. Right: representative cell of the periodic structure. The dark patterns represent the obstacles in the fluid.

4.1 Periodic Perforation and Acoustic Problem in the Transmission Layer

Let $\Gamma_0 \subset \mathbb{R}^2$ be an open bounded subdomain of the plane spanned by coordinates x_α, $\alpha = 1,2$ and containing the origin. Further let Γ_δ^+ and Γ_δ^- be equidistant to Γ_0 with the distance $\delta/2 = \mathrm{dist}(\Gamma_0, \Gamma_\delta^+) = \mathrm{dist}(\Gamma_0, \Gamma_\delta^-)$. We introduce *layer* $\Omega_\delta = \Gamma_0 \times]-\delta/2, \delta/2[\subset \mathbb{R}^3$, an open domain representing the transmission layer bounded by $\partial\Omega_\delta$ which is split as follows, see Fig. (6)

$$\begin{aligned} \partial\Omega_\delta &= \Gamma_\delta^+ \cup \Gamma_\delta^- \cup \partial\Omega_\delta^\infty , \\ \Gamma_\delta^\pm &= \Gamma_0 \pm \frac{\delta}{2}\vec{e}_3 , \\ \partial\Omega_\delta^\infty &= \partial\Gamma_0 \times]-\delta/2, \delta/2[, \end{aligned} \tag{48}$$

where $\delta > 0$ is the layer thickness and $\vec{e}_3 = (0,0,1)$, see Fig. (6). The acoustic medium occupies domain $\Omega_\delta^\varepsilon = \Omega_\delta \setminus \overline{S_\delta^\varepsilon}$, where S_δ^ε is the solid *rigid* obstacle which in a simple layout has a form of the periodically perforated sheet with the thickness $s\delta$, $s < 1$ and with ε characterizing the scale of the periodic perforation; thus, S_δ^ε is obtained by the usual *periodic lattice* extension of the solid unit structure. For passing to the limit $\varepsilon \to 0$ we consider a proportional scaling between the period length and the thickness, so that $\delta = h\varepsilon$, where $h > 0$ is fixed.

4.1.1 Acoustic Problem in the Layer

We assume a monochromatic wave propagation in layer Ω^δ. The total acoustic pressure, $p^{\varepsilon\delta}$ satisfies the Helmholtz equation in $\Omega_\delta^\varepsilon$ and Neumann condition on $\partial\Omega_\delta$

$$\begin{aligned} c^2\nabla^2 p^{\varepsilon\delta} + \omega^2 p^{\varepsilon\delta} &= 0 \quad \text{in } \Omega_\delta^\varepsilon , \\ c^2\frac{\partial p^{\varepsilon\delta}}{\partial n^\delta} &= -\mathrm{j}\omega g^{\varepsilon\delta\pm} \quad \text{on } \Gamma_\delta^\pm , \\ \frac{\partial p^{\varepsilon\delta}}{\partial n^\delta} &= 0 \quad \text{on } \partial S_\delta^\varepsilon \cup \partial\Omega_\delta^\infty , \end{aligned} \tag{49}$$

where $c = \omega/\kappa$ is the speed of sound propagation and by n^δ we denote the normal vector outward to

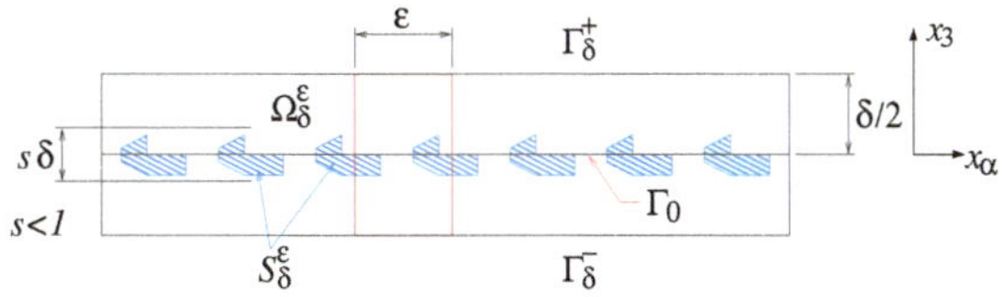

Fig. (**6**): Layer Ω_δ embedding the rigid obstacles periodically distributed. Obstacles should not approach the fictitious boundaries $\Gamma_\delta^\pm$, thus $s << 1$.

Ω_δ.

4.2 Homogenized Transmission Conditions

The asymptotic analysis of system (49) results in an equation which describes an acoustic wave propagating in the layer as a response to the incident wave acoustic momentum $g^{\varepsilon\pm}$. The following assumption is important.

Let us introduce *shifted* fluxes $\hat{g}^{\varepsilon\pm} \in L^2(\Gamma_0)$ such that $\hat{g}^{\varepsilon\pm}(\bar{x}) = g^{\varepsilon\pm}(x^\pm)$ where $x^\pm \in \Gamma^\pm$ are homologous points associated to $\bar{x} \in \Gamma_0$, i.e. $\bar{x} = (\bar{x}_\alpha, 0)$ and $x^\pm - \bar{x} = (0,0,\pm 1/2)$. We assume

$$\hat{g}^{\varepsilon\pm} \rightharpoonup g^{0\pm} \quad \text{weakly in } L^2(\Gamma_0)\,, \tag{50}$$

$$\frac{1}{\varepsilon}\left(\hat{g}^{\varepsilon+} + \hat{g}^{\varepsilon-}\right) \rightharpoonup 0 \quad \text{weakly in } L^2(\Gamma_0)\,, \tag{51}$$

consequently $g^0 \equiv g^{0+} = -g^{0-}$. This equality means continuity of the normal momentum, which is consistent with the consequence of (47).

The homogenized coefficients governing the acoustic transmission are introduced below using so called corrector functions defined in the reference periodic cell $Y =]0,1[^2 \times]-1/2,+1/2[\subset \mathbb{R}^3$. The acoustic medium occupies the domain $Y^* = Y \setminus S$, where $S \subset Y$ is the solid (rigid) obstacle. For clarity we use notation $I_y =]0,1[^2$ and $I_z =]-1/2,+1/2[$. The upper and lower boundaries are translations of $(I_y, 0)$; we define $I_y^+ = \{y \in \partial Y : z = 1/2\}$ and $I_y^- = \{y \in \partial Y : z = -1/2\}$. By $H^1_{\#(1,2)}(Y)$ we denote the space of $H^1(Y)$ functions which are "1-periodic" in coordinates y_α, $\alpha = 1,2$; in this paper such functions will be called "transversely Y-periodic".

In [38] the homogenization of problem (49) was considered in detail. As the result, the homogenized transmission conditions were obtained, being expressed in terms of the *interface mean acoustic pressure* $p^0 \in H^1(\Gamma_0)$, and the *fictitious acoustic transverse velocity* $g^0 \in L^2(\Gamma_0)$; these quantities satisfy the following PDE system in weak form:

$$\begin{aligned}
&\int_{\Gamma_0} A_{\alpha\beta}\partial^x_\beta p^0 \partial^x_\alpha q - f^*\omega^2 \int_{\Gamma_0} p^0 q + \mathrm{j}\omega \int_{\Gamma_0} B_\alpha \partial^x_\alpha q g^0 = 0\,,\\
&-\mathrm{j}\omega \int_{\Gamma_0} D_\beta \partial^x_\beta p^0 \psi + \omega^2 \int_{\Gamma_0} F g^0 \psi = \\
&\qquad -\mathrm{j}\omega \frac{1}{\varepsilon_0} \int_{\Gamma_0} (p^+ - p^-)\psi\,,
\end{aligned} \tag{52}$$

for all $q \in H^1(\Gamma_0)$ and $\psi \in L^2(\Gamma_0)$, where $f^* = \frac{|Y^*|}{|Y|}$ is the porosity related to the layer thickness. We remark that while $(52)_1$ is the direct consequence of (49) for $\varepsilon \to 0$, additional constraint $(52)_2$ arises due to coupling the "outer acoustic problem" imposed in $\Omega^G \setminus \Omega_\delta$ with the one imposed in the layer. A quite analogous treatment is employed in the electromagnetic transmission problem described in Section 5.2.4. Equations (52) involve the homogenized coefficients $A_{\alpha\beta}, B_\alpha, D_\alpha$ and F expressed in terms of the local corrector functions π^β and ξ.

The homogenized coefficients, A, B, F are determined by the solution of the local corrector problems. To simplify the notation, we introduce

$$\begin{aligned}
\hat{\nabla} q &= (\partial^y_\alpha q, h^{-1}\partial_z q),\\
a^*_Y(\pi,\xi) &= \int_{Y^*} \hat{\nabla}\pi \cdot \hat{\nabla}\xi \\
&= \int_{Y^*} \left(\partial^y_\alpha \pi \partial^y_\alpha \xi + \frac{1}{h^2}\partial_z \pi \partial_z \xi\right)\,,\\
\gamma^\pm(\xi) &= \int_{I_y^+} \xi - \int_{I_y^-} \xi\,.
\end{aligned} \tag{53}$$

The two following local corrector problems are defined: Find $\pi^\beta, \xi \in H^1_{\#(1,2)}(Y)/\mathbb{R}$ such that

$$\begin{aligned}
a^*_Y\left(\pi^\beta + y_\beta, \phi\right) &= 0\,,\ \forall \phi \in H^1_{\#(1,2)}(Y),\ \beta = 1,2\,,\\
a^*_Y(\xi,\phi) &= -\frac{|Y|}{hc^2}\gamma^\pm(\phi)\,,\ \forall \phi \in H^1_{\#(1,2)}(Y)\,,
\end{aligned} \tag{54}$$

see Fig. (**9**) where function ξ is displayed for three different microstructures. The homogenized coefficients are expressed in terms of π^α and ξ, as follows:

$$\begin{aligned}
A_{\alpha\beta} &= \frac{c^2}{|Y|} a^*_Y\left(\pi^\beta + y^\beta, \pi^\alpha + y^\alpha\right)\,,\\
h^{-1}D_\alpha = B_\alpha &= \frac{c^2}{|Y|} a^*_Y(\xi, y_\alpha)\,,\\
F &= \frac{1}{|I_y|}\gamma^\pm(\xi)\,.
\end{aligned} \tag{55}$$

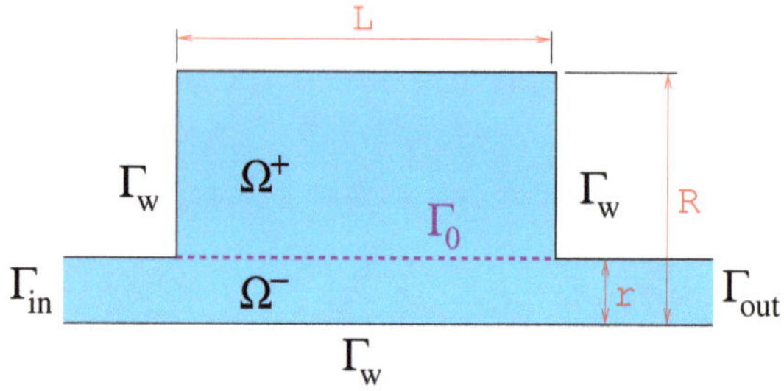

Fig. (7): The domain and boundary decomposition of the global acoustic problem considered. This layout is inspired by [8]

4.3 Structure of the Global Problem

The coupled system (52) described above constitute the transmission condition in a global problem considered. As an example, we shall present a model of an acoustic duct with perforated (rigid) plate.
Let us consider the domain of Ω^G, as in (46), where the outer boundary $\partial\Omega^G = \Gamma_{in} \cup \Gamma_{out} \cup \Gamma_w$ consists of the planar surfaces Γ_{in}, Γ_{out} and the channel walls Γ_w, see Fig. (7). On Γ_{in} we assume an incident wave of the form $\tilde{p}(x,t) = \bar{p}e^{-jkn_l \cdot x_l}e^{j\omega t}$, where (n_l) is the outward normal vector of Ω, on Γ_{out} we impose the radiation condition of the Sommerfeld type, so that

$$\begin{aligned} j\omega p + c\frac{\partial p}{\partial n} &= 2j\omega\bar{p} \quad \text{on } \Gamma_{in}\,, \\ j\omega p + c\frac{\partial p}{\partial n} &= 0 \quad \text{on } \Gamma_{out}\,, \\ \frac{\partial p}{\partial n} &= 0 \quad \text{on } \Gamma_w\,. \end{aligned} \tag{56}$$

The interface condition has the following form, see illustration in Fig. (8),

$$\left\{ \begin{array}{l} c^2\frac{\partial p}{\partial n^+} = j\omega g_0 \\ c^2\frac{\partial p}{\partial n^-} = -j\omega g_0 \end{array} \right. \quad \text{on } \Gamma_0\,, \tag{57}$$

where $\frac{\partial p}{\partial n^\pm} = n^\pm \cdot \nabla p$ are the normal derivatives on Γ_0 w.r.t. normals outward to Ω^+ and Ω^-, respectively. Thus, transmission conditions on the interface Γ_0 involve the transversal acoustic momentum g_0 satisfying

$$\begin{aligned} -\partial_\alpha(A_{\alpha\beta}\partial_\beta p^0) + \omega^2 f^* p^0 - \partial_\alpha(B_{\alpha}g^0) &= 0 \quad \text{on } \Gamma_0\,, \\ -jh\omega B_\beta + \omega^2 F g^0 &= -j\omega\frac{1}{\varepsilon_0}(p^+ - p^-) \quad \text{on } \Gamma_0\,, \\ A_{\alpha\beta}\partial_\beta p^0 &= 0 \quad \text{on } \partial\Gamma_0\,, \end{aligned} \tag{58}$$

where $\partial\Gamma_0$ is the edge of the obstacle Γ_0 and $f^* = |Y^*|/|Y|$ is the layer porosity (depending on parameter h). This is the differential form of integral identities (52) that were developed in [38] using asymptotic analysis.

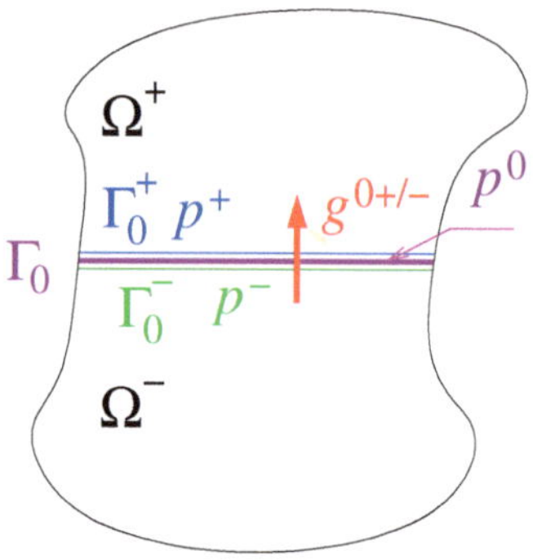

Fig. (8): Illustration of the transmission condition obtained by the homogenization of the perforated interface. Normal derivatives of the acoustic pressure are continuous, being proportional to g_0.

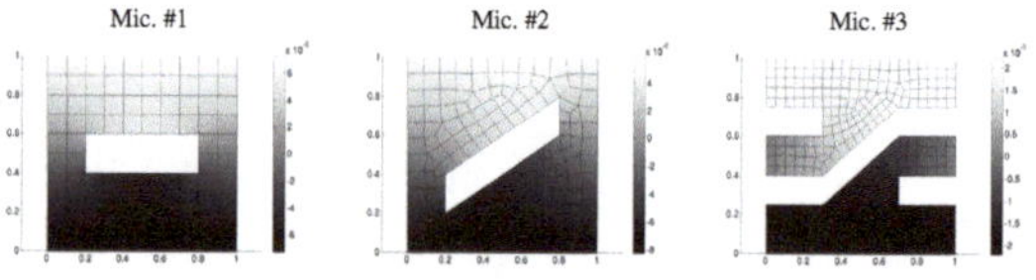

Fig. (9): Distribution of ξ in Y^*.

4.4 Numerical Illustration

In Table 1 we introduce homogenized transmission parameters A, B, F for 2D microstructures #1, #2 and #3 displayed in Fig. (9); whenever the microstructure is symmetric w.r.t. the vertical axis of Y, coefficient B vanishes and, as the consequence, the surface wave is decoupled from the transversal momentum.
We shall now illustrate that the global macroscopic response is very sensitive to the specific geometry of the perforation. The following numerical example shows the global response of a waveguide containing the homogenized transmission layer. The geometry of the waveguide is depicted in Figs. (7). The global response can be characterized by the trans-

Mic.	$A[(\text{m/s})^2]$	$B[\text{m}]$	$F[\text{s}^2]$
#1	$1.155 \cdot 10^5$	0	$1.391 \cdot 10^{-5}$
#2	$1.704 \cdot 10^5$	-0.251	$1.324 \cdot 10^{-5}$
#3	$2.186 \cdot 10^5$	-0.897	$4.265 \cdot 10^{-5}$

Table 1: Comparison of homogenized transmission parameters for different microstructures.

mission loss $TL = 20log\left(|\bar{p}_{|\Gamma_{\text{in}}}|/|p_{|\Gamma_{\text{out}}}|\right)$, where $\bar{p}$ is the incident plane wave, see (56). The transmission losses for the waveguide with perforations #1, #2 and #3 are shown in Fig. (**10**). On the horizontal axis there is the wave number κ ($\kappa = \omega/c$) multiplied by length L of the "expansion chamber" (see Fig. (**7**)). The resulting acoustic pressures in the waveguide are displayed in Fig. (**11**). The numerical results were obtained for acoustic speed $c = 343\,\text{m/s}$ and scale parameter $\varepsilon_0 = 0.035$, which e.g. for the microstructure type #1 means that the thickness of the perforated plate is 7mm. According to this study the perforation design seems to have quite important influence on the global behaviour of the acoustic pressure field, as viewed by the transmission losses. This is a motivation for an optimal perforation problem, see [29, 24].

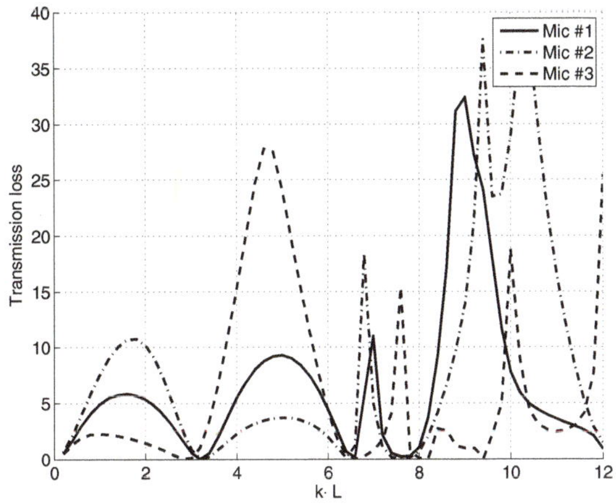

Fig. (**10**): Transmission losses for different perforation types.

5. ELECTROMAGNETIC WAVES IN PHOTONIC CRYSTALS

In analogy with the photonic crystals (materials) treated in Section 3, homogenization was employed to describe dispersion of optical waves in strongly heterogeneous periodic materials, cf.

5.1 Helmholtz Equation for Harmonic Waves

Here we recall the possible description of electromagnetic fields in heterogeneous materials using the Hertz potential (cf. [2]).

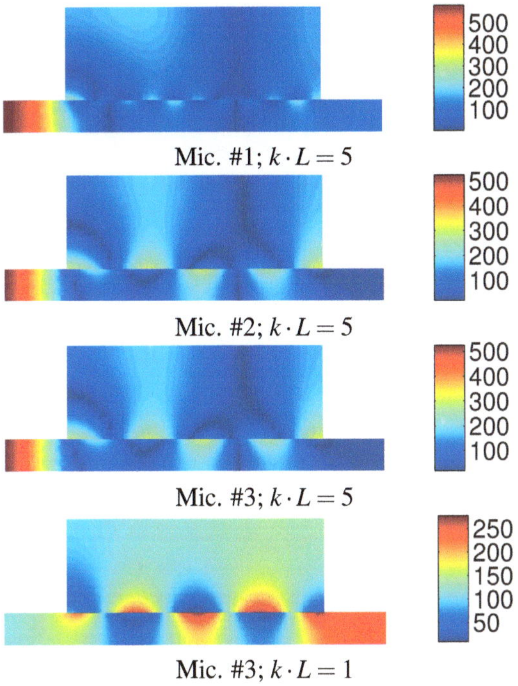

Fig. (**11**): Modulus of the acoustic pressure in Ω for $k \cdot L = 5$ (1 in the last picture). For this 2D computation a finite element mesh comprising 820 quadrilateral elements was used.

5.1.1 Maxwell Equation for Harmonic Waves

We assume monochromatic wave of frequency ω and amplitudes $\boldsymbol{H}$ and $\boldsymbol{E}$ standing for magnetic and electric Fields, respectively, which satisfy the Maxwell equations:

$$\begin{aligned} \nabla \times \boldsymbol{H} &= (-\mathrm{j}\omega\varepsilon + \sigma)\boldsymbol{E} + \boldsymbol{J}_e\,, \\ \nabla \times \boldsymbol{E} &= \mathrm{j}\omega\mu\boldsymbol{H}\,, \\ \nabla \cdot (\varepsilon\boldsymbol{E}) &= \rho\,, \\ \nabla \cdot (\mu\boldsymbol{H}) &= 0\,, \end{aligned} \tag{59}$$

where $\boldsymbol{J}_e$ is the current associated with external sources of electromagnetism, ρ is the volume electric charge density, ε is the electric permittivity (a real number), μ is the magnetic permeability (a real number) and σ is conductivity which is zero in vacuum (a real number).

Let us assume for a while, that the material is homogeneous, i.e. $(\varepsilon, \mu, \sigma)$ are constants. Then either $\boldsymbol{E}$, or $\boldsymbol{H}$ can be eliminated from system (59), so that the Helmholtz equations hold

$$\begin{aligned} \nabla^2\boldsymbol{E} + \kappa^2\boldsymbol{E} &= \varepsilon^{-1}\nabla\rho - \mathrm{j}\omega\mu\boldsymbol{J}_e\,, \quad \nabla\cdot\boldsymbol{E} = \rho/\varepsilon\,, \\ \nabla^2\boldsymbol{H} + \kappa^2\boldsymbol{H} &= -\nabla\times\boldsymbol{J}_e\,, \quad \nabla\cdot\boldsymbol{H} = 0\,, \end{aligned} \tag{60}$$

where κ is the wave number characterized by the material:

$$\kappa^2 = \omega^2 \mu\beta = \omega^2 \mu(\varepsilon + \mathrm{j}\sigma/\omega) . \qquad (61)$$

The vectorial Helmholtz equations (60) present three independent scalar "componentwise" equations, however they are coupled by the divergence conditions, which makes the analysis more difficult. To simplify construction of the solutions to (60), the *vector potentials* are introduced. Two standard cases can be treated:

1. **Electric Hertz potential.** Let us consider the special case $\boldsymbol{J}_e = 0$, thereby $\rho = 0$. Then by $(60)_1$ it follows that $\nabla \cdot \boldsymbol{E} = 0$. The electric Hertz potential $\boldsymbol{E} = \nabla \times \boldsymbol{A}^E$ then satisfies $(60)_1$, which yields

$$\nabla^2 \boldsymbol{A}^E + \kappa^2 \boldsymbol{A}^E = \nabla\phi , \qquad (62)$$

   where $\nabla\phi$ is any scalar differentiable function.

2. **Magnetic Hertz potential** of the magnetic field. Let $\boldsymbol{H} = \nabla \times \boldsymbol{A}^H$, where $\boldsymbol{A}^H$ is the Hertz potential. Then $(59)_2$ yields

$$\nabla^2 \boldsymbol{A}^H + \kappa^2 \boldsymbol{A}^H = -\boldsymbol{J}_e + \nabla\psi , \qquad (63)$$

   where ψ is any scalar differentiable function.

5.1.2 Transmission Conditions

Let Γ be the interface separating two subdomains Ω_1 and Ω_2 where in each the material parameters are constant. From the integral form of the Maxwell equations the following transmission conditions can be derived, see e.g. [2],

$$[\boldsymbol{n} \times \boldsymbol{E}]_\Gamma = 0 , \quad [\boldsymbol{n} \times \boldsymbol{H}]_\Gamma = 0 , \qquad (64)$$

where $[\bullet]_\Gamma$ is the jump of $\bullet$ on Γ and $\boldsymbol{n}$ is normal vector to Γ.

5.1.3 Two-Dimensional Model for a Heterogeneous Medium

Let us consider $\boldsymbol{A}^E = v\vec{e}_3$, so that $\vec{e}_3$ is the normal of the plane transversal to the fibres aligned with $\vec{e}_3$ and characterizing the heterogeneities, and $v = v(x_1, x_2)$ is the scalar potential of the transversal electric H-mode (TE-H-mode). Now (62) reduces to the scalar Helmholtz equation

$$\nabla^2 v + \kappa^2 v = \partial_3 \phi . \qquad (65)$$

In what follows we may put $\nabla\phi = 0$, thus $\partial_3\phi = 0$, (cf. [2]). Further we consider two materials occupying two disjoint domains Ω_1 and Ω_2, separated by interface Γ, so that $\Omega = \Omega_1 \cup \Gamma \cup \Omega_2$. For this special case we rewrite $(64)_1$, noting that $\boldsymbol{n} \cdot \vec{e}_3 = 0$ and also $\vec{e}_3 \cdot (\nabla \boldsymbol{n}) = 0$:

$$\begin{aligned} [\boldsymbol{n} \times \boldsymbol{E}]_\Gamma &= [\boldsymbol{n} \times \nabla \times \boldsymbol{A}^E]_\Gamma \\ &= [\nabla(\boldsymbol{n} \cdot \boldsymbol{A}^E) - (\nabla \boldsymbol{n}) \cdot \boldsymbol{A}^E - \partial_n \boldsymbol{A}^E]_\Gamma \\ &= -\vec{e}_3 [\partial_n v]_\Gamma , \end{aligned} \qquad (66)$$

where ∂_n is the normal derivative. Then we employ $(59)_2$ in $(64)_2$:

$$\begin{aligned} [\boldsymbol{n} \times \boldsymbol{H}]_\Gamma &= \frac{1}{\mathrm{j}\omega} [\frac{1}{\mu} \boldsymbol{n} \times \nabla \times \boldsymbol{E}]_\Gamma \\ &= \frac{-1}{\mathrm{j}\omega} [\frac{1}{\mu} \boldsymbol{n} \times \nabla^2 \boldsymbol{A}^E]_\Gamma \\ &= \vec{e}_3 \frac{-1}{\mathrm{j}\omega} [\frac{\kappa^2}{\mu} v]_\Gamma = 0 , \end{aligned} \qquad (67)$$

where (64) was employed. Thus, for the time-harmonic response featured by the frequency ω and the TE-mode, the Maxwell equations yields the following system

$$\begin{aligned} \nabla^2 v + \kappa^2 v &= 0 \quad \text{in } \Omega_k,\ k = 1,2 , \\ \text{some b.c. on } &\partial\Omega , \\ \text{transmission cond.:} \quad [\partial_n v]_\Gamma &= 0 \quad \text{on } \Gamma , \\ [\frac{\kappa^2}{\mu} v]_\Gamma &= 0 \quad \text{on } \Gamma , \end{aligned} \qquad (68)$$

where ∂_n denotes the co-normal derivative, i.e. $\partial_n = \boldsymbol{n} \cdot \nabla$. The *complex wave number* κ is defined locally by the material parameters; we consider them piecewise constant in Ω, in particular

$$(\mu, \varepsilon, \sigma)(x) = \left\{ \begin{array}{ll} (\mu_1, \varepsilon_1, \sigma_1) & x \in \Omega_1 \\ (\mu_2, \varepsilon_2, \sigma_2) & x \in \Omega_2 \end{array} \right. , \qquad (69)$$

where $(\mu_k, \varepsilon_k, \sigma_k)$, $k = 1,2$ are constants.

Meanwhile the boundary conditions on $\partial\Omega$ are not specified; importantly, when a part of $\partial\Omega$ is attached to a perfect conductor, then $\partial_n v = 0$ on this part.

It is worth noting that solutions to (68) have continuous co-normal derivative on Γ, but the traces of v on Γ are discontinuous. In the next section we modify the formulation represented by (68) to get rid of these discontinuities.

By virtue of the piecewise constant material properties (69) piecewise-defined rescaling of v restricted to Ω_k can be introduced. We shall see that there exists a continuous field u such that

$$v = \frac{\mu_k}{\kappa_k^2} u = \frac{1}{\varepsilon_k \omega^2 + \mathrm{j}\sigma_k \omega} u = \frac{1}{\omega^2 \beta_k} u \text{ in } \Omega_k \quad (70)$$

where $\beta_k = \varepsilon_k + \mathrm{j}\sigma_k/\omega$ and v satisfies (68). Substitution (70) is well defined provided $\omega > 0$ and $\varepsilon_k \neq 0$. Now we are allowed to apply this substitution in (68) to obtain the following modified system

$$\nabla \cdot \left(\frac{1}{\beta_k} \nabla u \right) + \omega^2 \mu_k u = \partial_3 g \text{ in } \Omega_k,\ k = 1,2\,,$$

$$\text{some b.c. on } \partial\Omega\,,$$

$$\text{transmission cond.: } [\frac{1}{\beta} \partial_n u]_\Gamma = 0 \quad \text{on } \Gamma\,,$$

$$[u]_\Gamma = 0 \quad \text{on } \Gamma\,, \quad (71)$$

where in $(71)_3$ $\beta = \beta_k$ on $\Gamma \cap \partial\Omega_k$. Obviously, continuity on Γ follows by $(71)_3$ and $(71)_4$ preserves continuity of the co-gradients.

Remark 1. Notation: Alternatively we can rewrite (71) using the relative permittivity and permeability. Let ε_0, μ_0 be the permittivity and permeability of the vacuum, then $\mu_k = \mu_k^r \mu_0$, $\varepsilon_k = \varepsilon_k^r \varepsilon_0$ and $\beta_k = \beta_k^r \varepsilon_0$, where $\beta_k^r(\omega) = \varepsilon_k^r + \mathrm{j}\sigma_k/(\omega\varepsilon_0)$. On introducing the wave number $\kappa_0 = \omega\sqrt{\varepsilon_0 \mu_0}$, $(71)_1$ can be rewritten (assuming $g = 0$)

$$\nabla \cdot \left(\frac{1}{\beta_k^r} \nabla u \right) + \kappa_0^2 \mu_k^r u = 0 \quad \text{in } \Omega_k,\ k = 1,2\,. \quad (72)$$

For magnetically inactive materials $\mu_k^r \approx 1$, therefore alternatively

$$\nabla \cdot \left(\frac{1}{(n_k^r)^2} \nabla u \right) + \omega^2 \mu_0 u = 0 \quad \text{in } \Omega_k,\ k = 1,2\,, \quad (73)$$

where $n_k^r = \sqrt{\beta_k^r/\varepsilon_0}$ is the refraction index.

△

Remark 2. Alternatively one can consider the so called transversal magnetic E-mode (TM-E-mode), on introducing $\boldsymbol{A}^H = w\vec{e}_3$, in analogy with the TE-H-mode. This applies in particular for $\boldsymbol{J}_e = j_e\vec{e}_3$, thus

$$\nabla^2 w + \kappa^2 w = -j_e + \partial_3 \psi\,.$$

The transmission conditions on Γ are

$$[\partial_n w]_\Gamma = 0, \quad [\mu w]_\Gamma = 0\,,$$

so that for μ constant in whole domain the solution w is smooth and continuous on Γ; typically this is satisfied by a class of optical materials where $\mu = \mu_0$.

△

5.2 Photonic Crystals

Photonic crystals and magnetically active materials became a quite interesting field of material science due to vast applications in optical technologies (waveguides, optical fibres, special lens...). There is a rich literature facing this subject, see e.g. [9][28][45].

In this section we aim to demonstrate the modelling analogy between acoustic waves in phononic materials and the electromagnetic waves in the photonic ones. Therefore, we shall focus on the homogenisation approach which consists in replacing a composite with a large number of periodic microstructures by a limit homogeneous material. Such a treatment is relevant for the modelling of the periodic structures presented by photonic crystals. As Bouchitté and Felbacq proposed [9] in the case of periodic photonic crystals made of "strongly heterogeneous composites" (i.e., with permittivity coefficients strongly different in the inclusions and in the matrix), the limit *homogenized permeability* is negative for certain wavelengths, thus yielding the existence of band gaps. More precisely, they showed that when the ratio between permeability of the inclusions and permeability of the background is of the order of the square of the size of the microstructures, then the band-gaps phenomenon appears. Historically this observation motivated the homogenization approach applied to elastic waves, as reported above.

5.2.1 Periodic Structure with Large Contrasts in Permittivity

Let us consider a periodic structure, as generated in (16), characterized by permeability $\mu^\varepsilon(x)$ and complex permittivity $\beta^\varepsilon(x)$ given as piecewise constant functions

$$\mu^\varepsilon(x) = \begin{cases} \mu^1 & \text{in } \Omega_1^\varepsilon, \\ \mu^2 & \text{in } \Omega_2^\varepsilon, \\ \mu^0 & \text{in } \mathbb{R}^2 \setminus \Omega, \end{cases}$$

$$\beta^\varepsilon(x) = \begin{cases} \beta^1 & \text{in } \Omega_1^\varepsilon, \\ \varepsilon^2\beta^2 & \text{in } \Omega_2^\varepsilon, \\ \beta^0 & \text{in } \mathbb{R}^2 \setminus \Omega \end{cases} \quad (74)$$

and assume that for $\varepsilon < \varepsilon_0$ no inclusion intersects $\partial\Omega$. Further we may assume that the heterogeneous medium occupying domain Ω is subject to an incident wave imposed in $\mathbb{R}^2 \setminus \Omega$ with the Sommerfeld radiation condition applied on the scattered field in the infinity, see [9]. Note that at any interface separating the inhomogeneities the standard interface condition of the type $(71)_3$ applies.

In [9] it was proved mathematically that the artificial magnetism can be obtained by homogenization (i.e. by asymptotic analysis) of the following problem

$$\begin{aligned} \nabla \cdot \left(\frac{1}{\beta^\varepsilon} \nabla u^\varepsilon \right) + \omega^2 \mu^\varepsilon u^\varepsilon = 0 \quad \text{in } \mathbb{R}^2, \\ \frac{1}{\beta^0} \partial_r u^{\mathrm{sc}\varepsilon} - \mathrm{j}\omega\mu^0 u^{\mathrm{sc}\varepsilon} = O(1/\sqrt{\kappa^0 r}) \\ \text{when } r \to +\infty \,, \end{aligned} \tag{75}$$

where u^{inc} is the incident wave and $u^{\mathrm{sc}\varepsilon} = u^\varepsilon - u^{\mathrm{inc}}$ is the scattered field. We shall here recall the model of homogenized material (*metamaterial* which will allow us to see the analogies between the homogenization of the phononic crystals (acoustic waves) and the photonic ones (electromagnetic waves).

5.2.2 Homogenized Coefficients

In analogy with the construction of mass tensor M^*_{ij} in (29) using eigensolutions of (27), the *effective permeability* is expressed in terms of eigensolutions of the problem: find couples $(\lambda^k, w^k) \in \mathbb{R} \times H^1_0(Y_2)$, $k = 1, 2, \ldots$

$$\begin{aligned} \int_{Y_2} \nabla w^k \cdot \nabla \phi = \lambda^k \int_{Y_2} w^k \phi \,, \quad \forall \phi \in H^1_0(Y_2), \\ \int_{Y_2} w^k w^l = \delta_{kl} \,. \end{aligned} \tag{76}$$

Now the effective permeability is computed as follows:

$$\begin{aligned} \mu^*(\omega) = \frac{\mu^1 |Y_1| + \mu^2 |Y_2|}{|Y|} + \\ + \mu^2 \frac{1}{|Y|} \sum_{k \in I_+} \frac{\omega^2}{\lambda^k / (\beta^2 \mu^2) - \omega^2} \left(\int_{Y_2} w^k \right)^2 \\ \text{where} \quad I_+ = \{k | \quad \left| \int_{Y_2} w^k \right| > 0\} \,. \end{aligned} \tag{77}$$

The *effective permittivity* becomes a 2×2 symmetric tensor:

$$A^*_{ij} = \frac{1}{\beta^1} \fint_{Y_1} \nabla_y (\eta^i + y_i) \cdot \nabla_y (\eta^j + y_j) \,, \tag{78}$$

where $\eta^i = H^1_\#(Y_1)$, being Y-periodic, satisfies the following identities:

$$\fint_{Y_1} \nabla_y (\eta^i + y_i) \cdot \nabla_y \psi = 0 \quad \forall \psi \in H^1_\#(Y_1) \,, \quad i = 1, 2 \,,. \tag{79}$$

5.2.3 Homogenized Photonic Materials

The limit analysis of the heterogeneous medium leads to the model of homogenized medium which is characterized by effective (homogenized) material parameters. One can show that $u^\varepsilon(x)$ in (75) *two-scale converges* (cf. the unfolding method of homogenization [13]) to $u(x) + \chi_2(y)\hat{u}(x,y)$, where χ_2 is the characteristic function of Y_2 and $\hat{u}(x,y)$ are the non-vanishing oscillations in the inclusions. u is the "macroscopic" solution satisfying

$$\begin{aligned} \nabla_x \cdot \boldsymbol{A}^* \cdot \nabla_x u + \omega^2 \mu^*(\omega) u = 0 \,, \quad \text{in } \Omega \,, \\ \frac{1}{\beta^0} \nabla^2 u + \omega^2 \mu^0 u = 0 \,, \quad \text{in } \mathbb{R}^2 \setminus \Omega \,, \\ \boldsymbol{n} \cdot \boldsymbol{A}^* \cdot \nabla_x u_- - \boldsymbol{n} \cdot \frac{1}{\beta^0} \nabla u_+ = 0 \quad \text{on } \partial\Omega \,, \\ u_+ - u_- = 0 \quad \text{on } \partial\Omega \,, \\ u^{\mathrm{sc}} \equiv u - u^{\mathrm{inc}} \quad \text{satisfies (75)} \,, \end{aligned} \tag{80}$$

where $\boldsymbol{n}$ is a normal vector on $\partial\Omega$ and u_-, u_+ are the interior and exterior values on $\partial\Omega$, respectively. Thus the solution is continuous on $\partial\Omega$.

5.2.4 Photonic Band Gaps

The homogenized medium represented by $\mu^*(\omega)$ and A^*_{ij} is the magnetic active metamaterial with possibly negative permeability $\mu^*(\omega) < 0$ for some ω. This effect features occurrence of band gaps, in analogy with the *phononic* material described above in the text, where the acoustic band gaps are indicated by negative effective mass $M^*(\omega)$.

6. ELECTROMAGNETIC WAVE TRANSMISSION ON HETEROGENEOUS LAYERS AND CLOAKING

In analogy with the acoustic transmission problem reported in Section 4.1, we discus the electromagnetic wave transmission through periodically heterogeneous layer.

We consider a strip $\Omega_\delta \subset \mathbb{R}^3$ with the thickness $\delta > 0$ generated by a planar surface Γ_0 and bounded by Γ^+_δ and Γ^-_δ, see Fig. (**12**); the same notation is used as that introduced in Section 3.4. In general, the strip may contain perfect conducting material; we denote by $S^{\varepsilon}_{\delta} \subset \Omega^{\varepsilon}_{\delta}$ union of all such conductor (e.g. realized by fibrous graining) which also constitute the periodic pattern in the strip; length of the period in x_α, $\alpha = 1$ is ε, see Remark 3; the pattern is defined by the 2D section spanning coordinates x_1, x_3, so that interfaces of the graining between different materials have the form of general infinite cylinders. he dielectric material with finite conductivity occupies domain $\Omega^{\varepsilon}_{\delta} = \Omega_\delta \setminus \overline{S^{\varepsilon}_{\delta}}$. The problem of the TE-mode radiation will be imposed in the perforated domain $\Omega^{\varepsilon}_{\delta}$.

Remark 3. Here we consider the TE-H-mode, i.e. the two-dimensional restriction of the electromagnetic wave propagation (65), which is characterized by scalar function $v = v(x_1, x_3)$, thus $\partial_2 v \equiv 0$. Such a situation is relevant whenever the heterogeneous structure is generated in 3D independently of coordinate x_2 (e.g. by fibrous graining aligned with x_2-axis). For generality we shall keep 3D description w.r.t. coordinates $(x_1, x_2, x_3) = (x_\alpha, x_3)$, where $\alpha = 1,2$ refers to the in-plane position in Γ_0 only. However, due to the TE-H-mode restriction, only gradients w.r.t. x_1 and x_3 coordinates do not vanish, therefore in the sequel one may consider $\alpha = 1$.

In the "ad hoc 2D" treatment, Γ_0 is just a line, whereas Ω_δ is a two-dimensional domain spanned by coordinates x_1, x_3.

△

From similar studies of elliptic problems in thin layers having a periodic microstructure it is well known that different limit models are obtained when commuting $\varepsilon \to 0$ (the period of heterogeneities) and $\delta \to 0$ (the thickness). Here we consider fixed proportion $\delta = h\varepsilon$, $h > 0$.

6.1 Non-Homogenized Layer – Problem Formulation

We can define the boundary value problem for the rescaled potential, see (71), and consider the Neumann conditions on $\Gamma^{\pm}_{\delta}$:

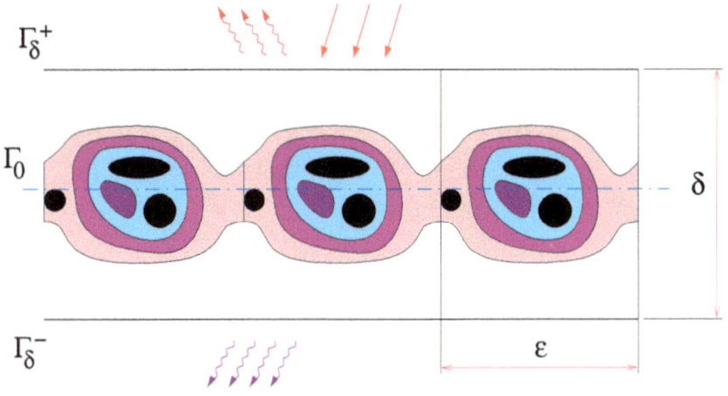

Fig. (**12**): Illustration of a section through the fictitious layer in which the heterogeneous structure is embedded. The black parts represent perfect conductors, in the "void" part the material coefficients are the same as those outside the layer; The colour (grey) regions are occupied by different materials.

$$
\begin{aligned}
\nabla \cdot \left(\frac{1}{\beta^2_{\varepsilon\delta}} \nabla u^{\varepsilon\delta} \right) + \mu\omega^2 u^{\varepsilon\delta} &= 0 \quad \text{in } \Omega^{\varepsilon}_{\delta} , \\
\frac{1}{\beta_0} \partial^{\pm}_n u^{\varepsilon\delta} &= \mathrm{j}\omega g^{\pm\delta} \quad \text{on } \Gamma^{\pm}_{\delta} , \\
\text{where } g^{\pm\delta} &= \pm g^0(x_\alpha) + \varepsilon g^{1\pm}(x_\alpha, x/\varepsilon) , \\
\text{so that } \fint_{I^+_y \cup I^-_y} g^{\pm\delta} &\approx \delta , \\
\partial_n u^{\varepsilon\delta} &= 0 \quad \text{on } \partial S^{\varepsilon}_{\delta} , \\
u^{\varepsilon\delta}, \partial_n u^{\varepsilon\delta} & \text{ periodic on opposite} \\
& \text{sides of } \partial\Omega^{\infty}_{\delta} ,
\end{aligned}
\tag{81}
$$

where $g^{1\pm}$ is the fluctuation part. The perfect conductor in S^{ε}_{δ} results in the zero Neumann condition on the associated perforation boundary. It is worth recalling that $\beta_{\varepsilon\delta}$ is piecewise constant in $\Omega^{\varepsilon}_{\delta}$ and ε-periodic in x_1 (for fibrous structure relevant to the TE-mode analysis $\beta_{\varepsilon\delta}(x_1, x_3)$ is independent of x_2). In any case we assume that material on $\Gamma^{\pm}_{\delta}$ is homogeneous, thus $\beta_{\varepsilon\delta} = \beta_0$ is a *constant* (whatever possibly a complex number). Due to $(71)_{3,4}$ the solution $u^{\varepsilon\delta}$ is smooth and the transmission conditions are satisfied automatically.

6.2 Induction Law Constraint

For stating the boundary conditions on $\Gamma^{\pm}_{\delta}$, as explained below, the induction low is needed to define a suitable scaling of the Neumann fluxes.

Let $\mathscr{S} \in \mathbb{R}^2$ be a planar surface spanned by coordinates x_1, x_3, bounded by $\partial\mathscr{S}$, and let us consider decomposition $\mathscr{S} = \bigcup_k \mathscr{S}_k$ using a finite number of mutually non-overlapping subdomains $\mathscr{S}_k$, $k = 1, 2, \ldots$; in each $\mathscr{S}_k$ the medium is assumed to be homogeneous. For zero external current, i.e. $\boldsymbol{J}_e = 0$, and using the electric Hertz potential $\boldsymbol{A}^E$ the Maxwell equations $(59)_{1,2}$ yield $\boldsymbol{H} = (\sigma - j\omega\varepsilon)\boldsymbol{A}^E$ and $\nabla \times \boldsymbol{E} = j\omega\mu(\sigma - j\omega\varepsilon)\boldsymbol{A}^E$ in each $\mathscr{S}_k$. Further let $\boldsymbol{t}^k$ be the tangent unit vector associated with closed oriented curve $\partial\mathscr{S}_k$ and let $\boldsymbol{E}^k$ be the trace on $\mathscr{S}_k$ of $\boldsymbol{E}$ defined in $\mathscr{S}_k$. On integrating in $\mathscr{S}_k$ and then using the summation over all subdomains, one obtains subsequently ($\mu^k, \varepsilon^k, \sigma^k$ are local material constants valid in $\mathscr{S}_k$):

$$\bigcup_k \int_{\partial\mathscr{S}_k} \boldsymbol{t}^k \cdot \boldsymbol{E}^k d\Gamma = \bigcup_k \mu^k (j\omega\sigma^k + \omega^2\varepsilon^k) \int_{\mathscr{S}_k} \boldsymbol{A}^E ,$$
$$\int_{\partial\mathscr{S}} \boldsymbol{t}\cdot\boldsymbol{E} d\Gamma = \int_{\mathscr{S}} \mu(j\omega\sigma + \omega^2\varepsilon)\boldsymbol{A}^E . \tag{82}$$

Above the equivalence between the l.h.s. expressions follows from the general transmission condition (66) which in 2D situation of the TE-H-mode yields $[\boldsymbol{t}\cdot\boldsymbol{E}]_\Gamma = 0$. Let $k \neq l$ and consider the integral over $\Gamma_{kl} = \partial\mathscr{S}_k \cap \partial\mathscr{S}_l$ which appears in the l.h.s. of $(82)_1$: due to the opposite curve orientation, $\boldsymbol{t}^k = -\boldsymbol{t}^l$ on Γ_{kl}, the following holds:

$$\int_{\partial\mathscr{S}_k\cap\Gamma_{kl}} \boldsymbol{t}^k \cdot \boldsymbol{E}^k d\Gamma + \int_{\partial\mathscr{S}_l\cap\Gamma_{kl}} \boldsymbol{t}^l \cdot \boldsymbol{E}^l d\Gamma = $$
$$= \int_{\Gamma_{kl}} [\boldsymbol{t}\cdot\boldsymbol{E}]_{\Gamma_{kl}} d\Gamma = 0 , \tag{83}$$

which yields the equivalence between the l.h.s. in $(82)_1$ and $(82)_2$.

In the 2D situation, due to the TE-mode assumption, (82) yields the following constraint

$$\int_{\partial\mathscr{S}} (-t_1\partial_3 v + t_3\partial_1 v) d\Gamma = -\int_{\mathscr{S}} \kappa^2 v , \tag{84}$$

where $(t_1, 0, t_3)$ is the tangent of $\partial\mathscr{S}$ and $v\vec{e}_2$ is the electric Hertz potential for the TE-mode. Note that (84) holds also on "perforated" domains $\mathscr{S}^* \subset \mathscr{S}$ when the perforation represents perfect conductors; this is the simple consequence of the homogeneous Neumann conditions on the part of $\partial\mathscr{S}^*$ attached to the conductors (the "holes").

We now consider $\Omega_\delta \supset \mathscr{S} = \Omega_{\delta L} = (\underline{x} +] - L/2, L/2[\,) \times] - \delta/2, \delta/2[$ where $\underline{x} \in \Gamma_0$ is such that $(\underline{x} +] - L/2, L/2[\,) \subset \Gamma_0$. Boundary of $\Omega_{\delta L}$ is as follows, see Fig. (**13**):

$$\partial\Omega_{\delta L} = \Gamma^+_{\delta L} \cup \Gamma^-_{\delta L} \cup \Xi^-_\delta \cup \Xi^+_\delta ,$$
$$\Gamma^\pm_{\delta L} \subset \Gamma^\pm_\delta , \tag{85}$$
$$\Xi^\pm_\delta = (\underline{x} \pm L/2) \times] - \delta/2, \delta/2[.$$

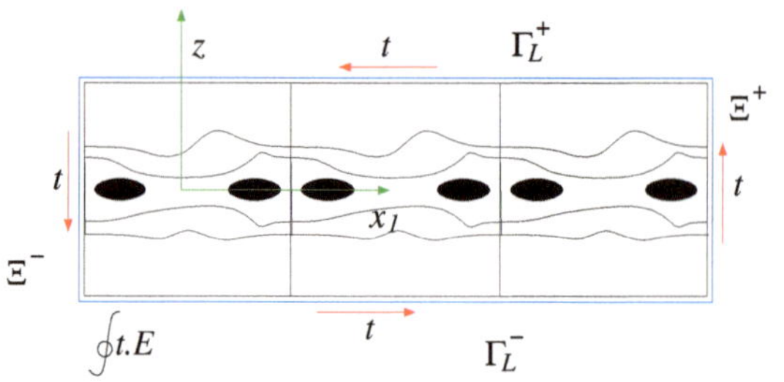

Fig. (**13**): Illustration of the integral form of the induction law.

Using substitution (70) in (84) we obtain

$$\frac{1}{\beta_0} \int_{\Gamma^\pm_{\delta L}} \partial^\pm_n u + \int_{\Xi^\pm_\delta} \frac{1}{\beta_{\varepsilon\delta}} \partial^\pm_n u = -\mu\omega^2 \int_{\Omega^{\varepsilon*}_{\delta L}} u , \tag{86}$$

where $\Omega^{\varepsilon*}_{\delta L} = \Omega_{\delta L} \cap \Omega^{\varepsilon*}_\delta$ and where $\pm$ sign matches the integration over $\Gamma^\pm_{\delta L}$ or $\Xi^\pm_\delta$. It is important to note, that $|\Omega^{\varepsilon*}_{\delta L}|$ and $|\Xi^\pm_\delta|$ are proportional to δ; this observation was respected in the definition of $g^\pm_\delta$ in (81).

6.3 The Homogenized Transmission Condition

The homogenized transmission condition is defined in terms of the homogenized coefficients which involve the corrector functions in the integral form. In what follows we explain, how the transmission condition can be evaluated, for its detailed derivation we refer to [37]. Here we shall just summarize the main steps of the homogenization procedure which is quite analogous to the result obtained for the acoustic problem reported above.

An important ingredient of the analysis is the dilation procedure, the affine mapping transforming domain Ω_δ on $\Omega = \Gamma_0 \times] - 1/2, 1/2[$ which, thus, is independent of $\delta = h\varepsilon$. The material structure in the layer is periodic being generated by representative cell Y in analogy with the acoustic problem discussed in Section 3.4 where the role of the fluid is now played by the dielectric material situated in Y^*, whereas the obstacles now represent the superconducting material.

Based on the *a priori estimate* of the solution to (81), one obtains the convergence result (in the sense of the two-scale convergence). There exist $u^0 \in L^2(\Gamma_0)$ and $u^1 \in L^2(\Gamma_0) \times H_{\#(1,2)}(Y)$ such that (denoting u^ε the solution of (81) on the dilated domain Ω) the following two-scale limits hold:

$$
\begin{aligned}
& u^\varepsilon \xrightarrow{2} u^0 \\
& \partial_\alpha u^\varepsilon \xrightarrow{2} \partial_\alpha^x u^0 + \partial_\alpha^y u^1 \,, \quad \alpha = 1,2 \\
& \frac{1}{\varepsilon}\partial_z u^\varepsilon \xrightarrow{2} \partial_z u^1
\end{aligned} \tag{87}
$$

Below we introduce the corrector basis functions which enable to express the "microscopic" function u^1 in terms of the "macroscopic" quantities $\partial_\alpha u^0$ and g^0; these are involved in the homogenized Helmholtz equation arising from $(81)_1$.

6.3.1 Coupling the Interface Layer Response with Outer Fields

In the limit situation the domain Ω_δ degenerates into the "mid-surface" (plane) Γ_0. Let the layer Ω_δ is embedded in Ω' where the scattered field can be observed,

$$
\Omega' = \Omega_\delta^+ \cup \overline{\Omega_\delta} \cup \Omega_\delta^- \,, \quad \Omega_\delta^\pm \cap \Omega_\delta = \emptyset \,, \tag{88}
$$

where also Ω_δ^+ and Ω_δ^- are disjoint. In order to be able to couple the exterior problem in $\Omega' \setminus \Omega_\delta$ with that in the homogenized layer represented by Γ, it is necessary to derive the relationship between the limit traces u^+ and u^- of the bulk field in $\Omega_\delta^\pm$ on $\Gamma^\pm$ for $\delta \to 0$ on one hand and the corresponding limit traces on $\Gamma_\delta^\pm$ on the other hand. Let $\widetilde{u^{\varepsilon\delta}}$ be the smooth extension over all perforations due to the perfect conductors. The traces from $\Omega_\delta^\pm$ satisfy

$$
\begin{aligned}
\int_{\Omega^\delta} \phi \partial_3 \widetilde{u^{\varepsilon\delta}} &= \int_{\Gamma_\delta^+} \phi\, u^\delta|_{\Gamma_\delta^+} d\Gamma - \int_{\Gamma_\delta^-} \phi\, u^\delta|_{\Gamma_\delta^-} d\Gamma \\
&\overset{\delta,\varepsilon\to 0}{\longrightarrow} \int_{\Gamma_0} \phi (u^+ - u^-) d\Gamma \,,
\end{aligned} \tag{89}
$$

for any $\phi \in L^2(\Omega')$ constrained by $\partial_3 \phi = 0$. We shall now consider a finite thickness $\delta_0 > 0$ of the layer. The l.h.s. in (89) can also be written as $\delta_0 \int_\Omega \phi \partial_3 \widetilde{u^{\varepsilon_0\delta_0}}$ (we recall the use of smooth extension $\widetilde{u^{\varepsilon_0\delta_0}}$ to entire Ω_{δ_0}). We consider the following approximation for $\varepsilon < \varepsilon_0$:

$$
\begin{aligned}
\delta_0 \int_\Omega \phi \partial_3 \widetilde{u^{\varepsilon_0\delta_0}} &\approx \varepsilon_0 \int_\Omega \frac{1}{\varepsilon} \phi \frac{\partial \widetilde{u^\varepsilon}}{\partial z} \\
&\overset{\varepsilon\to 0}{=\!=\!\longrightarrow} \varepsilon_0 \int_{\Gamma_0} \phi \fint_Y \left(\frac{\partial \widetilde{u^1}}{\partial z} \right) \\
&= \varepsilon_0 \int_{\Gamma_0} \phi \frac{1}{|I_y|} \left[\int_{I_y^+} u^1 \, d\Gamma_y - \int_{I_y^-} u^1 \, d\Gamma_y \right] ,
\end{aligned} \tag{90}
$$

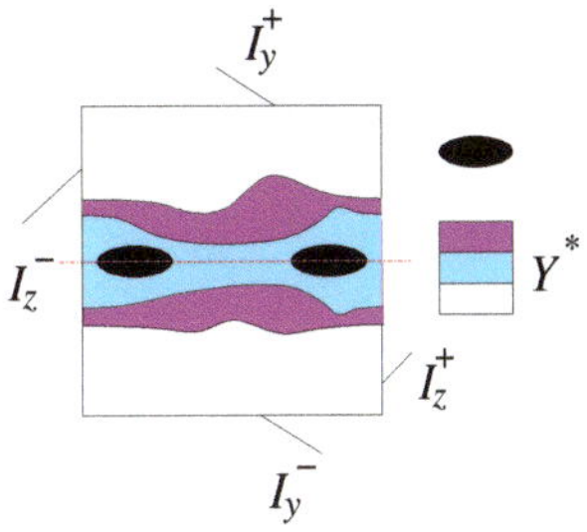

Fig. (**14**): Reference cell Y.

for all $\phi \in L^2(\Gamma_0)$, see $(87)_3$, hence using (89)

$$
\int_{\Gamma_0} \phi (u^+ - u^-) d\Gamma = \varepsilon_0 \int_{\Gamma_0} \phi \frac{1}{|I_y|} \left[\int_{I_y^+} u^1 \, d\Gamma_y - \int_{I_y^-} u^1 \, d\Gamma_y \right] . \tag{91}
$$

6.3.2 Continuity or a Jump of Potential Normal Derivative on Γ_0?

In the limit situation, $\varepsilon \to 0$, one can prove using the induction law constrain (86) that

$$
\frac{\partial u^+}{\partial n^+} + \frac{\partial u^-}{\partial n^-} = [\partial_n u]_{\Gamma_0} = 0 \,, \tag{92}
$$

where $\frac{\partial u^\pm}{\partial n^\pm}$ are traces from $\Omega^\pm$ of the normal derivatives on interface Γ_0.

However, an alternative treatment is possible. We may adapt the spirit of handling the potential jump $[u]_{\Gamma_0}$. For this we divide (86) by δ and approximate for a small $\delta_0 > 0$, which yields

$$
\frac{1}{\delta_0 \beta_0} \int_{\Gamma_L^\pm} \frac{\pm 1}{\delta} \frac{\partial}{\partial z} u + \int_{\Xi^\pm} \frac{\pm 1}{\beta_\varepsilon} \partial_1 u \approx -\mu \omega^2 \int_{\Omega_L^{\varepsilon *}} u \,. \tag{93}
$$

Above domains $\Gamma^\pm, \Xi^\pm, \Omega_L^{\varepsilon *}$ are obtained by the thickness dilatation (cf. [15],[38])of $\Gamma_\delta^\pm, \Xi_\delta^\pm, \Omega_{\delta L}^{\varepsilon *}$. Since for $\varepsilon \to 0$ the second l.h.s. term vanishes, he limit of (93) results in

$$
\frac{1}{\delta_0 \beta_0} \left[\frac{\partial u^+}{\partial n^+} + \frac{\partial u^-}{\partial n^-} \right] = \frac{1}{\delta_0 \beta_0} [\partial_n u]_{\Gamma_0} = -\rho^* \mu \omega^2 u^0 \quad \text{for a.a. } x \in \Gamma_0 \,, \tag{94}
$$

where $\rho^* = |Y^*|/|Y|$.

6.3.3 Corrector Basis Functions

We employ notation introduced in Section 3.4, however now the bilinear form a_Y^* is modified:

$$
a_Y^*(u, v) = \int_{Y^*} \frac{1}{\bar{\beta}} \hat{\nabla} u \cdot \hat{\nabla} v \, dy \,, \tag{95}
$$

where $\tilde{\beta}(y)$ is defined piecewise constant in Y^*. Due to linearity, we may define $\pi^\alpha, \xi \in H^1_{\#}(Y)$ such that

$$u^1 = \pi^\alpha \partial^x_\alpha u^0 + \mathrm{j}\omega \xi g^0 \,, \tag{96}$$

and they satisfy the following auxiliary problems:

$$\begin{aligned} a^*_Y\left(\pi^\beta + y_\beta, \phi\right) &= 0\,, \quad \forall \phi \in H^1_{\#}(Y)\,, \\ a^*_Y(\xi, \phi) &= \frac{1}{h}\gamma^{\pm}(\phi)\,, \quad \forall \phi \in H^1_{\#}(Y)\,. \end{aligned} \tag{97}$$

6.3.4 Macroscopic Wave Equation on Γ_0

The macroscopic equation governs the surface wave propagation. The limit of the Helmholtz equation reads as

$$\begin{aligned} &\partial_\alpha A_{\alpha\beta}\partial_\beta u^0 + \mathrm{j}\omega\partial_\alpha(B_\alpha g^0) + \\ &\qquad + \mu\omega^2\rho^* u^0 = \frac{\mathrm{j}\omega}{h}\sum_{s=+,-} \fint_{I^s_y} g^{1s} \end{aligned} \tag{98}$$

in Γ, where the homogenized coefficients are

$$\begin{aligned} A_{\alpha\beta} &= \frac{1}{|Y|}a^*_Y\left(\pi^\beta + y_\beta, \pi^\alpha + y_\alpha\right)\,, \\ B_\alpha &= \frac{1}{|Y|}a^*_Y(\xi, y_\alpha)\,. \end{aligned} \tag{99}$$

As the consequence of Remark 3, in fact $\alpha, \beta = 1$ and A, B are only scalar values. Also $\partial_2 u^0 = 0$ due to the TE-H-mode restriction.

6.3.5 Jump Condition

Using decomposition (96), from (90) for a.a. $x \in \Gamma_0$ we obtain

$$\begin{aligned} u^+ - u^- &= \varepsilon_0 \frac{1}{|I_y|}\gamma^{\pm}(u^1) \\ &= \varepsilon_0 \frac{1}{|I_y|}\left(\gamma^{\pm}(\pi^\alpha)\partial^x_\alpha u^0 + \mathrm{j}\omega\gamma^{\pm}(\xi) g^0\right) \\ &= \delta_0\left(-B_\alpha \partial^x_\alpha u^0 + \mathrm{j}\omega F g^0\right)\,, \end{aligned} \tag{100}$$

where (note $|Y| = |I_z||I_y|$ and $|I_z| = 1$)

$$F = \frac{1}{|Y|}a^*_Y(\xi, \xi) = \frac{1}{h|Y|\gamma^{\pm}(\xi)}\,. \tag{101}$$

Using auxiliary problems (97) one can verify that

$$\begin{aligned} a^*_Y(\xi, y_\alpha) &= -a^*_Y(\xi, \pi^\alpha) = -\frac{1}{h}\gamma^{\pm}(\pi^\alpha)\,, \\ \text{hence} \quad B_\alpha &= \frac{1}{|Y|}a^*_Y(\xi, y_\alpha) = -\frac{1}{h|I_y||Y_z|}\gamma^{\pm}(\pi^\alpha) \\ &= -\frac{1}{h|I_y|}\gamma^{\pm}(\pi^\alpha)\,, \end{aligned} \tag{102}$$

which was employed in (100).

6.3.6 Complete Homogenized Interface Conditions

They involve the in-plane limit electric Hertz potential u^0 (see the transformation (70)), the transformed tangential electric field components, $g^+ = g^0 + \varepsilon_0 g^{1+}$ and $g^- = -g^0 + \varepsilon_0 g^{1-}$ related to faces Γ^+ and Γ^-, respectively, where the fluctuating part is relevant for a given layer thickness $\delta_0 = \varepsilon_0 h > 0$. There is now discussion concerning the fluctuation parts $\varepsilon_0 g^{1s}$, $s = +, -$.

1. Let us consider the perfect continuity of normal derivatives according to (92). This is satisfied (in the sense of weak limits in $L^2(\Gamma_0)$) for the following two situations:

   a) for "the true limit case", $\varepsilon_0 = 0$, so that (92) holds for any $g^{1\pm}$ (since $\hat{g}^{\delta\pm} \rightharpoonup \pm g^0$ weakly in $L^2(\Gamma_0)$). In this case $g^{1\pm}$ is to be defined in (98).

   b) for the zero average in (98), i.e. assuming

   $$G^{\pm} \equiv \sum_{s=+,-} \fint_{I^s_y} g^{1s} = 0\,. \tag{103}$$

   In this case functions $g^{1\pm}$ are not present in the limit model.

2. Let us now consider (94). Since the boundaries $\Gamma^{\pm}_\delta$ are not related to any structural (material) discontinuity, the normal derivatives must be continuous. Thus, for $\varepsilon_0 > 0$, the external field gradients represented by $\mathscr{T}_\varepsilon(g^+) = g^0(x_\alpha) + \varepsilon_0 g^{1+}(x_\alpha, y)$ are related to $\partial^{\pm}_n u^{\pm}$ by

$$\begin{aligned} \partial^+_n|_{\Gamma^+} u^+ &= \mathrm{j}\omega\beta_0^2(g^0 + \varepsilon_0 \fint_{I^+_y} g^{1+})\,, \\ \partial^-_n|_{\Gamma^-} u^- &= \mathrm{j}\omega\beta_0^2(-g^0 + \varepsilon_0 \fint_{I^-_y} g^{1-})\,, \end{aligned} \tag{104}$$

These "external field boundary conditions" can be substituted in (94), therefore

$$\frac{1}{\delta_0 \beta_0^2}\left[\frac{\partial u^+}{\partial n^+}+\frac{\partial u^-}{\partial n^-}\right]= \\ =\frac{\mathrm{j}\omega}{\delta_0}\left(g^0+\varepsilon_0 \fint_{I_y^+} g^{1+}-g^0+\varepsilon_0 \fint_{I_y^-} g^{1-}\right) \\ \stackrel{!}{=} -\rho^* \mu \omega^2 u^0 , \tag{105}$$

hence the constraint

$$G^{\pm} \equiv \fint_{I_y^+} g^{1+} + \fint_{I_y^-} g^{1-} = \mathrm{j}\omega\rho^*\mu h u^0 \text{ a.e. on } \Gamma_0 . \tag{106}$$

We shall consider either (103) holds, so that the fluctuating parts are irrelevant in the limit situation, or (106) holds, which is an additional constraint. Therefore, the following problem is meaningful:
Let u^+ and u^- are given on faces Γ^+ and Γ^- of thin heterogeneous interface (with the thickness $\delta_0 << 1$) which is represented by surface (line in 2D – the relevant case) Γ_0 in the homogenized form. Denoting by $U_\#(\Gamma_0)$ the space of periodic functions on Γ_0, which is the consequence of periodic conditions $(81)_5$, we find $u^0 \in U_\#(\Gamma_0)$ and fluxes $g^0, G^{\pm} \in L^2(\Gamma_0)$ such that:

$$\int_{\Gamma_0} A_{\alpha\beta}\partial^x_\beta u^0 \partial^x_\alpha v^0 + \mathrm{j}\omega \int_{\Gamma_0} B_\alpha g^0 \partial^x_\alpha v^0 - \\ -\mu\omega^2 \int_{\Gamma_0} \rho^* u^0 v^0 - \frac{\mathrm{j}\omega}{h}\int_{\Gamma_0} G^{\pm} v^0 = 0 \quad \forall v^0 \in U_\#(\Gamma_0) , \\ \int_{\Gamma_0} \theta\left(-B_\alpha \partial^x_\alpha u^0 + \mathrm{j}\omega F g^0\right) = \frac{1}{\delta_0}\int_{\Gamma_0} \theta(u^+ - u^-) \\ \forall \theta \in L^2(\Gamma_0) , \\ G^{\pm} - \mathrm{j}\omega\zeta_0 \rho^* \mu h u^0 = 0 \quad \text{a.e. on } \Gamma_0 , \tag{107}$$

where $\zeta_0 = 0, 1$ in $(107)_3$, according to the case (103) and (106), respectively.

6.4 Cloaking Problem

The cloaking problem consists in finding model parameters related to some subdomain $\Omega^- \subset \Omega^G$ such that an object $\Omega_c \subset \Omega^-$ is not visible outside Ω^-, i.e. the *incident wave* imposed in $\Omega^+ = \Omega^G \setminus \Omega^-$ is not perturbed by a refracted field on $\Gamma_s \subset \partial\Omega^G$. The medium parameters in Ω^G are defined as piecewise constant functions (pcw. const. func.):

domain:	parameters of the medium:	description:
$\Omega_\delta^+, \Omega^+$	β_0^+, μ_0^+	const.
$\Omega_\delta^- \setminus \Omega_c$	β_0^-, μ_0^-	const.
$\Omega^- \setminus \Omega_c$	β_0^-, μ_0^-	const.
Ω_c	β, μ	pcw. const. func.
$\Omega_\delta^\varepsilon$	$\beta^{\varepsilon\delta}, \mu^{\varepsilon\delta}$	pcw. const. func.

We shall discus the following alternative definition of the cloaking problem with heterogeneous transmission layer:

1. **the δ-formulation – the layer is not homogenized**, $\Omega^G = \Omega_\delta^- \cup \Omega_\delta \cup \Omega_\delta^+$ (disjoined subdomains) and the observation manifold $\Gamma_s \subset \partial\Omega^G$ is located far away from Ω_δ^-.
2. **the homogenized formulation with the far-field cloaking effect**, i.e. the layer is represented by homogenized material distributed on $\Gamma_0 = \partial\Omega^+ \cap \partial\Omega^-$ and the manifold Γ_s is defined as above.
3. **the homogenized formulation with the strong cloaking effect**, in this case the cloaking effect is examined on the "exterior surface" of Γ^+, thus no scattered field component is observed in Ω^+.

In general there is the scattered field in Ω^+ given as $u^{\rm sc} = u - u^{\rm inc}$, i.e as the subtraction of the total and the incident field. A physically reasonable measure of the cloaking effect is the extinction function defined for a cylindric particle of unit length as:

$$Q^{\rm ext}_{\partial\Omega} = \frac{1}{d}\mathrm{Re}\left\{\int_{\partial\Omega}\left(\mathbf{n}\cdot\mathbf{d}u^{\rm inc}\overline{u^{\rm sc}} + \frac{\mathrm{j}\gamma}{k^{\rm inc}}u^{\rm sc}\overline{u^{\rm inc}}\right)\,\mathrm{d}l\right\}. \tag{108}$$

where $k^{\rm inc}$ is the incidence wavenumber, d is the effective diameter of the cross-sectional area of the particle projected onto a plane perpendicular to the direction of propagation $\mathbf{d}$ and $\gamma = \mathrm{j}\kappa + \frac{1}{2R}$, $\mathbf{n}$ is the outer normal unit vector, R is the radius of $\partial\Omega^G$.
The extinction function will be derived and its structure explained in the next section.

6.4.1 Far Field Cloaking Observation for Non-Homogenized Layer

The cloaking structure is situated in domain Ω_δ which is locally periodic in the sense we discussed above. The global domain, Ω^G, consists of three disjoint parts: $\Omega^G = \Omega_\delta^+ \cup \Omega_\delta \cup \Omega_\delta^-$, see Fig (**15**). The objects to conceal are located in Ω_δ^-, whereas on Γ_s the cloaking effect is evaluated using extinction function (108).

We assume that in Ω_δ^+ the material is homogeneous (material parameters labeled by subscript 0), whereas in $\Omega_\delta \cup \Omega_\delta^-$ the material is heterogeneous in general. However, to be consistent with the assumption considered in the next paragraph, we require that $\mu = \mu_0^\pm$, $\beta = \beta_0^\pm$ and $\sigma = \sigma_0^\pm$ on the respective interfaces $\Gamma_\delta^\pm$. The state problem has the following structure:

$$\frac{1}{\beta_0^+}\nabla^2 u^{\delta+} + \omega^2 \mu_0 u^{\delta+} = 0 \quad \text{in } \Omega_\delta^+ ,$$
$$\nabla\cdot\left(\frac{1}{\beta}\nabla u^{\delta-}\right) + \omega^2 \mu u^{\delta-} = 0 \quad \text{in } \Omega_\delta^- ,$$
$$\nabla\cdot\left(\frac{1}{\beta}\nabla u^{\delta}\right) + \omega^2 \mu u^{\delta} = 0 \quad \text{in } \Omega_\delta ,$$

standard transmission conditions:

$$\partial_n(u^{\delta+} - u^\delta) = 0 \text{ on } \Gamma_\delta^+ ,$$
$$\partial_n(u^{\delta-} - u^\delta) = 0 \text{ on } \Gamma_\delta^- ,$$
$$u^{\delta+} - u^\delta = 0 \text{ on } \Gamma_\delta^+ ,$$
$$u^{\delta-} - u^\delta = 0 \text{ on } \Gamma_\delta^- ,$$

boundary conditions:

$$\partial_n u^{\text{sc}} - \gamma u^{\text{sc}} = 0 \text{on} \, \partial\Omega^G,$$

where

$$u^{\text{sc}} = u^{\delta+} - u^{\text{inc}}. \tag{109}$$

The cloaking effect can be achieved by minimization of $Q_{\Gamma_s}^{\text{ext}}(u^{\text{inc}}, u^{\text{sc}})$.

The corresponding optimization problem can be treated as a *free material optimization problem* as follows.

$$\begin{cases} \min\limits_{\beta,\mu} Q_{\Omega_s}(u^{\text{inc}}, u^{\text{sc}}) \text{ s.t.} \\ (u^{\delta+}, u^{\delta-}, u^\delta) \text{ satisfies (109)} \\ (\beta, \mu) \in \mathscr{U}_{\text{ad}} , \end{cases} \tag{110}$$

where $\mathscr{U}_{\text{ad}}$ has to be specified. In particular, β, μ are fixed on the object to be cloaked (Ω_c) an can be chosen out of a set of materials in $\Omega_\delta^- \setminus \Omega_c =: \Omega_{\delta,c}^-$. The

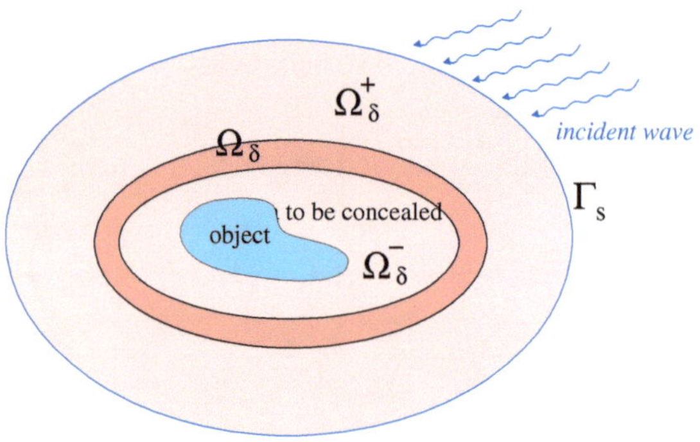

Fig. (**15**): Illustration to the cloaking problem formulation: for finite thickness layer Ω_δ, (109). Domain Ω^- contains the object to be cloaked by surface Γ_0 containing the metamaterial. Cloaking effect is evaluated on Γ_s.

so-called free material optimization problem would amount to require

$$\mathscr{U}_{\text{ad}} := \{a \in L^\infty(\Omega_{\text{ design}}; \mathbb{S}^3) | a_l \le a \le a_u, \text{tr } a \le V\}$$

for positive semi-definite matrices $a_0, a_u \in \mathbb{S}^3$. Existence of solutions and approximation properties with respect to H-convergence have been shown in a different context by Haslinger, Kocvara, Leugering, Stingl[18]. However, the realization of H-limits is well known to be a nontrivial problem. See however [18] for a numerical approximation analysis. The application of free material optimization to the cloaking problem (110) is under way. An alternative to treat the cloaking problem for (109) is to parametrize the material properties as well as the shapes of the inclusions and possible holes in the layer $\Omega_{\delta,c}^-$ and view the problem as a nonlinear finite dimensional constrained optimization problem in reduced form, in which the the problem (109) is solved for the given data and parameter set. In particular on the level of a suitable finite-element-discretization one can derive sensitivities of the cost-function with respect to the parameters by fairly standard means. Again, the numerical treatment is under way.

6.4.2 Far Field Cloaking Observation for Homogenized Layer

We consider the domain $\Omega^G = \Omega^+ \cup \Omega^- \cup \Gamma_0$, where $\Gamma_0 = \partial\Omega^+ \cap \partial\Omega^-$ can be curved as the straightforward generalization of the transmission layer model. Therefore, we shall introduce the the local coordinate system $(\tau, \nu)_X$ for any $X \in \Gamma_0$ where τ and ν are, respectively, the coordinates in the tangential and normal directions w.r.t. curve Γ_0 at position

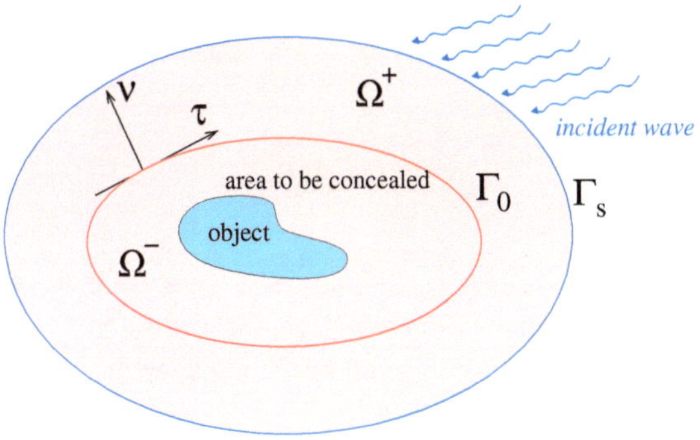

Fig. (**16**): Illustration to the cloaking problem formulation: for the homogenized layer represented by Γ_0, (111)

X. As above, the objects to conceal are located in Ω^-, see Fig. (**15**). On the rest of $\partial\Omega^G$, the radiation condition can be prescribed. The total field in $\Gamma_s \subset \partial\Omega^G$ is obtained by solving the following problem (we assume that in Ω^+ the medium is homogeneous, possibly air):

$$\begin{aligned} &\frac{1}{\beta_0^+}\nabla^2 u^+ + \omega^2\mu_0 u^+ = 0 \quad \text{in } \Omega^+ , \\ &\nabla\cdot\left(\frac{1}{\beta}\nabla u^-\right) + \omega^2\mu u^- = 0 \quad \text{in } \Omega^- , \end{aligned} \tag{111}$$

transmission conditions – Neumann type:

$$\begin{aligned} -\partial_\nu u^+ &= \partial_\nu^+ u^+ = -j\omega\beta_0^+ g^0 \quad \text{on } \Gamma_0 , \\ \partial_\nu u^- &= \partial_\nu^- u^- = j\omega\beta_0^- g^0 \quad \text{on } \Gamma_0 , \end{aligned} \tag{112}$$

wave transmission through the layer – jump control:

$$\begin{aligned} &\partial_\tau\left(A\partial_\tau u^0 + j\omega B g^0\right) + \omega^2\mu\rho^* u^0 = 0 \quad \text{on } \Gamma_0 , \\ &j\omega B\partial_\tau u^0 + \omega^2 F g^0 = -\frac{j\omega}{\delta_0}(u^+ - u^-) \\ &\qquad \text{on } \Gamma_0 , \end{aligned} \tag{113}$$

boundary conditions:

$$\partial_n u^{sc} - \gamma u^{sc} = 0 \,\text{on}\, \partial\Omega^G. \tag{114}$$

Above in the wave transmission condition we employed (107) with $G^\pm = 0$, i.e. $\zeta_0 = 0$.
As well as in the previous case, in this situation, the cloaking effect can be achieved by minimization of $Q_{\Gamma_s}^{ext}(u^{inc}, u^{sc})$. In other words, one is looking for the solutions of the following problem

$$\begin{cases} \min\limits_{\beta,\mu} Q_{\Gamma_s}^{ext}(u^{inc}, u^{sc}) \text{ s.t.} \\ (u^{\delta+}, u^{\delta-}, u^{\delta}) \text{ satisfies } (111)-(114) \\ (A, B, F, \beta, \mu) \in \mathscr{U}_{ad} , \end{cases} \tag{115}$$

where the optimization is with respect to a class of admissible functions A, B, F appearing in the transmission condition and μ, β as before. In order to understand in particular the transmission conditions along Γ_0 in (113) we focus on

$$\begin{aligned} &\partial_\tau\left(A\partial_\tau u^0 + j\omega B g^0\right) + \omega^2\mu\rho^* u^0 = 0 \quad \text{on } \Gamma_0 , \\ &j\omega B\partial_\tau u^0 + \omega^2 F g^0 = -\frac{j\omega}{\delta_0}(u^+ - u^-) \\ &\qquad \text{on } \Gamma_0 . \end{aligned} \tag{116}$$

The first equation contains a Laplace-Beltrami-Helmholtz equation on Γ_0. Indeed, we define the operator

$$\begin{aligned} &T_A : L^2(\Gamma_0) \to L^2(\Gamma_0), \\ &D(T_A) := \{u \in H^1_\#(\Gamma_0) | A\partial_\tau u \in H^1(\Gamma_0)\}, \\ &T_A u := -\partial_\tau A \partial_\tau u \end{aligned} \tag{117}$$

The operator T_A is self-adjoint and positive semi-definite with discrete spectrum. The equation to solve is now

$$-T_A u + \omega^2\mu\rho^* u = -j\omega\partial_\tau B g^0.$$

We introduce the resolvent $R(\lambda, T_A) := (\lambda I - T_A)^{-1}$ of T_A at a point $\lambda \in \rho(T_A)$. With this notation the first equation in (116) can be solved for u^0 as follows.

$$u^0 = -j\omega R(\omega^2\mu\rho^*, T_A)\partial_\tau B g^0, \tag{118}$$

while the second equation in (116) turns into

$$B\partial_\tau R(\omega^2\mu\rho^*)\partial_\tau B g^0 + F g^0 = \frac{1}{j\omega}(u^+ - u^-), \quad \text{on } \Gamma_0. \tag{119}$$

Equation (119) is an integral equation of the second kind which admits a unique solution g^0. If one then inserts g^0 into the Neumann conditions of (112) one obtains a nonlocal transmission condition along Γ_0 which contains the functions A, B, F, μ as material parameters to be used in the optimization. The optimization problem (115) has not yet been fully explored. This will be subject to a forthcoming publication.

6.4.3 Strong Form of the Cloaking Problem

We keep the domain $\Omega^G = \Omega^+ \cup \Omega^- \cup \Gamma_0$, the objects to conceal are located in Ω^-, as before. The incident wave is imposed in Ω^+. We impose the incident wave in Ω^+; let u^{inc} be the local amplitude of the plane wave, then

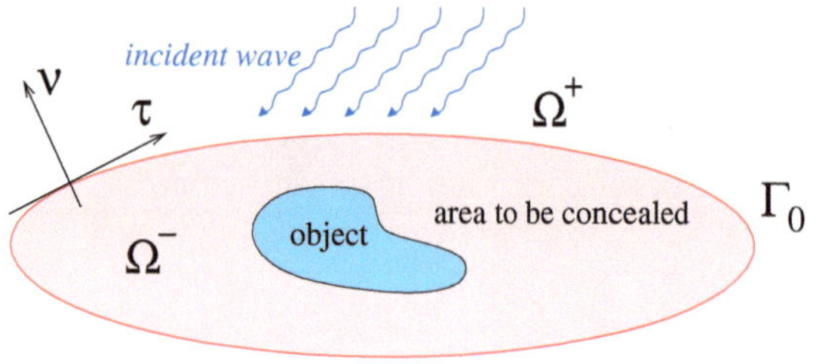

Fig. (**17**): Illustration to the problem formulation (121). Domain Ω^- contains the object to be cloaked by surface Γ_0 containing the metamaterial.

$$\mathrm{j}\omega\beta_0 g^0 = \partial_\nu^- u^- = \partial_\nu^+ u^+ = -k_\nu^+ u^{\mathrm{inc}} , \qquad (120)$$

where $u^\pm$ is the trace of u on $\partial\Omega^\pm \cap \Gamma_0$ and k_ν is the projection of the wave vector to the unit outward normal ν. Above $\int_{I_y^+} g^{1+} = 0$ applies due to the form of the incident wave. As the consequence, $\int_{I_y^-} g^{1-} = 0$ results by $G^\pm \equiv 0$, see (106) and (107). We consider the problem imposed in Ω^-, being defined in terms of triplet (u, u^0, g^0) which satisfies the following coupled system: wave in cloaked region:

$$\begin{aligned} \nabla\left(\frac{1}{\beta}\nabla u\right) + \omega^2\mu u = 0 \text{in } \Omega^- \\ \partial_\nu u = \mathrm{j}\omega\beta_0^- g^0 \text{on } \Gamma_0, \end{aligned} \qquad (121)$$

wave transmission through the layer:

$$\begin{aligned} \partial_\tau\left(A\partial_\tau u^0 + \mathrm{j}\omega B g^0\right) + \omega^2\mu\rho^* u^0 = 0 \quad &\text{on } \Gamma_0 , \\ \mathrm{j}\omega B\partial_\tau u^0 + \omega^2 F g^0 = -\frac{\mathrm{j}\omega}{\delta_0}(u^{\mathrm{inc}} - u) \\ &\text{on } \Gamma_0 . \end{aligned} \qquad (122)$$

In fact the *cloaking condition* (120) can be viewed as an exact controllability constraint with variables (A, B, F), the coefficients of the homogenized transmission through the heterogeneous layer, as controls. This exact controllability problem can be solved for special scenarios. However, in general we cannot expect exact controllability, and therefore the controllability constraint has to be relaxed by an appropriate optimization with penalty.
In general, the flux g^0 obtained by solving (121), (122) i.e. as the *State Problem* solution, is not consistent with the incident wave assumed in Ω^+; it fits the assumption of "no reflection", when

$$0 = k_\nu^+ u^{\mathrm{inc}} + \mathrm{j}\omega\beta_0 g^0 , \qquad \text{a.e. on } \Gamma_0 ,$$

therefore, the *cloaking effect* can be approached by the following minimization:

$$\min_{A,B,F} \Psi(g^0, A, B, F) , \qquad (123)$$

where $\Psi = \|k_\nu^+ u^{\mathrm{inc}} + \mathrm{j}\omega\beta_0 g^0\|_{\Gamma_0}$, s.t. g^0 solves the *State Problem* (121) with (122) for given u^{inc}. Coefficients (A, B, F) can be handled by designing the microstructure in cell Y.

Remark 4. In general, there is the jump on Γ_0, $[u]_{\Gamma_0} = u^+ - u^- \not\equiv 0$. u^0 involved in (122) is an internal variable which is relevant only if $B \not\equiv 0$ on Γ_0; otherwise (122) reduces to

$$\omega^2 F g^0 + \frac{\mathrm{j}\omega}{\delta_0}(u^{\mathrm{inc}} - u) = 0 \quad \text{on } \Gamma_0 .$$

In this case the problem (121), (122) reduces to a Helmholtz-problem

$$\begin{cases} \nabla\frac{1}{\beta}\nabla u + \omega^2\mu u = 0 \ \text{ in } \Omega^- \\ \partial_\nu u + \alpha u = \alpha u^{\mathrm{inc}} \ \text{ on } \Gamma_0 \end{cases}$$

with local Robin-type boundary condition on Γ_0. The cloaking constraint then also reduces to just another boundary condition on Γ_0. This leads to an overdetermined boundary value problem which may or not may have a solution.

△

7. TOPOLOGY OPTIMIZATION FOR THE CLOAKING PROBLEM

In this section we would like to demonstrate the topology optimization method to design a cloaking layer such that the given object will become less visible.
Let us consider a small object (i.e. a nanoparticle composed from a given material). Our aim is to design a topology of a cloaking layer (composite of the matrix medium and a medium with a low refractive index) in such a way that for an observer (sensor) present behind the particle, the particle becomes in some sense (specified by a cost function) invisible. Propagation of the electromagnetic waves in the composite is described by the Helmholtz equation (as defined in the previous sections). The geometry of the problem is described by figure (**18**). The state equation is considered in a circular domain $\overline{\Omega} = \cup_{i=1}^3 \overline{\Omega}_i$ with the boundary $\partial\Omega$. We place a particle (characterized by a complex refractive index) in the middle of the computational domain. Its body

is included in the set $\overline{\Omega}_1$. The particle is coated by a shell (Ω_2). And the core-shell is in turn embedded into a matrix medium (Ω_3).

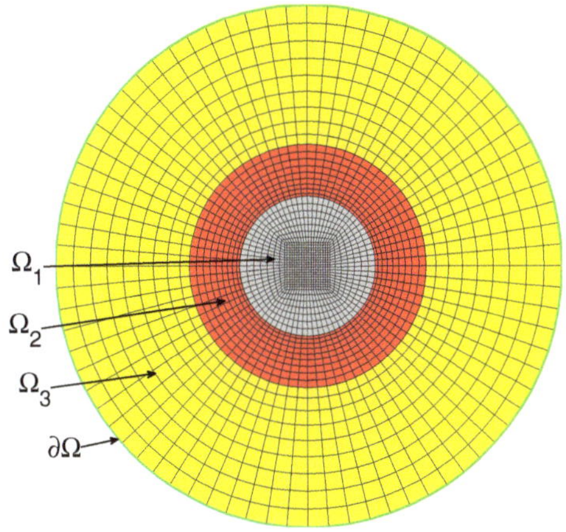

Fig. (**18**): Description of geometry for the strong form of cloaking problem

The refractive index is supposed to be constant in subdomains Ω_1, Ω_3, but is changing across interfaces and may vary in Ω_2. Since we will solve the Helmholtz equation on a finite computational domain we have to define appropriate boundary conditions. These conditions should prevent occurrence of non-physical reflections from the artificial boundary (i. e. the outer boundary should be transparent for the scattered field or the boundary conditions should absorb the scattered wave, that's why in the following we will call them absorbing boundary conditions). There are various ways in which such conditions can be chosen, we have used a.b.c. of first order for it's simplicity, these conditions retains sparsity of the finite element system matrix, on the other hand, they do not prevent reflections for all directions of incidence. The total rescaled electric Hertz potential u may be decomposed into the incident and the scattered field

$$u = u^{\text{inc}} + u^{\text{sc}}, u^{\text{inc}} = e^{-\text{j}\kappa^{\text{inc}}\mathbf{d}\cdot\mathbf{x}}, \tag{124}$$

where $\mathbf{d}$ is the direction of propagation of the incident wave. Furthermore we observe

$$\nabla u^{\text{inc}} = -\text{j}\kappa^{\text{inc}}\mathbf{d}u^{\text{inc}}. \tag{125}$$

The absorbing b.c. give the relation between the scattered field and its derivative in the direction of the outer normal on the boundary

$$\partial_n u^{\text{sc}} - \gamma u^{\text{sc}} = 0 \,\text{on}\, \partial\Omega, \tag{126}$$

where $\gamma = \text{j}\kappa + \frac{1}{2R}$. The Helmholtz equation has than the following form

$$\left\{\begin{aligned} \nabla\cdot\left(\frac{1}{\beta_r}\nabla u\right) + \kappa_0^2\mu_r u &= 0 \,\text{in}\, \Omega, \\ \left[\frac{1}{\beta_r}\partial_n u\right]_\Gamma &= 0 \,\text{on}\, \Gamma, \\ [u]_\Gamma &= 0 \,\text{on}\, \Gamma, \\ \partial_n u^{\text{sc}} - \gamma u^{\text{sc}} &= 0 \,\text{on}\, \partial\Omega, \end{aligned}\right. \tag{127}$$

where $\beta_r = n^2$ is the complex relative permittivity (square of the refractive index).

Remark 5. The total (or also scattered) potential $u(u^{\text{sc}})$ depends generally on the frequency $\omega \in \Lambda$ and on the direction of propagation $\mathbf{d} = (\cos\alpha, \sin\alpha), \alpha \in \Sigma$, where $\Lambda = \{\omega_1, \dots, \omega_n\}$ is a set of given frequencies, $\Sigma = \{\alpha_1, \dots, \alpha_m\}$ is the set of angles of incidence.

△

To obtain the weak form of Helmholtz equation we multiply $(127)_1$ by the test function $v \in H(\Omega)$, where $H(\Omega)$ is the standard Sobolev space

$$H(\Omega) = W^{1,2} = \left\{v | v, \frac{\partial v}{\partial x_i} \in L^2(\Omega), i = 1, 2\right\}. \tag{128}$$

We apply the Green's theorem, further we use (124), $(127)_4$ and (125). Then the weak formulation may be written as follows

$$\left\{\begin{array}{l} \text{Find } u^{\text{sc}} \in H(\Omega) \text{ such that for all } v \in H(\Omega) \text{ holds} \\ a(u^{\text{sc}}, v) = f(v). \end{array}\right. \tag{129}$$

where a sesquilinear form $a : H \times H \to \mathbf{C}$ is defined as

$$a(u^{\text{sc}}, v) = -\int_\Omega \frac{1}{\beta_r}\nabla u^{\text{sc}}\overline{\nabla v}\,\text{d}S + \int_\Omega \kappa_0^2\mu_r u^{\text{sc}}\overline{v}\,\text{d}S + \int_{\partial\Omega}\frac{1}{\beta_r}\gamma u^{\text{sc}}\overline{v}\,\text{d}l \tag{130}$$

and the operator $f(\cdot)$ is the operator of the right hand side $f : H \to \mathbf{C}$

$$f(v) = \int_\Omega \frac{1}{\beta_r}\nabla u^{\text{inc}}\overline{\nabla v}\,\text{d}S - \int_\Omega \kappa_0^2\mu_r u^{\text{inc}}\overline{v}\,\text{d}S + \int_{\partial\Omega}\frac{1}{\beta_r}\mathbf{n}\cdot\mathbf{d}\text{j}\kappa^{\text{inc}}u^{\text{inc}}\overline{v}\,\text{d}l \quad \forall v \in H(\Omega). \tag{131}$$

7.1 Cost Functional

Our aim is to minimize the so-called extinction efficiency. That is a function that reflects energy loss due to the inserted particle.
Energy flux at any point of space is represented by the Poynting vector

$$\mathbf{S} = \frac{1}{2}\mathrm{Re}\left\{\boldsymbol{E} \times \overline{\boldsymbol{H}}\right\}. \quad (132)$$

In the following we will define the energy that is scattered, absorbed and extincted per unit length of the cylinder L. We will ignore effects of the ends of the cylinder. Now imagine a fictive cylinder around the particle (in our concept it will be represented by the boundary of the computational domain $\partial\Omega$). We define net rate W^{abs} at which the electromagnetic energy crosses $\partial\Omega$

$$W^{\mathrm{abs}} = -L\int_{\partial\Omega} \mathbf{S}\cdot\mathbf{n}\,\mathrm{d}l. \quad (133)$$

If $W^{\mathrm{abs}} > 0$ energy is absorbed in Ω, if $W^{\mathrm{abs}} < 0$ energy is created in Ω (not considered in the following).
The absorbed energy rate W^{abs} may be decomposed into the incident energy rate (identically zero), extincted and scattered energy rates

$$W^{\mathrm{abs}} = W^{\mathrm{inc}} + W^{\mathrm{ext}} - W^{\mathrm{sc}} \quad (134)$$

Extinction efficiency is then defined as

$$Q^{\mathrm{ext}} = \frac{1}{GI^{\mathrm{inc}}}W^{\mathrm{ext}}, \quad (135)$$

where I^{inc} is incident irradiance - magnitude of the Poynting vector of the incident wave

$$I^{\mathrm{inc}} = |\mathbf{S}^{\mathrm{inc}}| = \frac{1}{2}\left|\mathrm{Re}\left\{\boldsymbol{E}^{\mathrm{inc}} \times \overline{\boldsymbol{H}}^{\mathrm{inc}}\right\}\right| \quad (136)$$

and $G = Ld$ is cross-sectional area of the particle projected onto a plane perpendicular to the direction of propagation (d is the diameter of the shelled particle).
In the following we will formulate the extinction efficiency in terms of the state variable u^{sc}. The magnetic end electric field intensities for a homogeneous and non-absorbing medium ($\beta = const > 0$) may be rewritten as follows

$$\boldsymbol{E} = \frac{1}{\omega^2\beta}\nabla\times(u\mathbf{e}_3) = -\frac{1}{\omega^2\beta}\mathbf{e}_3\times\nabla u, \quad (137)$$

$$\boldsymbol{H} = (\sigma - \mathrm{j}\varepsilon\omega)\frac{1}{\omega^2\beta}u\mathbf{e}_3 = -\frac{\mathrm{j}}{\omega}u\mathbf{e}_3. \quad (138)$$

Then the Poynting vector may be rewritten as follows (noting that $\mathbf{e}_3\cdot\nabla u = 0$)

$$\begin{aligned}\mathbf{S} &= -\frac{1}{2\omega^3\beta}\mathrm{Re}\left\{\mathrm{j}\,(\mathbf{e}_3\times\nabla u)\times\overline{u}\mathbf{e}_3\right\},\\ &= -\frac{1}{2\omega^3\beta}\mathrm{Re}\left\{\mathrm{j}\overline{u}\nabla u\right\}.\end{aligned} \quad (139)$$

The incident irradiance is then given by (using (125))

$$I^{\mathrm{inc}} = \frac{1}{2}\left|\mathrm{Re}\left\{\boldsymbol{E}^{\mathrm{inc}}\times\overline{\boldsymbol{H}^{\mathrm{inc}}}\right\}\right| = \frac{k^{\mathrm{inc}}}{2\omega^3\beta}. \quad (140)$$

Using (139) also extinction energy rate is obtained as (using $(127)_4$ and again (125))

$$\begin{aligned}W^{\mathrm{ext}} &= \frac{L}{2\omega^3\beta}\int_{\partial\Omega}\mathrm{Re}\left\{\mathrm{j}\overline{u^{\mathrm{sc}}}\nabla u^{\mathrm{inc}} + \mathrm{j}\overline{u^{\mathrm{inc}}}\nabla u^{\mathrm{sc}}\right\}\cdot\mathbf{n}\,\mathrm{d}l,\\ &= \frac{L}{2\omega^3\beta}\int_{\partial\Omega}\mathrm{Re}\left\{\mathbf{n}\cdot\mathbf{d}k^{\mathrm{inc}}u^{\mathrm{inc}}\overline{u^{\mathrm{sc}}} + \mathrm{j}\gamma u^{\mathrm{sc}}\overline{u^{\mathrm{inc}}}\right\}\mathrm{d}l.\end{aligned} \quad (141)$$

Using (135) the final formula for the extinction efficiency is obtained as

$$Q^{\mathrm{ext}} = \frac{1}{d}\mathrm{Re}\left\{\int_{\partial\Omega}\left(\mathbf{n}\cdot\mathbf{d}u^{\mathrm{inc}}\overline{u^{\mathrm{sc}}} + \frac{\mathrm{j}\gamma}{k^{\mathrm{inc}}}u^{\mathrm{sc}}\overline{u^{\mathrm{inc}}}\right)\mathrm{d}l\right\}. \quad (142)$$

7.2 Min-Max Problem

The aim of the optimization is to minimize values of the cost functional for a selected interval of frequencies. It can be achieved by the worst scenario approach: we shall minimize the cost functional value for the worst case frequency.
We would like to find an optimal distribution of two isotropic materials characterized with refractive indices n_0, n_1. This leads to the discrete optimization, which is generally a very difficult problem. One possibility to handle this problem is to introduce relaxation of the material (the SIMP method, [5]). We define pseudo density function $\rho(\mathbf{x}) \in \mathscr{U}_{ad}$

$$\begin{aligned}n(\rho(\mathbf{x}),\omega) &= n_0(\omega) + (n_1(\omega) - n_0(\omega))\rho(\mathbf{x})^p,\\ &\quad p > 1,\\ \mathscr{U}_{ad} &= \left\{\frac{1}{|\Omega_2|}\int_{\Omega_2}\rho(\mathbf{x})\,\mathrm{d}S \le \rho^*,\right.\\ &\qquad \left. 0 \le \rho(\mathbf{x}) \le 1, \mathbf{x}\in\Omega_2\right\},\end{aligned} \quad (143)$$

where $\mathscr{U}_{ad}$ is the admissible set, ρ^* is the maximal fraction of the material with refractive index n_1 that may be included in the design layer. The worst scenario approach may be formulated as follows

$$\min_{\rho\in\mathscr{U}_{ad}} \max_{\omega\in[\omega_1,\omega_n],\alpha\in[\alpha_1,\alpha_m]} \Psi(u^{sc}_{\omega,\alpha}), \tag{144}$$

where Ψ is the cost functional depending on the state variable.

For the finite element analysis we have to define the discrete form of the previous problem. Let E be a set of indices of finite elements in the design subdomain Ω_2. Then the refractive index for every finite element in E is defined as follows

$$\begin{aligned} n_e(\omega) &= n_0(\omega)+(n_1(\omega)-n_0(\omega))\rho_e^p,\ p>1,\\ \hat{\rho}(\mathbf{x}) &= \sum_{e\in E}\rho_e\chi_e(\mathbf{x}), \hat{\rho}\in\widetilde{\mathscr{U}_{ad}}\\ \widetilde{\mathscr{U}_{ad}} &= \Big\{card(E)\sum_{e\in E}\rho_e\le\rho^*,\\ &\quad 0\le\rho_e\le 1\,\forall e\in E\Big\}, \end{aligned} \tag{145}$$

where χ_e is a characteristic function of the finite element e in Ω_2, $card(E)$ is the amount of finite elements in the design layer. The problem (144) is then in then reformulated as follows

$$\min_{\hat{\rho}\subset\widetilde{\mathscr{U}_{ad}}} \max_{\omega\in\Lambda,\alpha\in\Sigma} \Psi(u^{sc}_{\omega,\alpha}). \tag{146}$$

The Method of Moving Asymptotes (MMA) is used to solve the preceding problem. One additional reformulation of (146) is necessary

$$\min_{\hat{\rho}\in\widetilde{\mathscr{U}_{ad}}} c \tag{147}$$

subject to:

$$\begin{aligned} h_{i,j} &\le 0, \quad i=1,\dots,n, j=1,\dots,m,\\ g &\le 0,\\ 0\le \rho_e &\le 1, \quad \forall e\in E, \end{aligned} \tag{148}$$

where

$$\begin{aligned} h_{i,j} &= \Psi(u^{sc}_{\omega_i,\alpha_j}), \omega_i\in\Lambda, \alpha_j\in\Sigma\\ &\quad \text{for } i=1,\dots,n, j=1,\dots,m,\\ g &= \frac{1}{card(E)}\Sigma_{e\in E}\rho_e-\rho^*. \end{aligned} \tag{149}$$

The MMA method requires knowledge of the gradient of the cost functional which is obtained via the sensitivity analysis. Sensitivity analysis of similar problems is provided in a detailed way in [40, 32].

The main task is the solution of the adjoint equations (that are in fact optimality conditions of the Lagrangian $\mathscr{L}$ of our problem), the equations are formally defined as follows

$$\begin{cases} \text{Find } w\in H(\Omega) \text{ such that for all } v\in H(\Omega) \text{ holds}\\ (\delta_{Re}\psi(u^{sc})-j\delta_{Im}\psi(u^{sc}))\cdot v+a(v,w)=0. \end{cases} \tag{150}$$

Then the final sensitivity of the cost functional for a given frequency ω and an angle of incidence α is

$$\begin{aligned} \delta\psi = \delta\mathscr{L}(\rho,u^{sc},w) &= \delta_\rho\left(a(u^{sc},w)-f(w)\right)\\ &= \delta_\rho\left(a(u,w)\right),\\ &= \int_{\Omega_D} -2n_e(\rho_e)^{-3}p(n_1-n_0)\rho_e^{p-1}\nabla u\overline{\nabla w}\,dS. \end{aligned} \tag{151}$$

7.3 Implementation and Results

The discretization of the state equations was done by the classical approach of the finite element method (for details we recommend the well known book of Zienkiewicz et al. [47]). The state equation is solved by the finite element method using isoparametric, bilinear, hexahedral finite elements (an introduction is given by Jianming Jin in [19]).

In all following examples the extinction efficiency was minimized ($\Psi = Q^{ext}$), although the scattering efficiency would be also a good alternative, since

$$Q^{ext} = Q^{abs} + Q^{sc} \tag{152}$$

and we observed the decrease of the extinction was mainly due to lower scattering than absorption.

On figure (**19**) we may observe a particle with higher refractive index (2.1) that is surrounded by the layer with refractive index given by the pseudo density $\rho = 0.3$. Dark blue color in the shell corresponds to the matrix material ($n = 1.31$), by the red color low refractive index material is represented ($n = 0.95$, that is more or less air). We see that the design evolves to two oval inclusions (**24**), which maintains more than 60 % decrease in extinction.

The extinction efficiency curves for particular iterations are displayed on figure (**25**). The pink interrupted curve corresponds to the bare particle.

The inclusions in the final design (**24**) have no clear interface with respect to the matrix medium. The production of such shell is out of reach of nowadays technology. Our suggestion is to use the optimal topology design as an initial guess for the shape optimization method. On Figures (**26**), (**27**) the contour lines and initial shape of 3 layers with piece-

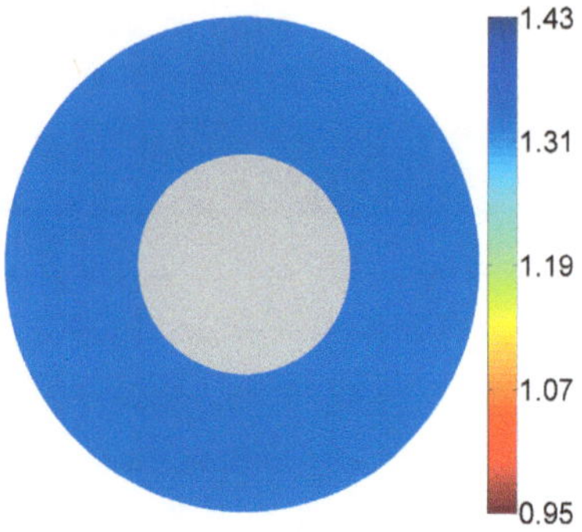

Fig. (**19**): Initial design - iteration 0.

Fig. (**20**): Design - iteration 6.

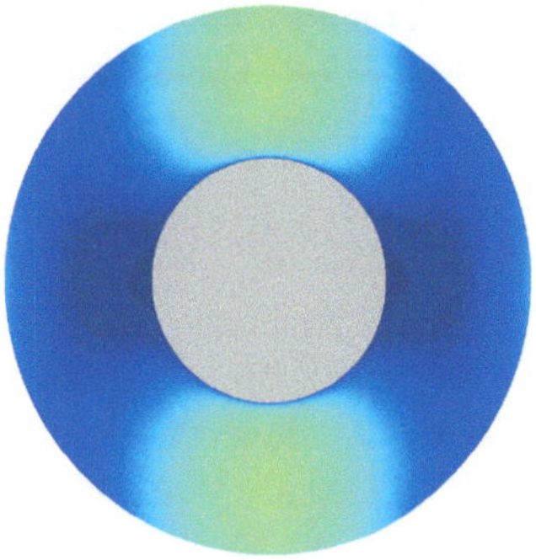

Fig. (**21**): Design - iteration 9.

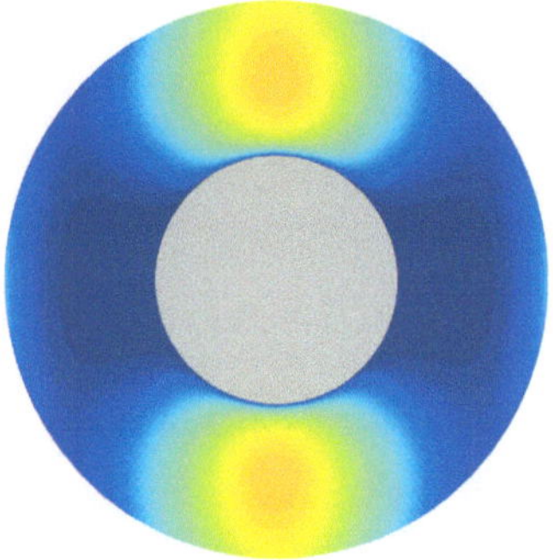

Fig. (**22**): Design - iteration 12.

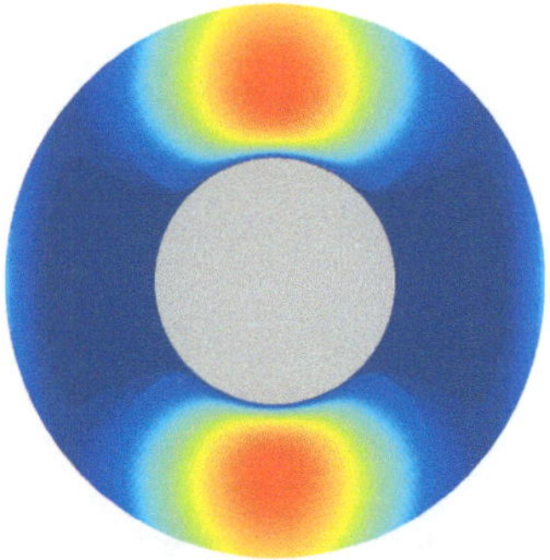

Fig. (**23**): Design - iteration 14.

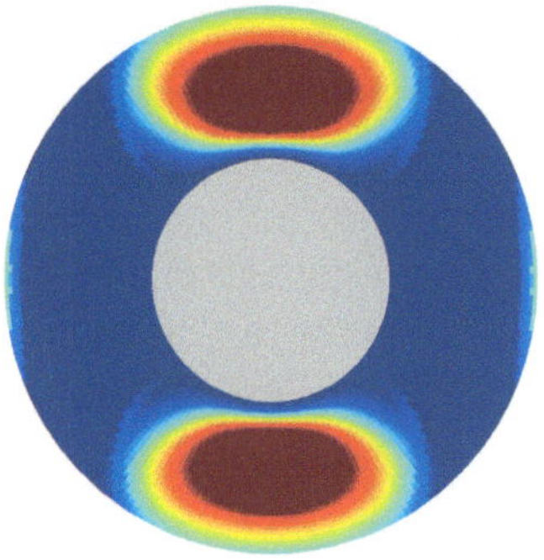

Fig. (**24**): Design - iteration 18.

wise constant refractive index are defined. The geometry of such structure could be parametrized and optimized in a similar way as in [39, 40].

The optimal design for 2, 3 and 4 angles of incidence is displayed in Figures (**28**), (**29**) and (**30**). The decrease of extinction is not so huge as in the previous simulation, but we still get improvement of approximately 20-40 %. The complicated structures that develop give us a hint back to the previous section (Fig. (**15**)). We believe that the optimally designed micro-structure would reduce the extinction even more significantly than shown in Fig. (**25**).

CONCLUSION

As we have amply demonstrated, meta-materials in the acoustic, electromagnetic, elastic and piezoelectric context can be approached by quite analogous mathematical methods. Therefore, a unifying theory of meta-materials for wave propagation is within reach. It turns out that micro- or nanostructured layers play an important role in obtaining meta-properties, like cloaking and band-gap phenomena. Similarly, micro-structures appear in auxetic elastic materials, like metallic or ceramic foams. In order to achieve results that lead to a an actual

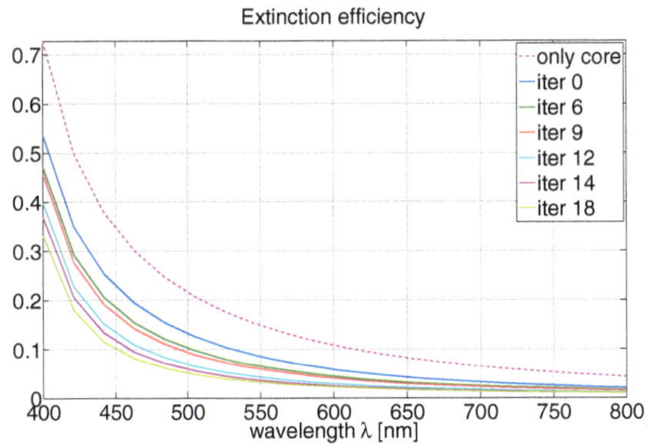

Fig. (**25**): Cost functional values for particular iterations

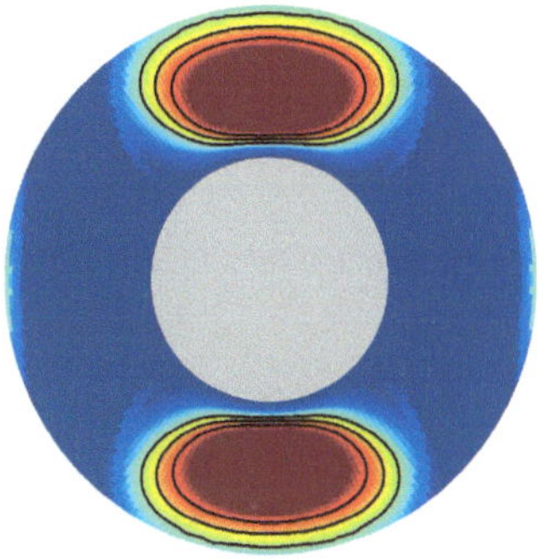

Fig. (**26**): Contour lines..

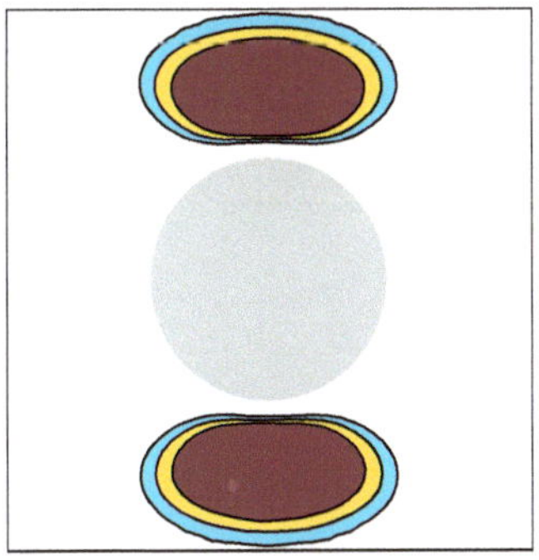

Fig. (**27**): Contour layers.

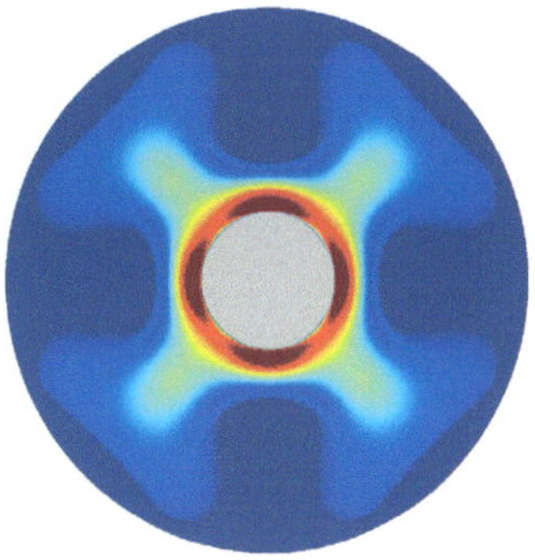

Fig. (**28**): Optimal design for 2 directions, $\Sigma = \{-1/4\pi, 1/4\pi\}$.

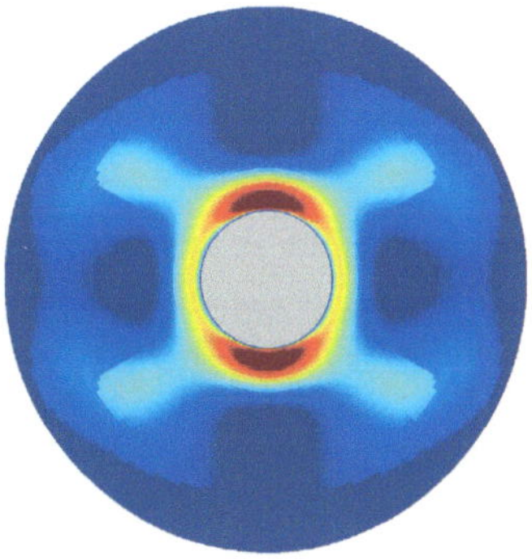

Fig. (**29**): Optimal design for 3 directions, $\Sigma = \{-1/4\pi, 0, 1/4\pi\}$.

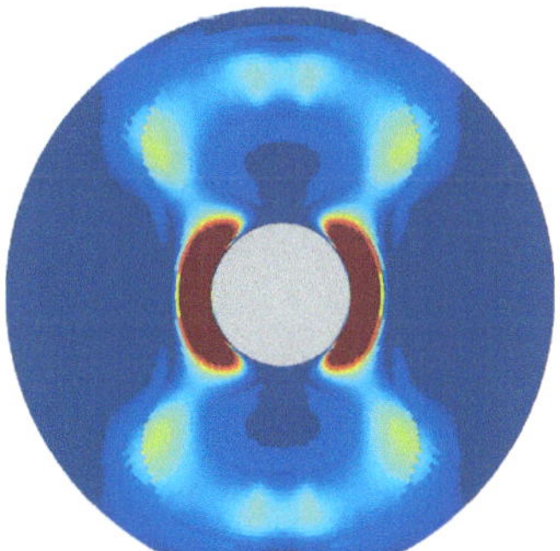

Fig. (**30**): Optimal design for 4 directions, $\Sigma = \{-1/2\pi, -1/3\pi, 1/3\pi, 1/2\pi\}$.

mechanical, acoustic or electromagnetical device, further research has to be conducted. In particular, post-processing and interpretation tools have to be developed in order to transfer the numerical results into practice.

ACKNOWLEDGEMENT

The notion 'Engineering of Advanced Materials' implicitly announcing a program all by itself, is also the title of a research cluster within the German Excellence Initiative funded by the Deutsche Forschungsgemeinschaft (DFG). The current contribution was initiated in the Center of 'Multiscale Modeling and Simulation' run by the first author. Second author has been visiting the institute a number of times, while the third author is employed there. The second author has been also supported by projects of the Czech Science Foundation 101/07/1417 and by project MSM 4977751301 of the Czech Ministry of Education.

REFERENCES

[1] Allaire G. Homogenization and two-scale convergence. SIAM J Math Anal 1992; 23: 1482-1518.

[2] Angell TS, Kirsch A. Optimization methods in electromagnetic radiation. Springer-Verlag, New York, 2004.

[3] Auriault JL, Bonnet G. Dynamique des composites elastiques periodiques. Arch Mech 1985; 37: 269-284.

[4] Ávila A, Griso G, Miara B, Rohan E. Multi-scale modelling of elastic waves – theoretical justification and numerical simulation of band gaps. SIAM Multi Model Simul 2008; 7: 1-21.

[5] Bendsoe MB, Sigmund O. Topology optimization, theory, methods and applications. Springer-Verlag, Berlin, Heidelberg, 2003 and 2004.

[6] Bohren CR, Huffman DR. Absorption and scattering of light by small particles. Wiley-VCH, Weinheim, 2004.

[7] Bonnet-Bendhia AS, Drissi D, Gmati N. Simulation of muffler's transmission losses by a homogenized finite element method. J Comput Acous 2004; 12: 447-474.

[8] Bonnet-Bendhia AS, Drissi D, Gmati N. Mathematical analysis of the acoustic diffraction by a muffler containing perforated ducts. Math Model Meth Appl Sci 2005; 15: 1059-1090.

[9] Bouchitte G, Felbacq D. Homogenization near resonances and artificial magnetism from dielectrics. CR Acad Sci, Ser 2004; 1339: 377-382.

[10] Cimrman R, Rohan E. Three phase phononic materials. Appl Comput Mech 2009; 3: 5-16.

[11] Cimrman R, Rohan E. Three-phase phononic materials. accepted : Appl. Comp. Mech., Univ. West Bohemia, Pilsen, 2009.

[12] Cioranescu D, Damlamian A, Griso G, Onofrei D. The periodic unfolding method for perforated domains and neumann sieve models. J Math Pure Appl. 2008; 89: 248-277.

[13] Cioranescu D, Damlamian A, Griso G. The periodic unfolding method in homogenization. SIAM J Math Anal 2008; 40: 1585-1620.

[14] Cioranescu D, Donato P. An introduction to homogenization. Oxford Lecture Series in Mathematics and its Applications 17. Oxford University Press, Oxford, 1999.

[15] Cioranescu D, Saint Jean Paulin J. Homogenization of reticulated structures. Applied Mathematical Sciences 136, Springer, New York, 1999.

[16] Greenleaf A, Kurylev Y, Lassas M, Uhlmann G. Approximate quantum and acoustic cloaking. arXiv:0812.1706v1[math-ph] 9 Dec 2008.

[17] Gurtin ME. Configurational forces as basic concept in continuum physics. Applied Mathematical Sciences 137, Springer-Verlag, 2000.

[18] Haslinger J, Kocvara M, Leugering G, Stingl M. Multidisciplinary free material optimization. submitted 2009.

[19] Jin J. The finite element method in electromagnetics. John Wiley & Sons, Inc., New York, 2002.

[20] Kogut P, Leugering G. S-homogenization of optimal controls problems in Banach spaces. Math Nachr 2002; 233-234: 141-169.

[21] Kogut P, Leugering G. Homogenization of optimal control problems in variable domains. Principle of the fictitious homogenization. Asympt Anal 2001; 26: 37-72.

[22] Kogut P, Leugering G. Homogenization of constrained optimal control problems for one-dimensional elliptic equations on periodic graphs. ESAIM, Control Optim Calc Var 2009;15: 471-498.

[23] Kogut P, Leugering G. Homogenization of Dirichlet optimal control problems with exact partial controllability constraints. Asympt Anal 2008; 57: 229-249.

[24] Lukeš V, Rohan E. Computational analysis of acoustic transmission through periodically perforated interfaces. Appl Comput Mech 2009; 3: 111-120.

[25] Miara B, Rohan E, Zidi M, Labat B. Piezomaterials for bone regeneration design – homogenization approach. J Mech Phys Solids 2005; 53: 2529-2556.

[26] Milton GW, Willis JR. On modifications of Newton's second law and linear continuum elastodynamics. Proc Royal Soc A 2007; 483: 855-880.

[27] Pendry JB, Schurig D, Smith DR. Calculation of of material properties and ray tracing in transformation optics. Opt Eng 2006; 14: 9794.

[28] Pendry JB. Negative refraction makes perfect lens. Phys. Rev. Letters 2000; 85: 3966-3969.

[29] Rohan E, Lukeš V. Sensitivity analysis for the optimal perforation problem in acoustic transmission. Appl Comput Mech 2009; 3: 163-176.

[30] Rohan E, Cimrman R. Dispersion properties in homogenized piezoelectric phononic materials. In: Proceedings of the conference Engineering Mechanics 2009, Svratka 2009.

[31] Rohan E, Miara B. Modelling and sensitivity analysis for design of phononic crystals. PAMM, Proc Appl Math Mech 2006; 6: 505-506.

[32] Rohan E, Miara B. Sensitivity analysis of acoustic wave propagation in strongly heterogeneous piezoelectric composite. In: Topics on Mathematics for Smart Systems, World Scientific Publishers, 2006: 139-207.

[33] Rohan E, Miara B. Shape sensitivity analysis for material optimization of homogenized piezo-phononic materials. In: Proceedings of the WCSMO-8, Lisbon, Portugal 2009.

[34] Rohan E, Miara B. Homogenization and shape sensitivity of microstructures for design of piezoelectric bio-materials. Mech Adv Materials Struct 2006; 13: 473-485.

[35] Rohan E, Miara B, Seifrt F. Numerical simulation of acoustic band gaps in homogenized elastic composites. Int J Eng Sci 2009; 47: 573-594.

[36] Rohan E et al. On nonsmooth sensitivity of phononic band gaps in elastic structures. in preparation.

[37] Rohan E, Leugering G. Homogenization of particle-cloaking layer in EM radiation. in preparation.

[38] Rohan E, Lukeš V. Homogenization of the acoustic transmission through perforated layer. J Appl Math Phys, 2009 (in press).

[39] Seifrt F, Leugering G, Rohan E. Shape optimization for the Helmholtz equation. PAMM, Proc Appl Math Mech 2008; 8: 10705-10706.

[40] Seifrt F, Leugering G, Rohan E. Tracking optimization for propagation of electromagnetic waves for multiple wavelengths. Appl Comp Mech 2008; 2: 369-378, Univ. of West Bohemia, Pilsen.

[41] Sanchez-Palencia E. Non-homogeneous media and vibration theory. Lecture Notes in Physics 127, Springer, Berlin, 1980.

[42] Seyranian AP, Lund E, Olhoff N. Multiple eigenvalues in structural optimization problems. Struct Opt 1994; 8: 207-227.

[43] Sigmund O, Jensen JS. Systematic design of phononic band-gap materials and structures by topology optimization. Phil Trans R Soc Lond, A 2003; 361: 1001-1019.

[44] Shalaev V. Optical negative-index metamaterials. Nature Photonics 2007; 1.

[45] Smith DR, Pendry JB. Homogenization of metamaterials by field averaging. J Opt Soc Am 2006; 23: 391-403.

[46] Wadbro E, Berggren M. Topology optimization of an acoustic horn. Comput Meth Appl Mech Engrg 2006; 196: 420-436.

[47] Zienkiewicz OC, Taylor RL, Zhu JZ. The finite elements method: its basis and fundamentals. Elsevier, Burlington, 2005.

Index

A

B

C

D

E

Matthias Ehrhardt (ED)

F

G

H

I

L

M

N

O

P

Q

R

S

T

www.ingramcontent.com/pod-product-compliance
Lightning Source LLC
Chambersburg PA
CBHW050834220326
41598CB00006B/365
* 9 7 8 1 6 0 8 0 5 6 9 1 0 *